This textbook deals with the theoretical basis of chemical equilibria and chemical changes and emphasizes the properties of phase diagrams. A feature of the book is the author's treatment of the field from a modern perspective, taking into account the fact that thermodynamics can now be applied using computers and databases.

Thermodynamics is an extremely powerful tool, applicable to a wide range of subjects in science and technology. In materials science, the main application has traditionally been through the use of phase diagrams. In recent years, however, there have been dramatic changes. These have come about as a result of the development both of computer-operated programs for sophisticated equilibrium calculations and of extensive databases containing thermodynamic parameter values for individual phases (from which all thermodynamic properties may be calculated). The present book is an attempt to give to students a proper thermodynamic basis in this new situation. Irreversible thermodynamics is introduced from the very beginning of the book by retaining the entropy production due to internal processes when the first and second laws are combined. Illustrative examples of thermodynamic modelling are given, and the book contains many exercises and worked solutions.

This modern text will be of value to undergraduate and graduate students in materials science, including ceramics and metallurgy, and also chemistry.

Phase Equilibria,
Phase Diagrams and
Phase Transformations

Their Thermodynamic Basis

Mats Hillert

Department of Materials Science and Engineering
KTH, Stockholm

Phase Equilibria, Phase Diagrams and Phase Transformations

Their Thermodynamic Basis

CAMBRIDGE
UNIVERSITY PRESS

PUBLISHED BY THE PRESS SYNDICATE OF THE UNIVERSITY OF CAMBRIDGE
The Pitt Building, Trumpington Street, Cambridge CB2 1RP, United Kingdom

CAMBRIDGE UNIVERSITY PRESS
The Edinburgh Building, Cambridge CB2 2RU, United Kingdom
40 West 20th Street, New York, NY 10011-4211, USA
10 Stamford Road, Oakleigh, Melbourne 3166, Australia

First published 1998

Printed in the United Kingdom at the University Press, Cambridge

Typeset in 9.5/13 pt Times [VN]

A catalogue record for this book is available from the British Library

Library of Congress Cataloguing in Publication data

Hillert, Mats, 1924–
    Phase equilibria, phase diagrams, and phase transformations: a
thermodynamic basis/Mats Hillert.
        p.   cm.
    Includes bibliographical references and index.
    ISBN 0 521 56270 8 (hc). – ISBN 0 521 56584 7 (pbk.)
    1. Phase rule and equilibrium. 2. Chemical equilibrium.
    3. Thermodynamics. I. Title.
    QD503.H554 1998
    541.3'63–dc21   97–12280 CIP

ISBN 0 521 56270 8 hardback
ISBN 0 521 56584 7 paperback

Contents

Preface

Thermodynamics is an extremely powerful tool applicable to a wide range of science and technology. However, its full potential has been utilized by relatively few experts and the practical application of thermodynamics has often been based simply on dilute solutions and the law of mass action. In materials science the main use of thermodynamics has taken place indirectly through phase diagrams. These are based on thermodynamic principles but, traditionally, their determination and construction have not made use of thermodynamic calculations, nor have they been used fully in solving practical problems. It is my impression that the role of thermodynamics in the teaching of science and technology has been declining in many faculties during the last few decades, and for good reasons. The students experience thermodynamics as an abstract and difficult subject and very few of them expect to put it to practical use in their future career.

Today we see a drastic change of this situation which should result in a dramatic increase of the use of thermodynamics in many fields. It may result in thermodynamics regaining its traditional role in teaching. The new situation is caused by the development both of computer operated programs for sophisticated equilibrium calculations and extensive databases containing assessed thermodynamic parameter values for individual phases from which all thermodynamic properties can be calculated. Experts are needed to develop the mathematical models and to derive the numerical values of all the model parameters from experimental information. However, once the fundamental equations are available, it will be possible for engineers with limited experience to make full use of thermodynamic calculations in solving a variety of complicated technical problems. In order to do this, it will not be necessary to remember much from a traditional course in thermodynamics. Nevertheless, in order to use the full potential of the new facilities and to avoid making mistakes, it is still desirable to have a good understanding of the basic principles of thermodynamics. The present book has been written with this new situation in mind. It does not provide the reader with much background in numerical calculation but should give him/her a solid basis for an understanding of the thermodynamic principles behind a problem, help him/her to present the problem to the computer and allow him/her to interpret the computer results.

The principles of thermodynamics were developed in an admirably logical way by Gibbs but he only considered equilibria. It has since been demonstrated, e.g. by Prigogine and Defay, that classical thermodynamics can also be applied to systems not at equilibrium whereby the affinity (or driving force) for an internal process is evaluated as an ordinary thermodynamic quantity. I have followed that approach by introducing a clear distinction between external variables and internal variables referring to entropy-producing internal processes. The entropy production is retained when the first and second laws are combined and the driving force for internal processes then plays a central role throughout the development of the thermodynamic principles. In this way, the driving force appears as a natural part of the thermodynamic application 'tool'.

Computerized calculations of equilibria can easily be directed to yield various types of diagram, and phase diagrams are among the most useful. The computer provides the user with considerable freedom of choice of axis variables and in the sectioning and projection of a multicomponent system, which is necessary for producing a two-dimensional diagram. In order to make good use of this facility, one should be familiar with the general principles of phase diagrams. Thus, a considerable part of the present book is devoted to the interrelations between thermodynamics and phase diagrams. Phase diagrams are also used to illustrate the character of various types of phase transformations. My ambition has been to demonstrate the important role played by thermodynamics in the study of phase transformations.

I have tried to develop thermodynamics without involving the special properties of particular kinds of phases, but have found it necessary sometimes to use the ideal gas or the regular solution to illustrate principles. However, even though thermodynamic models and derived model parameters are already stored in databases, and can be used without the need to inspect them, it is advantageous to have some understanding of thermodynamic modelling. The last few chapters are thus devoted to this subject. Simple models are discussed, not because they are the most useful or popular, but rather as illustrations of how modelling is performed.

Many sections may give the reader little stimulation but may be valuable as reference material for later parts of the book or for future work involving thermodynamic applications. The reader is advised to peruse such sections very quickly, but to remember that this material is available for future consultation.

Practically every section ends with at least one exercise and the accompanying solution. These exercises often contain material that could have been included in the text, but would have made the text too massive. The reader is advised not to study such exercises until a more thorough understanding of the content of a particular section is required.

This book is the result of a long period of research and teaching, centred on thermodynamic applications in materials science. It could not have been written without the inspiration and help received through contacts with numerous students and colleagues. Special thanks are due to my former students, Professor Bo Sundman and Docent Bo Jansson, whose development of the *Thermo-Calc* data bank system has inspired me to penetrate the underlying thermodynamic principles and has made me aware of many important questions. Thanks are also due to Dr Malin Selleby for producing a large number of diagrams by skilful operation of *Thermo-Calc*. All her diagrams in this book can be identified by the use of the *Thermo-Calc* logotype, .

Stockholm
Mats Hillert

1

Basic concepts of thermodynamics

External state variables

Thermodynamics is concerned with the state of a system when left alone, and when interacting with the surroundings. By 'system' we shall mean any portion of the world that can be defined for consideration of the changes that may occur under varying conditions. The system may be separated from the surroundings by a real or imaginary wall. The properties of the wall determine how the surroundings may interact with the system. The wall itself will not usually be regarded as part of the system but rather as part of the surroundings. We shall first consider two kinds of interactions, thermal and mechanical, and we may regard the name 'thermodynamics' as an indication that these interactions are of main interest. Secondly, we shall introduce interactions by exchange of matter in the form of chemical species. The name 'thermochemistry' is sometimes used as an indication of such applications. The term 'thermophysical properties' is sometimes used for thermodynamic properties which do not primarily involve changes in the amounts of various chemical species, e.g. heat capacity, thermal expansivity and compressibility.

One might imagine that the content of matter in the system could be varied in a number of ways equal to the number of species. However, species may react with each other inside the system. It is thus convenient instead to define a set of **independent components**, the change of which can accomplish all possible variations. By denoting the number of independent components as c and also considering thermal and mechanical interactions with the surroundings, we find by definition that the state of the system may vary in $c + 2$ independent ways. For metallic systems the components are usually identical to the elements. For systems with covalent bonds it may sometimes be convenient to regard a very stable molecular species as a component. For systems with a strongly ionic character it may be convenient to select the independent components from the neutral compounds rather than from the ions.

By waiting for the system to come to rest after making a variation we may hope to establish a **state of equilibrium**. A criterion that a state is actually a state of equilibrium, would be that the same state can be established from different starting points. After a system has reached a state of equilibrium we can, in principle, measure the values

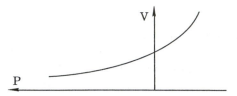

Figure 1.1 Property diagram for a constant amount of a solid material at a constant temperature showing the volume as a function of pressure. Notice that P has here been plotted in the negative direction. The reason will be explained later.

of many quantities which are uniquely defined by the state and independent of the history of the system. Examples are temperature T, pressure P, volume V and amount of each component N_i. We may call such quantities **state variables** or **state functions**, depending upon the context. It is possible to identify a particular state of equilibrium by giving the values of a number of state variables under which it is established. As might be expected, $c + 2$ variables must be given. The values of all other variables are fixed, provided that equilibrium has really been established. There are thus $c + 2$ independent variables and, after they have been selected and equilibrium has been established, the rest are dependent variables. As we shall see, there are many ways to select the set of independent variables. For each application a certain set is usually most convenient. For any selection of independent variables it is possible to change the value of each one, independent of the others, but only if the wall containing the system is open for exchange of $c + 2$ kinds, i.e. exchanges of mechanical work, heat and c components.

The equilibrium state of a system can be represented by a point in a $c + 2$ dimensional diagram. In principle, all points in such a diagram represent possible states of equilibrium although there may be practical difficulties in establishing the states represented by some region. One can use the diagram to define a state by specifying a point or a series of states by specifying a line. Such a diagram may be regarded as a **state diagram**. It does not give any information on the properties of the system under consideration unless such information is added to the diagram. We shall later see that some vital information on the properties can be included in the state diagram but in order to show the value of some dependent variable a new axis must be added. For convenience of illustration we shall now decrease the number of axes in the $c + 2$ dimensional state diagram by sectioning at constant values of $c + 1$ of the independent variables. All the states to be considered will thus be situated along a single axis which may now be regarded as the state diagram. We may then plot a dependent variable by introducing a second axis. That property is thus represented by a line. We may call such a diagram a **property diagram**. An example is shown in Fig. 1.1. Of course, we may arbitrarily choose to consider any one of the two axes as the independent variable. The shape of the line is independent of that choice and it is thus the line itself that represents the property of the system.

In many cases the content of matter in a system is kept constant and the wall is only open for exchange of mechanical work and heat. Such a system is often called a **closed** system and we shall start by discussing the properties of such a system. In other cases the content of matter may change and, in particular, the **composition** of the system by which we mean the relative amounts of the various components. In metallurgy such an open system is called an 'alloy system' and its behaviour is often shown in so-called **phase diagrams**, which are state diagrams with some additional information on what phases are present under various conditions. We shall later discuss the properties of phase diagrams in considerable detail.

The state variables are of two kinds which we shall call **intensive** and **extensive**. Temperature T and pressure P are intensive variables because they can be defined at each point of the system. As we shall see later, T must have the same value at all points in a system at equilibrium. An intensive variable with this property will be called **potential**. We shall later meet intensive variables which may have different values at different parts of the system. They will not be regarded as potentials.

Volume V is an extensive variable because its value for a system is equal to the sum of its values of all parts of the system. The content of component i, usually denoted by n_i or N_i, is also an extensive variable. Such quantities obey the law of additivity. For a homogeneous system their values are proportional to the size of the system.

One can imagine variables which depend upon the size of the system but do not always obey the law of additivity. The use of such variables is complicated and will not be further considered here.

If the system is contained inside a wall that is rigid, thermally insulating and impermeable to matter, then all the interactions mentioned are prevented and the system may be regarded as completely closed to interactions with the surroundings. It is left 'completely alone'. It is often called an **isolated** system. By changing the properties of the wall we can open the system to exchanges of mechanical work, heat or matter. By an **open** system one usually understands a system that is open to all these exchanges. We may thus control the values of $c + 2$ variables and we may regard them as **external variables**.

Exercise 1.1.1

A system consists of two subsystems with the values U_1, V_1, P_1 and U_2, V_2, P_2, where U is the internal energy. U and V are both extensive quantities. The law of additivity applies to them and we have for the complete system $U = U_1 + U_2$; $V = V_1 + V_2$. One often defines another function, enthalpy $H = U + PV$, where P is the external pressure. Evidently, H is also a state function and an extensive quantity. Discuss whether the law of additivity applies to H in a way that $H = U_1 + P_1V_1 + U_2 + P_2V_2$.

Hint

Before we discuss this question, it is necessary to define the situation better. How can the two different pressures be maintained? You may find two different cases.

Solution

In principle, one can change any one of V_1, V_2, U_1 and U_2 by making the opposite change in the surroundings. For the surroundings we get a change in volume of $-dV = -dV_1 - dV_2$ and in energy of $-dU = -dU_1 = dU_2$ and V and U are state variables of the complete system. For pressure there are two possibilities. One possibility is that both subsystems have direct contact with the surroundings and there are two independent external pressures, P_1 and P_2. We may then accept that H is additive but actually we have to derive a whole new set of thermodynamic equations using two external pressures. Another possibility is that one subsystem (1) is enclosed in the other (2) and the surface tension of the interface gives the pressure difference $P_1 - P_2 = 2\sigma/r$. In the calculation of the total enthalpy of the system one must also include the surface energy of the interface U_{int}: $H = U + P_2V = U_1 + U_2 + U_{int} + P_2(V_1 + V_2) = U_1 + P_2V_1 + U_2 + P_2V_2 + U_{int}$. One will thus obtain $H = H_1 + H_2 + U_{int}$ if one define $H_1 = U_1 + P_2V_1$ instead of $H_1 = U_1 + P_1V_1$ because P_2 represents an external variable but P_1 is an internal variable which cannot be controlled from the outside.

1.2   *Internal state variables*

After some or all of the $c + 2$ independent variables have been changed to new values and before the system has come to rest at equilibrium, it is also possible to describe the state of the system, at least in principle. For that description additional variables are required. We may call them **internal variables** because they will change due to internal processes as the system approaches the state of equilibrium under the new values of the $c + 2$ external variables.

An internal variable ξ (pronounced 'xeye') is illustrated in Fig. 1.2(a) where $c + 1$ of the independent variables are again kept constant in order to obtain a two-dimensional diagram. The equilibrium value of ξ for various values of the remaining independent variable T is represented by a curve. In that respect, the diagram is a property diagram. On the other hand, by a rapid change of the independent variable T the system may be brought to a point away from the curve. Any such point represents a possible non-equilibrium state and in that sense the diagram is a state diagram. In order to define such a point one must give the value of the internal variable in addition to T. The quantity ξ is thus an independent variable for states of non-equilibrium.

For such states of non-equilibrium one may plot any other property versus the value of the internal variable. An example of such a property diagram is given in Fig. 1.2(b). In this particular case we have chosen to show a property called Helmholtz energy F which will decrease by all spontaneous changes at constant T and V. Given sufficient time the system will approach the minimum of F which corresponds to a point on the curve to the left. That curve is the locus of all points of minimum of F, each one

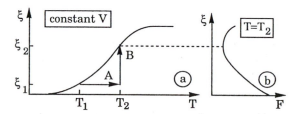

Figure 1.2 (a) Property diagram showing the equilibrium value of an internal variable, ξ, as a function of temperature. Arrow A represents a sudden change of temperature and arrow B the gradual approach to a new state of equilibrium. (b) Property diagram for non-equilibrium states at T_2, showing the change of Helmholtz energy F as a function of the internal variable, ξ. There will be a spontaneous change with decreasing F and a stable state will eventually be reached at the minimum of F.

obtained under its own constant value of T. Any state of equilibrium can thus be defined by giving T and the proper ξ value or by giving T and the requirement of equilibrium. Under that requirement ξ is a dependent variable and does not need to be given.

It is sometimes possible to imagine that a non-equilibrium state can be 'frozen-in' (see Section 1.4), i.e. by the temperature being so low that the non-equilibrium state does not change markedly during the time it takes to measure an internal variable. Under the given restrictions such a state may be regarded as a state of equilibrium, but one or several internal variables must be included in the set of independent variables which defines the state. There is a particular type of internal variable which can be controlled from outside the system under such restrictions. Such a variable can then be treated as an external variable. It can for instance be the number of O_3 molecules in a system, the rest of which is O_2. At high temperature the chemical reaction between these species will be rapid and the amount of O_3 may be regarded as a dependent variable. In order to define a state of equilibrium at high temperature it is sufficient to give the amount of oxygen. At a lower temperature the reaction may be frozen and the system has two independent variables, the amounts of O_2 and O_3.

Exercise 1.2.1

Consider a box of fixed volume containing a certain amount of a liquid which fills the box only partly. Some of the liquid thus evaporates. The equilibrium vapour pressure is $P = k \cdot \exp(-b/T)$. Calculate and show with a property diagram how the amount of gas varies as a function of T. Also, make a property diagram for non-equilibrium states showing the relation between the amount of gas and the pressure at two constant temperatures, i.e. under the assumption that condensation and evaporation are slow enough to be negligible during the time of studying the relation. Indicate where the equilibrium pressure can be found. Apply the ideal gas law, $PV = NRT$, where R is the gas constant and N the amount of gas measured as moles of gas molecules.

Hint

In order to simplify the calculations, neglect the volume of the liquid and assume that the volume of the gas is constant.

Solution

$N = PV/RT = (kV/RT)\exp(-b/T)$. Let us introduce dimensionless variables, $N/(kV/bR) = (b/T)\exp(-b/T)$. This function has a maximum at $T/b = 1$. For non-equilibrium at constant T/b: $P = NRT/V$; $P/k = N/(kV/bR)\cdot T/b$. In the following diagram we consider two temperatures which at equilibrium give the same content of gas, N, but different pressures.

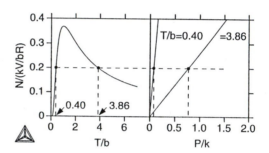

Solution 1.2.1

1.3   *The first law of thermodynamics*

The development of thermodynamics starts by the definition of Q, the amount of heat flown into a system, and W, the amount of work done on the system. The concept of work may be regarded as a useful device to avoid having to define what actually happens to the surroundings as a result of certain changes within the system. The first law of thermodynamics is related to the law of conservation of energy which says that energy cannot be created, nor destroyed. As a consequence, if a system receives an amount of heat, Q, and the work W is done on the system, then the energy of the system must have increased by $Q + W$. This must hold quite independent of what happened to the energy inside the system. In order to avoid such discussions, the concept of internal energy U has been invented, and the first law of thermodynamics is formulated as

$$\Delta U = Q + W$$

In differential form we have

$$dU = dQ + dW$$

It is rather evident that the internal energy of the system is uniquely determined by the state of the system and independent of how it has been established. U is a state variable. It should be emphasized that Q and W are not properties of the system but define different ways of interaction with the surroundings. A system can be brought from one state to another by different combinations of heat and mechanical work. It is possible to bring the system from one state to another by some route and then let it return to the initial state by a different route. It would thus be possible to get mechanical work out of the system by supplying heat but without any net change of the system. An examination of how efficient such a process can be resulted in the formulation of the second law of thermodynamics. It will be discussed in Sections 1.5 and 1.6.

The internal energy U is a variable which is not easy to vary experimentally in a controlled fashion. Thus, we shall usually regard U as a state function rather than a state variable. At equilibrium it may, for instance, be convenient to consider U as a function of temperature and pressure because those variables may be easily controlled in the laboratory

$$U = U(T,P)$$

However, we shall soon find that there are two other more natural variables for U. It is evident that U is an extensive property and obeys the law of additivity. The total value of U of a system is equal to the sum of U of the various parts of the system. Its value does not depend upon how the additional energy, due to added heat and work, is distributed within the system.

It should be emphasized that the absolute value of U is not defined through the first law, but only *changes* of U. Thus, there is no natural zero point for the internal energy. One can only consider changes in internal energy. For practical purposes one often chooses a point of reference, an arbitrary zero point.

For compressional work against a hydrostatic pressure P we have

$$dW = P(-dV) = -PdV$$
$$dU = dQ - PdV$$

The present discussion will be limited to cases where the work done on the system is hydrostatic. The treatment will be applicable to gases and liquids which cannot support shear stresses. It should be emphasized that a complete treatment of the thermodynamics of solid systems requires a consideration of non-hydrostatic stresses. We shall not discuss such problems.

Mechanical work against a hydrostatic pressure is so important that one needs to define a special state function called enthalpy H in the following way, $H = U + PV$

$$dH = dU + PdV + VdP = Q + VdP$$

Exercise 1.3.1

Consider a system under such conditions that there is no exchange of heat or work with the surroundings. Suppose there is some spontaneous reaction inside the system. What can we say about the change of the internal energy?

Solution

It follows directly from the definition that U does not change, $dU = dQ + dW = 0 + 0 = 0$.

Exercise 1.3.2

One mole of a gas at pressure P_1 is contained in a cylinder of volume V_1 which has a piston. The pressure is changed rapidly and the volume thus decreases to V_2, without time for heat conduction to or from the surroundings.
(a) Evaluate the change in internal energy of the gas if it behaves as a classical ideal gas for which $PV = RT$ and $U = A + BT$.
(b) Then evaluate the amount of heat flow until the temperature has returned to its initial value, assuming that the piston is locked in the new position, V_2.

Hint

The internal energy can change due to mechanical work and heat conduction. The first step is with mechanical work only, the second step with heat conduction only.

Solution

(a) Without heat conduction $dU = -PdV$ but we also know that $dU = BdT$. This yields $BdT = -PdV$. Elimination of P using $PV = RT$ gives $BdT/RT = -dV/V$ and by integration we then find $(B/R)\ln(T_2/T_1) = -\ln(V_2/V_1) = \ln(V_1/V_2)$ and $T_2 = T_1(V_1/V_2)^{R/B}$ where T_1 is the initial temperature, $T_1 = P_1V_1/R$. Thus: $\Delta U_a = B(T_2 - T_1) = (BP_1V_1/R)[(V_1/V_2)^{R/B} - 1]$.
(b) By heat conduction the system returns to the initial temperature and thus to the initial value of U, since U in this case depends only on T. Since the piston is now locked, there will be no mechanical work this time, so that $dU_b = dQ_b$ and, by integration, $\Delta U_b = Q_b$. Considering both steps we find because U depends only upon T:
$0 = \Delta U_a + \Delta U_b = \Delta U_a + Q_b$; $Q_b = -\Delta U_a = -(BP_1V_1/R)[(V_1/V_2)^{R/B} - 1]$

Exercise 1.3.3

Consider a system at T_1,P_1,V_1, which is composed of one mole of a classical ideal gas for which $PV = RT$ and $U = A + BT$. Let the system change in four steps. The first step is a very slow compression to P_2, V_2 at a constant temperature of T_1. The second step is a very quick compression to T_3, P_3, V_3. The third step is a very slow expansion to P_4, V_4 at a constant temperature of T_3. The fourth step is a very quick expansion back to T_1, P_1, V_1. The second and fourth steps are so quick that there is no heat exchange. Calculate the relations between the total work, the heat of the first step and the heat of the third step, expressing the result in terms of T_1 and T_3.

Hint

$\Sigma \Delta U = 0$ because the system returns to the initial state. Thus, $Q_1 + Q_3 + W_1 + W_2 + W_3 + W_4 = 0$. Furthermore, U is only a function of T and thus $\Delta_1 U = 0$ and $\Delta_3 U = 0$ giving $Q_1 + W_1 = 0$ and $Q_3 + W_3 = 0$ and thus $W_2 + W_4 = 0$. Consequently, the net work $W = W_1 + W_3$. The problem is to calculate W_1 $(= - Q_1)$ and W_3 $(= - Q_3)$ and to relate them using T_1 and T_3.

Solution

$W_1 = - \int P \, dV = - RT_1 \int dV/V = - R/T_1 \ln(V_2/V_1); \quad W_3 = - RT_3 \ln(V_4/V_3)$. The second step where $Q_2 = 0$ gives $dU = dW; \quad B dT = - P dV = - RT dV/V;$ $B\ln(T_3/T_1) = - R\ln(V_3/V_2)$ and the fourth step $B\ln(T_1/T_3) = - R\ln(V_1/V_4)$. Thus $V_3/V_2 = V_4/V_1$ or $V_1/V_2 = V_4/V_3$ and $W_1/T_1 = R\ln(V_1/V_2) = R\ln(V_4/V_3) = - W_3/T_3; \quad - W = - W_1 - W_3 = - W_1 + W_1 T_1/T_3 = W_1(T_3 - T_1)/T_1 = - W_3(T_3 - T_1)/T_3 = - Q_1(T_3 - T_1)T_1 = Q_3(T_3 - T_1)/T_3$. By drawing the heat Q_3 from a heat source at T_3 one may thus produce mechanical work in the amount of $Q_3(T_3 - T_1)/T_3$ if the rest of the energy, $Q_3 \cdot T_1/T_3$, can be disposed of as heat to a heat sink at T_1.

Exercise 1.3.4

One mole of a fluid is heated from T_1 to T_2 under constant volume. Show how one can calculate the work done by the surroundings and the heat absorbed from the surroundings. What state function of the fluid do we need to know?

Hint

Since nothing is said about the exchange of work, suppose there is only hydrostatic work.

Solution

The work $- P dV$ is zero if V is constant. The heat is $dQ = dU + P dV = dU$ and we get $Q = \int dU = U_2 - U_1 = U(T_2, V_1) - U(T_1, V_1)$. We must know U as a function of T at constant V.

Exercise 1.3.5

Two completely isolated containers are each filled with one mole of gas. They are at different temperatures but at the same pressure. The containers are then connected and can exchange heat and molecules freely but do not change their volumes. Evaluate the final temperature and pressure. Suppose that the gas is classical ideal for which $PV = RT$ and $U = A + BT$ if one considers one mole.

Hint

Use the fact that the containers are still completely isolated from the surroundings. Thus, the total internal energy has not changed.

Solution

$V = V_1 + V_2 = RT_1/P_1 + RT_2/P_1 = R(T_1 + T_2)/P_1$; $A + BT_1 + A + BT_2 = U = 2A + 2BT_3$; $T_3 = (T_1 + T_2)/2$; $P_3 = 2RT_3/V = R(T_1 + T_2)/[R(T_1 + T_2)/P_1] = P_1$.

1.4   *Freezing-in conditions*

As a continuation of our discussion on internal variables we may now consider heat absorption under two different conditions.

We shall first consider an increase in temperature slow enough to allow an internal process to adjust continuously to the changing conditions. If the heating is made under conditions where we can control the volume, we can regard T and V as the independent variables and write

$$dU = \left(\frac{\partial U}{\partial T}\right)_V dT + \left(\frac{\partial U}{\partial V}\right)_T dV$$

By combination with $dU = dQ - PdV$ we find

$$dQ = \left(\frac{\partial U}{\partial T}\right)_V dT + \left[\left(\frac{\partial U}{\partial V}\right)_T + P\right]dV$$

We thus obtain for the heat capacity under constant V,

$$C_V \equiv \left(\frac{\partial Q}{\partial T}\right)_V = \left(\frac{\partial U}{\partial T}\right)_V$$

Secondly, we shall consider an increase in temperature so rapid that an internal process is practically inhibited. Then we must count the internal variable as an additional independent variable which is kept constant. Denoting the internal variable as ξ we obtain

$$dU = \left(\frac{\partial U}{\partial T}\right)_{V,\xi} dT + \left(\frac{\partial U}{\partial T}\right)_{T,\xi} dV + \left(\frac{\partial U}{\partial \xi}\right)_{T,V} d\xi$$

$$dQ = \left(\frac{\partial U}{\partial T}\right)_{V,\xi} dT + \left[\left(\frac{\partial U}{\partial T}\right)_{T,\xi} + P\right] dV + \left(\frac{\partial U}{\partial \xi}\right)_{T,V} d\xi$$

Under constant V and ξ we now obtain the following expression for the heat capacity

$$C_{V,\xi} \equiv \left(\frac{\partial Q}{\partial T}\right)_{V,\xi} = \left(\frac{\partial U}{\partial T}\right)_{V,\xi}$$

Experimental conditions under which an internal variable ξ does not change will be called *freezing-in* conditions and an internal variable that does not change due to such conditions will be regarded as being *frozen-in*. We can find a relation between the two heat capacities by comparing the two expressions for dU at constant V,

$$\left(\frac{\partial U}{\partial T}\right)_V = \left(\frac{\partial U}{\partial T}\right)_{V,\xi} + \left(\frac{\partial U}{\partial \xi}\right)_{T,V} \left(\frac{\partial \xi}{\partial T}\right)_V$$

$$C_V \equiv C_{V,\xi} + \left(\frac{\partial U}{\partial \xi}\right)_{T,V} \left(\frac{\partial \xi}{\partial T}\right)_V$$

The two heat capacities will thus be different unless either $(\partial U/\partial \xi)_{T,V}$ or $(\partial \xi/\partial T)_V$ is zero, which may rarely be the case.

It is instructive to note that the last expression given for dQ allows the heat of the internal process to be evaluated,

$$\left(\frac{\partial Q}{\partial \xi}\right)_{T,V} = \left(\frac{\partial U}{\partial \xi}\right)_{T,V}$$

The processes may be melting, for instance. It is then convenient to define ξ as the amount of melt formed, expressed in moles.

Exercise 1.4.1

Suppose there is an internal reaction by which a system can adjust to a new equilibrium if the conditions change. There is a complete adjustment if the change is very slow and for a slow increase of T one measures $C_{V,\text{slow}}$. For a very rapid change there will be practically no reaction and one measures $C_{V,\text{rapid}}$. What value of C_V would one find if the change is intermediate and the reaction at each temperature is half-way between the initial value and the equilibrium value.

Hint

In every case

$$C_V \equiv (\partial Q/\partial T)_{\text{expt.cond.}} = (\partial U/\partial T)_{V,\xi} + (\partial U/\partial \xi)_{T,V} \cdot (\partial \xi/\partial T)_{\text{expt.cond.}}$$

Solution

$C_{V,\text{rapid}} = (\partial U/\partial T)_{V,\xi}$; $C_{V,\text{slow}} = (\partial U/\partial T)_{V,\xi} + (\partial U/\partial \xi)_{T,V} \cdot (\partial \xi/\partial T)_{\text{eq.}}$;
$C_{V,\text{interm.}} = (\partial U/\partial T)_{V,\xi} + (\partial U/\partial \xi)_{T,V} \cdot 0.5(\partial \xi/\partial T)_{\text{eq.}} = (C_{V,\text{rapid}} + C_{V,\text{slow}})/2.$
    It should be noticed that the values of the two derivatives of U may depend on ξ as well as T. They may thus change during heating and in different ways depending on whether and how ξ changes. The last step in the derivation is thus strictly valid only at the starting point.

1.5   *Reversible and irreversible processes*

Consider a cylinder filled with an ideal gas and with a frictionless piston which exerts a pressure P on the gas in the cylinder. By gradually increasing P we can compress the gas and perform the work $W = -\int P\,dV$ on the gas. If the gas is thermally insulated from the surroundings, its temperature will rise because $\Delta U = Q + W = -\int P\,dV$. By then decreasing P we can make the gas expand again and perform the same work on the surroundings through the piston. The initial situation has thus been restored without any net exchange of work or heat with the surroundings and no change of temperature or pressure of the gas. The whole process and any part of it are regarded as **reversible**.

    The process would be different if the gas were not thermally insulated. Suppose it were instead in thermal equilibrium with the surroundings during the compression. For an ideal gas the internal energy only varies with the temperature and would thus stay constant during the compression. We would thus find $Q + W = \Delta U = 0$ and $Q = -W = \int P\,dV$. It is evident that Q would be negative for the compression (negative dV) and the system (the gas) would thus give heat to the surroundings during the compression. By then decreasing P we could make the gas expand and, as it returns to the initial state, it would give back the work to the surroundings and take back the heat. Again there would be no net exchange with the surroundings. This process is also regarded as reversible and it may be described as a reversible **isothermal** process. The previous case may be described as a reversible **adiabatic** process.

    By combination of the above processes and with the use of two heat reservoirs,

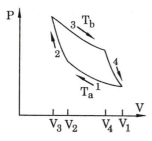

Figure 1.3 Schematic diagram of Carnot's cycle.

T_a and T_b, one can make the system go through a cycle which may be defined as reversible because all the steps are reversible. Fig. 1.3 illustrates a case with four steps where $T_b > T_a$.

1 Isothermal compression from V_1 to V_2 at a constant temperature T_a. The surroundings perform the work W_1 on the system and the system gives away heat, $-Q_1$, to the surroundings, i.e. to the colder heat reservoir, T_a. The heat received by the system, Q_1, is negative.

2 Adiabatic compression from V_2 to V_3 under an increase of the temperature inside the cylinder from T_a to T_b. The surroundings perform the work W_2 on the system but there is no heat exchange, $Q_2 = 0$.

3 Isothermal expansion from V_3 to V_4 after the cylinder has been brought into contact with a warmer heat reservoir, T_b. The system now gives back some work to the surroundings; W_3 is negative whereas Q_3 is positive. The warm heat reservoir, T_b, thus gives away heat to the system.

4 Adiabatic expansion from V_4 back to V_1 under a decrease of temperature inside the cylinder from T_b to T_a; W_4 is negative and $Q_4 = 0$.

The system has thus received a net heat of $Q = Q_1 + Q_3$ but it has returned to the initial state and for the whole process we obtain $Q + W = \Delta U = 0$ and $-W = Q = Q_1 + Q_3$ where W is the net work done on the system. According to Fig. 1.3 the inscribed area is positive and mathematically it corresponds to $\int P dV$. The net work, W, is equal to $-\int P dV$ and it is thus negative and the system has performed work on the surroundings. The net heat, Q, is positive and the system has thus received energy by heating. The system has performed work on the surroundings, $-W$, by transforming into mechanical energy some of the thermal energy, Q_3, received from the warm heat reservoir. The remaining part of Q_3 is given off to the cold heat reservoir, $-Q_1 < Q_3$. This cycle may thus be used for the construction of a heat engine that can produce mechanical energy from thermal energy. It was first discussed by Carnot (c. 1820, see translation, 1960) and is called Carnot's cycle. From a practical point of view the important question is how efficient that engine would be. The efficiency may be defined as the ratio between the mechanical work produced, $-W$, and the heat drawn from the warm heat reservoir, Q_3.

$$\eta = \frac{-W}{Q_3} = \frac{Q_1 + Q_3}{Q_3} = 1 + \frac{Q_1}{Q_3}$$

This is less than unity because Q_1 is negative and its absolute value is smaller than Q_3.

We can let the engine run in the reverse direction. It would then draw heat from the cold reservoir and deposit it in the warm reservoir by means of some mechanical work. It would thus operate as a heat pump or refrigerator.

Before continuing the discussion, let us consider the flow of heat through a wall separating two heat reservoirs. There is no method by which we could reverse this process. Heat can never flow from a cold reservoir to a warmer one. Heat conduction is an **irreversible** process.

Let us then go back to the Carnot cycle and examine it in more detail. It is clear that in reality it must have some irreversible character. The flow of heat in steps 1 and 3 cannot occur unless there is a temperature difference between the system and the heat reservoir. The irreversible character of the heat flow may be decreased by making the temperature difference smaller but then the process will take more time. A completely reversible heat transfer could, in principle, be accomplished by decreasing the temperature difference to zero but then the process would take an infinite time. A completely reversible process is always an idealization of reality which can never be attained. However, it is an extremely useful concept because it defines the theoretical limit. Much of thermodynamics is concerned with reversible processes.

We may expect that the efficiency would increase if the irreversible character of the engine could be decreased. However, it may also seem conceivable that the efficiency of a completely reversible engine could depend on the choice of temperatures of the two heat reservoirs and on the choice of fluid (gas or liquid) in the system. These matters will be considered in the next section.

Exercise 1.5.1

Discuss by what physical mechanisms the adiabatic steps of the Carnot cycle can get an irreversible character.

Solution

There may be heat conduction through the wall of the cylinder also during the adiabatic step, i.e. it would not be completely adiabatic. That effect would be less if the compression is very fast. However, for a very fast compression it is possible that there will be violent motions or oscillations inside the system. The damping of them is an irreversible process.

The second law of thermodynamics

Let us now compare the efficiency of two heat engines which are so close to the ideal case that they may be regarded as reversible. Let them operate between the same two heat reservoirs, T_a and T_b. Suppose one engine has a lower efficiency than the other and let it operate in the reverse direction, i.e. as a heat pump. Build the heat pump of such a size that it will give to the warm reservoir the same amount of heat as the heat engine will take. Thanks to its higher efficiency the heat engine will produce more work than needed to run the heat pump. The difference can be used for some useful purpose and the equivalent amount of thermal energy must come from the cold reservoir because the warm reservoir is not affected and could be disposed of.

The above arrangement would be a kind of perpetuum mobile. It would for ever produce mechanical work by drawing thermal energy from the surroundings without using a warmer heat source. This does not seem reasonable and one has thus formulated the second law of thermodynamics which states that this is not possible. It then follows that the efficiency of all reversible heat engines must be the same if they operate between the same two heat reservoirs. From the expression for the efficiency η it follows that the ratio Q_1/Q_3 can only be a function of T_a and T_b and the same function for all choices of fluid in the cylinder.

A heat engine which is not reversible will have a lower efficiency but, when used in the reverse direction, it will have different properties because it is not reversible. Its efficiency will thus be different in the reverse direction and it could not be used to make a perpetuum mobile.

It remains to examine how high the efficiency is for a reversible heat engine and how it depends on the temperatures of the two heat reservoirs. The answer could be obtained by studying any well-defined engine, for instance an engine built on the Carnot cycle using a classical ideal gas. It has already been examined in Exercise 1.3.2 and the result was

$$\eta = \frac{-W}{Q_3} = \frac{T_b - T_a}{T_b}$$

In line with Carnot's ideas, we can give a more general derivation by first considering the production of work when a body of mass ΔM is moved from a higher level to a lower one, i.e. from a higher gravitational potential, g_b, to a lower one, g_a.

$$-W = \Delta M \cdot (g_b - g_a)$$

The minus sign is added because $+W$ should be defined as mechanical energy received by the system (the body). With this case in mind, let us assume that the work produced by a reversible heat engine could be obtained by considering some appropriate thermal quantity which would play the same role as mass. That quantity is now called entropy and denoted by S. When a certain amount of that quantity is moved from a higher thermal potential (temperature T_b) to a lower one (temperature T_a) the production of work should be given in analogy to the above equation,

$$-W = \Delta S \cdot (T_b - T_a)$$

However, we already know that $-W$ is the sum of Q_1 and Q_3,

$$\Delta S \cdot T_b - \Delta S \cdot T_a = Q_3 + Q_1$$

We can find an appropriate quantity S to satisfy this equation by defining S as a state function, the change of which in a system is related to the heat received,

$$\Delta S = Q/T$$

The amount of S received by the system from the warm heat reservoir would then be

$$\Delta S = Q_3/T_b$$

The amount of S given by the system to the cold reservoir would be

$$\Delta S = -Q_1/T_a$$

The equation is satisfied and we also find

$$\frac{Q_1}{Q_3} = \frac{-\Delta S \cdot T_a}{\Delta S \cdot T_b} = -\frac{T_a}{T_b}$$

$$\eta = \frac{-W}{Q_3} = 1 + \frac{Q_1}{Q_3} = \frac{T_b - T_a}{T_b}$$

in agreement with the previous examination of the Carnot cycle.

Let us now look at entropy and temperature in a more general way. By adding a small amount of heat to a system by a reversible process we would increase its entropy by

$$dS = dQ/T$$

For a series of reversible changes that brings the system back to the initial state

$$\int dQ/T = \Delta S = 0$$

This can be demonstrated with the Carnot cycle,

$$\Delta S = \int dQ/T = Q_1/T_a + Q_3/T_b = 0$$

The quantity T is a measure of temperature but it remains to be discussed exactly how to define T. It is immediately evident that the zero point must be defined in a unique way because T_a/T_b would change if the zero point is changed. That is not allowed because it must be equal to $-Q_1/Q_3$. The quantity T is thus measured relative to an absolute zero point and one can say that T measures the absolute temperature.

It has already been demonstrated that by applying the ideal gas law $PV = RT$ to the fluid in the Carnot engine, one can derive the correct expression for the efficiency, $\eta = (T_b - T_a)/T_b$. One can thus define the absolute temperature as the temperature scale used in the ideal gas law and one can measure the absolute temperature with a gas thermometer. When this was done it was decided to express the difference between the boiling and freezing point of water at 1 atm as 100 units, in agreement with the Celsius scale. This unit is now called kelvin (K).

Let us now return to the irreversible process of heat conduction from a warm reservoir to a cold one. By transferring an amount dQ one would decrease the entropy of the warm reservoir by dQ/T_b and increase the entropy of the cold one by dQ/T_a. The net change of the entropy would thus be

$$dS = -dQ/T_b + dQ/T_a = dQ \cdot (T_b - T_a)T_bT_a$$

The irreversible process thus produces entropy. When an irreversible process occurs inside a system, one talks about **internal entropy production**

$$d_{ip}S > 0$$

The subscript 'ip' indicates that this change of the entropy is due to an internal process. If there is a simultaneous heat exchange with the surroundings, the total change in the system would be

$$dS = dQ/T + d_{ip}S > dQ/T$$

A spontaneous process can occur in one direction only, and the above criterion can be used to predict its direction. A spontaneous process is always an irreversible process, otherwise it would have no preferred direction and it would be reversible.

A reversible process can be defined by either one of the following criteria,

$$d_{ip}S = 0$$
$$dS = dQ/T$$

Exercise 1.6.1

Suppose a simple model for an internal reaction yields the following expression for the internal production of entropy under conditions of constant volume and temperature, $\Delta_{ip}S = -\xi K/T - R[\xi\ln\xi - (1 + \xi)\ln(1 + \xi)]$ where ξ is a measure of the progress of the reaction. Find the equilibrium value of ξ, i.e. the value of ξ where the reaction can no longer occur spontaneously.

Hint

The spontaneous reaction will stop when $d_{ip}S$ is no longer positive, i.e. when $d_{ip}S/d\xi = 0$.

Solution

$d_{ip}S/d\xi = d(\Delta_{ip}S)/d\xi = -K/T - R[1 + \ln\xi - 1 - \ln(1 + \xi)] = 0$; $\xi/(1 + \xi) = \exp(-K/RT)$; $\xi = 1/[\exp(K/RT) - 1]$

Exercise 1.6.2

In the preceding exercise suppose the heat of reaction, i.e. the absorption of heat necessary to keep the temperature constant, is ξK. Construct an expression for the total increase of entropy inside the system. Calculate the difference between the total increase of entropy of the system and the internal production of entropy, when ξ increases from zero to the equilibrium value at the experimental temperature.

Hint

Don't calculate each one. The difference can be calculated directly.

Solution

Under isothermal conditions the heat of reaction must be compensated by heat flow from the surroundings, $Q = \xi K$ and the total increase of entropy is $\Delta S = Q/T + \Delta_{ip}S = \xi K/T + \Delta_{ip}S = - R[\xi \ln \xi - (1 + \xi)\ln(1 + \xi)]$. However, we get the difference directly: $\Delta S - \Delta_{ip}S = Q/T = \xi K/T$. Inserting the equilibrium value of ξ given by the preceding exercise gives $\Delta S - \Delta_{ip}S = (K/T)/[\exp(K/RT) - 1]$.

Exercise 1.6.3

The fact that S is a state function is reflected by the fact that dS can be integrated for a reversible process and the resulting ΔS only depends upon the initial and final states. Show this for a classical ideal gas.

Hint

First, calculate dQ from the first law. Then use d$S = $ dQ/T. For one mole of a classical ideal gas $PV = RT$ and $U = A + BT$.

Solution

$dU = dQ + dW = dQ - PdV$; $dQ = dU + PdV = BdT + (RT/V)dV$; $dS = dQ/T = (B/T)dT + (R/V)dV$. It is interesting to note that by dividing dQ with T we have thus been able to separate the variables T and V and obtain a state function from Q which is not a state function itself. Integration yields $\Delta S = B\ln(T_2/T_1) + R\ln(V_2/V_1)$.

Exercise 1.6.4

Consider a Carnot cycle with a non-ideal gas and suppose that the process is somewhat irreversible. Use the second law to derive an expression for the efficiency.

Hint

For each complete cycle we have $\Sigma\Delta U = 0$ and $\Sigma\Delta S = 0$ because U and S are both state functions.

Solution

$\Sigma\Delta U = W + Q_1 + Q_3 = 0$; $\Sigma\Delta S = Q_1/T_a + Q_3/T_b + \Delta_{ip}S = 0$ where W is the sum of W_i over the four steps and $\Delta_{ip}S$ is the sum of the internal entropy production over the four steps. We seek $\eta = -W/Q_3$ and should thus eliminate Q_1 by combining the two equations:

$$-Q_1 = Q_3 T_a/T_b + \Delta_{ip}S \cdot T_a;$$
$$-W = Q_1 + Q_3 = -Q_3 T_a/T_b - \Delta_{ip}S \cdot T_a + Q_3 = Q_3(T_b - T_a)/T_b - \Delta_{ip}S \cdot T_a$$

and thus $\eta = -W/Q_3 = (T_b - T_a)/T_b - \Delta_{ip}S \cdot T_a/Q_3 < (T_b - T_a)T_b$ because $\Delta_{ip}S$, T_a and Q_3 are all positive.

Exercise 1.6.5

Examine if it would be possible to have a system with a spontaneous process such that dS and dQ have different signs under spontaneous isothermal conditions. Try to find such a system.

Hint

Start from the second law.

Solution

$TdS = dQ + Td_{ip}S$. Since $d_{ip}S > 0$ we get $TdS > dQ$. If dQ is positive, dS must also be positive. However, if dQ is negative, dS could be positive or negative. In this case it is easier to make dS positive if the process is only slightly exothermic and if it has a high internal production of entropy. A possible case would be the mixing of two elements to form a solution with a slight negative heat of reaction.

Exercise 1.6.6

Find a state function from which one could evaluate the heat absorption from the surroundings when a homogeneous material is compressed isothermally.

Hint

Heat is not a state function of a system. In order to solve the problem we must know how the change was made. Let us first assume that it was reversible.

Solution

For a reversible change $Q = \int T dS$. When the change is also isothermal $Q = T_1 \int dS = T_1(S_2 - S_1)$. However, this Q is negative because S decreases by isothermal compression, $S_2 - S_1 < 0$. The energy $-Q = T_1(S_1 - S_2)$ is thus extracted from the material as heat. For irreversible conditions $dQ < T dS$; $Q < T_1(S_2 - S_1)$ and the extracted heat is $-Q > T_1(S_1 - S_2)$, i.e. larger than before. However, if the final state is the same, ΔU must be the same because it is a state function and the higher value of $-Q$ must be compensated by a higher value of the work of compression W. How much higher $-Q$ and W will be cannot be calculated without detailed information on the factor making the compression irreversible.

Exercise 1.6.7

We have seen that a perpetuum mobile could be constructed from two reversible heat engines of different efficiencies (if they could be found) by running the less efficient one in reverse. However, this seems impossible and that is why we concluded that all reversible heat engines must have the same efficiency. On the other hand, an irreversible heat engine (i) will have a lower efficiency. It would thus seem tempting to run such an engine in reverse and couple it to a reversible engine (r). Evaluate the net work produced by such an arrangement.

Hint

The net work produced is $-W = -W^r - W^i = Q_1^r + Q_3^r + Q_1^i + Q_3^i = Q_1^r + Q_1^i$ because we should make the size of the irreversible engine such that the heat it delivers to the warm reservoir, $-Q_3^i$, is equal to the heat the reversible engine takes from the same reservoir, Q_3^r, i.e. $Q_3^r + Q_3^i = 0$. Furthermore, for each engine $\Delta S = 0$ for each whole cycle.

Solution

$\Delta S = 0 = Q_3^r/T_b + Q_1^r/T_a$; $\Delta S = 0 = Q_3^i/T_b + Q_1^i/T_a + \Delta_{ip}S$ where $\Delta_{ip}S$ is the sum of internal entropy production in all four steps. By adding the two equations we get: $(Q_3^r + Q_3^i)/T_b + (Q_1^r + Q_1^i)/T_a + \Delta_{ip}S = 0$. However, $Q_3^r + Q_3^i = 0$ and $-W = -W^r - W^i = Q_1^r + Q_1^i = -\Delta_{ip}S \cdot T_a < 0$ because $\Delta_{ip}S$ must be positive (second law). One cannot produce work by this arrangement.

Condition of internal equilibrium

The second law states that an internal process may continue as long as $d_{ip}S$ is positive. It must stop when for a continued process one would have

$$d_{ip}S \leq 0$$

This is the condition for internal equilibrium and a system is in such a state if this condition applies to all conceivable internal processes in the system. By integrating $d_{ip}S$ over an internal process we may obtain a measure of the net production of entropy by the process, $\Delta_{ip}S$. It has its maximum value at equilibrium. The maximum may be smooth, $d_{ip}S = 0$, or sharp, $d_{ip}S < 0$.

As an example of the first case, Fig. 1.4 shows a diagram for the formation of vacancies in a pure metal. The internal variable, generally denoted by ξ, is here the number of vacancies per mole of the metal.

As an example of the second case, Fig. 1.5 shows a diagram for the solid state reaction between two phases, graphite and $Cr_{0.7}C_{0.3}$, by which a new phase $Cr_{0.6}C_{0.4}$ is formed. The internal variable here represents the amount of $Cr_{0.6}C_{0.4}$. The curve only exists up to a point of maximum where one or both of the reactants have been consumed (in this case $Cr_{0.7}C_{0.3}$). From the point of maximum the reaction can only go in the reverse direction and that would give $d_{ip}S < 0$ which is not permitted for a spontaneous reaction. The sharp point of maximum thus represents a state of equilibrium. This case is often neglected and one usually treats equilibrium with the equality sign only, $d_{ip}S = 0$.

If $d_{ip}S = 0$ it is possible that the system is in a state of minimum $\Delta_{ip}S$ instead of maximum. By a small, finite change the system could then be brought into a state where $d_{ip}S > 0$ for a continued change. Such a system is thus at an **unstable equilibrium**. As a consequence, for a stable equilibrium we require that either $d_{ip}S < 0$, or $d_{ip}S = 0$ but then its second derivative must be negative.

It should be mentioned that instead of introducing the internal entropy production, $d_{ip}S$, one sometimes introduces dQ'/T where dQ' is called 'uncompensated heat'. It represents the extra heat which must be added to the system if the same change of the system were accomplished by a reversible process. Under the actual, irreversible conditions one has $dS = dQ/T + d_{ip}S$. Under the hypothetical, reversible conditions one has $dS = (dQ + dQ')/T$. Thus, $dQ' = Td_{ip}S$. In the actual process $d_{ip}S$ is produced

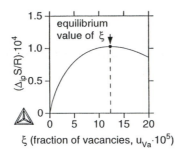

Figure 1.4 The internal entropy production due to the formation of thermal vacancies in one mole of a pure element at a temperature where the energy of formation of a vacancy is $9kT$, k being Boltzmann's constant. The initial state is a pure element without any vacancies. The internal variable is here the number of vacancies expressed as moles of vacancies per mole of metal, u_{Va}.

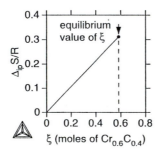

Figure 1.5 The internal entropy production due to the solid state phase transformation $C + Cr_{0.7}C_{0.3} \rightarrow Cr_{0.6}C_{0.4}$ at 1500 K and 1 bar. The initial state is 0.5 mole each of C (graphite) and $Cr_{0.7}C_{0.3}$. The internal variable ξ here represents the amount of $Cr_{0.6}C_{0.4}$, expressed as moles of this formula unit.

without the system being compensated by such a heat flow from the surroundings.

If the reversible process could be carried out and the system thus received the extra heat dQ', as compared to the actual process, then the system must also have delivered the corresponding amount of work to the surroundings in view of the first law. Because of the irreversible nature of the process, this work will not be delivered and that is why one sometimes talks about the '**loss of work**' in the actual process which is irreversible and produces some entropy instead of work, $dW = dQ' = Td_{ip}S$.   •

Exercise 1.7.1

Check the loss of work in a cyclic process working with a high-temperature heat source of T_b and a low-temperature heat sink of T_a and having some internal entropy production.

Hint

In exercise 1.6.4 we found $- W = Q_3(T_b - T_a)/T_b - \Delta_{ip}S \cdot T_a$. From this result we can calculate the 'loss of work', e.g. if the amount of heat extracted from the heat source is the same in the irreversible case as in the reversible one.

Solution

For a reversible cycle one would have $- W = Q_3(T_b - T_a)/T_b$. The 'loss of work' is thus $\Delta_{ip}S \cdot T_a$ if the amount of heat extracted from the heat source is the same. If, instead, the amount of heat given to the heat sink is kept constant, we should eliminate Q_3 from the two equations in the solution of exercise 1.6.4:
$$- Q_3 = \Delta_{ip}S \cdot T_b + Q_1 T_b/T_a; \quad - W = Q_1 + Q_3 = - Q_1(T_b - T_a)/T_a - \Delta_{ip}S \cdot T_b.$$
The 'loss of work' is thus $\Delta_{ip}S \cdot T_b$ in this case. The 'loss of work' is here found by multiplying the total internal entropy production with the temperature of the heat sink.

1.8 *Driving force*

Let the internal variable ξ represent the extent of a certain internal process. The internal entropy production should be a function of this variable and we may define $d_{ip}S/d\xi$ as a new state variable or state function. For convenience, we shall multiply by T to obtain

$$D = T \cdot \frac{d_{ip}S}{d\xi}$$

Classically, this quantity has been regarded as the **affinity** between reacting chemical species. However, it has a much wider applicability and will here be regarded as the **driving force** for any internal process. The symbol D, chosen here, may either be regarded as an abbreviation of driving force or as an honour to De Donder who explored its applicability in great detail. D is a state function because its value depends only on (i) the state of the system and (ii) the definition of the internal process, the extent of which is measured by ξ which is then also regarded as a new state variable. It is usually convenient to define the extent of an internal process ξ by a variable which is an extensive property, subject to the law of additivity. The driving force D is then an intensive variable.

   If a system is not in a state of equilibrium, there may be a spontaneous internal process for which the second law gives $d_{ip}S > 0$ and thus

$$Dd\xi > 0$$

It is evident that $d\xi$ and D must have the same sign in order for the process to proceed. By convention, $d\xi$ is given a positive value in the direction one wants to examine and D

must then be positive for a spontaneous process in that direction. In many applications one even attempts to predict the rate of a process from the magnitude of D. Simple models often predict a proportionality. This will be further discussed in Section 4.5.

If $D > 0$ for some internal process, then the system is not in a state of equilibrium. The process may proceed and it will eventually approach a state of equilibrium where $D = 0$. The equilibrium value of the variable ξ can, in principle, be evaluated from the condition $D = 0$, which is usually more directly applicable than the basic condition $d_{ip}S = 0$.

In the preceding section we connected an internal entropy production with the progress of an internal process. However, we can now see that it is possible, in principle, to change an internal variable without any entropy production. This can be done by changing the external variables in such a way that the driving force D is always zero. Since D is zero at equilibrium only, it is necessary to change the external variables so slowly that ξ can all the time adjust itself to the new value required by equilibrium. In practice, this cannot be completely achieved because the rate of the process should be zero if its driving force is zero. An infinitely slow change is thus necessary. Such an idealized change is identical to the reversible process mentioned in the preceding section and it is sometimes described as an 'equilibrium reaction'. It takes the system through a series of equilibrium states.

It may be convenient to consider a reversible process if one knows a state of equilibrium for a system and wants to find other states of equilibrium under some different conditions. This is the reason why one often applies 'reversible conditions'. As an example we may consider the heating of a system under constant volume, discussed in Section 1.4. The heat capacity under such conditions, C_V, was found to be different under slow and rapid changes. Both of these cases may be regarded as reversible because the internal entropy production is negligible when D is small for a very slow change and also when $d\xi$ is small for a frozen-in internal process. For both cases we may thus use $dS = dQ/T$ and we obtain two different quantities,

$$C_V \equiv \left(\frac{\partial Q}{\partial T}\right)_V = T\left(\frac{\partial S}{\partial T}\right)_V$$

$$C_{V,\xi} \equiv \left(\frac{\partial Q}{\partial T}\right)_{V,\xi} = T\left(\frac{\partial S}{\partial T}\right)_{V,\xi}$$

These expressions are equivalent to those given in Section 1.4 in terms of U. For intermediate cases, which are not reversible, one should consider U and not S, i.e. use the first law and not the second law.

Exercise 1.8.1

Consider an internal reaction which gives an entropy production under isothermal conditions, $\Delta_{ip}S = -\xi K/T - R[\xi \ln \xi - (1 + \xi)\ln(1 + \xi)]$. Derive the stability at equilibrium, defined as $B = - T \cdot d_{ip}^2 S/d\xi^2$. (See Section 5.1.).

Hint

In exercise 1.6.1 we have already calculated $d_{ip}S/d\xi$ and ξ at equilibrium.

Solution

$d_{ip}S/d\xi = - K/T - R[\ln\xi - \ln(1 + \xi)]$; $d_{ip}^2 S/d\xi^2 = - R[1/\xi - 1/(1 + \xi)]$.
However, at equilibrium $1/\xi = \exp(K/RT) - 1$; $1/(1 + \xi) = [\exp(K/RT) - 1]/\exp(K/RT)$. Thus, $B = + R[1/\xi - 1/(1 + \xi)] = R[1 - \exp(K/RT)]^2/\exp(K/RT)$.
This is always positive. The state of equilibrium must be stable.

1.9   *The combined first and second law*

For a non-insulated system the second law takes the following form if one internal process is considered

$$T dS = dQ + T d_{ip}S = dQ + D d\xi$$

By combination with the first law, applied to hydrostatic work $dU = dQ - PdV$, we obtain

$$T dS = dU + P dV + D d\xi$$

This may be regarded as the combined first and second law for the state variables U and V and the quantity S is regarded as the **characteristic state function** for this set of independent external variables. U and V may thus be regarded as the **natural variables** for S. Both U and V may be completely controlled by actions from the external world and are thus regarded as external state variables, whereas ξ is an internal state variable.

It is more common to write the combined law in the following form

$$dU = T dS - P dV - D d\xi$$

Here, U is the characteristic state function and its natural variables are S and V. One usually regards S as an external variable although its value is also influenced by internal processes and it is not possible to control its value by actions from the outside without an intimate knowledge of the properties of the system.

When there are i internal processes, one should replace $D d\xi$ by $\Sigma D^i d\xi^i$. For the sake of simplicity this will be done only when we actually consider more than one process.

We can write the combined law in a more general form by considering more types of work. By grouping together the products of the external variables we write

$$dU = \Sigma Y^a dX^a - D d\xi$$

Y^a represents potentials like T. It is evident that the pressure should be expressed as $-P$ in order to be comparable with other potentials. As a consequence, we shall plot P in the negative direction in many diagrams (see, for instance, Fig. 1.1). X^a represents extensive quantities like S and V. The pair of one potential and one extensive quantity, Y^a and X^a, is called a pair of **conjugate variables**, for instance T,S or $-P,V$. Other pairs of conjugate variables may be included through the first law by considering other types of work, for instance gravitational work. It is important to notice that the change in U is given in terms of the changes in variables all of which are extensive like S and V and all of them are subject to the law of additivity.

Since U is a state variable which is a function of all the external variables, X^a, X^b, etc., and the internal ξ variables, we have

$$Y^b = \left(\frac{\partial U}{\partial X^b}\right)_{X^c,\xi}$$

where X^c represents all the X variables except for X^b. It is interesting to note that all the Y variables are obtained as partial derivatives of an energy with respect to an extensive variable. That is why they are regarded as potentials. One may also regard $-D$ and ξ as a pair of conjugate variables where $-D$ is the potential and is obtained as

$$-D = \left(\frac{\partial U}{\partial \xi}\right)_{X^a}$$

where X^a represents all the X variables. It should be emphasized that the Y potentials have here been defined for a frozen-in state because ξ was treated as an independent variable which is kept constant. Under conditions of maintained equilibrium one should treat ξ as a dependent variable and the potentials are defined as

$$Y^b = \left(\frac{\partial U}{\partial X^b}\right)_{X^c}$$

We shall soon see that for equilibrium states the two definitions of Y^b give the same result.

In the following discussions we shall not want to be limited to frozen-in states $(\mathrm{d}\xi = 0)$, nor to equilibrium states or reversible changes $(D = 0)$ and we shall thus retain the $D\mathrm{d}\xi$ term in the combined law. It should again be emphasized that there are these two different cases for which the term $D\mathrm{d}\xi$ is zero and can be omitted.

The combined law can be expressed in several alternative forms depending upon the choice of independent external variables. These forms make use of new state functions which will soon be discussed.

Exercise 1.9.1

Formulate the combined law for a system which interacts with the surroundings by means of two pistons working at different pressures.

Hint

One may speculate on what happens inside such a system, i.e. on the nature of the internal process, the progress of which we represent by ξ. However, that is not a problem we need to concern ourselves with now.

Solution

The first law becomes $dU = dQ + dW_1 + dW_2 = dQ - P_1 dV_1 - P_2 dV_2$ and the combined law becomes $dU = TdS - P_1 dV_1 - P_2 dV_2 - Dd\xi$.

Exercise 1.9.2

Try to include the effect of electrical work in the combined law.

Hint

There are two cases. Firstly, consider the addition of an extra charge to the system. Secondly, consider the case where the system is made part of an electrical circuit.

Solution

In the first case, the first law gives $dU = dQ + dW + dW_{el}$ where we may write $dW_{el} = E \cdot d(\text{charge}) = -E\mathcal{F}dn_e$ where $\mathcal{F}$ is the Faraday constant (the negative of the charge of one mole of electrons) and n_e is the number of extra electrons (in mole). E is the electrical potential. The combined law becomes $dU = TdS - PdV - E\mathcal{F}dn_e - Dd\xi$. However, E increases very rapidly with n_e and reaches extremely high values before n_e is large enough to have a chemical effect. This form is thus of little practical interest.

   Let us then consider the case of a circuit. From the discussion of the first case, we understand that the charge entering a system through one lead must be practically equal to the charge leaving the system from the other lead, i.e. $dn_{e1} = dn_{e2}$. The first law becomes $dU = dQ + dW + dW_{e1} = dQ + dW - E_1\mathcal{F}dn_{e1} - E_2\mathcal{F}dn_{e2} = dQ + dW - (E_1 - E_2)\mathcal{F}dn_{e1}$, and the combined law becomes $dU = TdS - PdV - (E_1 - E_2)\mathcal{F}dn_{e1} - Dd\xi$. Right now we do not need to speculate on what happens inside the system.

1.10   *General conditions of equilibrium*

A system is in a state of equilibrium if the driving forces for all possible internal processes are zero. Many kinds of internal processes can be imagined in various types of systems but there is one class of internal process which should always be considered, the transfer of a quantity of an extensive variable from one part of the system to another part. In this section we shall examine the equilibrium condition for such a process.

Let us examine an internal process taking place in a system under constant values of the external extensive variables S and V, here collectively denoted by X^a, and let us not be concerned about the experimental difficulties encountered in performing such an experiment. We could then turn to the combined first and second law in terms of dU, which is reduced as follows

$$dU = \Sigma Y^a dX^a - D d\xi = - D d\xi$$

The driving force for the internal process will be

$$D = - (\partial U/\partial \xi)_{X^a}$$

The process can occur spontaneously until U has reached a minimum. The state of minimum in U at constant S and V is thus a state of equilibrium.

The internal process we shall now consider is the transfer of dX^b from one half of the system (') to the other half ("), keeping the remaining Xs constant in the two halves. It is convenient to measure the extent of this internal process by identifying $d\xi$ with $- dX^b$ for the first half of the system and $+ dX^b$ for the second half. We thus obtain, by applying the law of additivity to D,

$$- D = \left(\frac{\partial U}{\partial \xi}\right)_{X^a} = \left(\frac{\partial U}{- \partial X^b}\right)'_{X^c} + \left(\frac{\partial U}{\partial X^b}\right)''_{X^c} = - \left(\frac{\partial U}{\partial X^b}\right)'_{X^c} + \left(\frac{\partial U}{\partial X^b}\right)''_{X^c}$$

The derivative $\partial U/\partial X^b$ is identical to the conjugate potential Y^b and we thus find

$$D = Y^{b'} - Y^{b''}$$

The driving force for this process will be zero and the system will be in equilibrium with respect to the process if the potential Y^b has the same value in the two parts of the system. We have thus proved that each potential must have the same value in the whole system at equilibrium. This applies to T, and to P with an exception to be treated in Section 14.6. It also applies to chemical potentials μ_i which have not yet been introduced.

Exercise 1.10.1

One may derive a term $- E\mathcal{F}dn_e$ for the electrical contribution to dU. Here E is the electrical potential and $- \mathcal{F}dn_e$ the electrical charge because dn_e is the number of moles of extra electrons and $- \mathcal{F}$ is the charge of one mole of electrons. Evaluate the driving force for the transfer of electrons from one half of the system to the other if their electrical potentials are E' and E''. Define $d\xi$ as dn_e.

Solution

$$-D = (\partial U/\partial \xi) = -(\partial U/\partial n_e)' + (\partial U/\partial n_e)'' = E'\mathscr{F} - E''\mathscr{F}; D = (E'' - E')\mathscr{F}.$$

1.11   *Characteristic state functions*

We have seen that a combination of the first and second laws allows us to calculate how the internal energy changes as a result of variations in S and V and also as a result of an internal process

$$dU = TdS - PdV - Dd\xi$$

Under experimental conditions of constant S and V this equation yields

$$dU = -Dd\xi$$

and the equilibrium condition, $D = 0$, can then be written as

$$-D = (\partial U/\partial \xi)_{S,V} = 0$$

for all conceivable internal processes. If this condition is not fulfilled, then the internal energy may decrease spontaneously by the internal process and eventually approach a minimum.

From an experimental point of view it is not very easy to control S but relatively easy to control T. A change of independent variable may thus be desirable and it can be performed by subtracting $d(TS)$ which is equal to $TdS + SdT$. The combined law is thus modified to

$$d(U - TS) = -SdT - PdV - Dd\xi$$

We may regard this as the combined law for the variables T and V, and $U - TS$ is regarded as the characteristic state function for these variables, whereas U is regarded as the characteristic state function for the variables S and V. The new function $(U - TS)$ has been given its own name and symbol, Helmholtz energy, F,

$$F = U - TS$$

This state function can change as a result of changes in T and V and also as a result of an internal process

$$dF = -SdT - PdV - Dd\xi$$

Under experimental conditions of constant T and V we obtain

$$dF = -Dd\xi$$

and the equilibrium condition can then be written as

$$-D = (\partial F/\partial\xi)_{T,V} = 0$$

for all conceivable processes. In an experiment under constant T and V there may be spontaneous changes until F has approached a minimum.

In the same way we may modify the combined law by changing from V to P as independent variable by adding $d(PV)$,

$$d(U + PV) = TdS + VdP - Dd\xi$$

This may be regarded as the combined law for the variables S and P, and $(U + PV)$ is regarded as the characteristic state function for these variables. It is called enthalpy, H,

$$H = U + PV$$

The equilibrium condition under constant S and P is

$$-D = (\partial H/\partial\xi)_{S,P} = 0$$

In an experiment under constant S and P there may be spontaneous internal changes until H has approached a minimum.

It is important to notice that the change in enthalpy in a particular experiment can be obtained directly from the first law

$$dH = d(U + PV) = dQ - PdV + PdV + VdP = dQ + VdP$$

In fact, the enthalpy H has already been introduced in connection with the first law in Section 1.3.

By applying both modifications we obtain

$$d(U - TS + PV) = -SdT + VdP - Dd\xi$$

This may be regarded as the combined law for the variables T and P and the characteristic state function for these variables, $(U - TS + PV)$, is called Gibbs energy, G,

$$G = U - TS + PV$$
$$dG = -SdT + VdP - Dd\xi$$

This characteristic state function is of particular interest because T and P are the variables which are mot easily controlled experimentally and they are both potentials. In experiments under constant T and P the equilibrium conditions is

$$-D = (\partial G/\partial\xi)_{P,T} = 0$$

G may decrease spontaneously to a minimum under constant T and P.

Finally, it may be mentioned that the mathematical operation, we have used in order to introduce a potential instead of an extensive variable, is called **Legendre**

transformation. An important aspect is that no information is lost during such a transformation, as will be discussed in Section 2.1.

Exercise 1.11.1

Suppose a system, when it transforms between two states, can perform work in addition to $\int Pd V$. Show that the maximum amount of work is obtained as the change in Gibbs energy if the two states are at the same pressure and temperature.

Hint

Start with the first law and introduce two kinds of work. If the additional work done on the system is dW_a, then the additional work performed by the system is $-dW_a$. Combine with the second law.

Solution

$dU = dQ + dW = dQ - PdV + dW_a$; $dU = TdS - PdV + dW_a - Dd\xi$ where ξ measures the progress of the transformation. Introducing T and P as independent variables yields: $dU - d(TS) + d(PV) = dG = -SdT + VdP + dW_a - Dd\xi$. At constant T,P: $-dW_a = -dG - Dd\xi$. The work performed, $-dW_a$, would be larger the smaller $Dd\xi$ is. The system would thus perform the maximum work if one could arrange that $Dd\xi = 0$ because $Dd\xi$ cannot be negative for a spontaneous process. Then we get: $-dW_a = -dG$; $-\Delta W_a = G_1 - G_2$.

Exercise 1.11.2

Suppose that it would be practically possible to keep H and P constant during an internal reaction in a system. What state function should then be used in order to predict the state of equilibrium?

Hint

Find a form of the combined law which has dH and dP on the right-hand side.

Solution

$dH = TdS + VdP - Dd\xi$; $TdS = dH - VdP + Dd\xi$. We thus obtain $D = T(\partial S/\partial \xi)_{H,P} > 0$ for spontaneous reactions. Equilibrium is where $S(\xi)$ is at its maximum.

Exercise 1.11.3

In two exercises on the second law we considered an internal reaction giving the following internal production of entropy under isothermal conditions. $\Delta_{ip}S = -\xi K/T - R[\xi\ln\xi - (1 + \xi)\ln(1 + \xi)]$ and giving a heat of reaction $\Delta Q = \xi K$ under constant volume. Derive an expression for the Helmholtz energy F and calculate the equilibrium value of ξ. Compare with the previous result obtained by minimizing the internal production of entropy.

Hint

Use $\Delta F = \Delta U - \Delta(TS)$ where ΔU is obtained from the first law and ΔS from the second law.

Solution

Under isothermal conditions the heat of reaction must be compensated by heat flow from the surroundings, $\Delta Q = \xi K$. Since the volume is constant $\Delta U = \Delta Q = \xi K$. The total increase of entropy is $\Delta S = \Delta Q/T + \Delta_{ip}S = \xi K/T + \Delta_{ip}S = - R[\xi\ln\xi - (1 + \xi)\ln(1 + \xi)]$ and thus $\Delta F = \Delta U - \Delta(TS) = \xi K + RT[\xi\ln\xi - (1 + \xi)\ln(1 + \xi)]$. This is identical to $- T\Delta_{ip}S$. We thus get the same result if we minimize ΔF at constant temperature or maximize $\Delta_{ip}S$.

Exercise 1.11.4

A system is contained inside an elastic wall such that $V = V_0(1 - \alpha P)$. Derive the equilibrium condition for an internal process in such a container if the temperature is kept constant.

Hint

Since V_0 is a constant and we are going to keep T constant, we should formulate the combined law with T and V_0 as the variables and then keep them constant. (Since T is to be constant we do not need to consider the fact that the properties of the wall may depend on T.) First, we introduce V_0 instead of P in the expression for dU. Then we must find a function $f(V, V_0)$ such that the dV term in dU vanishes when we subtract df. This function is found by integrating the dV term under constant V_0.

Solution

$P = 1/\alpha - V/\alpha V_0$; $dU = TdS - (1/\alpha - V/\alpha V_0)dV - Dd\xi$. We now want to subtract df from dU where f is a function such that dV will not appear in $d(U - f)$. It is evident that we can choose $f = -V/\alpha + V^2/2\alpha V_o$; $df = -(1/\alpha - V/\alpha V_o)dV + (V^2/2\alpha)d(1/V_o)$. We also subtract $d(TS)$ and obtain $d(U - TS - f) = TdS - TdS - SdT - (1/\alpha - V/\alpha V_o)dV + (1/\alpha - V/\alpha V_o)dV - (V^2/2\alpha)d(1/V_o) - Dd\xi = -SdT - (V^2/2\alpha)d(1/V_o) - Dd\xi$. Equilibrium at constant T,V_o is given by $D = -(\partial[U - TS - f]/\partial\xi)_{T,V_o}$. The equilibrium condition is thus the minimum of a new function $U - TS + V/\alpha - V^2/2\alpha V_o$.

1.12   *Entropy*

Before finishing the present discussion of basic concepts of thermodynamics, a few words regarding entropy should be added. No attempt will be made to explain the nature of entropy. However, it is important to realize that there is a fundamental difference between entropy and volume in spite of the fact that these two extensive state variables appear in equivalent places in many thermodynamic equations, for instance in the forms of the combined law defining dU or dG. For volume there is a natural zero point and one can give absolute values of V. As a consequence, the change of G due to a variation of P, $(\partial G/\partial P)_T$, is a well-defined quantity because it is equal to V. One may thus compare the values of G of two systems at different pressures.

As for internal energy or enthalpy, there is no natural zero point for entropy although it is quite common to put $S = 0$ for a well-crystallized substance at absolute zero. That is only a convention and it does not alter the fact that the change of the Gibbs energy G due to a variation of T, $(\partial G/\partial T)_P$, cannot be given an absolute value because it is equal to $-S$. As a consequence, it makes no sense to compare the values of G of two systems at the same pressure but different temperatures. The interaction between such systems must be based upon kinetic consideration, not upon the difference in G values. The same is true for the Helmholtz energy F at the same volume but different temperatures because $(\partial F/\partial T)_V$ is also equal to $-S$.

The convention to put $S = 0$ at absolute zero is useful because the entropy difference between two crystalline states of a system of fixed composition goes to zero there according to Nernst's heat theorem, sometimes called the **third law**. It should be emphasized that the third law defined in this way only applies to states which are not frozen in a disordered arrangement.

Statistical thermodynamics can provide answers to some questions which are beyond classical thermodynamics. It is based upon the Boltzmann relation

$$S = k\ln W$$

where k is the Boltzmann constant ($= R/N^A$ where N^A is Avogadro's number) and W is the number of different ways in which one can arrange a state. $1/W$ is thus a measure of the probability that a system in this state will actually be arranged in a particular way. Boltzmann's relation is a very useful tool in developing thermodynamic models for various types of phases and it will be used extensively in Chapters 16–19. It will there be applied to one physical phenomenon at a time. The contribution to the entropy from such a phenomenon will be denoted by ΔS or more specifically by ΔS_i and we can write Boltzmann's relation as

$$\Delta S_i = k \ln W_i$$

W_i and ΔS_i are evaluated for this phenomenon alone. Such a separation of the effects of various phenomena is permitted because $W = W_1 \cdot W_2 \cdot W_3 \cdots$

$$S = k \ln W = k \ln(W_1 \cdot W_2 \cdot W_3 \cdots) = k(\ln W_1 + \ln W_2 + \ln W_3 + \ldots)$$
$$= \Delta S_1 + \Delta S_2 + \Delta S_3 + \ldots$$

Finally, we should mention here the possibility of writing the combined law in a form which avoids using entropy as a variable and instead treats entropy as the characteristic state function, although this will be discussed in much more detail in Chapters 3 and 5.

$$- T\mathrm{d}S = -\mathrm{d}U - P\mathrm{d}V - D\mathrm{d}\xi$$
$$- \mathrm{d}S = -(1/T)\mathrm{d}U - (P/T)\mathrm{d}V - (D/T)\mathrm{d}\xi$$

This formalism is sometimes called the **entropy scheme** and the formalism based upon $\mathrm{d}U$ is then called the **energy scheme**. With the entropy scheme we have here introduced new pairs of conjugate variables, $(-1/T, U)$ and $(-P/T, V)$. We may also introduce H into this formalism by using $\mathrm{d}U = \mathrm{d}H - P\mathrm{d}V - V\mathrm{d}P$, whence

$$- \mathrm{d}S = -(1/T)\mathrm{d}H + (V/T)\mathrm{d}P - (D/T)\mathrm{d}\xi$$
$$- \mathrm{d}(S + PV/T) = -(1/T)\mathrm{d}H - P\mathrm{d}(V/T) - (D/T)\mathrm{d}\xi$$

Two more pairs of conjugate variables appear here, $(-1/T, H)$ and $(-P, V/T)$. It is evident that S is the characteristic state function for H and P as well as for U and V.

By subtracting $\mathrm{d}S$ from $\mathrm{d}(H/T)$ we further obtain

$$\mathrm{d}(H/T - S) = H\mathrm{d}(1/T) + (V/T)\mathrm{d}P - (D/T)\mathrm{d}\xi$$

and this is equal to $\mathrm{d}(G/T)$ because $G = U - TS + PV = H - TS$. This form of the combined law has the interesting property that it yields directly the expression for the enthalpy,

$$H = \left(\frac{\partial(G/T)}{\partial(1/T)} \right)_P$$

It should again be emphasized that enthalpy has no natural zero point and $(\partial[G/T]/\partial[1/T])_P$ cannot be given an absolute value, just as $(\partial G/\partial T)_P$ cannot.

Exercise 1.12.1

Consider two pieces of Zn, one at each end of a sealed silica tube, filled with argon at 1 bar pressure. The ends are kept at different temperatures. One piece may shrink by evaporization and the other grow by condensation. Which one would grow if the principle governing the process were the minimization of G? Discuss the conclusion.

Hint

Suppose the vapour pressure of Zn is small enough to be neglected in comparison with 1 bar and thus $dG/dT = -S$.

Solution

With the usual convention, S is always positive and thus G is lower at the high-T end. If Zn would go from higher G to lower G, it seems that it would evaporate from the cold end and condense at the hot end. However, by experience we know that condensation will occur at the cold end. Actually, there is no basis for the suggestion that G should be minimized by a non-isothermal reaction because the difference in G depends upon the arbitrary choice of reference for S. ΔG between two systems at different temperatures has no physical significance. Instead, the reaction is governed by kinetic factors and depends upon the properties of the substance separating the two pieces, in our case the gas. The vapour pressure of Zn is greater at the higher temperature and Zn will diffuse through the gas from the higher temperature to the lower one.

Exercise 1.12.2

We know that S can be calculated from $G(T,P)$ as $-(\partial G/\partial T)_P$. It may be tempting to try to derive this relation as follows: $G = H - TS$; $\partial G/\partial T = -S$. However, that derivation is not very satisfactory. Show that a correction should be added and then prove that the correction is zero under some conditions.

Hint

Remember that H and S may depend upon T.

Solution

$G = H - TS$ gives strictly $(\partial G/\partial T)_P = (\partial H/\partial T)_P - T(\partial S/\partial T)_P - S$. However, the sum of the first two terms (the contributions from the T-dependence of H and S) is zero under reversible conditions and constant P because then $dH = TdS + VdP - Dd\xi = TdS$.

Exercise 1.12.3

We have discussed the consequences for G and F of the fact that S has no natural zero point. In fact, nor does U. Find a quantity for which this has a similar consequence.

Hint

Find a form of the combined law which has U as a coefficient just as S is a coefficient in $\mathrm{d}G = -S\mathrm{d}T + V\mathrm{d}P - D\mathrm{d}\xi$.

Solution

$\mathrm{d}S = (1/T)\mathrm{d}U + (P/T)\mathrm{d}V + (D/T)\mathrm{d}\xi$; $\mathrm{d}(-F/T) = \mathrm{d}(S - U/T) = -U\mathrm{d}(1/T) + (P/T)\mathrm{d}V + (D/T)\mathrm{d}\xi$; $(\partial[F/T]/\partial[1/T])_V = U$. Thus, one cannot compare the values of F/T at the same volume but different temperatures.

Manipulation of thermodynamic quantities

Evaluation of one characteristic state function from another

In the first chapter we have defined some characteristic state functions, U, F, H and G in addition to S. Each one was introduced through a particular form of the combined law. The independent variables in each form may be regarded as the natural variables for the corresponding characteristic state function. In integrated form these functions can thus be written as

$$U = U(S,V,\xi)$$
$$H = H(S,P,\xi)$$
$$F = F(T,V,\xi)$$
$$G = G(T,P,\xi)$$

According to Gibbs, these integrated forms may all be regarded as **fundamental equations** because, if any one of them is known for a substance, then all thermodynamic properties of the substance can be evaluated. This is because the values of all the dependent variables can be calculated for any given set of independent variables. As an example, from the combined law for the variables S, V and ξ we get

$$T = (\partial U/\partial S)_{V,\xi}$$
$$- P = (\partial U/\partial V)_{S,\xi}$$
$$- D = (\partial U/\partial \xi)_{S,V}$$

As a consequence, we can now calculate the value for any other of the characteristic state functions at a given set of S, V and ξ for instance

$$G = U - TS + PV = U - S(\partial U/\partial S)_{V,\xi} - V(\partial U/\partial V)_{S,\xi}$$

It should be noted that the calculation of G from U is only possible because U is known as a function of its natural variables. As a consequence of the same principle, even if G can thus be obtained as an analytical expression from $U(S,V,\xi)$ by the use of the above relation, the result is not a fundamental equation because $G(S,V,\xi)$ does not allow the dependent variables to be calculated. They can only be calculated from G as $G(T,P,\xi)$ through $S = -(\partial G/\partial T)_{P,\xi}$ and $V = (\partial G/\partial P)_{T,\xi}$. It is thus necessary first to replace S and V by T and $-P$, which is seldom possible to do analytically. If that replacement is not done, then some information has been lost in the calculation of G from U.

It should again be emphasized that a characteristic state function will be a fundamental equation only if expressed as a function of its natural variables.

Exercise 2.1.1

Show $G = (\partial[H/S]/\partial[1/S])_P$.

Hint

Evidently, no internal reaction is considered. The natural variables of H are S and P. It should thus be possible to express G in terms of H and its derivatives.

Solution

Without any internal reaction we have, from Section 1.11, $dH = TdS + VdP$; $T = (\partial H/\partial S)_P$; $G = H - TS = H - S(\partial H/\partial S)_P$. This is equal to the expression given for G because it can be transformed: $(\partial[H/S]/\partial[1/S])_P = H + (1/S)(\partial H/\partial[1/S])_P = H - S(\partial H/\partial S)_P$.

2.2   *Internal variables at equilibrium*

We have already emphasized that ξ is a dependent variable if the system is to remain in internal equilibrium. Since $D = 0$ in such a state, the equilibrium value of ξ can be evaluated from any one of the fundamental equations, for instance from $U(S,V,\xi)$,

$$-D = (\partial U/\partial \xi)_{S,V} = 0$$

In principle, this condition yields a relation for the equilibrium value

$$\xi = \xi(S,V)$$

which may be inserted in $U(S,V,\xi)$ in order to yield

$$U = U(S,V)$$

for states of internal equilibrium.

For a state of equilibrium, $D = 0$ and $dU = TdS - PdV - Dd\xi = TdS - PdV$. One can thus evaluate T and $-P$ from

$$Y^b = \left(\frac{\partial U}{\partial X^b}\right)_{X^c}$$

where X^b and X^c represent S and V, or V and S. Compared to the expressions given in Section 2.1, the variable ξ is now omitted because at equilibrium it is regarded as a dependent variable. The fact that ξ is omitted from the subscript should thus be interpreted as meaning that the derivative is evaluated under conditions of internal equilibrium. However, since

$$\left(\frac{\partial U}{\partial \xi}\right)_{X^a} = -D = 0$$

for a state of internal equilibrium, this new definition gives the same result as the definition for a state frozen-in at some ξ value, when that definition is applied to the state of equilibrium. As in Section 1.9, X^a as a subscript represents all X variables.

$$\left(\frac{\partial U}{\partial X^b}\right)_{X^c} = \left(\frac{\partial U}{\partial X^b}\right)_{X^c,\xi} + \left(\frac{\partial U}{\partial \xi}\right)_{X^a}\left(\frac{\partial \xi}{\partial X^b}\right)_{X^c} = \left(\frac{\partial U}{\partial X^b}\right)_{X^c,\xi}$$

When calculating the value of a partial derivative at equilibrium, one may use either expression. The new definition, i.e. the left-hand side of this equation, is preferable if ξ can be easily eliminated analytically from $U(S,V,\xi)$ which requires that $\xi(S,V)$ is an analytical function. However, this is sometimes laborious or even impossible and then the first definition, i.e. the right-hand side, may be used. Of course, in order to obtain a numerical value of the potential at equilibrium it is then necessary to evaluate and insert the equilibrium value of ξ as the frozen-in value.

This method of calculation can be applied to derivatives of all the characteristic state functions and one may thus use a mixed set of potentials and extensive variables as independent variables. However, it is necessary to use the particular characteristic state function which has the chosen set of state variables as natural variables. For the variables S and V this function is U. At equilibrium $(\partial U/\partial \xi)_{S,V} = 0$ but other derivatives are generally not zero. As an example, in general

$$\left(\frac{\partial U}{\partial \xi}\right)_{T,V} \neq 0$$

not even for a state of equilibrium. That is why C_V, the heat capacity at constant volume, is different when ξ is frozen-in and when ξ has time to adjust to internal equilibrium (see Section 1.4 on 'freezing-in' conditions).

Furthermore, it should be emphasized that this method of calculation can only be applied to the first derivatives of the characteristic state function, and not to higher-order derivatives, since, in general

$$\frac{\partial^2 U}{\partial X^b \partial \xi} \neq 0$$

In the preceding section, ξ was included as one of the variables in the fundamental equations. They are still called fundamental equations when applied to internal equilibrium conditions and thus without showing explicitly how they vary with ξ. Of course, such a fundamental equation only contains the properties at internal equilibrium.

Exercise 2.2.1

At high temperatures, H_2 gas may partially dissociate into H atoms. The following expression can be used for the Gibbs energy of a system which initially contained one mole of H_2 molecules: $G = (1 + \xi)RT\ln[P/(1 + \xi)] + (1 - \xi)[^{\circ}G_{H_2} + RT\ln(1 - \xi)] + 2\xi[^{\circ}G_H + RT\ln(2\xi)]$. Here, ξ is the fraction of H_2 molecules which have dissociated into free H atoms, and $^{\circ}G_{H_2}$ and $^{\circ}G_H$ are the Gibbs energies of one mole of pure H_2 and H, respectively, at a pressure equal to the unit used for pressure. Show how V can be calculated under freezing-in conditions at constant T and P. Then derive an expression for the equilibrium value of ξ and show how V can be calculated under equilibrium conditions.

Hint

It is convenient to introduce the equilibrium constant for the reaction $H_2 \Leftrightarrow 2H$, $K = \exp[(2^{\circ}G_H - ^{\circ}G_{H_2})/RT]$.

Solution

Freezing-in conditions: $V = (\partial G/\partial P)_{T,\xi} = (1 + \xi)RT/P$ for any constant value of ξ.
Equilibrium conditions: $- D = (\partial G/\partial \xi)_{T,P} = RT\ln P - RT(1 + \xi)/(1 + \xi) -$
$RT\ln(1 + \xi) - ^{\circ}G_{H_2} - RT\ln(1 - \xi) + RT(1 - \xi)/(1 - \xi)(- 1) + 2^{\circ}G_H +$
$2RT\ln(2\xi) + 2RT\xi/\xi = 0$ gives $(1 - \xi^2)4P\xi^2 = K$ and $\xi = 1/\sqrt{1 + 4PK}$ at
equilibrium. This ξ value could be inserted in the expression for G and
$V = (\partial G/\partial P)_T$ could then be evaluated. However, since $(\partial G/\partial \xi)_{T,P} = 0$ we can
directly use the expression derived for the frozen-in case, with the equilibrium value
of ξ inserted. We thus obtain $V = (1 + 1/\sqrt{1 + 4PK})RT/P$.

Exercise 2.2.2

A very simple model for the magnetic disordering of a ferromagnetic element gives the following expression at a constant pressure, $\Delta G = \xi(1 - \xi)K + RT[\xi\ln\xi + (1 - \xi)\ln(1 - \xi)]$ where K is a constant and ξ is the fraction of spins being disordered. The degree of magnetic disorder varies with T, $\xi = \xi(T)$. Derive an expression for the contribution to the enthalpy due to magnetic disordering, ΔH. Then calculate, at the temperature where $\xi = 1/4$, the corresponding contribution to the heat capacity at constant P which is defined as $\Delta C_P = (\partial \Delta H / \partial T)_P$.

Hint

The magnetic state cannot be frozen-in. ξ will always have its equilibrium value. It can be found from $(\partial \Delta G / \partial \xi)_{T,P} = 0$ which yields $(1 - 2\xi)K + RT\ln[\xi/(1 - \xi)] = 0$. Unfortunately, this does not give ξ as an analytical function of T and we cannot replace ξ by T in ΔG. Thus, it is convenient to make use of the fact that $(\partial \Delta G / \partial \xi)_{T,P} = 0$, which gives $- \Delta S = (\partial \Delta G / \partial T)_P = (\partial \Delta G / \partial T)_{P,\xi}$. We can then evaluate ΔH from $\Delta G + T\Delta S$.

Solution

$\Delta S = - (\partial \Delta G / \partial T)_{P,\xi} = - R[\xi\ln\xi + (1 - \xi)\ln(1 - \xi)];$
$\Delta H = \xi(1 - \xi)K + RT[\xi\ln\xi + (1 - \xi)\ln(1 - \xi)] - RT[\xi\ln\xi + (1 - \xi)\ln(1 - \xi)]$
$= \xi(1 - \xi)K$. However, $\Delta C_P = (\partial \Delta H / \partial T)_P$ cannot be calculated as $(\partial \Delta H / \partial T)_{P,\xi}$ because $(\partial \Delta H / d\xi)_{P,T} \neq 0$. The natural variables for H are P and S and the fact that $(\partial \Delta H / d\xi)_{P,S} = 0$ at equilibrium is not very useful in the present problem. Thus, we must use the basic equation $\Delta C_P = (\partial \Delta H / \partial T)_P = (\partial \Delta H / \partial T)_{P,\xi} + (\partial \Delta H / \partial \xi)_{T,P}(\partial \xi / \partial T)_P$. From the relation between ξ and T, given in the hint, we get: $(\partial T / \partial \xi)_P = - (K/R)\{ - 2/\ln[\xi/1 - \xi)] - (1 - 2\xi)[1/\xi + 1/(1 - \xi)]/\ln[\xi/(1 - \xi)])^2\}; \Delta C_P = 0 + (1 - 2\xi)KR/K\{2/\ln[\xi/(1 - \xi)] + (1 - 2\xi)/\xi(1 - \xi)(\ln[\xi/(1 - \xi)])^2\} = 1.3R$ for $\xi = 1/4$.

Exercise 2.2.3

According to a simple theory, the internal energy of a solid metal at absolute zero can be described with an expression of the form $U/N = k_1(V/N)^{-2/3} - k_2(V/N)^{-1/3}$ where N is the number of moles in the volume V. The equilibrium volume is then calculated from the minimum of U obtained from $d(U/N)/d(V/N) = (2/3)k_1(V/N)^{-5/3} + (1/3)k_2(V/N)^{-4/3} = 0$ giving $V/N = (2k_1/k_2)^3$. This procedure may look incorrect because S and V are supposed to be constant when equilibrium is calculated by minimizing U. Discuss whether it may be correct, nevertheless.

Hint

Equilibrium should be calculated with respect to an *internal* variable, ξ, using $(\partial U / \partial \xi)_{S,V} = 0$ and V (or V/N) is regarded as ξ in the above calculation. It must be different from the V quantity which is kept constant. Try to define the nature of the two volumes by considering one mole of the metal contained in a constant volume V. What will happen?

Solution

We have no problem with S, which we can put to zero at absolute zero. Identifying ξ with V in the expression for U and denoting the volume of the container by V_c, we can rewrite $(\partial U/\partial \xi)_{S,V}$ as $(\partial U/\partial V)_{S,V_c}$. Suppose there is no atmosphere in the container and neglect the vapour pressure of the metal itself. The metal is then free to adjust its volume V (i.e. ξ) without affecting the volume of the whole system, V_c (i.e. an external state variable) and without affecting the number of atoms in the system. N should be interpreted as the number of atoms in the volume V_c, and N is thus constant. The calculation is then correct.

2.3   *Equations of state*

If a characteristic state function for a particular substance is given with a different set of variables than the natural one, then it describes some of the properties but not all of them. Such an equation is often regarded as a state equation and not a fundamental equation. As an example

$$U = U(T,P)$$

is often called the **caloric** equation of state. Some of the quantities which are usually regarded as variables may also be represented with an equation between other variables, for instance

$$V = V(T,P)$$

This is sometimes called the **thermal** equation of state. The practical importance of some state equations stems from the fact that they can be evaluated fairly directly from measurable quantities and can thus be used to rationalize the results of measurements on a particular substance. As an example, the derivatives of V with respect to T and P can be obtained by measuring the thermal expansivity and the isothermal compressibility, respectively. It is much more laborious to evaluate the fundamental equation for a substance.

A major problem in the evaluation of the fundamental equation or an equation of state is the choice of mathematical form. The form is not specified by thermodynamics but must be chosen from a knowledge of the physical character of the particular substance under consideration. A considerable part of the present text will be concerned with the modelling of the fundamental equation for various types of substances.

Exercise 2.3.1

A classical ideal gas is defined by two equations of state. For one mole, they are $PV = RT$ and $U = A + BT$ where A and B are two constants. Try to derive a fundamental equation.

Hint

Try to find $F(T,V)$. Use $U = F - T(\partial F/\partial T)_V$ which yields $(\partial U/\partial T)_V = -T(\partial^2 F/\partial T^2)_V$. Also use $P = -(\partial F/\partial V)_T$.

Solution

$(\partial^2 F/\partial T^2)_V = -(\partial U/\partial T)_V/T = B/T$. Integration yields $(\partial F/\partial T)_V = -B\ln T + K_1$; $F = -BT\ln T + K_1 T + K_2$ where K_1 and K_2 are independent of T but may depend upon V. That dependency is obtained from $RT/V = P = -(\partial F/\partial V)_T = -T(\partial K_1/\partial V)_T - (\partial K_2/\partial V)_T$. Thus, $(\partial K_2/\partial V)_T = 0$ and $(\partial K_1/\partial V)_T = -R/V$. K_2 is independent of V and $K_1 = -R\ln V + K_3$. We get $F = -BT\ln T - RT\ln V + K_3 T + K_2$. To determine K_2: $A + BT = U = F + TS = F - T(\partial F/\partial T)_V = BT + K_2$. We thus find $K_2 = A$. We cannot determine K_3 from the information given. Our result is $F = A + K_3 T - BT\ln T - RT\ln V$ which is a fundamental equation, $F(T,V)$. We can also derive the Gibbs energy $G = F - V(\partial F/\partial V)_T = F + RT$ but in order to have a fundamental equation in G we must obtain $G(T,P)$ by replacing V with P and T which is possible in the present case where $V = RT/P$. We thus get $G = A + (K_3 + R - R\ln R)T - (B + R)T\ln T + RT\ln P$.

2.4 *Experimental conditions*

By experimental conditions we here mean the way an experiment is controlled from the outside. It primarily concerns variables which we may regard as external. Let us first consider the pair of conjugate variables $-P$ and V. Either one of them can be controlled from the outside without any knowledge of the properties of the system. In the pair of conjugate variables T and S one can control the value of T from the outside but the control of S requires knowledge of the properties of the system or extremely slow changes. In practice, it may even be difficult to keep S constant when another variable is changed. On the other hand, one can control the change in two other state variables, U and H, by controlling the heat flow

$$dU = dQ - PdV$$
$$dH = dU + d(PV) = dQ + VdP$$

For a reaction under constant V we find $dU = dQ$ and the heat of reaction is also the energy of reaction. For a reaction under constant P we find $dH = dQ$ and the heat of reaction is also the enthalpy of reaction. Of course, Q itself is not a state variable, because it does not concern the system itself but its interaction with another system, usually the so-called surroundings. An important experimental technique is to keep the system thermally insulated from the surroundings, i.e. to make $dQ = 0$, which is called **adiabatic** conditions. Experimental conditions under which various state variables are kept constant are often given special names,

constant P	**isobaric**
constant V	**isochoric**
constant T	**isothermal**
constant P and T	**isobarothermal**
constant H	**isenthalpic**
constant S	**isentropic**
constant U	**isoenergetic**
constant composition	**isoplethal**
constant potential	**equipotential**

From the above equations for dU and dH, it is evident that an isenthalpic reaction can be accomplished under a combination of isobaric and adiabatic conditions and an isoenergetic reaction can be realized under a combination of isochoric and adiabatic conditions.

Let us now turn to the internal variables which we have represented by the general symbol ξ. At equilibrium ξ has reached a value where the driving force for its change, D, is zero. If the conditions are changed very slowly by an action from the outside, ξ may vary slowly but all the time be very close to its momentary equilibrium value. In Section 1.8 we have already concluded that D is then very low and the internal entropy production

$$d_{ip}S = Dd\xi/T$$

is very low. In the limit, one talks about a reversible reaction where $d_{ip}S = 0$. In view of the relation

$$TdS = dQ + Dd\xi$$

we see that a reversible reaction, $D = 0$, which is carried out under adiabatic conditions, $dQ = 0$, is isentropic. By examining the combined law in the variables T and V we see that a reversible reaction which is carried out under isothermal and isochoric conditions takes place under constant F. If it is carried out under isothermal and isobaric conditions, it takes place under constant G. It is usual to consider such conditions and they may be called isobarothermal conditions.

The heat flow into a system on heating is often studied experimentally under conditions which may not approach reversible ones. The heat capacity is defined as follows, independent of the reversible or irreversible character of the process.

$$C = dQ/dT$$

For isochoric conditions

$$C_V = \left(\frac{\partial U}{\partial T}\right)_V$$

because dU is always equal to $dQ - PdV$ according to the first law. For isobaric conditions we obtain

$$C_P = \left(\frac{\partial H}{\partial T}\right)_P$$

because dH is always equal to $dQ + VdP$ according to the first law.

We have already seen that for heat capacity the result will be different if ξ is kept at a constant value or is allowed to be adjusted to its equilibrium value which varies with T. In many experiments with molecular species, their amounts are frozen-in at reasonably low temperatures and ξ is thus kept constant. At higher temperatures, the amounts may be adjusted by molecular reactions and ξ may thus be adjusted to its equilibrium value. When discussing C_P and C_V it may sometimes be wise to specify the conditions regarding ξ. Usually it is assumed that the ξs for all possible internal processes are adjusted to their equilibrium values but it is not unusual to consider some process as frozen-in.

Exercise 2.4.1

Find the conditions for which $T(\partial S/\partial T)_V$ is equivalent to $C_V = (\partial U/\partial T)_V$.

Hint

Start from the basic form of the combined law.

Solution

$dU = TdS - PdV - Dd\xi$. Under constant V we get $dU = TdS$ and $(\partial U/\partial T)_V = T(\partial S/\partial T)_V$ if $Dd\xi = 0$. This occurs under two different conditions. One is $D = 0$, so-called reversible conditions. The change of T is slow enough to make ξ adjust to equilibrium all the time. The other is $d\xi = 0$, so-called freezing-in conditions. The change is so rapid that there is no internal reaction. For clarity we could write this case as $(\partial U/\partial T)_{V,\xi} = T(\partial S/\partial T)_{V,\xi}$. Of course, it is always true if there is no possible internal process in the system.

Exercise 2.4.2

The isothermal compressibility κ_T is defined as $- (\partial V/\partial P)_T/V$ where the derivative is evaluated under reversible, isothermal conditions, i.e. a very slow compression, $D = 0$. Show a similar way of defining the adiabatic compressibility.

Hint

Suppose the adiabatic compression is so rapid that there can be no internal process, i.e. $d\xi = 0$.

Solution

The second law gives $TdS = dQ + Dd\xi$ and both terms are now zero. We thus have $dS = 0$ and can use the following definition of the adiabatic compressibility $- (\partial V/\partial P)_S/V$. This is why this quantity is usually denoted by κ_S. Note that this is justified only if the compression is much faster than all internal reactions, including heat condition.

Exercise 2.4.3

It is well known that S can increase spontaneously towards a maximum if U and V are kept constant. Find another condition under which S can increase spontaneously towards a maximum. Then, discuss how the two kinds of conditions can be realized experimentally.

Hint

We can use any form of the combined law which contains dS, then use the first law in order to examine the experimental conditions.

Solution

$dH = TdS + VdP - Dd\xi$ can be written as $TdS = dH - VdP + Dd\xi$ and S may thus increase spontaneously towards a maximum if H and P are constant as well as if U and V are kept constant. The first law gives $dU = dQ - PdV$ and the condition of constant U and V may thus be realized experimentally by keeping dQ and dV equal to zero, i.e. by using adiabatic and isochoric conditions. By introducing H into the first law we get $dH = d(U + PV) = dQ - PdV + PdV + VdP = dQ + VdP$. The condition of constant H and P can thus be realized experimentally by keeping dQ and dP equal to zero, i.e. by using adiabatic and isobaric conditions.

In fact, S will increase spontaneously towards a maximum under all adiabatic conditions. The second law gives directly $dS = dQ + Dd\xi = Dd\xi > 0$ for all spontaneous processes.

Exercise 2.4.4

One mole of a fluid at T_1, P_1 is compressed adiabatically to P_2. Discuss how one can calculate the work done on the system if one knows the properties of the system.

Hint

The problem may look like a simple first-law problem. However, it is not well defined. We must make some assumption regarding the process of compression in addition to its being adiabatic. Make the assumption which would give the simplest calculation.

Solution

Suppose the experiment is made under reversible conditions in addition to adiabatic conditions. Then the second law gives $TdS = dQ + Dd\xi = 0$ and S is constant. If we had an equation of state $S = S(T,P)$, then we could calculate the new T from the new P and the old T and P. If we also had an equation of state $U = U(T,P)$, then we could calculate ΔU between the two states and that would be equal to the work done on the system because the first law gives $dU = dQ + dW = dW$.

2.5   *Notation for partial derivatives*

Since there are many alternative sets of independent variables it is necessary to indicate which variables are to be kept constant in the evaluation of a particular partial derivative. In order to simplify the notation for characteristic state functions, we can omit this information when we use the natural variables, i.e. the particular set of independent variables characteristic of the state function under consideration. Since G is the characteristic state function for T,P we could then write $\partial G/\partial T$ instead of $(\partial G/\partial T)_P$. Furthermore, we may introduce a short-hand notation for these derivatives, say G_T. Second-order derivatives can be denoted by two subscripts and G_{TP} would thus mean

$$G_{TP} = \left(\frac{\partial}{\partial T} \left(\frac{\partial G}{\partial P} \right)_T \right)_P = \left(\frac{\partial}{\partial P} \left(\frac{\partial G}{\partial T} \right)_P \right)_T$$

Full information must be given as soon as a set of variables, different from the natural one, is used.

In the preceding section, C_V and C_P were given as derivatives which cannot be expressed by this short-hand notation. Their values will be compared in the next section.

The short-hand notation can be used for frozen-in conditions, $d\xi = 0$, and for equilibrium conditions where ξ is regarded as a dependent variable. When there is any doubt as to what conditions are considered, such information should be given.

Exercise 2.5.1

How should H_{TT} be interpreted?

Hint

Study the combined law in the form $dH = TdS + VdP - Dd\xi$.

Solution

The natural variables of H are S and P. Thus H_{TT} is an illegal notation because T is not one of the natural variables of H. We conclude that H_{TT} should not be used.

Exercise 2.5.2

How should H_{PP} be interpreted?

Solution

S and P are the natural variables of H and H_{PP} thus means $(\partial^2 H/\partial P^2)_{S,\xi}$ or $(\partial^2 H/\partial P^2)_S$. In the latter case ξ is a dependent variable which is continuously adjusted to equilibrium. In order to interpret H_{PP} one would need to know the conditions.

2.6   *Use of various derivatives*

Of course, C_P and C_V can both be related to any one of the characteristic state functions but in each case a certain choice gives a shorter derivation. C_P is defined with T and P as independent variables and we should thus use G which has T and P as its natural variables. The fundamental equation $G = G(T,P)$ gives $S = -(\partial G/\partial T)_P$ and thus

$$H(T,P) = G + TS = G - T(\partial G/\partial T)_P$$
$$C_P = (\partial H/\partial T)_P = -T(\partial^2 G/\partial T^2)_P = T(\partial S/\partial T)_P$$

For C_V we should use $F(T,V)$ and, since $S = -(\partial F/\partial T)_V$, we find in an analogous way

$$C_V = (\partial U/\partial T)_V = -T(\partial^2 F/\partial T^2)_V = T(\partial S/\partial T)_V$$

However, we may wish to compare the two heat capacities and must then be prepared to derive both from the same characteristic state function, say G. For C_P we already have an expression $-TG_{TT}$, and C_V will now be derived from U through G as a function of T and P.

$$U = G - TG_T - PG_P$$
$$dU = (G_T - G_T - TG_{TT} - PG_{PT})dT + (G_P - TG_{TP} - G_P - PG_{PP})dP$$
$$= -(TG_{TT} + PG_{PT})dT - (TG_{TP} + PG_{PP})dP$$

However, in order to evaluate C_V which is equal to $(\partial U/\partial T)_V$ we must know U as a function of T and V instead of T and P. We need a relation between dV, dT and dP. Starting with $V = (\partial G/\partial P)_T = G_P$ we obtain

$$dV = G_{PT}dT + G_{PP}dP$$
$$dP = dV/G_{PP} - G_{PT}dT/G_{PP}$$

This gives dP as a function of dV and dT which can be inserted in the above equation

$$dU = -(TG_{TT} + PG_{PT})dT - (TG_{TP} + PG_{PP})(dV/G_{PP} - G_{PT}dT/G_{PP})$$

Remembering that G_{PT} is identical to G_{TP}, we thus obtain

$$C_V = (\partial U/\partial T)_V = -TG_{TT} - PG_{TP} + T(G_{TP})^2/G_{PP} + PG_{TP}$$
$$= T(G_{TP})^2/G_{PP} - TG_{TT}$$

so that

$$C_V = C_P + T(G_{TP})^2/G_{PP}$$

Using the same method we can derive an expression for any quantity in terms of the derivatives of G with respect to T and P.

It should be pointed out that, by tradition, one instead relates various quantities in terms of the following three quantities which are directly measurable.

Heat capacity at constant pressure $C_P = (\partial H/\partial T)_P = -TG_{TT}$
Thermal expansivity $\alpha = (\partial V/\partial T)_P/V = G_{TP}/G_P$
Isothermal compressibility $\kappa_T = -(\partial V/\partial P)_T/V = -G_{PP}/G_P$

These three quantities are thus closely related to the three second-order derivatives G_{TT}, G_{TP} and G_{PP}. The two schemes of relating quantities can easily be translated into each other. It is interesting to note that through experimental information on the three quantities C_P, α, and κ_T one has information on all the second-order derivatives of G.

Together, they thus form a good basis for an evaluation of the fundamental equation $G(T,P)$.

Exercise 2.6.1

Derive an expression for C_V for a substance with $G = a + bT + cT\ln T + dT^2 + eP^2 + fTP + gP^2$

Hint

Use either one of the equations given for C_V but remember first to make sure that the proper variables are used.

Solution

Let us use $C_V = T(\partial S/\partial T)_V$ but then we must evaluate $S(T,V)$ from G. First, we get $- S(T,P) = (\partial G/\partial T)_P = b + c + c\ln T + 2dT + fP$. In order to replace P by V we need $V = (\partial G/\partial P)_T = e + fT + 2gP$, which gives $- S = b + c + c\ln T + 2dT + f(V - e - fT)/2g$; $C_V = T(\partial S/\partial T)_V = - T(c/T + 2d - f^2/2g) = - c - 2dT + f^2T/2g$.

Exercise 2.6.2

Show how one can calculate the heat absorption on reversible isothermal compression from easily measured quantities. In exercise 1.6.6 we obtained the result $Q = T_1(S_2 - S_1)$ which is very convenient but only if the properties of the substance have already been evaluated from the experimental information.

Hint

Since T and P are most easily controlled experimentally, we should use these variables.

Solution

Under reversible, isothermal conditions: $dQ = TdS = T(\partial S/\partial P)_T dP = - T(\partial^2 G/\partial T\partial P)dP = - T(\partial V/\partial T)_P = - TV\alpha dP$; $Q = - T\int V\alpha dP$ where $\alpha = (\partial V/\partial T)_P/V$.

Exercise 2.6.3

Prove the following relation which is known to be a very useful equation $(\partial U/\partial V)_T = T(\partial P/\partial T)_V - P$.

Hint

Since V and T are used as variables, express U in terms of F which has V and T as its natural variables. Remember that $(\partial P/\partial T)_V = -F_{VT} = (\partial S/\partial V)_T$.

Solution

$U = F + ST;\ \ (\partial U/\partial V)_T = (\partial F/\partial V)_T + T(\partial S/\partial V)_T = -P + T(\partial P/\partial T)_V.$

Exercise 2.6.4

It is well known that the heat capacity at constant volume is defined as dQ/dT at constant V. Another important quantity is defined as dQ/dV at constant T. It is sometimes called the latent heat of volume change. Derive a general expression for this new quantity in terms of state variables.

Hint

Since the variables are V and T, it is convenient to use F and its derivatives. Start with the first law, then use $U = F + ST = F - TF_T$.

Solution

$dQ = dU + PdV;\ \ (\partial Q/\partial V)_T = (\partial U/\partial V)_T + P = F_V - TF_{TV} + P = -TF_{TV}.$
Using the hint in the preceding problem, the result can also be expressed as
$T(\partial S/\partial V)_T$ or $T(\partial P/\partial T)_V.$

Exercise 2.6.5

It is well known that G has a minimum at equilibrium under constant T and P but F has a minimum at equilibrium under constant T and V. That being so, what is wrong in the following derivation?

Consider V as a function of P and T,

$dV = (\partial V/\partial P)_T dP + (\partial V/\partial T)_P dT = -V\kappa_T dP + V\alpha dT$
$dF = -PdV - SdT - Dd\xi = PV\kappa_T dP - (S + PV\alpha)dT - Dd\xi$

If this derivation were correct, it would seem that there could be a spontaneous process under constant P and T until F has reached a minimum.

Hint

We are considering a system with a possible internal process. We must then include ξ among the state variables and realize that the volume depends upon the ξ value.

Solution

We want to replace dV in $dF = -PdV - SdT - Dd\xi$ and instead introduce dP. It is evident that we must then consider V as a function of the new set of variables P,T,ξ: $dV = (\partial V/\partial P)_{T,\xi}dP + (\partial V/\partial T)_{P,\xi}dT + (\partial V/\partial \xi)_{T,P}d\xi = -V\kappa_T dP + V\alpha dT + (\partial V/\partial \xi)_{T,P}d\xi$ and here κ_T and α must be determined at constant ξ. We now get $dF = PV\kappa_T dP - (S + PV\alpha)dT - [D + P(\partial V/\partial \xi)_{T,P}]d\xi$. At constant T,P: $D + P(\partial V/\partial \xi)_{T,P} = -(\partial F/\partial \xi)_{T,P}$ and $D = -[\partial(F + PV)/\partial \xi]_{T,P} = -(\partial G/\partial \xi)_{T,P}$.

2.7   *Comparison between C_V and C_P*

Let us now examine the relation between C_V and C_P in more detail. It is usually given in the following form

$$C_P = C_V(1 + \gamma\alpha T)$$

where γ is a dimensionless quantity called Grüneisen's constant. By comparison with the relation between C_V and C_P given in Section 2.6 we can express γ in terms of the directly measurable quantities

$$\gamma = -(G_{TP})^2/\alpha C_V G_{PP} = V\alpha/\kappa_T C_V$$

κ_T and C_V are both positive and, with few exceptions, α is also positive and it is never strongly negative. The γ quantity often has a value of about 2. Note that C_P is always larger than C_V, independent of the sign of α, because $\gamma\alpha$ is equal to $V\alpha^2/\kappa_T C_V$, which is always positive.

The quantity γ can be expressed in many ways, some of which are given here without proof

$$\gamma = -\frac{1}{T}\cdot\frac{G_P G_{TP}}{G_{TP}^2 - G_{TT}G_{PP}} = \frac{V}{T}\frac{F_{VT}}{F_{TT}} = \frac{V}{C_V}\left(\frac{\partial S}{\partial V}\right)_T = V\left(\frac{\partial P}{\partial U}\right)_V$$

In all these forms γ is proportional to V which in turn varies with T if P is kept constant. It is evident that one cannot discuss how γ for a particular substance varies with T without specifying if P or V is kept constant. If $C_V(T)$ is evaluated from $C_P(T)$ using values of γ and α measured at 1 bar, then the resulting values hold for different volumes at different temperatures. C_V may be regarded either as a function of T,V or T,P and it is

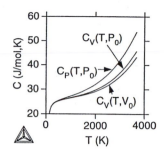

Figure 2.1 The heat capacity of tungsten evaluated in three different ways. Even though C_V is defined as the heat capacity at constant V, it may be regarded as a function of T,V or T,P. At each temperature $C_V(T,P_0)$ is evaluated at the volume given by the actual temperature and a selected constant pressure P_0.

evident that the functions $C_V(T,V_0)$ and $C_V(T,P_0)$ are different. A comparison of the two functions, evaluated from $C_P(T,P_0)$ for tungsten is given in Fig. 2.1. It should be emphasized that all other quantities, such as α and κ_T, can also be treated as functions of either T,V or T,P.

Exercise 2.7.1

Calculate the heat flow into a system in two cases:

    (a) Heating from T_1 to T_2 under constant V.
    (b) Heating from T_1 to T_2 under constant P and then compression back to the initial volume under constant T_2.

       Suppose the system is an ideal classical gas for which $PV = RT$ and $U = A + BT$.

Hint

Use the first law in all cases.

Solution

    (a) $dQ = dU + PdV = dU; \quad Q = \Delta U = B(T_2 - T_1)$.
    (b_1) $dQ = dU + PdV; \quad Q_1 = \Delta U + P_1\Delta V = B(T_2 - T_1) + P_1(V_2 - V_1)$.
    (b_2) $dQ = dU + PdV; \quad Q_2 = \Delta U + \int PdV = 0 + \int (RT_2/V)dV =$
        $RT_2\ln(V_1/V_2)$

where $V_1 = RT_1/P_1; \quad V_2 = RT_2/P_1$. Therefore, $\Sigma Q = B(T_2 - T_1) + R(T_2 - T_1) + RT_2\ln(T_1/T_2)$. For small $T_2 - T_1$: $T_2\ln(T_1/T_2) = T_2\ln[1 + (T_1 - T_2)/T_2]$ $\cong T_1 - T_2$, and $\Sigma Q \cong (B + R - R)(T_2 - T_1) = B(T_2 - T_1)$. (Same as for (a).)

2.8 *Change of independent variables*

One often wants to change the set of independent variables. An example was given in Section 2.6 where C_V was first given as $(\partial U/\partial T)_V$ and was then evaluated as a function of the derivatives of G with respect to T and P. Such changes can be made by the following automatic procedure which is based upon the properties of Jacobians. It is here given without mathematical proof. We start by a definition of the Jacobian

$$\frac{\partial(u,v)}{\partial(x,y)} = \begin{vmatrix} \dfrac{\partial u}{\partial x} & \dfrac{\partial u}{\partial y} \\ \dfrac{\partial v}{\partial x} & \dfrac{\partial v}{\partial y} \end{vmatrix}$$

It obeys the following rule

$$\frac{\partial(u,v)}{\partial(x,y)} = \frac{\partial(u,v)}{\partial(r,s)} \bigg/ \frac{\partial(x,y)}{\partial(r,s)}$$

We can thus introduce r and s as new independent variables instead of x and y.

The derivative of a thermodynamic quantity can be expressed by a Jacobian because

$$\frac{\partial(u,y)}{\partial(x,y)} = \begin{vmatrix} \dfrac{\partial u}{\partial x} & \dfrac{\partial u}{\partial y} \\ \dfrac{\partial y}{\partial x} & \dfrac{\partial y}{\partial y} \end{vmatrix} = \begin{vmatrix} \dfrac{\partial u}{\partial x} & \dfrac{\partial u}{\partial y} \\ 0 & 1 \end{vmatrix} = \left(\frac{\partial u}{\partial x}\right)_y$$

and the new independent variables, r and s, can thus be introduced in the following way

$$\left(\frac{\partial u}{\partial x}\right)_y = \frac{\partial(u,y)}{\partial(x,y)} = \frac{\partial(u,y)}{\partial(r,s)} \bigg/ \frac{\partial(x,y)}{\partial(r,s)} = \begin{vmatrix} \dfrac{\partial u}{\partial r} & \dfrac{\partial u}{\partial s} \\ \dfrac{\partial y}{\partial r} & \dfrac{\partial y}{\partial s} \end{vmatrix} \bigg/ \begin{vmatrix} \dfrac{\partial x}{\partial r} & \dfrac{\partial x}{\partial s} \\ \dfrac{\partial y}{\partial r} & \dfrac{\partial y}{\partial s} \end{vmatrix}$$

It should be realized that $\partial u/\partial x$ actually means $(\partial u/\partial x)_y$ and $\partial u/\partial r$ means $(\partial u/\partial r)_s$. This equation contains the following relations as special cases. They can of course be proved in a much simpler way.

$$\left(\frac{\partial u}{\partial x}\right)_y = 1 \bigg/ \left(\frac{\partial x}{\partial u}\right)_y$$

$$\left(\frac{\partial u}{\partial x}\right)_y = - \left(\frac{\partial y}{\partial x}\right)_u \bigg/ \left(\frac{\partial y}{\partial u}\right)_x$$

Exercise 2.8.1

Express $(\partial G/\partial T)_V$ in terms of functions usually measured and tabulated.

Hint

Most measurements are made by controlling T and P. Change to these variables.

Solution

$$(\partial G/\partial T)_V = \begin{vmatrix} \partial G/\partial T & \partial G/\partial P \\ \partial V/\partial T & \partial V/\partial P \end{vmatrix} \bigg/ \begin{vmatrix} \partial T/\partial T & \partial T/\partial P \\ \partial V/\partial T & \partial V/\partial P \end{vmatrix}$$

$$= (\partial G/\partial T)_P - (\partial G/\partial P)_T (\partial V/\partial T)_P/(\partial V/\partial P)_T = -S + V\alpha\kappa_T.$$

Exercise 2.8.2

A condensed phase is compressed adiabatically and reversibly from a pressure 0 to P. Derive an equation for the temperature change in terms of easily measured quantities.

Hint

Adiabatic and reversible conditions are also isentropic. We want $(\partial T/\partial P)_S$. Change the variables to T and P.

Solution

$$(\partial T/\partial P)_S = \begin{vmatrix} \partial T/\partial T & \partial T/\partial P \\ \partial S/\partial T & \partial S/\partial P \end{vmatrix} \bigg/ \begin{vmatrix} \partial P/\partial T & \partial P/\partial P \\ \partial S/\partial T & \partial S/\partial P \end{vmatrix}$$

$$= -(\partial S/\partial P)_T/(\partial S/\partial T)_P = -G_{TP}/G_{TT} = -V\alpha/(-C_P/T) = TV\alpha/C_P.$$

We thus get $dT = (TV\alpha/C_P)dP$. In order to integrate, we must know V, α and C_P as functions of T,P.

Exercise 2.8.3

From $\gamma = \alpha V/C_V\kappa_T$, show that $\gamma = \alpha V/C_P\kappa_S$, i.e. show that $C_V\kappa_T = C_P\kappa_S$.

Hint

$\kappa_S = -(\partial V/\partial P)_S/V$. Change all the derivatives to G of T and P. Use the expressions of C_P and C_V in terms of such derivatives. C_V and κ_T were given in Section 2.6.

Solution

$$-V\kappa_S = (\partial V/\partial P)_S = \begin{vmatrix} \partial V/\partial P & \partial V/\partial T \\ \partial S/\partial P & \partial S/\partial T \end{vmatrix} \Big/ \begin{vmatrix} \partial P/\partial P & \partial P/\partial T \\ \partial S/\partial P & \partial S/\partial T \end{vmatrix}$$

$$= (\partial V/\partial P)_T - (\partial V/\partial T)_P(\partial S/\partial P)_T/(\partial S/\partial T)_P = G_{PP} - G_{PT}G_{TP}/G_{TT}.$$

But $C_P = -TG_{TT}$ and $V = G_P$. Thus $C_P\kappa_S = T(G_{TT}G_{PP} - G_{TP}^2)/G_P$. From expressions for C_V and κ_T we get $C_V\kappa_T = T(G_{TP}^2 - G_{TT}G_{PP})(1/G_{PP})(-G_{PP}/G_P) = C_P\kappa_S$.

2.9   *Maxwell relations*

Some partial derivatives can be transformed in a very special way. The requirement is that the variable to be kept constant is a conjugate variable to one of the quantities in the derivative. The method may be illustrated by the following example

$$dG = -SdT + VdP; \quad -S = (\partial G/\partial T)_P; \quad V = (\partial G/\partial P)_T$$

$$\left(\frac{\partial V}{\partial T}\right)_P = \left(\frac{\partial(\partial G/\partial P)_T}{\partial T}\right)_P = \frac{\partial^2 G}{\partial P\partial T} = \frac{\partial^2 G}{\partial T\partial P} = \left(\frac{\partial(\partial G/\partial T)_P}{\partial P}\right)_T = -\left(\frac{\partial S}{\partial P}\right)_T$$

or by the short-hand notation

$$\left(\frac{\partial V}{\partial T}\right)_P = G_{PT} = G_{TP} = -\left(\frac{\partial S}{\partial P}\right)_T$$

The relations obtained in this way are called Maxwell relations. We can use any form of the combined law and thus obtain a large number of such relations. It should be noticed that all derivatives related by Maxwell relations are constructed in such a way that the variable to be kept constant is conjugate to the quantity in the numerator but the relations may be inverted, of course.

Exercise 2.9.1

Transform $(\partial V/\partial T)_S$ using a Maxwell relation.

Hint

In the inverse quantity, $(\partial T/\partial V)_S$, T and S are conjugate quantities. Use the characteristic state function with V and S as natural variables.

Solution

$(\partial V/\partial T)_S = 1/(\partial T/\partial V)_S = 1/(\partial^2 U/\partial V\partial S) = -1/(\partial P/\partial S)_V = -(\partial S/\partial P)_V$

Exercise 2.9.2

Show that $dS = (C_V/T)dT + (\alpha/\kappa_T)dV$.

Hint

We want to show that $C_V/T = (\partial S/\partial T)_V$ and $\alpha/\kappa_T = (\partial S/\partial V)_T$. The first relation is obtained from the second law under reversible conditions. The second relation is obtained from the definitions of α and κ_T in derivatives of G after some manipulation involving the last equation derived in Section 2.8 and a Maxwell relation.

Solution

Under reversible conditions we get $dQ = TdS$; $C_V = (dQ/dT)_V = T(\partial S/\partial T)_V$. From the definitions of α and κ_T we get $\alpha/\kappa_T = (\partial V/\partial T)_P/[-(\partial V/\partial P)_T] = (\partial P/\partial T)_V = (\partial S/\partial V)_T$ and thus $dS = (C_V/T)dT + (\alpha/\kappa_T)dV$.

Exercise 2.9.3

Prove the identity $T(\partial^2 P/\partial T^2)_V = (\partial C_V/\partial V)_T$.

Hint

Since T and V are the variables, it is convenient to base the solution on F and its derivatives. $dF = -SdT - PdV$ gives the following Maxwell relation: $(\partial S/\partial V)_T = (\partial P/\partial T)_V$. Also, use $C_V = T(\partial S/\partial T)_V$.

Solution

Using the Maxwell relation we get $T(\partial^2 P/\partial T^2)_V = T(\partial(\partial P/\partial T)_V/\partial T)_V = T(\partial(\partial S/\partial V)_T/\partial T)_V = T\partial^2 S/\partial V\partial T$. From $C_V = T(\partial S/\partial T)_V$ we get $(\partial C_V/\partial V)_T = T\partial^2 S/\partial T\partial V$ which is the same result.

Exercise 2.9.4

Evaluate the thermal expansivity at absolute zero for a substance obeying the third law.

Hint

According to the third law, S approaches the same value at absolute zero for all ordered states of a substance. The limiting S value is not only independent of the crystalline structure but also of the pressure, $(\partial S/\partial P)_{T=0} = 0$.

Solution

$\alpha = (\partial V/\partial T)_P/V$. Since V and P are conjugate variables, we can use a Maxwell relaion, $\alpha = (\partial^2 G/\partial P\partial T)/V = -(\partial S/\partial P)_T/V = 0$ at $T = 0$.

Exercise 2.9.5

Transform $(\partial H/\partial P)_T$ using a Maxwell relation.

Hint

In order to introduce H as a conjugate variable to T (or rather $1/T$) in the combined law, one should use the entropy scheme, presented in Section 1.12.

Solution

From the entropy scheme we get $\mathrm{d}(G/T) = H\mathrm{d}(1/T) + (V/T)\mathrm{d}P$. Thus, $(\partial H/\partial P)_T = (\partial H/\partial P)_{1/T} = (\partial[V/T]/\partial[1/T])_P = V - T(\partial V/\partial T)_P$.

Systems with variable composition

Chemical potential

We have considered a system which can exchange heat and work with the surroundings as well as being capable of one or more internal processes. In Section 1.9 each one of these several changes gave a term in the combined first and second law. If the system is also open to exchange of a certain type of matter, j, with the surroundings, there would be a corresponding term because the internal energy U must also be a function of the content of matter. A more general form of the combined law should thus be

$$dU = TdS - PdV + \Sigma\mu_i dN_i - Dd\xi$$

where we have introduced the notation N_i for the amount of component i and μ_i for a new intensive state variable, which must be a potential, like T, $-P$ and $-D$. This was first introduced by Gibbs (1876, 1948) and is called **chemical potential**. μ_i and N_i are conjugate variables and the terms $\mu_i dN_i$ may thus be included in $\Sigma Y^a dX^a$ in the generalized form of the combined law, introduced in Section 1.9. It contains $c + 2$ conjugate pairs of external variables and U is a function of the $c + 2$ extensive variables. For any component j of the system the chemical potential may be defined as

$$\mu_j = (\partial U/\partial N_j)_{S,V,N_k,\xi}$$

The subscript 'N_k' indicates that all N_i are kept constant except for N_j. At equilibrium with respect to the internal process, ξ is a dependent variable and then we have

$$\mu_j = (\partial U/\partial N_j)_{S,V,N_k}$$

The summation $\Sigma\mu_i dN_i$ is taken over all components in a chosen set of independent components. In chemical thermodynamics one often takes the summation over all molecular species but then one must also define a set of independent reactions. That procedure is less general and will be avoided in the present text.

From the new, more general form of the combined law introduced above, it is evident that the number of external variables which can be varied independently of each other under equilibrium conditions is $c + 2$, if there are c independent components. As U is a function of S, V, N_i and ξ, so too is μ_j and we can write

$\mu_j = \mu_j(S,V,N_i,\xi)$

or under conditions of internal equilibrium

$\mu_j = \mu_j(S,V,N_i)$

This is another equation of state but it is not a fundamental equation and, thus, it does not contain all the properties.

When considering systems with variable composition it is useful to define many new quantities. A large part of the present chapter is devoted to discussions of such quantities.

Exercise 3.1.1

Examine how μ_i would enter into a conjugate pair of variables according to the entropy scheme.

Hint

Rearrange the combined law already given.

Solution

$dS = (1/T)dU + (P/T)dV - \Sigma(\mu_i/T)dN_i + (D/T)d\xi$. It is evident that μ_i/T and N_i form a conjugate pair.

Exercise 3.1.2

Derive an expression for μ_j as a derivative of S instead of U.

Hint

Use the entropy scheme.

Solution

$-dS = (-1/T)dU + (-P/T)dV + \Sigma(\mu_i/T)dN_i - (D/T)d\xi$ yields:
$(\partial S/\partial N_j)_{U,V,N_k,\xi} = -\mu_j/T$.

Exercise 3.1.3

By $\mu_j = (\partial U/\partial N_j)_{S,V,N_k}$ we mean a quantity evaluated under internal equilibrium, i.e. under the equilibrium value of ξ which varies with N_j. Show that this quantity has the same value as $(\partial U/\partial N_j)_{S,V,N_k,\xi}$ with the equilibrium value of ξ inserted as a constant value.

Hint

The kind of problem was treated in Section 2.2.

Solution

$(\partial U/\partial N_j)_{S,V,N_k,eq} = (\partial U/\partial N_j)_{S,V,N_k,\xi} + (\partial U/\partial \xi)_{S,V,N_i}\cdot(\partial \xi/\partial N_j)_{S,V,N_k,eq}$. However, at equilibrium $(\partial U/\partial \xi)_{S,V,N_i} = -D = 0$ and thus $(\partial U/\partial N_j)_{S,V,N_k,eq} = (\partial U/\partial N_j)_{S,V,N_k,\xi}$.

3.2   *Molar quantities*

Let us consider a homogeneous system at equilibrium and define a part of it as a subsystem. The size of the subsystem may be expressed by the value of any extensive variable. The most natural way may be to use the content of matter because from the experimental point of view it is easier to control the content of matter than the volume or entropy. One usually uses the total content of matter, N, defined by

$$N = \Sigma N_i$$

Sometimes we shall use the content of a particular component, N_j, instead of the total content of matter, N.

As a measure of the content of matter Gibbs used the mass, but today it is more common to use the number of atoms or species. We shall use the latter method but it should be emphasized that it is often necessary to specify what species are considered, which Gibbs did not have to do. On the other hand, thermodynamic models of special kinds of substances are often based upon considerations of atoms and it is then convenient to interpret N and N_i as the number of atoms (or groups of atoms). The number is usually expressed in units of moles, i.e. approximately $6\cdot10^{23}$ pieces (Avogadro's number, N^A).

The volume V is proportional to N in a homogeneous system and we may define a new quantity, the molar volume

$$V_m = V/N$$

This quantity has a defined value at each point of a system. It is thus an **intensive** variable like T, $-P$ and μ_i. However, its properties are quite different, a fact which becomes evident if we consider a system consisting of more than one phase, i.e. regions exhibiting different properties. In each phase V_m has a different, constant value but T, $-P$ and μ_i must have the same value in the whole system at equilibrium (with one exception which we shall deal with later). This is the property of a potential as noted in Section 1.10 and we may conclude that V_m is not a potential. It is very important to distinguish between two kinds of intensive variables, **potentials** and **molar quantities**. One should never use the word 'intensive variable' without specifying what kind one is considering.

In the same way we may define the **molar content** of component i. Usually it is denoted by x_i and is called mole fraction

$$x_i = N_i/N$$

However, it is sometimes essential to stress its close relation to other molar quantities. 'Molar content' is thus preferable.

Since entropy S is also an extensive quantity, subject to the law of additivity, the molar entropy can be defined in the same way

$$S_m = S/N$$

The molar quantities have been defined for a homogeneous system or for a homogeneous part of a system. The definition may very well be extended to the whole of a system with more than one phase but such a molar quantity is not strictly an intensive quantity but may be regarded as an average of an intensive quantity.

Let us return to a homogenous system at equilibrium, i.e. with $D = 0$, and define a very small subsystem enclosed inside an imaginary wall. We shall let the subsystem grow in size by expanding the wall but without making any real changes in the system, i.e. without changing P or T. During this process we have

$$dS = S_m dN$$
$$dV = V_m dN$$
$$dN_i = x_i dN$$

but $d\xi = 0$ because no real process is going on. We may thus evaluate the change in U as follows

$$dU = TdS - PdV + \Sigma \mu_i dN_i = (TS_m - PV_m + \Sigma \mu_i x_i)dN$$

where the value of the expression in parentheses is constant. By integrating over the expansion we obtain

$$U = (TS_m - PV_m + \Sigma \mu_i x_i)N = TS - PV + \Sigma \mu_i N_i$$

It is evident that U is also an extensive quantity for which we may define a molar quantity, U_m,

$$U/N = U_m = TS_m - PV_m + \Sigma \mu_i x_i$$

In order to be distinguished from molar quantities, extensive quantities like U are sometimes called **integral quantities**.

When defining N as the mass, as Gibbs did, one usually calls the quantities, obtained by dividing with N, **specific** instead of molar. Most of the thermodynamic relations are valid independent of how N is defined.

In many cases it is convenient to consider one mole of formula units or groups of atoms and the molar quantities are then defined by dividing with the number of formula units or groups of atoms, expressed as moles.

It can be easily shown that all the relations between extensive quantities also apply to molar quantities, e.g.

$$S_m = \frac{1}{N}S = \frac{-1}{N}\left(\frac{\partial G}{\partial T}\right)_{P,N_i} = \frac{-1}{N}\left(\frac{\partial NG_m}{\partial T}\right)_{P,N_i}$$

$$= -\left(\frac{\partial G_m}{\partial T}\right)_{P,N_i} = -\left(\frac{\partial G_m}{\partial T}\right)_{P,x_i}$$

because all x_i are constant if all N_i are kept constant.

Exercise 3.2.1

Cuprous oxide has a density of $6000 \, kg/m^3$. Give its molar volume in two different ways and also its specific volume.

Hint

The atomic mass is for Cu 63.546 and for O 15.9994.

Solution

The mass of one mole of the formula unit Cu_2O is 143.09 g or 0.14309 kg and the molar volume is thus $0.14309/6000 = 24 \cdot 10^{-6} \, m^3$ per mole of Cu_2O or $8 \cdot 10^{-6} \, m^3$ per mole of atoms or mole of $Cu_{0.67}O_{0.33}$. The specific volume is $1/6000 = 167 \cdot 10^{-6} \, m^3/kg$, whether one considers Cu_2O or $Cu_{0.67}O_{0.33}$.

Exercise 3.2.2

Consider a real change in a system with a constant amount of one mole of atoms, $N = 1$. Formulate the combined law for that system.

Hint

Start from an ordinary form of the combined law and divide by N. Since N is constant, we get $(1/N)dU = d(U/N) = dU_m$, etc.

Solution

$$dU_m = TdS_m - PdV_m + \Sigma\mu_i dx_i - Dd(\xi/N)$$

3.3   *More about characteristic state functions*

We have seen that

$$U = TS - PV + \Sigma\mu_i N_i$$

in a multicomponent system. It is evident that we get the following relation for the Gibbs energy

$$G = U - TS + PV = \Sigma\mu_i N_i$$

The relations between the characteristic state functions can be summarized as follows

$$\Sigma\mu_i N_i = G = H - TS = U + PV - TS = F + PV$$

It should be emphasized that the extensive quantities introduced from the beginning, U, S and V, obey the law of additivity and they can be applied to non-equilibrium states. The derived quantities G, H and F are usually applied to systems with constant values of T and P. Under such conditions, they also obey the law of additivity and we may define the corresponding molar quantities for homogeneous systems

$$\Sigma\mu_i x_i = G_m = H_m - TS_m = U_m + PV_m - TS_m = F_m + PV_m$$

In Section 1.11 we discussed various forms of the combined law obtained by changing from S to T and from V to P in the set of independent variables. We can now generalize them as follows

$$dU = TdS - PdV + \Sigma\mu_i dN_i - Dd\xi$$
$$d(U - TS) = dF = -SdT - PdV + \Sigma\mu_i dN_i - Dd\xi$$
$$d(U + PV) = dH = TdS + VdP + \Sigma\mu_i dN_i - Dd\xi$$
$$d(U - TS + PV) = dG = -SdT + VdP + \Sigma\mu_i dN_i - Dd\xi$$

It is evident that the chemical potentials for a substance can be evaluated from any one of these characteristic state functions if it is given in terms of its natural variables.

$$\mu_j = (\partial U/\partial N_j)_{S,V,N_k,\xi} = (\partial F/\partial N_j)_{T,V,N_k,\xi} = (\partial H/\partial N_j)_{S,P,N_k,\xi} = (\partial G/\partial N_j)_{T,P,N_k,\xi}$$

We may consider ξ as a dependent variable under equilibrium conditions but, in view of Section 2.2, that fact does not change the value of a partial derivative. We could thus omit ξ and write

$$\mu_j = (\partial U/\partial N_j)_{S,V,N_k} = (\partial F/\partial N_j)_{T,V,N_k} = (\partial H/\partial N_j)_{S,P,N_k} = (\partial G/\partial N_j)_{T,P,N_k}$$

The remaining extensive variables, N_i, can also be replaced by their conjugate potentials, μ_i, and we can get four new forms of the combined law,

$$d(U - \Sigma\mu_i N_i) = d(TS - PV) = TdS - PdV - \Sigma N_i d\mu_i - Dd\xi$$
$$d(U - TS - \Sigma\mu_i N_i) = d(-PV) = -SdT - PdV - \Sigma N_i d\mu_i - Dd\xi$$
$$d(U + PV - \Sigma\mu_i N_i) = d(TS) = TdS + VdP - \Sigma N_i d\mu_i - Dd\xi$$
$$d(U - TS + PV - \Sigma\mu_i N_i) = 0 = -SdT + VdP - \Sigma N_i d\mu_i - Dd\xi$$

The first three forms define new characteristic state functions. The fourth form is unique because it defines a function which is identically equal to zero since $U - TS - PV = \Sigma\mu_i N_i$. For reversible conditions, $D = 0$, or in the absence of internal processes, $d\xi = 0$, it yields a direct relation between the $c + 2$ potentials, the so-called **Gibbs–Duhem relation**. Consequently, one of the potentials is no longer an independent variable.

$$SdT - VdP + \Sigma N_i d\mu_i = 0$$

This relation is often given in terms of molar quantities

$$S_m dT - V_m dP + \Sigma x_i d\mu_i = 0$$

The second form gives a characteristic state function which is equal to $(-PV)$ and is particularly interesting in statistical thermodynamics. This characteristic state function is sometimes denoted by Ω and is called 'grand potential'. It can be evaluated from the so-called grand partition function, Ξ,

$$\Omega = -kT\ln\Xi$$

The grand partition function is defined for a so-called grand canonical ensemble for which T, V and μ_i are the independent variables and $\Omega = \Omega(T,V,\mu_i)$. It is sometimes useful in calculations of equilibrium states because it may yield relatively simple relationships. The fact that it applies under constant values of μ_i, which may be difficult to control experimentally, does not limit its usefulness in such calculations.

In this connection it may be mentioned that the ordinary partition function Z is defined for an ordinary canonical ensemble for which T, V and N_i are the independent variables. It can be used to evaluate the Helmholtz energy

$$F = -kT\ln Z$$

Furthermore, for a microcanonical ensemble one keeps U, V and N_i constant and can evaluate $S(U,V,N_i)$.

The remaining two new forms of the combined law and their characteristic state functions have not found much direct use. However, in the next section they will prove useful in some thermodynamic derivations. It should finally be emphasized that a large number of additional forms may be derived by selecting as independent variables some of the N_i and some of the μ_i. We shall discuss one such example in Section 13.5.

Exercise 3.3.1

Express μ_A in terms of molar quantities and insert the expression in the combined law for a pure element A.

Solution

$\Sigma\mu_i x_i = G_m$ gives $\mu_A = G_m$ (pure A). For pure A: $dU = TdS - PdV + G_m dN$.

Exercise 3.3.2

Prove the well-known equality $(\partial G/\partial N_A)_{T,P,N_j} = (\partial F/\partial N_A)_{T,V,N_j}$ by changing variables using Jacobians.

Hint

T and all N_j are not to be changed. Simplify the notation by omitting them from the subscripts. Change from N_A,P to N_A,V. Then express all derivatives of G in terms of F using $P = -(\partial F/\partial V)_{T,N_i} = -F_V$ and $G = F + PV = F - VF_V$.

Solution

$$\left(\frac{\partial G}{\partial N_A}\right)_P = \begin{vmatrix} \partial G/\partial N_A & \partial G/\partial V \\ \partial P/\partial N_A & \partial P/\partial V \end{vmatrix} \bigg/ \begin{vmatrix} \partial N_A/\partial N_A & \partial N_A/\partial V \\ \partial P/\partial N_A & \partial P/\partial V \end{vmatrix}$$

$$= (\partial G/\partial N_A)_V - (\partial G/\partial V)_{N_A}(\partial P/\partial N_A)_V/(\partial P/\partial V)_{N_A}.$$

But, $(\partial G/\partial N_A)_V = F_{N_A} - VF_{VN_A}$; $(\partial G/\partial V)_{N_A} = F_V - F_V - VF_{VV} = -VF_{VV}$; $(\partial P/\partial N_A)_V = -F_{VN_A}$; $(\partial P/\partial V)_{N_A} = -F_{VV}$. Inserting these we get $(\partial G/\partial N_A)_P = (F_{N_A} - VF_{VN_A} - (-VF_{VV})(-F_{VN_A})/(-F_{VV})) = F_{N_A} = (\partial F/\partial N_A)_V$.

3.4   *Various forms of the combined law*

In Section 3.3 the discussion was based on the energy scheme which starts from the combined law in the form

$$dU = TdS - PdV + \Sigma\mu_i dN_i - Dd\xi$$

where all the independent variables are extensive ones. It defines the following set of conjugate pairs of variables (T,S), $(-P,V)$ and (μ_i,N_i). However, there are many more possibilities to express the combined law in terms of only extensive quantities as independent variables. Using the new characteristic state functions, obtained in Section 3.3, we can change variables in the combined law. For example, let us replace S by $(TS - PV)/T + PV/T$, obtaining

$$\mathrm{d}S = \mathrm{d}[(TS - PV)/T] + (P/T)\mathrm{d}V + V\mathrm{d}(P/T)$$

By inserting this expression we get

$$
\begin{aligned}
\mathrm{d}U &= T\mathrm{d}[(TS - PV)/T] + P\mathrm{d}V + TV\mathrm{d}(P/T) - P\mathrm{d}V + \Sigma\mu_i\mathrm{d}N_i - D\mathrm{d}\xi \\
&= T\mathrm{d}[(TS - PV)/T] + TV\mathrm{d}(P/T) + \Sigma\mu_i\mathrm{d}N_i - D\mathrm{d}\xi
\end{aligned}
$$

By subtracting $(P/T)\cdot TV$ (which is equal to PV) from U, we can form a new characteristic state function with only extensive quantities as independent variables

$$\mathrm{d}(U - PV) = \mathrm{d}[U - (P/T)\cdot TV] = T\mathrm{d}[(TS - PV)/T] - (P/T)\mathrm{d}(TV) + \Sigma\mu_i\mathrm{d}N_i - D\mathrm{d}\xi$$

This form of the combined law defines a new set of conjugate pairs, $\{T,[(TS - PV)/T]\}$, $(-P/T,TV)$ and (μ_i, N_i).

We may instead replace V by $[(PV - TS)/P + TS/P]$ and after some manipulations we obtain a new characteristic state function with only extensive variables as independent variables

$$\mathrm{d}(U + TS) = \mathrm{d}[U + (T/P)\cdot PS] = -P\mathrm{d}[(PV - TS)/P] + (T/P)\mathrm{d}(PS) + \Sigma\mu_i\mathrm{d}N_i - D\mathrm{d}\xi$$

which yields a new set of conjugate pairs.

We may also rearrange the terms in the combined law before introducing new functions. The entropy scheme uses

$$-\mathrm{d}S = -(1/T)\mathrm{d}U - (P/T)\mathrm{d}V + \Sigma(\mu_i/T)\mathrm{d}N_i - (D/T)\mathrm{d}\xi$$

It immediately defines a new set of conjugate pairs and two more alternatives are obtained by replacing U or V in the way demonstrated above. One may also rearrange the terms in the combined law as follows

$$\mathrm{d}V = (T/P)\mathrm{d}S - (1/P)\mathrm{d}U + \Sigma(\mu_i/P)\mathrm{d}N_i - (D/P)\mathrm{d}\xi$$

This may be called the **volume scheme** and it yields three more alternatives. We have thus obtained the sets of conjugate pairs of variables given in Table 3.1. In each pair the potential is given first and between them one can formulate a Gibbs–Duhem relation. In each case the characteristic state function for the extensive variables is given to the left.

We may also define a number of **content schemes** by the following arrangement of terms, but they will probably have very limited use and will not be discussed further.

$$-\mathrm{d}N_j = (T/\mu_j)\mathrm{d}S - (1/\mu_j)\mathrm{d}U - (P/\mu_j)\mathrm{d}V + \sum_k (\mu_k/\mu_j)\mathrm{d}N_k - (D/\mu_j)\mathrm{d}\xi$$

Table 3.1 *Sets of conjugate pairs of state variables.*

From the energy scheme			
U:	T,S	$-P,V$	μ_i,N_i
$U-PV$:	$T,(TS-PV)/T$	$-P/T,TV$	μ_i,N_i
$U+TS$:	$T/P,PS$	$P,(TS-PV)/P$	μ_i,N_i
From the entropy scheme			
$-S$:	$-1/T,U$	$-P/T,V$	$(\mu_i/T),N_i$
$-(TS+PV)/T$:	$-1/T,H$	$-P,V/T$	$(\mu_i/T),N_i$
$-(TS+U)/T$:	$-P/T,H/P$	$-1/P,PU/T$	$(\mu_i/T),N_i$
From the volume scheme			
V:	$T/P,S$	$-1/P,U$	$(\mu_i/P),N_i$
$(PV-U)/P$:	$-1/T,TU/P$	$-T/P,F/T$	$(\mu_i/P),N_i$
$(PV+TS)/P$:	$T,S/P$	$-1/P,F$	$(\mu_i/P),N_i$

Exercise 3.4.1

Suppose one would like to consider U as an independent variable. What would be its conjugate potential?

Solution

From the entropy scheme we find $-1/T$ and from the volume scheme $-1/P$. Evidently, the choice depends on what other conjugate pairs one would like to consider at the same time.

Exercise 3.4.2

Formulate the Gibbs–Duhem relation based on the basic volume scheme. Then show that it is equivalent to the ordinary form.

Solution

$Sd(T/P) + Ud(-1/P) + \Sigma N_i d(\mu_i/P) = 0$; $(S/P)dT - (TS/P^2)dP + (U/P^2)dP + \Sigma(N_i/P)d\mu_i - \Sigma(N_i\mu_i/P^2)dP = 0$. Multiply by P and replace $\Sigma N_i\mu_i$ by $U + PV - TS$: $SdT - (TS/P)dP + (U/P)dP + \Sigma N_i d\mu_i - (U + PV - TS)/P \cdot dP = 0$; $SdT + \Sigma N_i d\mu_i - VdP = 0$

Exercise 3.4.3

Find the characteristic state function for the variables T/P, $1/T$ and N_i. Then change to T, P and N_i and check that the function changes to the Gibbs energy G.

Hint

In order to replace dX^a by dY^a, subtract $X^a Y^a$ from the characteristic state function.

Solution

Line 8 in Table 3.1 yields $d[(PV - U)/P] = -(1/T)d(TU/P) - (T/P)d(F/T) + \Sigma(\mu_i/P)dN_i$. Subtract $-(1/T)\cdot(TU/P) = -U/P$ and $-(T/P)\cdot(F/T) = -F/P$ from $(PV - U)/P$: $d[(PV - U + U + F)/P] = (TU/P)d(1/T) + (F/T)d(T/P) + \Sigma(\mu_i/P)dN_i$. But $(PV - U + U + F)/P = G/P$. This is the new characteristic state function (for the independent variables T/P, $1/T$ and N_i). Thus $d(G/P) = (1/P)dG - (G/P^2)dP = -(TU/PT^2)dT + (F/TP)dT - (TF/TP^2)dP + \Sigma(\mu_i/P)dN_i$; i.e. $dG = (1/P)dP\cdot(G - F) - (1/T)dT\cdot(U - F) + \Sigma\mu_i dN_i = VdP - SdT + \Sigma\mu_i dN_i$.

Exercise 3.4.4

Consider a solid system with no volatile components but in intimate contact with a large controllable volume of an ideal gas. The state of the system can be changed by altering its temperature and the volume of the gas, V_g. What characteristic state function would be the natural one to use?

Hint

The ideal gas law relates some variables.

Solution

At constant volume, the ideal gas will relate T and P by $T/P = V_g/NR$ and the value of T/P will thus be controlled by the volume of the gas. The two state variables will thus be T and T/P. We find no such alternative in Table 3.1 but we find $-1/T$ and $-T/P$ on line 8. By changing variables in order to introduce the potentials as independent variables we get $d[(PV - U)/P] - d[(-1/T)\cdot(TU/P)] - d[(-T/P)\cdot(F/T)] = -(TU/P)d(-1/T) - (F/T)d(-T/P) = d[V - U/P + U/P + F/P] = d(G/P)$. We could also have used line 2 in Table 3.1 with T and $-P/T$ or line 4 with $-1/T$ and $-P/T$.

Exercise 3.4.5

In Chapter 8 we shall find that the two axes in a phase diagram should be taken from the same set of conjugate pairs. Suppose one would like to use U, F or G as one of the axes. How should the other axis be chosen?

Solution

From the entropy scheme we find that U could be combined with $-P/T$ (which may not be very practical) or V. From the volume scheme we find that U could be combined with T/P (which again may not be very practical) or S. We find F in the volume scheme only, and it can be combined with T or S/P (which is not very practical). We do not find G in any scheme except in the form of μ_i for a unary system.

3.5 *Partial quantities*

It is most common to control T, P and N_i experimentally and then to regard the other external variables as functions of them. For any extensive quantity, A, we have, under equilibrium conditions,

$$A = A(T,P,N_i)$$

Alternatively, one may (i) introduce the molar contents x_i as variables by substituting Nx_i for N_i, and (ii) introduce A_m which is defined as A/N. Thus,

$$A = A(T,P,N_i) = A(T,P,Nx_i) = NA_m(T,P,x_i)$$

Since A_m is defined for one mole, $N = 1$, we can write x_i instead of Nx_i when giving the variables for A_m. We see that the molar quantity, A_m, being an intensive quantity, can be expressed as a function of a set of other intensive quantities T,P,x_i.

It is common to keep constant T and P but vary the amount of some component, N_i. It is interesting to examine what happens to various thermodynamic quantities under such conditions and we shall thus define a new kind of quantity called a **partial quantity**.

partial quantity of j: $A_j \equiv (\partial A/\partial N_j)_{T,P,N_k}$

Such partial quantities appear in the expression for the differential of A

$$dA = (\partial A/\partial T)_{P,N_i}dT + (\partial A/\partial P)_{T,N_i}dP + \Sigma A_i dN_i$$

In Section 3.3 we saw that the chemical potential μ_j can be derived as a partial derivative of any one of the characteristic state functions U, F, H and G. However, it is important to notice that only one of these partial derivatives is a partial quantity with the definition used here, $(\partial G/\partial N_j)_{T,P,N_k}$, because it is evaluated under constant T and P. We can thus write

$$\mu_j = (\partial G/\partial N_j)_{T,P,N_k} = G_j$$

With the short-hand notation introduced for partial derivatives in Section 2.5, this quantity could also be denoted by G_{N_j}.

Since the chemical potential μ_j is identical to the partial Gibbs energy G_j one may wonder if both names or symbols are necessary. However, we shall find it useful sometimes to use one and sometimes the other. When we are interested in the variation of properties of a homogeneous system consisting of a single phase with variable composition, and employ an analytical function $G_m(T,P,x_i)$, then G_j is the most natural term to use. When we are concerned with a more complex system, where G_j of a small part cannot be defined because the composition of that part cannot vary gradually, then μ_j is the most natural term to use.

In order to distinguish the notation for a partial quantity A_j at any composition from the notation for the same quantity in pure j, the latter one will be identified by a small superscript circle in front, $^\circ A_j$. It should be noticed that $^\circ A_j$ is actually identical to A_m of pure j, because for a system with one component we have $N_j = N$ and obtain

$$A = NA_m(T,P)$$
$$^\circ A_j = (\partial A/\partial N_j)_{T,P,N_k} = (\partial A/\partial N)_{T,P} = A_m$$

It is evident that A_j is also an intensive quantity and this can be demonstrated by the fact that it is related to the intensive quantity A_m and can be calculated from it. Using the following relations: $N = \Sigma N_i$; $x_j = N_j/\Sigma N_i$; $\partial x_j/\partial N_j = (N - N_j)/N^2 = (1 - x_j)/N$; $x_k = N_k/\Sigma N_i$; $\partial x_k/\partial N_j = -N_k/N^2 = -x_k/N$, we obtain

$$A_j = \partial A/\partial N_j = \partial(NA_m)/\partial N_j = 1 \cdot A_m + N \cdot \frac{\partial A_m}{\partial x_j}\frac{\partial x_j}{\partial N_j} + N \cdot \sum_{k \neq j} \frac{\partial A_m}{\partial x_k}\frac{\partial x_k}{\partial N_j}$$

$$= A_m + (1 - x_j)\frac{\partial A_m}{\partial x_j} - \sum_{k \neq j} x_k \frac{\partial A_m}{\partial x_k}$$

All the partial derivatives of A_m are here taken under constant T and P and molar contents of the other components. x_j is excluded from the summation. We can modify the equation by including x_j in the summation

$$A_j = A_m + \frac{\partial A_m}{\partial x_j} - \Sigma x_i \frac{\partial A_m}{\partial x_i}$$

When the set of N_i variables was replaced by the set of x_i variables, N was also introduced but the number of independent variables was not changed because there is a relation between all the x_i, their sum being unity. It should be emphasized that the derivatives with respect to x_i were then evaluated at constant values of all the other x quantities. Since this is a physical impossibility, these derivatives cannot be used alone.

The difference between two such quantities can be expressed in a simple way using a derivative which has a physical meaning

$$A_j - A_k = \frac{\partial A_m}{\partial x_j} - \frac{\partial A_m}{\partial x_k} = \left(\frac{\partial A_m}{\partial x_j}\right)_{x_j + x_k}$$

An example is the so-called **diffusion potential**, which represents the driving force for diffusion of j in exchange for k

$$G_j - G_k = \left(\frac{\partial G_m}{\partial x_j}\right)_{x_j + x_k}$$

The equations derived here for calculating partial quantities from molar quantities are frequently used for calculating chemical potentials as partial Gibbs energies.

Exercise 3.5.1

For substitutional solutions one often defines an activity coefficient for a component i as $\gamma_i = \exp[(G_i - {}^\circ G_i - RT\ln x_i)/RT]$. Show that for low contents of B and C in A one has the following approximate relation under constant T and P, if x_A is not included in the set of independent composition variables, $\partial\ln\gamma_B/\partial x_C = \partial\ln\gamma_C/\partial x_B$.

Hint

Start from a Maxwell relation $\partial G_B/\partial N_C = \partial^2 G/\partial N_C\partial N_B = \partial G_C/\partial N_B$. Then go from derivatives with respect to N_i to derivatives with respect to x_i by using $x_i = N_i/N$; $\partial x_i/\partial N_i = (N - N_i)/N^2 = (1 - x_i)/N$; $\partial x_i/\partial N_j = -N_i/N^2 = -x_i/N$.

Solution

$G_i = {}^\circ G_i + RT\ln x_i + RT\ln\gamma_i$; $\partial G_B/\partial N_C = RT[(1/x_B)(-x_B/N) + (\partial\ln\gamma_B/\partial N_C)] = \partial G_C/\partial N_B = RT[(1/x_C)(-x_C/N) + \partial\ln\gamma_C/\partial N_B]$ and thus $\partial\ln\gamma_B/\partial N_C = \partial\ln\gamma_C/\partial N_B$, exactly. However, we should examine derivatives with respect to x_i and not N_i. Notice that an analytical expression for γ_B may contain x_B and x_C. For small x_B and x_C we get approximately:

$\partial\ln\gamma_B/\partial N_C = (\partial\ln\gamma_B/\partial x_B)(-x_B)/N + (\partial\ln\gamma_B/\partial x_C)(1 - x_C)/N \cong (\partial\ln\gamma_B/\partial x_C)/N$;
$\partial\ln\gamma_C/\partial N_B = (\partial\ln\gamma_C/\partial x_B)(1 - x_B)/N + (\partial\ln\gamma_C/\partial x_C)(-x_C)/N \cong (\partial\ln\gamma_C/\partial x_B)/N$.

Thus, $\partial\ln\gamma_B/\partial x_C \cong \partial\ln\gamma_C/\partial x_B$.

Exercise 3.5.2

Rearrange the derivatives in the last equation of this section in such a way that it only contains terms of physical significance.

Hint

First multiply the first derivative by Σx_i which is equal to 1.

Solution

$$A_j = A_m + \Sigma x_i[\partial A_m/\partial x_j - \partial A_m/\partial x_i]$$

Exercise 3.5.3

Suppose one has the following expression for a ternary solution phase: $G_m = \Sigma x_i(^\circ G_i + RT\ln x_i) + x_A x_B x_C L_{ABC}$, where L is a ternary interaction parameter.
(a) Evaluate G_A directly from the last equation of this section.
(b) Evaluate G_A after an elimination of x_A using $x_A = 1 - x_B - x_C$.

Solution

(a) $G_A = {}^\circ G_A(x_A + 1 - x_A) + {}^\circ G_B(x_B - x_B) + {}^\circ G_C(x_C - x_C) + RT\{x_A\ln x_A + \ln x_A + (x_A/x_A) - x_A[\ln x_A + (x_A/x_A)] + x_B\ln x_B - x_B[\ln x_B + (x_B/x_B)] + x_C\ln x_C - x_C[\ln x_C + (x_C/x_C)]\} + L_{ABC}(x_A x_B x_C + x_B x_C - x_A x_B x_C - x_B x_A x_C - x_C x_A x_B) = {}^\circ G_A + RT\ln x_A + x_B x_C(1 - 2x_A)L_{ABC}$;
(b) $G_m = (1 - x_B - x_C)^\circ G_A + x_B{}^\circ G_B + x_C{}^\circ G_C + RT[(1 - x_B - x_C)\ln(1 - x_B - x_C) + x_B\ln x_B + x_C\ln x_C] + (1 - x_B - x_C)x_B x_C \cdot L_{ABC}$.
Now we get $G_A = {}^\circ G_A(1 - x_B - x_C + x_B + x_C) + {}^\circ G_B(x_B - x_B) + {}^\circ G_C(x_C - x_C) + RT\{1 - x_B - x_C)\ln(1 - x_B - x_C) - x_B[-\ln(1 - x_B - x_C) - (1 - x_B - x_C)/(1 - x_B - x_C)] - x_C[-\ln(1 - x_B - x_C) - (1 - x_B - x_C)/(1 - x_B - x_C)] + x_B\ln x_B - x_B[\ln x_B + (x_B/x_B)] + x_C\ln x_C - x_C[\ln x_C + (x_C/x_C)]\} + L_{ABC}\{(1 - x_B - x_C)x_B x_C - x_B[-x_B x_C + (1 - x_B - x_C)x_C] - x_C[-x_B x_C + (1 - x_B - x_C)x_B]\} = {}^\circ G_A + RT\ln(1 - x_B - x_C) + L_{ABC}[x_B{}^2 x_C + x_B x_C{}^2 - x_B x_C(1 - x_B - x_C)] = {}^\circ G_A + RT\ln(1 - x_B - x_C) + x_B x_C(2x_B + 2x_C - 1)L_{ABC} = {}^\circ G_A + RT\ln x_A + x_B x_C(1 - 2x_A)L_{ABC}$.

3.6   *Relations for partial quantities*

In Section 3.2 we saw how an expression for the integral internal energy could be derived by integration over a homogeneous system. It will now be demonstrated that the same method can be applied to any other extensive quantity, A. Consider a homogeneous system with constant T, P and x_i. Then all the partial quantities A_i are also constant. We select an infinitely small subsystem and allow it to grow by simply extending its imaginary wall. The growth in size may be represented by dN and the increase of the i content is obtained as

$$dN_i = x_i dN$$

By integrating the differential of A under constant T and P and remembering that all A_i and x_i are constant, we obtain

$$A = \int dA = \int \Sigma A_i dN_i = \int \Sigma A_i x_i dN = \Sigma A_i x_i \int dN = \Sigma A_i x_i N$$

$$A = \Sigma A_i N_i \text{ or } A_m = \Sigma A_i x_i$$

It may again be emphasized that the partial quantities are always defined with T and P as independent variables. If we were to define a corresponding quantity under constant T and V, for instance, it would not have the same properties because V is an extensive variable.

By differentiating $A = \Sigma A_i N_i$ we obtain

$$dA = \Sigma A_i dN_i + \Sigma N_i dA_i$$

Comparison with the expression for dA derived in Section 3.5,

$$dA = (\partial A/\partial T)_{P,N_i} dT + (\partial A/\partial P)_{T,N_i} dP + \Sigma A_i dN_i$$

yields

$$\Sigma N_i dA_i - (\partial A/\partial T)_{P,N_i} dT - (\partial A/\partial P)_{T,N_i} dP = 0$$

This expression is most useful when applied to the Gibbs energy, giving

$$\Sigma N_i dG_i + S dT - V dP = 0$$

This is identical to the Gibbs–Duhem relation since G_i is identical to μ_i. For other quantities it may be most useful under conditions of constant T and P. As an example, for volume we would obtain, under constant T and P,

$$\Sigma N_i dV_i = 0 \text{ or } \Sigma x_i dV_i = 0$$

Since all the partial quantities are defined as the partial derivatives with respect to some content under constant T and P, it is evident that the following relations hold between them

$$G_i = H_i - TS_i = U_i + PV_i - TS_i = F_i + PV_i$$

It is also evident that the expressions for other extensive state variables as derivatives of the characteristic state functions can be applied to partial quantities as well. As an example, we can start from an expression for S in terms of G and derive a similar expression for S_j in terms of G_j,

$$S = -\left(\frac{\partial G}{\partial T}\right)_{P,N_i}$$

$$S_j = \left(\frac{\partial S}{\partial N_j}\right)_{T,P,N_k} = -\left(\frac{\partial}{\partial N_j}\left(\frac{\partial G}{\partial T}\right)_{P,N_i}\right)_{T,P,N_k}$$

$$= -\left(\frac{\partial}{\partial T}\left(\frac{\partial G}{\partial N_j}\right)_{T,P,N_k}\right)_{P,N_i} = -\left(\frac{\partial G_j}{\partial T}\right)_{P,N_i}$$

Exercise 3.6.1

Derive the relation $H_j = (\partial[G_j/T]/\partial[1/T])_{P,N_i}$ from $H = (\partial[G/T]/\partial[1/T])_{P,N_i}$.

Hint

Start from the basic definition of H_j.

Solution

$H_j = (\partial H/\partial N_j)_{P,N_k} = (\partial(\partial[G/T]/\partial[1/T])_{P,N_i}/\partial N_j)_{T,P,N_k} = (\partial(\partial[G/T]/\partial N_j)_{T,P,N_k}/\partial[1/T])_{P,N_i} = (\partial[G_j/T]/\partial[1/T])_{P,N_i}$.

3.7   *Alternative variables for composition*

By composition we mean the *relative* amounts of various components, preferably the set of molar contents, x_i. We shall now examine different ways of expressing the molar contents in a ternary system. The same methods may be applied in higher-order systems. In order to distinguish the methods we shall use a number of different notations.

(i) $x_j = N_j/N = N_j/\Sigma N_i$
(ii) $z_j = N_j/N_1 = x_j/x_1$
(iii) $u_j = N_j/(N_1 + N_2 + \cdots + N_k) = N_j/(\Sigma N - N_{k+1} - \cdots) = x_j/(1 - x_{k+1} - \cdots)$

The size of the system is thus measured by N, N_1 and $(N_1 + N_2 + \cdots + N_k)$, respectively.

     The characteristics of the three methods for a ternary system (with $k = 2$ in the third method) are compared in Fig. 3.1 where the regular triangle introduced by Gibbs is shown in Fig. 3.1(a) to the left. Isopleths (lines along which some composition variable is held constant) according to the other schemes are shown in Fig. 3.1(b) and (c). It should be noticed that the isopleths for u_2 are also isopleths for u_1, since $u_1 + u_2 = 1$. In Fig. 3.1(c) it should be noticed that $z_1 = 1$ everywhere. The three diagrams are redrawn with linear scales for each kind of variable in Fig. 3.2. Here, the isopleths with arrows go to infinity. It should be emphasized that any line which is straight in the Gibbs triangle is

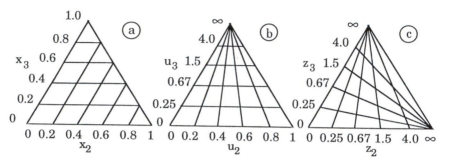

Figure 3.1 The Gibbs triangle showing three different methods of representing composition. The corners represent pure component 1, 2 and 3, respectively.

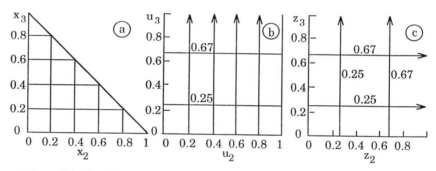

Figure 3.2 The diagrams from Fig. 3.1 drawn with linear scales for the composition variables. An arrow indicates that the component is at infinity and parallel lines with arrows point to the same pure component at infinity.

still a straight line in the modified diagrams.

When these new composition variables are used, the calculation of partial quantities is changed. Before turning to these calculations, it should be realized that the definition of all the molar quantities to be used in one context should be modified in the same way. Taking the Gibbs energy as an example, its molar quantity should be defined as G/N_1 when discussed in connection with z_i (case (ii)) and $G/(N_1 + N_2)$ when discussed in connection with u_i (case (iii)) if $k = 2$. We shall denote these molar quantities by G_{m1} and G_{m12}.

With the method used in deriving an expression for A_j in Section 3.5 we obtain, for case (ii),

$$G = N_1 G_{m1}(z_1,z_2,z_3) \text{ with } z_1 = 1$$

$$\mu_1 \equiv G_1 = G_{m1} - z_2 \frac{\partial G_{m1}}{\partial z_2} - z_3 \frac{\partial G_{m1}}{\partial z_3}$$

$$\mu_2 \equiv G_2 = \frac{\partial G_{m1}}{\partial z_2}$$

$$\mu_3 \equiv G_3 = \frac{\partial G_{m1}}{\partial z_3}$$

For case (iii) we have

$$G = (N_1 + N_2)G_{m12}(u_1, u_2, u_3) \text{ with } u_1 + u_2 = 1$$

$$\mu_1 \equiv G_1 = G_{m12} + \frac{\partial G_{m12}}{\partial u_1} - \Sigma u_i \frac{\partial G_{m12}}{\partial u_i}$$

$$\mu_2 \equiv G_2 = G_{m12} + \frac{\partial G_{m12}}{\partial u_2} - \Sigma u_i \frac{\partial G_{m12}}{\partial u_i}$$

$$\mu_3 \equiv G_3 = \frac{\partial G_{m12}}{\partial u_3}$$

On the other hand, it should be emphasized that many equations derived with the ordinary way of expressing the size of the system, will hold without further modification, if one simply replaces all molar contents x by the corresponding z or u and all other molar quantities by the corresponding molar quantities which may be denoted by A_{m1} or A_{m12}. The following relations are useful

$$u_j = N_j/(N_1 + N_2) = x_j/(x_1 + x_2)$$
$$A_{m12} = A/(N_1 + N_2) = A/N(x_1 + x_2) = A_m/(x_1 + x_2)$$
$$A_{m12} = A/(N_1 + N_2) = \Sigma N_i A_i/(N_1 + N_2) = \Sigma u_i A_i$$
$$z_j = N_j/N_1 = x_j/x_1$$
$$A_{m1} = A/N_1 = A/Nx_1 = A_m/x_1$$
$$A_{m1} = A/N_1 = \Sigma N_i A_i/N_1 = \Sigma z_i A_i = A_1 + z_2 A_2 + z_3 A_3$$

It should be observed that A_i is the usual partial quantity $(\partial A/\partial N_i)_{T,P,N_j}$.

Sometimes it may be convenient to use the notations A_m and x_i for all these quantities. It is then necessary always to specify how one mole is defined, i.e. whether one considers one mole of 1, one mole of $1 + 2$ or one mole total. In higher-order systems one may measure the size of the system in several ways. It may be convenient to use the notations $u_{i(1\ldots k)}$ and $A_{m(1\ldots k)}$ where $1\ldots k$ are the components used to measure the size.

Exercise 3.7.1

Show the details of the derivation of the expression for μ_2 from G_m as a function of the u variables in a ternary system where component 3 has a special character.

Hint

Since components 1 and 2 have similar character, it may be convenient to define $u_i = N_i/(N_1 + N_2)$. Derive the partial derivatives of u_j with respect to N_2. Derive μ_2 from $(\partial G/\partial N_2)_{N_1,N_3}$ where G is given by $G = (N_1 + N_2)G_{m12}$.

Solution

$u_j = N_j/(N_1 + N_2)$; $\partial u_1/\partial N_2 = -N_1/(N_1 + N_2)^2 = -u_1/(N_1 + N_2)$; $\partial u_2/\partial N_2 = (N_1 + N_2 - N_2)/(N_1 + N_2)^2 = (1 - u_2)/(N_1 + N_2)$; $\partial u_3/\partial N_2 = -N_3/(N_1 + N_2)^2 = -u_3/(N_1 + N_2)$. Now we can calculate the potentials from $G = (N_1 + N_2)G_{m12}(u_1, u_2, u_3)$: $\mu_2 = G_2 = (\partial G/\partial N_2)_{N_1, N_3} = 1 \cdot G_{m12} + (N_1 + N_2)[(\partial G_{m12}/\partial u_1)(-u_1)/(N_1 + N_2) + (\partial G_{m12}/\partial u_2)(1 - u_2)/(N_1 + N_2) + (\partial G_{m12}/\partial u_3)(-u_3)/(N_1 + N_2)] = G_{m12} + \partial G_{m12}/\partial u_2 - \Sigma u_i \partial G_{m12}/\partial u_i$.

Exercise 3.7.2

Show that $\mu_2 = G_m + (1 - x_2)(\partial G_m/\partial x_2)_{x_3/x_1}$ in a ternary system.

Hint

Replace variables N, N_2 and N_3 using x_2 $(= N_2/N)$, x_3/x_1 $(= N_3/N_1)$ and N $(= N_1 + N_2 + N_3)$.

Solution

Let G_m be a function of x_2 and x_3/x_1: $G = NG_m(x_2, x_3/x_1)$; $\mu_2 = (\partial G/\partial N_2)_{N_1, N_3} = G_m + N(\partial G_m/\partial x_2)_{x_3/x_1}(\partial x_2/\partial N_2)_{N_1, N_3} + N(\partial G_m/\partial[x_3/x_1])_{x_2}(\partial[x_3/x_1]/\partial N_2)_{N_1, N_3} = G_m + N(\partial G_m/\partial x_2)_{x_3/x_1}(N - N_2)/N^2 + N(\partial G_m/\partial[x_3/x_1])_{x_2} \cdot 0 = G_m + (1 - x_2)(\partial G_m/\partial x_2)_{x_3/x_1}$.

Exercise 3.7.3

One can easily derive $(\partial G_C/\partial z_B)_{z_C} = (\partial G_B/\partial z_C)_{z_B}$ from $\partial G_C/\partial N_B = \partial^2 G/\partial N_B \partial N_C = \partial G_B/\partial N_C$. Derive the same relation by using the Gibbs–Duhem relation.

Hint

Formulate the Gibbs–Duhem relation using the z variables by dividing with N_A. First consider a variation dz_B, then take the derivative with respect to z_C. Next, do it the other way around and compare the two results.

Solution

For any variation, $dG_A + z_B dG_B + z_C dG_C = 0$.
If z_B is varied: $\partial G_A/\partial z_B + z_B \partial G_B/\partial z_B + z_C \partial G_C/\partial z_B = 0$; Take the derivative with

respect to z_C: $\partial^2 G_A/\partial z_B \partial z_C + z_B \partial^2 G_B/\partial z_B \partial z_C + \partial G_C/\partial z_B + z_C \partial^2 G_C/\partial z_B \partial z_C = 0$ (a).
If z_C is varied: $\partial G_A/\partial z_C + z_B \partial G_B/\partial z_C + z_C \partial G_C/\partial z_C = 0$; take the derivative with
respect to z_B: $\partial^2 G_A/\partial z_C \partial z_B + \partial G_B/\partial z_C + z_B \partial^2 G_B/\partial z_C \partial z_B + z_C \partial^2 G_C/\partial z_C \partial z_B = 0$ (b).
Then, (a) − (b): $\partial G_C/\partial z_B - \partial G_B/\partial z_C = 0$.

3.8   *The lever rule*

Let us consider some molar quantity A_m in two homogeneous subsystems (phases), α
and β, with different properties, and then evaluate the average of the molar quantity, A_m^{av},
in the total system. By definition we have

$$A_m^\alpha = A^\alpha/N^\alpha$$
$$A_m^\beta = A^\beta/N^\beta$$

Using the law of additivity we obtain

$$A_m^{av} = (A^\alpha + A^\beta)/(N^\alpha + N^\beta) = \frac{N^\alpha}{N^\alpha + N^\beta} A_m^\alpha + \frac{N^\beta}{N^\alpha + N^\beta} A_m^\beta = f^\alpha A_m^\alpha + f^\beta A_m^\beta$$

The fractions of atoms present in each subsystem, i.e. the relative sizes of the two
subsystems, are denoted by f^α and f^β. The terms can be rearranged because $f^\alpha + f^\beta = 1$.

$$f^\alpha(A_m^{av} - A_m^\alpha) = f^\beta(A_m^\beta - A_m^{av})$$

This is often called the **lever rule** and is often used when A_m is a molar content x_i. That
case is illustrated in Fig. 3.3(a).

The terms can be rearranged in another way

$$A_m^{av} - A_m^\alpha = f^\beta(A_m^\beta - A_m^\alpha)$$

and this equation can be illustrated by two balancing forces, each of which tries to turn
the lever around the point representing the α subsystem (see Fig. 3.3(b)).

The lever rule can be extended to more subsystems. It is easy to see that

$$A_m^{av} = f^\alpha A_m^\alpha + f^\beta A_m^\beta + f^\gamma A_m^\gamma + \cdots$$

For three subsystems in a diagram with two molar quantities one obtains a triangle and
the total system will be represented by a point placed at its centre of gravity. This case is
illustrated in Fig. 3.4.

When the positions of the three subsystems and the total system are known,
then one can evaluate the fractions by several graphical methods, as illustrated in Fig.
3.5.

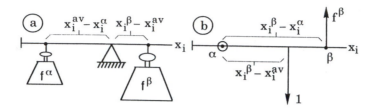

Figure 3.3 Two ways of applying the lever rule.

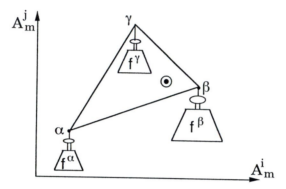

Figure 3.4 The lever rule applied to a system with three subsystems α, β and γ. The triangle is regarded as capable of rotating around the point representing the total system.

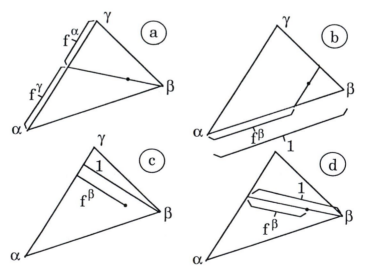

Figure 3.5 (a)–(d) Four methods of evaluating the fraction of a subsystem, f^β, or the ratio of the fractions of two subsystems, f^α/f^γ.

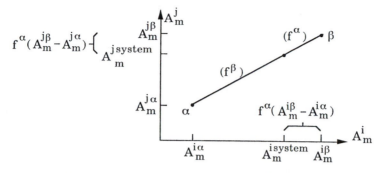

Figure 3.6 The tie-line rule in a diagram with two molar axes. A mixture falls on the straight line joining the points representing the two subsystems, α and β.

3.9   *The tie-line rule*

Suppose two molar quantities are plotted in a two-dimensional diagram. The lever rule can be used to find the average value in the system of the two quantities if the fractions are known. This is illustrated on the sides of the diagram in Fig. 3.6. We have thus proved the rule that the point representing the total system falls on the straight line connecting the two points representing the subsystems and the lever rule applies to that line as well. Such a line may be called a **tie-line** or **conode** and we shall call the rule the **tie-line rule**. It is sometimes called the chord rule. It applies to all diagrams with two molar axes, subject to a restriction to be mentioned soon.

The most common applications of the tie-line rule are found in isobarothermal sections of ternary phase diagrams and the two axes then represent molar contents of two components. In a binary system one may plot another molar quantity as a function of a molar content and thus obtain a property diagram in which the tie-line rule applies. As an example, a plot of the molar volume is shown in Fig. 3.7.

In Fig 3.7 a tangent has been drawn to a point representing an alloy with a B content of x'_B. The two end-points represent the partial molar volumes of A and B, V_A and V_B, and the tie-line rule says that the value of the alloy falls on the straight line connecting those points. Then we find

$$V_m(x'_B) = (1 - x'_B)V_A(x'_B) + x'_B V_B(x'_B)$$

The fact that the end-points of the tangent actually represent the partial volumes can be shown by interpreting geometrically the mathematical expression obtained by applying the formula for partial quantities from Section 3.5. For V_B we obtain for the point of tangency in the diagram

$$V_B = V_m + \frac{\partial V_m}{\partial x_B} - x_A \frac{\partial V_m}{\partial x_A} - x_B \frac{\partial V_m}{\partial x_B}$$

If V_m is described as a function of x_B only, then

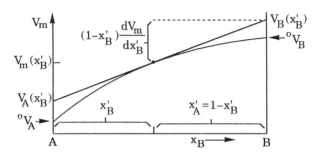

Figure 3.7 Property diagram for a binary system showing the molar volume as a function of composition at constant T and P. The intercepts made by the tangent give the partial quantities.

$$V_B = V_m + (1 - x_B) \frac{dV_m}{dx_B}$$

$$V_B - V_A = \frac{dV_m}{dx_B}$$

It should be noted that dV_m/dx_B is here an ordinary derivative, not a partial one. This kind of property diagram is often applied to the molar Gibbs energy and the tie-line rule plays an essential role in many of those applications. This will be demonstrated in Chapter 6.

It is worth emphasizing that the lever rule and the tie-line rule only hold when the molar quantities are plotted with linear scales. As an example, in a diagram with logarithmic scales one should not draw the tie-lines as straight lines.

In our derivation of the lever rule and the tie-line rule we have used the content of atoms, N, as the measure of the size of the system. We could just as well have used the content of atoms of a particular element, N_j, or of all elements of a particular type, ΣN_j, or the mass, which was used by Gibbs. However, in order for the tie-line rule to apply it is necessary that all the molar quantities have been defined with the same measure of the size, otherwise the fractions f^α, f^β, etc., will be different for the quantities plotted on the two axes and the line representing mixtures between α and β will not be a straight line. Figs. 3.8 and 3.9 show two different ways of plotting the same information on the Gibbs energy in a binary system. Fig. 3.8 corresponds to Fig. 3.7 for the molar volume where the size is measured by the total content of atoms, N. In Fig. 3.9, the size is measured by the content of A, N_A, and the molar Gibbs energy is thus the quantity denoted by G_{m1} in Section 3.7. In both diagrams the intercept of the tangent on the left-hand side gives the partial Gibbs energy of A but in the second case the partial Gibbs energy of B is given by the slope of the tangent dG_{m1}/dz_B, as required by the expression for μ_2 in Section 3.7.

On the other hand, Fig. 3.10(a) and (b) demonstrate the effect of defining the size in two different ways by plotting $z_{Cr} = N_{Cr}/N_{Fe}$ versus $x_C = N_C/(N_{Fe} + N_{Cr} + N_C)$ in the Fe–Cr–C system. In Fig. 3.10(a) one has incorrectly drawn the sides of the

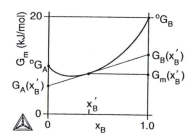

Figure 3.8 Property diagram for a binary system showing the molar Gibbs energy as a function of composition at constant T and P, using the total content of atoms, N, as a measure of the size.

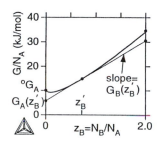

Figure 3.9 The diagram of Fig. 3.8 drawn with a different composition variable and a related modification of G_m.

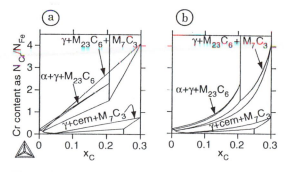

Figure 3.10 z_{Cr}, x_C diagram for Fe–Cr–C at 1 bar and 1200 K. (a) is drawn under the incorrect assumption that the tie-line rule applies; (b) is correct.

three-phase triangles as straight lines and obtained overlapping triangles. In Fig. 3.10(b) they have been drawn in a correct way. One must realize that the tie-line rule does not apply for this choice of axes and that choice is thus inconvenient.

The tie-line rule also holds in three dimensions. As an example, Fig. 3.11 shows a molar Gibbs energy diagram for a ternary system and the intercepts of the tangent plane on the component axes represent the partial Gibbs energies. According to the tie-line rule, the molar Gibbs energy of the alloy falls on the plane through these points. This is also in accordance with the well-known equation $G_m = \Sigma x_i \mu_i$ where μ_i is identical to G_i (see Section 3.3).

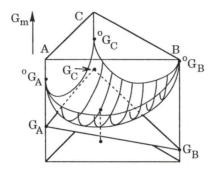

Figure 3.11 Molar Gibbs energy diagram for a ternary solution.

Exercise 3.9.1

Suppose one has measured the lattice parameter a of face centred cubic (fcc)-Fe as a function of the carbon content. What composition variable should be most convenient in a diagram showing the volume of the unit cell, a^3?

Hint

It would be most convenient if the tie-line rule could be applied. Then one could, for instance, see immediately if the volume of a system would be different when carbon is distributed uniformly and non-uniformly. Use the fact that carbon dissolves interstitially in fcc-Fe.

Solution

The unit cell contains a fixed number of Fe atoms and a variable number of C atoms. Its volume is thus proportional to V_m/x_{Fe}. The composition should thus be expressed as x_C/x_{Fe}.

3.10   *Different sets of components*

When considering a system open to exchange of matter with the surroundings in Section 3.1, we introduced the terms $\Sigma \mu_i dN_i$ in the expression for dU. These terms were subsequently carried over into the expression for dG and a chemical potential for any component j can thus be defined as

$$\mu_j = \left(\frac{\partial G}{\partial N_j}\right)_{T,P,N_k,\xi}$$

The quantity N_j represents the amount of component j. The quantities N_i and N are often measured as the number of atoms or groups of atoms, whether the corresponding molecule exists or not. However, the set of independent components can be chosen in different ways and it is self-evident that whatever choice is made it cannot be allowed to affect the total value of $dG = \Sigma\mu_i dN_i$. As a consequence, there is a relation between the chemical potentials defined for different sets of components. Let us compare two sets. As the first set we shall take the elements i, j, k, etc., and as the second set we shall take formula units denoted by d, e, f, etc. Let a_i^d be the number of i atoms in a formula unit of d. It is interesting to note that the set of a_i^d values for a new component d defines its position in the i, j, k compositional space. If the formula unit of d is defined for one mole of atoms then a_i^d is equal to the molar contents of the elements in the new component, x_i^d. Further, let N_d be the total number of formula units of the new component d. The total number of i atoms in the system is then obtained by a summation over all the new components.

$$N_i = \sum_d a_i^d N_d$$

We thus obtain

$$\Sigma\mu_i dN_i = \sum_i \mu_i \left(\sum_d a_i^d dN_d \right) = \sum_d \left(\sum_i a_i^d \mu_i \right) dN_d$$

This result can be inserted into the expression for dG instead of $\Sigma\mu_i dN_i$ and we thus find for the chemical potential of component d,

$$\mu_d = \left(\frac{\partial G}{\partial N_d} \right)_{T,P,N_e} = \sum_i a_i^d \mu_i$$

It is interesting to note that the final expression for μ_d is independent of how the other components in the new set were selected. The expression can thus be used to calculate the chemical potential of any compound or species or combination of atoms, whether it is used in a set of independent variables or not.

Exercise 3.10.1

It is difficult to measure the oxygen potential when it has low values. Suppose one has instead measured the chemical potentials of Ce_2O_3 and CeO_2 in a complicated multicomponent system. Show how one can calculate the oxygen potential.

Solution

$2\mu_{CeO_2} = 2\mu_{Ce} + 4\mu_O$ and $\mu_{Ce_2O_3} = 2\mu_{Ce} + 3\mu_O$ give $2\mu_{CeO_2} - \mu_{Ce_2O_3} = \mu_O$

Exercise 3.10.2

Consider a solution phase with two sublattices and the same number of sites on each. If A and B can occupy the first one and C and D the second one, then we can use the chemical formula $(A_xB_{1-x})_1(C_yD_{1-y})_1$. It may seem reasonable to use the following expressions for the properties in a simple case where all the ternary solutions behave as ideal solutions between two compounds each, e.g. $(A_xB_{1-x})_1C_1$ as xA_1C_1 and $(1-x)B_1C_1$ yielding $\mu_{AC} = {}^\circ G_{AC} + RT\ln(xy)$ and $\mu_{AD} = {}^\circ G_{AD} + RT\ln[x(1-y)]$, etc. However, this would be reasonable only under an additional condition. Accept the expressions given here and find the condition.

Hint

The four μs are related.

Solution

By definition $\mu_{AC} = \mu_A + \mu_C$, etc. Thus, $\mu_{AC} + \mu_{BD} - \mu_{AD} - \mu_{BC} = 0$ and ${}^\circ G_{AC} + RT\ln(xy) + {}^\circ G_{BD} + RT\ln[(1-x)(1-y)] - {}^\circ G_{AD} - RT\ln[x(1-y)] - {}^\circ G_{BC} - RT\ln[(1-x)y] = 0$ and thus ${}^\circ G_{AC} + {}^\circ G_{BD} = {}^\circ G_{AD} + {}^\circ G_{BC}$. This requires that the pair AC + DB has the same stability as AD + BC, which must be an unusual case. Better expressions for μ_A, etc., are discussed in Chapters 18 and 19.

3.11   *Constitution and constituents*

The composition of a system together with the condition of equilibrium defines the state of the system but it gives no direct information on how the atoms are arranged. In order to understand the properties and to make a realistic model of the thermodynamic properties as a function of composition, it is necessary to have some idea about the arrangement of the atoms. The modelling should be based on the **constitution** of the system, i.e. the detailed description of the arrangement of the atoms. The occurrence of regions of different structures and compositions, so-called **phases**, is of primary importance. The arrangement of atoms within each phase may also be important, for instance their arrangement on different sublattices or in groups like molecules, ions or complexes. Groups of atoms, including ions and single atoms, are often called **species**. They may be so stable that they can be transferred from one phase to another and even from the system to the surroundings.

Another useful concept is **constituent** by which one understands a certain kind of species on a certain sublattice in a certain phase. In the following discussion of constituents we shall only consider single atoms. However, the results can be generalized easily to molecular or ionic species.

Let us consider a phase with several sublattices in a higher-order system. The sublattices may be identified by superscripts, s, t, u, etc., their numbers of sites may be denoted by a^s, a^t, a^u, etc., the number of j atoms in the t sublattice by N_j^t and the corresponding **site fraction** by y_j^t. By definition

$$y_j^t = N_j^t / \sum_i N_i^t$$

The site fraction is thus a kind of molar content (mole fraction), evaluated for each sublattice separately. The molar contents in the whole phase can be evaluated from the site fractions

$$x_j = a^t y_j^t / \sum_s a^s$$

where t represents the sublattice in which j resides. In simple cases the relation can be inverted and the site fractions can be evaluated from the composition of the phase

$$y_j^t = x_j \sum_s a^s / a^t$$

However, in the general case an element may enter into more than one sublattice. One can still evaluate the composition from the site fractions

$$x_j = \sum_t a^t y_j^t / \sum_s a^s$$

but it is not certain that this relation can be inverted, i.e. that the site fractions can be evaluated from the composition. Instead there may now be one or more internal variables, describing the distribution of the elements on the various sublattices. Such internal variables will be discussed further in Chapter 19. Together with the external x_j parameters they define the state. An alternative way of defining the state is by only giving the site fractions. A site fraction may thus have a mixed character of internal and external variable.

The total number of formula units can be obtained by considering any sublattice or the whole phase,

$$N = \sum_i N_i^t / a^t = \sum_i N_i^u / a^u = \cdots = \sum_s \sum_i N_i^s / \sum_s a^s$$

Exercise 3.11.1

For a so-called 'reciprocal' system with two elements on each sublattice, $(A,B)_a(C,D)_c$, one often uses a solution model yielding the expression $G_m = y_A y_C \mu_{A_a C_c} + y_B y_C \mu_{B_a C_c} + y_A y_D \mu_{A_a D_c} + y_B y_D \mu_{B_a D_c}$. Examine if this expression is correct.

Hint

Use $\mu_{B_aC_c} = a\mu_B + c\mu_C$, etc., and $x_A = ay_A/(a + c)$, etc.

Solution

$y_Ay_C\mu_{A_aC_c} + y_By_C\mu_{B_aC_c} + y_Ay_D\mu_{A_aD_c} + y_By_D\mu_{B_aD_c} = a\mu_A(y_Ay_C + y_Ay_D) + a\mu_B(y_By_C + y_By_D) + c\mu_C(y_Ay_C + y_By_C) + c\mu_D(y_Ay_D + y_By_D) = a\mu_Ay_A + a\mu_By_B + c\mu_Cy_C + c\mu_Dy_D = (a + c)(x_A\mu_A + x_A\mu_B + x_C\mu_C + x_D\mu_D) = G_m$ for one formula unit.

Exercise 3.11.2

For a reciprocal system, $(A,B)_a(C,D)_c$, prove that $G_m = y_C\mu_{B_aC_c} + y_A\mu_{A_aD_c} + (y_D - y_A)\mu_{B_aD_c}$.

Hint

Use $\mu_{B_aC_c} = a\mu_B + c\mu_C$, etc. $x_A = ay_A/(a + c)$, etc., and $y_A + y_B = 1 = y_C + y_D$.

Solution

$y_C\mu_{B_aC_c} + y_A\mu_{A_aD_c} + (y_D - y_A)\mu_{B_aD_c} = y_Ca\mu_B + y_Cc\mu_C + y_Aa\mu_A + y_Ac\mu_D + (y_D - y_A)a\mu_B + y_Dc\mu_D - y_Ac\mu_D = y_Cc\mu_C + y_Aa\mu_A + y_Dc\mu_D + (y_C + y_D - y_A)a\mu_B = y_Cc\mu_C + y_Aa\mu_A + y_Dc\mu_D + y_Ba\mu_B = (a + c)(x_A\mu_A + x_B\mu_B + x_C\mu_C + x_D\mu_D) = G_m$ for one mole of formula units.

3.12   *Chemical potentials in a phase with sublattices*

When trying to evaluate a chemical potential of a component in a phase with two or more sublattices, we run into difficulties because we cannot vary the content of one component alone unless it is present in all sublattices. The reason is the fixed total amount of atoms in each sublattice relative to the total amounts in the other sublattices. This kind of restriction on the contents of a phase may be called **stoichiometric constraint** and this kind of phase is called a **stoichiometric compound**. The word stoichiometric actually means that the coefficients in the chemical formula are small integers but the word is often used to mean 'fixed composition'. Usually, one follows from the other.

If we were to neglect the difficulty with the stoichiometric constraint and calculate the chemical potential of a constituent j in sublattice s with the method used in Section 3.5, we should get the following formal result

$$\mu_j^s = G_m + \frac{1}{a^s}\left[\frac{\partial G_m}{\partial y_j^s} - \sum_i y_i^s \frac{\partial G_m}{\partial y_i^s}\right]$$

where G_m is defined for 1 mole of atoms and $\Sigma a^s = 1$. The factor $1/a^s$ comes from the fact that $\Sigma N_i^s = a^s N$. It must be emphasized that the expression for μ_j^s cannot be used alone. It can only be used in combinations obeying the stoichiometric constraint. Two methods of obeying the constraint should be considered. In the first method one considers the addition of balanced amounts of atoms to all sublattices, corresponding to the addition of a compound $j_{a^t}k_{a^u}l_{a^v}$. For the chemical potential of that compound we obtain

$$\mu_{j_{a^t}k_{a^u}l_{a^v}} = a^t\mu_j^t + a^u\mu_k^u + a^v\mu_l^v = G_m + \frac{\partial G_m}{\partial y_j^t} + \frac{\partial G_m}{\partial y_k^u} + \frac{\partial G_m}{\partial y_l^v} - \sum_s\sum_i y_j^s\frac{\partial G_m}{\partial y_j^s}$$

It is evident that we can here drop the restriction $\Sigma a^s = 1$ and redefine G_m to hold for 1 mole of formula units.

If an element A appears in all sublattices, then one could consider a compound which is a form of the pure element A and with $\Sigma a^s = 1$ its chemical potential would be

$$\mu_A = \mu_{A_{a^t}A_{a^u}A_{a^v}} = a^t\mu_A^t + a^u\mu_A^u + a^v\mu_A^v = G_m + \frac{\partial G_m}{\partial y_A^t} + \frac{\partial G_m}{\partial y_A^u} + \frac{\partial G_m}{\partial y_A^v} - \sum_s\sum_i y_j^s\frac{\partial G_m}{\partial y_j^s}$$

This calculation cannot be performed if the element is not present in all sublattices and in such a case μ_A has no unique physical meaning in that phase.

The other method of obeying the stoichiometric constraint is to substitute an element for another one in a certain sublattice. The result will be

$$\mu_j - \mu_k = \mu_j^t - \mu_k^t = \frac{1}{a^t}\left[\frac{\partial G_m}{\partial y_j^t} - \frac{\partial G_m}{\partial y_k^t}\right] = \frac{1}{a^t}\left(\frac{\partial G_m}{\partial y_j^t}\right)_{y_j^t + y_k^t}$$

The difference $\mu_j - \mu_k$ is sometimes called the diffusion potential because its gradient may be regarded as the driving force for diffusion of element j in one direction in exchange of another element k, diffusing in the opposite direction. It was mentioned in Section 3.5 where it was denoted $G_j - G_k$.

For a system in internal equilibrium the calculation of $\mu_j - \mu_k$ must give the same result independent of what sublattice is used in the calculation, otherwise there would be a driving force for an exchange of atoms between the sublattices. Thus,

$$\mu_j^t - \mu_k^t = \mu_j^u - \mu_k^u$$

If there are vacancies in one of the sublattices, then one can evaluate the chemical potential of any element present in that sublattice at equilibrium because the vacancies may be treated as an additional element with chemical potential μ_{Va} which can be defined as zero at equilibrium. It would also be possible to calculate the chemical potential of an element not present in that sublattice but present in all the other ones.

It should again be emphasized that the quantities μ_j^t, etc., which refer to a specified sublattice, in general have no unique meaning by themselves and they do not have the same value in different sublattices, not even at equilibrium. There may be several methods of calculating the μ_j^t quantities and they may give different results. But in the combinations obeying the stoichiometric constraint the results must be the same. This question is connected with the fact that one cannot add just one element to a phase with a stoichiometric constraint. It follows that the set of independent components contains less components than there are elements. In such a case one may define the set of components by using compounds and talk about **component compounds**. On the other hand, there may also be too many possible component compounds and for the set of independent components one must make a selection.

Exercise 3.12.1

Consider a system with two sublattices and $a^t = a^u = 0.5$. The elements A and B can dissolve in both sublattices and one can thus evaluate the quantities $\mu_{A_{0.5}B_{0.5}}$ and $\mu_{B_{0.5}A_{0.5}}$. Under what conditions are they equal.

Hint

At internal equilibrium $\mu_A^t - \mu_B^t = \mu_A^u - \mu_B^u$.

Solution

$2\mu_{A_{0.5}B_{0.5}} = \mu_A^t + \mu_B^u = \mu_B^t + \mu_A^u = 2\mu_{B_{0.5}A_{0.5}}$. They are equal at internal equilibrium and under such conditions one can evaluate the chemical potential of a compound between one atom of each of A and B in two ways, either as $2\mu_{A_{0.5}B_{0.5}}$ or as $2\mu_{B_{0.5}A_{0.5}}$.

Exercise 3.12.2

We have seen that the chemical potential of an element A in a system with more than one sublattice can be evaluated under two different conditions. In one case the element is present in all sublattices and in the other case it is present in a sublattice, t, which has vacancies. In the latter case, consider another element B which is only present in a second sublattice, u, which has no vacancies. Can its chemical potential also be evaluated?

Hint

Use the fact that the chemical potential of the first element, A, can be evaluated.

Solution

Using the second sublattice we can evaluate $\mu_A - \mu_B = \mu_A^u - \mu_B^u$. Under equilibrium conditions, μ_A is known from the first sublattice $\mu_A = \mu_A^t - \mu_{Va}^t = \mu_A^t$ and thus $\mu_B = \mu_A^t - (\mu_A^u - \mu_B^u)$ but only under equilibrium.

Exercise 3.12.3

It is natural to start the expression for G_m of a solution phase with a term characteristic of a so-called 'mechanical mixture' $\Sigma x_i {}^\circ G_i$ where ${}^\circ G_i$ is the molar Gibbs energy of pure i and the mole fraction of i can be expressed with site fractions $x_i = \sum_s a^s y_i^s$. Consider a binary system with two sublattices and examine how the corresponding term in μ_A^t would look.

Hint

Apply $\mu_j^t = G_m + (1/a^t)\cdot[(\partial G_m/\partial y_j^t) - \Sigma y_i^t\cdot(\partial G_m/\partial y_i^t)]$ to $G_m = \Sigma x_i {}^\circ G_i$.

Solution

$\mu_A^t = (a^t y_A^t + a^u y_A^u){}^\circ G_A + (a^t y_B^t + a^u y_B^u){}^\circ G_B + (1/a^t)\cdot(a^t\cdot{}^\circ G_A - y_A^t a^t\cdot{}^\circ G_A - y_B^t a^t\cdot{}^\circ G_B) = {}^\circ G_A + [a^u y_A^u - (1 - a^t)y_A^t]{}^\circ G_A + [a^u y_B^u - (1 - a^t)y_B^t]{}^\circ G_B = {}^\circ G_A + a^u(y_A^u - y_A^t)({}^\circ G_A - {}^\circ G_B)$. It may seem strange that the result is not simply ${}^\circ G_A$ (see next exercise).

Exercise 3.12.4

One would normally expect the first term in an expression for μ_A to be just ${}^\circ G_A$, the molar Gibbs energy of pure A. In the preceding exercise the result for μ_A^t was ${}^\circ G_A + a^u(y_A^u - y_A^t)({}^\circ G_A - {}^\circ G_B)$. This difference is not completely surprising because in a phase with a stoichiometric constraint it should not be possible to define the chemical potential of an element in a unique way and the formula given in this section is just one alternative. However, it is necessary that such a difference has no effect in combinations satisfying the stoichiometric constraint. Check this by forming the corresponding term in G_m by the combination $\Sigma\Sigma a^s y_i^s \mu_i^s$.

Hint

For a binary system $y_A^t + y_B^t = 1 = y_A^u + y_B^u$.

Solution

$\Delta\mu_A^t = a^u(y_A^u - y_A^t)(^\circ G_A - {}^\circ G_B)$; $\Delta\mu_B^t = a^u(y_B^u - y_B^t)(^\circ G_B - {}^\circ G_A)$; $\Delta\mu_A^u = a^t(y_A^t - y_A^u)(^\circ G_A - {}^\circ G_B)$; $\Delta\mu_B^u = a^t(y_B^t - y_B^u)(^\circ G_B - {}^\circ G_A)$; $\Delta G_m = a^t y_A^t \Delta\mu_A^t + a^t y_B^t \Delta\mu_B^t + a^u y_A^u \Delta\mu_A^u + a^u y_B^u \Delta\mu_B^u = a^t a^u(y_A^u - y_A^t)(^\circ G_A - {}^\circ G_B)(y_A^t + y_B^t - y_A^u - y_B^u) = 0$.

Evaluation and use of driving force

4.1 Irreversible thermodynamics

Many authors emphasize that classical thermodynamics is strictly applicable to states of equilibrium, only, and Gibbs only considered equilibria. Thermodynamics of irreversible processes has been defined as a separate branch of science. It is an extension of strict thermodynamics based on several steps. The first two steps are the assumption that the internal entropy production can be used to derive the driving force for an irreversible process and the definition of driving force (or 'affinity') as a thermodynamic property of a state not in equilibrium. In the present text these two steps are treated as integral parts of basic thermodynamics. The third step is the assumption that the rate of a process is proportional to the driving force, i.e. a linear dependence between flux and force. This is certainly not generally true but seems to apply to many processes at low values of the driving force. The fourth step is defined by Onsager's reciprocity relations, concerned with the coupling between two or more simultaneous processes. A very brief demonstration of these features of irreversible thermodynamics will be given in the present chapter. It will start with a short discussion of how to calculate the value of an internal variable in a state of equilibrium under different external conditions because that is the first step in the calculation of the driving force.

4.2 Calculation of equilibrium

Suppose a state has been established in a system under some experimental conditions and one knows the values of all the external variables as well as an internal variable, ξ. Then one would like to test whether it is a state of equilibrium, i.e. whether ξ has its equilibrium value. If it has, then the system is at rest and it does not matter what the experimental conditions are, i.e. what external variables are used for controlling the experiment. One could evaluate the driving force from any of the following relations or similar ones and test if it is zero, as it should be at equilibrium.

$$- D = (\partial U/\partial \xi)_{S,V,N_i} = (\partial F/\partial \xi)_{T,V,N_i} = (\partial H/\partial \xi)_{S,P,N_i} = (\partial G/\partial \xi)_{T,P,N_i} = (\partial \Omega/\partial \xi)_{T,V,\mu_i}$$

The question is simply what fundamental equation is available. In most cases the Gibbs energy is used because $G = G(T,P,N_i,\xi)$ is available.

Suppose one finds that the system is not at equilibrium and then would like to know what the state of equilibrium would be, i.e. the equilibrium value of ξ. Then it is essential to know the experimental conditions because one wants to find a state of equilibrium under the initial values of a particular set of external variables. Suppose one is going to keep T and V constant during the experiment. Then one would primarily like to use $F = F(T,V,N_i,\xi)$, derive an expression for $-D = (\partial F/\partial \xi)_{T,V,N_i} = 0$ and solve for the equilibrium value of ξ.

However, suppose that one has only $G = G(T,P,N_i,\xi)$. The calculation is then carried out by iteration, starting with the prescribed T value and guessing what P value would yield an equilibrium at the desired T and V values. After the equilibrium values of ξ under various P values have been calculated, one can evaluate V from $(\partial G/\partial P)_{T,N_i,\xi}$ and compare with the desired value. Then the result should be improved by trying new P values.

Suppose the initial state is defined in some way and the experimental conditions are adiabatic, $dQ = dU + PdV = 0$. Then the experimental conditions must be defined in more detail. Suppose V is kept constant. Then U is also kept constant because the conditions are adiabatic. First, one should evaluate U, V and ξ for the initial state using whatever fundamental equation is available. Secondly, one should calculate the equilibrium, preferably with $S = S(U,V,N_i,\xi)$, if available. If only $G = G(T,P,N_i,\xi)$ is available, one must find the desired U and V values by iteration.

If, instead, P is kept constant under adiabatic conditions, we find

$$dH = dU + PdV + VdP = dQ = 0$$

and would prefer to use the fundamental equation with the variables P, H and N_i. By rearranging $dH = TdS + VdP - Dd\xi$ we find

$$-dS = (-1/T)dH + (V/T)dP + \Sigma\mu_i dN_i - (D/T)d\xi$$

We would thus need the fundamental equation $S(H,P,N_i,\xi)$. In practice, one may have to use $G = G(T,P,N_i,\xi)$ and find a state of equilibrium with the initial values of H, P and N_i by iteration.

When a thermodynamic model for a certain kind of system is based on basic physical properties, it may not result in an explicit expression for $G(T,P,N_i,\xi)$. It may then be advantageous to use the grand potential $\Omega(T,V,\mu_i,\xi)$ and find an equilibrium for the desired N_i values by iteration.

Next, consider an $\alpha + \beta$ two-phase system where the relative amounts and compositions of the phases can vary but not the content of the whole system. The internal variable can be defined as $\xi = N^\alpha = N - N^\beta$, where N is the total content, but it is not immediately evident how the equilibrium compositions of α and β can be related to ξ. However, the compositions can be calculated directly from the two-phase equilibrium if T and P of the equilibrium state are known, using $G_m^\alpha = G_m^\alpha(T,P,N_i^\alpha)$ and $G_m^\beta = G_m^\beta(T,P,N_i^\alpha)$. Finally, ξ can be calculated from a mass balance. If, instead T and V

of the equilibrium are known, then the fundamental equation $F_m^\alpha = F_m^\alpha(T,V_m^\alpha,N_i^\alpha)$ and $F_m^\beta = F_m^\beta(T,V_m^\beta,N_i^\alpha)$ would be of little use because the molar volumes of the phases are not known before P and the phase compositions have been calculated. One would have to guess the final P value, carry out a calculation based on G_m^α and G_m^β as already described, and finally evaluate the total volume V and compare with the required value. By iteration one could eventually find the P value that gives the correct V value. For the calculation of a phase equilibrium it is evident that $G_m^\alpha(T,P,N_i^\alpha)$ is the most useful fundamental equation for all experimental conditions.

Exercise 4.2.1

Examine what would be the most convenient way of calculating a two-phase equilibrium under given values of T, V and N in a pure element.

Hint

We have already seen that it is not practical to use $F_m^\alpha(T,V_m^\alpha)$ and $F_m^\beta = F_m^\beta(T,V_m^\beta)$ for a two-phase equilibrium at given T and V because V_m^α and V_m^β are not defined directly by the experimental conditions.

Solution

Using the molar Gibbs energy for each phase we get for the whole system
$G(T,P,\xi) = \xi G_m^\alpha(T,P) + (N - \xi)G_m^\beta(T,P)$ where $\xi = N^\alpha = N - N^\beta$. Equilibrium requires that $-D = (\partial G/\partial \xi)_{T,P} = G_m^\alpha(T,P) - G_m^\beta(T,P) = 0$. For the given T we may thus calculate a particular P for the two-phase equilibrium without iteration. Then we can calculate $V_m^\alpha = (\partial G_m^\alpha/\partial P)_T$ and $V_m^\beta = (\partial G_m^\beta/\partial P)_T$ for these T and P values. Finally, we calculate the ξ value satisfying $\xi V_m^\alpha + (N - \xi)V_m^\beta = V$.

4.3   *Evaluation of the driving force*

In the preceding section we discussed the calculation of the equilibrium value of an internal variable, ξ, under various conditions. The calculation of the driving force for the corresponding reaction is simpler because the system does not 'feel' which variables are to be kept constant until the reaction is under way. One could use $-D = (\partial G/\partial \xi)_{T,P,N_i}$ as well as any other expression for D. On the other hand, as the reaction gets under way there will be changes in the variables that are not controlled and the result will depend upon the experimental conditions. Then one must either use the appropriate fundamental equation or an iteration technique similar to the one described in the preceding section. For example, when using $G(T,P,N_i,\xi)$ for an experiment under constant T and V, one can make a series of calculations along the reaction path by selecting a number of ξ values. For each value one can use iteration to evaluate the P value yielding the experimental value of $V = (\partial G/\partial P)_{T,N_i,\xi}$. Using that pair of ξ,P values one can calculate $-D$ from $(\partial G/\partial \xi)_{T,P,N_i}$.

There are many cases where one knows the initial and final states for a process but does not know or is not interested in the '**reaction path**' in detail. In such cases it may be interesting to evaluate the total production of entropy due to internal processes

$$\Delta_{ip}S = \int d_{ip}S = \int (D/T)d\xi$$

For isothermal reactions T is constant and

$$\Delta_{ip}S = (1/T)\int Dd\xi$$

The quantity $\int Dd\xi$ could be called the **integrated driving force** but unfortunately it is often called simply 'driving force'. It could also be identified with the integrated value of the 'loss of work' discussed in Section 1.7. Anyway, it should only be applied to isothermal reactions because T initially appears in the integrand.

Under constant T, P and N_i we obtain, using the combined law expressed for G,

$$Dd\xi = -SdT + VdP + \Sigma\mu_i dN_i - dG = -dG$$

Under these conditions, the integrated driving force is thus equal to the decrease in Gibbs energy,

$$\int Dd\xi = -\Delta G$$

Since G is a state function it is evident that ΔG is here independent of the reaction path and so is the integrated driving force, whereas the driving force at any value of ξ, i.e. at any stage of the reaction, depends critically upon the reaction path.

If the reaction occurs under other conditions, the integrated driving force will be given by the change in the characteristic state function for which the natural variables are constant during the reaction. For instance, under constant T, V and composition, $\int Dd\xi = -\Delta F$. However, suppose $G(T,P,N_i,\xi)$ is the only fundamental equation available, then one must first find the final equilibrium by iteration, as described in the preceding section. Then one can use $G(T,P,N_i,\xi)$ to evaluate

$$\int Dd\xi = -\Delta F = -\Delta[G - P(\partial G/\partial P)_{T,N_i,\xi}]$$

Exercise 4.3.1

Consider a unary system at constant T, V and μ_A. It is in a metastable state of β.

Show how one can calculate the integrated driving force for the transformation to a more stable phase α.

Hint

$D\mathrm{d}\xi$ is present in all forms of the combined law. It is most convenient to use the form where T, V and μ_A are the independent variables.

Solution

Choose $\mathrm{d}(-PV) = -S\mathrm{d}T - P\mathrm{d}V - \Sigma N_i\mathrm{d}\mu_i - D\mathrm{d}\xi$. In our case $\mathrm{d}(-PV) = -D\mathrm{d}\xi$; $\int D\mathrm{d}\xi = \int \mathrm{d}(PV) = (PV)_2 - (PV)_1 = V(P_2 - P_1)$. It is evident that P must increase during the spontaneous transformation. It should be noticed that the content N_A is not constant under these experimental conditions.

4.4

Evaluation of integrated driving force as function of T or P

According to the preceding section, the integrated driving force for an $\alpha \to \beta$ phase transformation, which takes place under constant T, P and N_i, should be equal to $-\Delta G = G^\alpha - G^\beta$. One is sometimes interested in evaluating the variation of $-\Delta G$ with T or P. The following procedure can be used close to equilibrium.

For constant P it is convenient to evaluate the effect of a change of T on the relative stability of the two phases by starting from the following equation, obtained by applying $G = H - TS$ to both phases under the same T,

$$\Delta G(T) = \Delta H - T\Delta S$$

If the two phases are in equilibrium with each other at T_o for the P value under consideration, we have

$$0 = \Delta H - T_o\Delta S$$

Suppose the difference $T - T_o$ is so small that ΔH and ΔS have practically the same values at both temperatures. By eliminating ΔS or ΔH we obtain

$$\int D\mathrm{d}\xi = -\Delta G = \Delta H(T - T_o)/T_o = \Delta S(T - T_o)$$

For constant T it is convenient to evaluate the effect of a change of P by starting from the following equation

$$\Delta G(P) = \Delta F + P\Delta V$$

By the same procedure we now obtain

$$\int Dd\xi = -\Delta G = \Delta F(P - P_o)/P_o = \Delta V(P_o - P)$$

Again, this equation can only be used so close to P_o that the variation of ΔF and ΔV with P is negligible.

Exercise 4.4.1

Consider two phases of pure A, α and L, which are in equilibrium at T_o, P_o. At $T = T_o + \Delta T$ and $P = P_o$ there is a driving force for the transformation $\alpha \rightarrow$ L. How much should P be changed in order to restore the equilibrium? To get a numerical value, use the 'typical' values $\Delta S_m = R$ and $\Delta V_m = 0.2 \cdot 10^{-6}\, \mathrm{m^3/mol}$.

Hint

The driving forces due to the two changes must eliminate each other.

Solution

$\Delta S(T - T_o) + \Delta V(P_o - P) = 0;\ (P - P_o)/(T - T_o) = \Delta S/\Delta V = \Delta S_m/\Delta V_m = R/\Delta V_m = 8.3/(0.2 \cdot 10^{-6})\, \mathrm{Pa/K} = 400\, \mathrm{bar/K}$

Exercise 4.4.2

Consider 1 mole of a substance having a transformation point from α to β at T_o under 1 bar. Use the relation $dS = dQ/T + d_{ip}S$ to calculate the change of entropy at $T_1 > T_o$ and 1 bar. Suppose ΔH_m and ΔS_m are independent of T.

Hint

The internal entropy production is obtained from the driving force. The heat of reaction is equal to ΔH_m under constant P.

Solution

$\Delta_{ip}S = (1/T)\int Dd\xi = -\Delta G_m/T_1 = -\Delta H_m(T_o - T_1)/T_oT_1 = -\Delta H_m/T_1 + \Delta H_m/T_o;\ \Delta S = \Delta Q/T_1 + \Delta_{ip}S = \Delta H_m/T_1 - \Delta H_m/T_1 + \Delta H_m/T_o = \Delta H_m/T_o = \Delta S_m$, as one might have expected.

4.5   ## *Driving force for molecular reactions*

Many kinds of system contain aggregates of atoms, so-called molecules. Even though there may be reactions between the molecular species (often called 'chemical reactions') the individual molecule often has a long lifetime, not only inside a phase but also with respect to exchange of matter between phases or between a system and the surroundings.

In a study of the rate of molecular reactions, it may be interesting to evaluate their driving forces. By denoting the extent of a reaction by ξ^j they may be included in the combined law through the terms $\Sigma D^j \mathrm{d}\xi^j$. Let v_i^j be the reaction coefficients of reaction j, yielding $\mathrm{d}N_i^j = v_i^j \mathrm{d}\xi^j$. The driving force for reaction j would be

$$D^j = -\left(\frac{\partial G}{\partial \xi^j}\right)_{T,P,\xi^k} = -\sum_i \left(\frac{\partial G}{\partial N_i^j}\right)_{T,P,\xi^k} \cdot \frac{\mathrm{d}N_i^j}{\mathrm{d}\xi^j} = -\sum_i \mu_i v_i^j$$

The net effect of all reactions would be

$$\mathrm{d}N_i = \sum_j \mathrm{d}N_i^j = \sum_j v_i^j \mathrm{d}\xi^j$$

$$\sum D^j \mathrm{d}\xi^j = -\sum_j \sum_i \mu_i v_i^j \mathrm{d}\xi^j = -\sum_i \mu_i \sum_j v_i^j \mathrm{d}\xi^j = -\sum_i \mu_i \mathrm{d}N_i$$

where $\mathrm{d}N_i$ is the change due to all the reactions. This looks just like the term $\Sigma \mu_i \mathrm{d}N_i$ which is always part of the combined law. However, in that case $\mathrm{d}N_i$ represents the exchange with the surroundings and it only concerns the independent components. It is evident that the two expressions can be combined and $\mathrm{d}N_i$ should then be defined as the total change of N_i due to internal reactions and the exchange with the surroundings. This is often done and all species are then regarded as components. It results in complications in the application of the Gibbs phase rule, to be derived in Section 7.1, and necessitates the definition of the number of independent reactions. One may just as well avoid this by instead making an effort to define the number of independent components. In the present discussion we do not need to consider this question further because we shall be concerned only with internal reactions.

The equilibrium conditions for an internal chemical reaction is given by

$$D^j = -\sum_i v_i^j \mu_i = 0$$

As an example, for an ideal gas mixture one can write the chemical potential as a function of the partial pressure

$$\mu_i = {}^o\mu_i + RT\ln P_i$$

By inserting this in the equilibrium condition we get

$$\Pi(P_i)_{\mathrm{eq.}}^{vi} = \exp(-\Sigma v_i^o \mu_i / RT)$$

This is the law of mass action. The value of the right-hand side is regarded as the equilibrium constant and may be denoted by K. When the left-hand side is not equal to K, then the system is not in equilibrium and the driving force for the reaction is

$$D^j = - \sum_i v_i^j \mu_i = RT\ln[K/\Pi(P_i)^v{}_i]$$

Exercise 4.5.1

Suppose one has evaluated an equilibrium constant K' for a gas reaction in terms of the mole fractions. How could one evaluate the driving force for the reaction from the set of initial values of the mole fractions?

Hint

For an ideal gas mixture the partial pressures are defined from the mole fractions y_i by $P_i = y_i P$.

Solution

$$K = \Pi(P_i)^{v_i}_{eq.} = P^{\Sigma v_i} \cdot \Pi(y_i)^{v_i}_{eq.} = P^{\Sigma v_i} \cdot K';$$
$$D = RT\ln[K/\Pi(P_i)^{v_i}] = RT\ln[P^{\Sigma v_i}K'/P^{\Sigma v_i}\Pi(y_i)^{v_i}] = RT\ln[K'/\Pi(y_i)^{v_i}]$$

Exercise 4.5.2

For dilute, condensed solutions one can express the chemical potential with Henry's law, $\mu_i = {}^o\mu_i + RT\ln f_i + RT\ln x_i$, where x_i is the molar content of component i and f_i is the activity coefficient. Show how one can express the equilibrium with a compound having the stoichiometric coefficients v_i. The compound is present in another, coexistent phase. Derive an expression for the driving force for the dissolution of the compound in the solution.

Hint

Suppose the chemical potential of the compound in the other phase is μ_c. The reaction would be compound $\rightarrow \Sigma v_i \cdot I_{in\,sol.}$.

Solution

$$D = 1 \cdot \mu_c - \Sigma v_i \mu_i = \mu_c - \Sigma v_i^o \mu_i - RT\ln[\Pi(f_i)^{v_i}] - RT\ln[\Pi(x_i)^{v_i}] = RT\ln[\exp(\Delta_f G^c/RT)/\Pi(f_i)^{v_i}\Pi(x_i)^{v_i}]$$ where $\Delta_f G_c$ denotes Gibbs energy of formation of the compound from the elements in their reference states. Of course, one may define $(\Delta_f G_c/RT)/\Pi(f_i)^{v_i}$ as an equilibrium constant K. At equilibrium, $D = 0$, one would then have $K = \Pi(x_i)^{v_i}_{eq.}$ and in general $D = RT\ln[K/\Pi(x_i)^{v_i}]$.

Onsager's reciprocity relations

It is generally agreed that the rate of physical processes is proportional to the driving force, at least for low values of the driving force. Let us measure the rate of a reaction by $d\xi^j/dt$,

$$v^j = d\xi^j/dt = K^j \cdot D^j$$

and the rate of entropy production

$$T \cdot d_{ip}S/dt = D^j \cdot d\xi^j/dt = D^j \cdot v^j = K^j \cdot (D^j)^2$$

If there are two simultaneous but quite independent reactions

$$L \to M \text{ (i)}$$
$$L \to N \text{ (ii)}$$

then the total entropy production will be the sum of the two. However, let us now assume that the two reactions are coupled by a third reaction

$$M \to N \text{ (iii)}$$

The rate of production of M and N would then be

$$v_M = v^i - v^{iii} = K^i D^i - K^{iii} D^{iii}$$
$$v_N = v^{ii} + v^{iii} = K^{ii} D^{ii} + K^{iii} D^{iii}$$

However, the three driving forces are related

$$D^{ii} = D^i + D^{iii}$$

Eliminating D^{iii} we find

$$v_M = (K^i + K^{iii})D^i - K^{iii} D^{ii}$$
$$v_N = -K^{iii} D^i + (K^{ii} + K^{iii})D^{ii}$$

We note that the coefficients in the cross terms are equal, $(-K^{iii})$. The question is how we should have treated this case if we had no information on the detailed mechanism but simply knew from experimental observation that L reacts and forms M and N simultaneously. The answer is that we should be prepared for the possibility of a coupling reaction and write the kinetic equations as

$$v_M = K_{MM}D_M + K_{MN}D_N$$
$$v_N = K_{NM}D_M + K_{NN}D_N$$

where the driving forces are the same as before, $D^i = D_M$ and $D^{ii} = D_N$. A coupling reaction would result in cross terms and in the present case we have found that the coefficients in the cross terms are equal,

$$K_{MN} = K_{NM} = -K^{iii}$$

This is called Onsager's reciprocity law and it has a very general applicability although it was here derived only for a very simple case. It is of particular importance when one has no, or insufficient, information on coupling reactions.

Let us now examine what happens to the production of entropy if there is a coupling reaction. If we have information on all three reactions in the above case, then we would write

$$T \cdot d_{ip}S/dt = \Sigma D^j v^j = K^i(D^i)^2 + K^{ii}(D^{ii})^2 + K^{iii}(D^{iii})^2$$

If we only know about the production of M and N molecules but were aware of the possibility of a coupling reaction, resulting in cross terms, we should write

$$
\begin{aligned}
T \cdot d_{ip}S/dt &= D_M v_M + D_N v_N \\
&= D_M(K_{MM}D_M + K_{MN}D_N) + D_N(K_{NM}D_M + K_{NN}D_N) \\
&= (K_{MM} + K_{MN})(D_M)^2 + (K_{NN} + K_{NM})(D_N)^2 - K_{MN}D_M(D_M - D_N) \\
&\quad - K_{NM}D_N(D_N - D_M)
\end{aligned}
$$

In order to get agreement with the first expression for the entropy production, it is necessary to have $K_{MN} = K_{NM}$. Then, by comparing terms, we obtain

$$K_{MN} = K_{NM} = -K^{iii}$$
$$K_{MM} = K^i + K^{iii}$$
$$K_{NN} = K^{ii} + K^{iii}$$

We have thus found again that the coefficients of the cross terms must be equal. This is a demonstration of a rule we may formulate as follows: 'Suppose the set of kinetic equations, describing simultaneous reactions, is transformed in such a way that the entropy production can still be evaluated as $\Sigma D^j v^j$. Then the coefficients of the cross terms must be equal.'

Exercise 4.6.1

Demonstrate that Onsager's reciprocity relations apply to the reactions between CO, CO_2 and O_2 in a gas and solid graphite.

Hint

First we should decide how many independent reactions there are. We have four species, CO, CO_2, O_2 and C but only two components, C and O. Thus, there must be $4 - 2 = 2$ independent reactions, only. Start by choosing them and denoting their driving forces D_1 and D_2. Then define as many dependent reactions as possible but, for each one, relate its driving force to D_1 and D_2.

Solution

We may, for instance, choose the following reactions as independent: $2CO \rightarrow CO_2 + C$ (1) and $2CO_2 \rightarrow 2CO + O_2$ (2). Then, $CO_2 \rightarrow C + O_2$ (3) is a dependent reaction. It can be obtained as (1) + (2) and thus $D_3 = D_1 + D_2$.

Furthermore, $2CO \rightarrow 2C + O_2$ (4) is also a dependent reaction. It can be obtained as $2 \cdot (1) + (2)$ and thus $D_4 = 2D_1 + D_2$. The rates of formation of C and O_2 will be
$v_C = K_1 D_1 + K_3 D_3 + 2K_4 D_4 = K_1 D_1 + K_3 (D_1 + D_2) + 2K_4 (2D_1 + D_2) =$
$(K_1 + K_3 + 4K_4)D_1 + (K_3 + 2K_4)D_2$; and
$v_{O_2} = K_2 D_2 + K_3 D_3 + K_4 D_4 = K_2 D_2 + K_3 (D_1 + D_2) + K_4 (2D_1 + D_2) =$
$(K_3 + 2K_4)D_1 + (K_2 + K_3 + K_4)D_2$. Both cross coefficients are thus equal to $K_3 + 2K_4$.

Exercise 4.6.2

Examine if it is possible that the amount of a species decreases although its driving force is positive. To make the discussion more specific, consider the case of L, M and N in the text.

Hint

Consider the formation of N. For a single reaction D and $d\xi$ must have the same sign (usually ' $+$ ' by definition) because $D d\xi > 0$ for a spontaneous reaction. There is no such restriction on the cross coefficients.

Solution

$v_N = K^{ii} D^{ii} + K^{iii} D^{iii}$. The second term represents the rate of formation of N by the reaction $M \rightarrow N$. Its value and even its sign depends on its driving force. By starting with very little M this reaction would go backwards and consume more N than is produced by $L \rightarrow N$ if the amounts of L and N are initially close to their equilibrium.

4.7   *Driving force and entropy production in diffusion*

The kind of chemical reaction considered in Sections 4.5 and 4.6 occurs in a system that is homogeneous at every instant. It occurs in every point and with the same rate everywhere. It is classified as a homogeneous reaction. The opposite case is a heterogeneous reaction occurring at the interfaces in a system of different regions, usually regarded as phases. A heterogeneous reaction often results in the growth of some phase and the shrinkage of another. Such a reaction is also called phase transformation. If the phase transformation does not result in a change of the local composition, then the driving force is easily obtained as the difference in the value of the appropriate characteristic function, $\Delta\Theta$, between the parent phases and the product phases, counted per mole of atoms, for instance. This case and related ones will be further discussed in Section 6.8.

In most cases, a heterogeneous reaction is accompanied by a change in composition and occurs under diffusion. Diffusion itself is yet another type of reaction. It may occur in a system of a single phase which initially has differences in composition. In the general case, it occurs everywhere where there is a composition difference but not with the same rate. The result will be that local differences decrease and eventually disappear. In this case one should discuss the progress of the reaction in each point. At any particular point there may be a flux J of diffusing material and the derivative of the flux results in a change dc of the local composition. Phenomenologically this situation is described with Fick's laws. His first law states that

$$J_1 = -\mathscr{D}_1 \cdot \frac{dc_1}{dy}$$

For simplicity we shall introduce the molar content, $x_1 = c_1/V_m$ and assume that the molar volume V_m is constant. Then we obtain

$$J_1 = -\frac{\mathscr{D}_1}{V_m}\frac{dx_1}{dy}$$

We should now like to interpret this equation thermodynamically by starting with our basic expression, $Dd\xi$. In this case it is convenient to consider two large reservoirs, separated by a layer of thickness Δy of the phase in which we are interested. The layer is our system and in this unusual case we have two surroundings and twice as many independent variables as usual. However, if T and P have the same values and are constant in the two reservoirs, then we can formulate the change in a characteristic state function, Θ, related to Gibbs energy,

$$d\Theta = \Sigma\mu_i' dN_i' + \Sigma\mu_i'' dN_i'' - Dd\xi$$

where ' and '' identify the two reservoirs. Of course, there is only one $Dd\xi$ term because it represents the effect of the process inside the system. There may be diffusion through the system if μ_i' and μ_i'' are different in the two reservoirs. After a stationary state of diffusion has been established, there will be no further changes inside the system and the value of the characteristic state function will not change with time, $d\Theta = 0$, and $dN_i' + dN_i'' = 0$. If we further assume that there are only two components, 1 and 2, and they diffuse with the same rate but in opposite directions, $J_1 + J_2 = 0$, then we get $dN_1' = -dN_2' = -dN_1'' = dN_2''$ and

$$[(\mu_1 - \mu_2)' - (\mu_1 - \mu_2)''] \cdot dN_1' = Dd\xi$$

The quantity $\mu_1 - \mu_2$ was mentioned in Section 3.5 as the diffusion potential. dN_1' can be expressed as $AJ_1 dt$ where J_1 is the flux of 1 through the layer and A is the area of the cross section.

$$D \cdot d\xi/dt = AJ_1 \cdot \Delta(\mu_2 - \mu_1)$$

where Δ means the difference between the two reservoirs. We could now formulate a kinetic equation using the same scheme as before

$$d\xi/dt \equiv AJ_1 = K \cdot \Delta(\mu_2 - \mu_1)$$

However, if we apply this equation to a thin slice of the layer, the result would depend on the thickness. Thus it does not seem very useful to define $d\xi/dt$ as AJ_1. We would prefer to define it as $AJ_1\Delta y$, yielding the following kinetic equation

$$J_1 = K \cdot \frac{d(\mu_2 - \mu_1)}{dy}$$

It could be brought into the form of Fick's law by writing

$$\frac{d(\mu_2 - \mu_1)}{dy} = \frac{d(\mu_2 - \mu_1)}{dx_1} \cdot \frac{dx_1}{dy} = -\frac{d^2 G_m}{dx_1^2} \cdot \frac{dx_1}{dy}$$

Inserting this and assuming that K is proportional to $x_1 x_2$, i.e. $K = L x_1 x_2$, where L is a constant, because $x_1 x_2$ gives the probability that 1 and 2 are available for an exchange of positions, we get

$$-J_2 = J_1 = -L x_1 x_2 \cdot \frac{d^2 G_m}{dx_1^2} \cdot \frac{dx_1}{dy}$$

Comparison with Fick's law gives

$$\frac{\mathscr{D}_1}{V_m} = L x_1 x_2 \cdot \frac{d^2 G_m}{dx_1^2}$$

The derivative $d^2 G_m / dx_1{}^2$ is often regarded as the thermodynamic factor in diffusion.

The interesting question is now to see if we can express the entropy production with the flux and force used in the kinetic equation. For a thin layer we can transform our previous expression, to yield

$$\frac{T \cdot d_{ip} S}{dt} = D \frac{d\xi}{dt} = AJ_1 \cdot \Delta(\mu_2 - \mu_1) = AJ_1 \cdot \frac{d(\mu_2 - \mu_1)}{dy} \cdot \Delta y$$

But $A \cdot \Delta y$ is the volume of the system, ΔV. We may thus express the rate of entropy production per volume with the flux and force from the kinetic equation

$$\frac{T \cdot d_{ip}^2 S}{dt dV} = D \cdot \frac{d^2 \xi}{dt dV} = J_1 \cdot \frac{d(\mu_2 - \mu_1)}{dy}$$

We may thus identify J_1 with $d^2\xi/dt dV$ and D with $d(\mu_2 - \mu_1)/dy$. Note that K in our equation has different dimensions than in previous kinetic equations.

The present derivation was made under two assumptions that we should now discuss. Firstly, we assumed that $J_1 + J_2 = 0$, which is always satisfied if one uses a so-called 'number-fixed frame of reference' for diffusion. However, one can mathematically transform Fick's law to other frames of reference, e.g. the 'lattice-fixed frame of reference' and one gets different expressions for J_1 and $-J_2$. However, that is just a mathematical operation and the assumption of $J_1 + J_2 = 0$ does not make our derivation less general.

Secondly, we assumed stationary conditions. However, the final equation applies to a thin layer and it is no longer limited to stationary conditions. If the composition profile in a non-stationary case is known, then one could evaluate the entropy production by integration. The equation could then be used in three different forms,

$$\frac{T \cdot d_{ip}^2 S}{dt\, dV} = D \cdot \frac{d^2 \xi}{dt\, dV} = J_1 \cdot \frac{d(\mu_2 - \mu_1)}{dy} = K \cdot \left(\frac{d(\mu_2 - \mu_1)}{dy} \right)^2 = \frac{1}{K} \cdot (J_1)^2$$

For stationary states the integration would simply yield the initial equation containing $\Delta(\mu_2 - \mu_1)$, i.e. the difference between the two reservoirs.

Exercise 4.7.1

Derive Fick's law for diffusion of B in an A–B phase assuming that the driving force is $d\mu_B/dy$.

Hint

In this case it would seem reasonable to assume that the rate constant contains a factor x_B, representing the chance that a B atom is in the right place for jumping. Also, remember that $\mu_B = G_m + x_A dG_m/dx_B$.

Solution

$d\mu_B/dx_B = dG_m/dx_B - 1 \cdot dG_m/dx_B + x_A d^2 G_m/dx_B^2 = x_A \cdot d^2 G_m/dx_B^2;$
$J_B = - K_B x_B (x_A d^2 G_m/dx_B^2) dx_B/dy.$ In the same way we would find $J_A = K_A x_A (x_B d^2 G_m/dx_B^2) dx_B/dy.$ This is in complete agreement with the case $J_A + J_B = 0$ if A and B have the same mobility, i.e. $K_A = K_B = K.$

4.8   *Effective driving force*

When a phase transformation occurs under diffusion it often happens that the processes occurring at the phase interfaces are rapid compared to the rate of diffusion. The transformation will then be diffusion controlled and the boundary conditions governing the rate of diffusion can be evaluated by assuming that, whenever two phases meet at an interface, their compositions right at the interface are very close to those required by equilibrium. This is called the **local equilibrium** approximation. That approximation will be used in the following, except when other conditions are clearly defined. For such exceptions, see Section 6.8 and Chapter 13.

So far, we have chosen to regard $T \cdot d_{ip} S/d\xi$ as the driving force for the process, the progress of which is measured by ξ, and it thus seemed natural to assume that the rate of the process is proportional to $D = T \cdot d_{ip}S/d\xi$, at least as a first approximation, yielding $d\xi/dt = KD$. However, one should be aware of the possibility that a process may be accompanied by an entropy production that does not contribute to the rate of the process. This possibility may be best explained by an example from a very simple type of transformation.

Let us first consider particles of pure A immersed in liquid B. The component A may dissolve in the liquid to a small but measurable extent, but B does not dissolve in the solid. It is well known that smaller particles will dissolve and larger ones will grow, so-called coarsening or 'Ostwald ripening'. The driving force comes from the increased pressure inside the smaller particles. Next, suppose that B can dissolve in solid A but the temperature is so low that diffusion can be neglected. We would still expect that the pressure difference makes the smaller particles go into solution and the larger ones grow. However, the growing layer of a large particle should now be a solid solution of B in A and the process could be written as: solid A + liquid B → solid A–B alloy. The chemical driving force for such a reaction can be evaluated from $-\Delta G_m$ assuming that all the phases are under the same pressure but then we should add the effect of the pressure difference. It would seem that the chemical driving force should give a drastic increase of the driving force for the process and make it possible even without the pressure effect, at least after the process has started. Such a process has actually been observed in sintering in the presence of a liquid.

However, in this description of the process we did not consider the local equilibrium conditions at the solid A/liquid interface. Even though the rate of diffusion inside solid A is negligible, the rate of transfer of atoms between the solid and liquid may be appreciable. Under ordinary conditions the net rate of dissolution of A is obtained as the difference between opposite fluxes that are much larger. We should thus recognize that there is a very localized reaction at the interface by which a monolayer of an A + B solid solution forms. The chemical driving force will drive that reaction but it will soon slow down if B does not diffuse into the interior of the A particle. Only the pressure difference may remain and cause material from the monolayer to go into solution and diffuse to a larger, growing particle. B from the liquid will then again react with the fresh A surface and the monolayer will he healed.

This example has demonstrated that the Gibbs energy may decrease as a result of the progress of a process without actually driving the process. One might say that the decrease of Gibbs energy depends on the progress of the process but the process does not depend on the decrease of Gibbs energy. The effective driving force, from which one may estimate the rate of reaction, has to come from another source, in our example from a pressure difference.

In the above example, it was fairly easy to identify the various steps in the whole process and thus to identify what part of the total driving force actually contributes to the rate. The example gets more complicated if we replace the liquid by a grain boundary which has contact with a B-rich reservoir outside the A material. Even in that case it has been observed experimentally that an A–B solid solution can grow at the expense of pure A, a phenomenon called DIGM (diffusion-induced grain boundary migration). This is actually the case for which Cahn, Pan and Balluffi (1979) first emphasized that the whole driving force does not necessarily contribute to driving a process. For DIGM they argued that the chemical driving force does not contribute at all and they proposed that the effective driving force comes from the process of diffusion of B down the grain boundary. Later, it was proposed that a part of the chemical driving force is not dissipated, as described above, thanks to the action of coherency stresses, and that this undissipated part thus is able to drive the main process. There may also be a deviation from the local-equilibrium approximation by the sluggishness of processes occurring at or in the interface. Contrary to the natural expectation that this would cause further dissipation of the driving force, it may, in fact, result in less dissipation and thus again leave some of the driving force to drive the main process.

This kind of complication has not attracted much attention and it will not be further considered in this book. Thus, we shall regard chemical driving forces as forces actually contributing to the rate of processes and the local-equilibrium approximation will be applied in most cases.

Exercise 4.8.1

Consider the mechanical device shown here. The left hand piston is supposed to have a friction against the cylinder, but not the right-hand piston. Between the two chambers there is a thin tube and the rate of transfer of gas from left to right is governed by the flow of gas through that tube. Examine how close an analogue this device would be to the example discussed in the text?

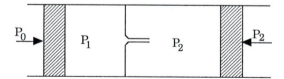

Exercise 4.8.1

Hint

Suppose the friction force is constant, i.e. independent of the rate of movement.

Solution

If we compare $P_0 - P_1$ with the chemical driving force and $P_1 - P_2$ with the effect of the pressure difference between small and large particles, then the analogue is very close. The friction between piston and cylinder corresponds to the dissipation of chemical driving force by the interface reaction.

5 Stability

Introduction

For the internal entropy production due to an internal process we formulated the following expression in Section 1.8

$$T \cdot d_{ip}S = D d\xi$$

D was identified with the driving force for the process. At equilibrium we have $D_{eq} = 0$ but in order to discuss whether it is a stable or an unstable equilibrium we must consider the second derivative because close to equilibrium a series expansion yields

$$T \cdot d_{ip}S = 0.5T \cdot (d_{ip}^2 S/d\xi^2)(d\xi)^2$$

For a stable equilibrium $d_{ip}S < 0$ for any internal process, preventing the process from proceeding if it has started by a small fluctuation. We may regard $-T \cdot (d_{ip}^2 S/d\xi^2)$ as the stability of the equilibrium state. It will be denoted by B and must always be positive for a stable state of equilibrium. It can be illustrated by the mechanical analogues in Fig. 5.1 which shows two bodies with different profiles, in contact with a flat floor. Their potential energy depends upon the angle θ.

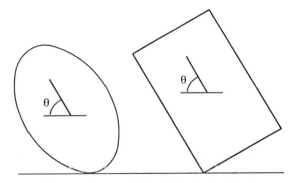

Figure 5.1 Mechanical analogues of two cases of thermodynamic stability or instability.

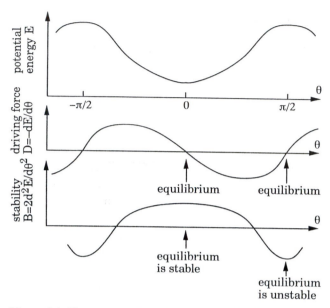

Figure 5.2 The energy, driving force and stability for the elliptical body in Fig. 5.1 as functions of the angle of rotation.

Only very slow changes will be considered, and it will be assumed that any release of potential energy goes into frictional losses. Kinetic energy will thus be neglected. The extent of the process, ξ, will be expressed by the angle θ and the potential energy will be denoted by E.

The variation of E, D and B with θ is illustrated in Fig. 5.2 for the body with an elliptical profile. It has an energy minimum at $\theta = 0$ and a maximum at $\theta = \pi/2$. In both these positions the driving force for a further rotation is zero, $D = - dE/d\theta = 0$, and they both represent equilibria. The quantity $d^2E/d\theta^2 = - dD/d\theta$, may there be regarded as the stability and the lower part of the diagram shows that for $\theta = 0$ it is positive and the equilibrium is thus a stable one. For $\theta = \pi/2$ it is negative and the equilibrium is unstable. A small fluctuation of θ away from $\pi/2$ in any direction will here give a force for a further growth of the fluctuation.

Fig. 5.3 is for the body with a rectangular cross-section. It has two equilibria, at $\theta = 0$ and $\pi/2$, which are both stable because a small fluctuation of θ will give a force for rotation back to the initial position. This case corresponds to Fig. 1.5 where $\Delta_{ip}S$ has a sharp maximum. In order to decide that this equilibrium is stable it is not only unnecessary but even incorrect to look at the value of $d^2E/d\theta^2$ because it represents the stability *only when the driving force is zero*, $D = - dE/d\theta = 0$, which is not the case for $\theta = 0$ or $\pi/2$. It would give an incorrect prediction. On the other hand, $D = - dE/d\theta = 0$ at some angle between 0 and $\pi/2$ and there $d^2E/d\theta^2 < 0$ and that equilibrium is thus unstable.

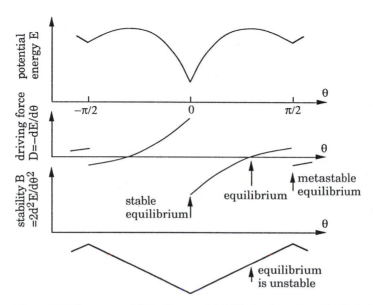

Figure 5.3 The energy, driving force and stability for the rectangular body in Fig. 5.1 as functions of the angle of rotation.

Of the two stable equilibria, one ($\theta = \pi/2$) has a higher energy than the other ($\theta = 0$). For thermodynamic systems such a state is called **metastable**.

5.2   *Some necessary conditions of stability*

In the discussion of general conditions of equilibrium in Section 1.10 we saw that a system is in a state of internal equilibrium with respect to the extensive variables if each one of the potentials has the same value in the whole system. It remains to be tested if it is a stable or unstable equilibrium. We thus return to the combined law according to the energy scheme and apply dU to the whole system, but we replace $Dd\xi$ by $-0.5B(d\xi)^2$ because we shall only consider a state of equilibrium where $D_{eq} = 0$.

$$dU = \Sigma Y^a dX^a + 0.5B(d\xi)^2$$

We shall soon see that the stability B has different values depending on what variables are kept constant in the partial derivative $\partial D/\partial \xi$. At present we shall consider only one internal process at a time, the transfer of dX^b from one half of the system, denoted by ', to the other, denoted by ", $d\xi - dX^{b"} - - dX^{b'}$. All the dX^a of the total system are zero and we have

$$dU = 0.5B(d\xi)^2$$

$$B = \frac{\partial^2 U}{\partial \xi^2} = \left(\frac{\partial^2 U}{(\partial X^b)^2}\right)'_{X^c} + \left(\frac{\partial^2 U}{(\partial X^b)^2}\right)''_{X^c}$$

Here we have used the fact that the change of U in the total system must be equal to the sum of the changes in the two subsystems. The two terms are equal if the system consists of a homogeneous substance. By introducing the potential $Y^b = (\partial U/\partial X^b)_{X^c}$ we obtain

$$B = \frac{\partial^2 U}{\partial \xi^2} = 2\left(\frac{\partial^2 U}{(\partial X^b)^2}\right)'_{X^c} = 2\left(\frac{\partial Y^b}{\partial X^b}\right)'_{X^c}$$

The value of this derivative depends upon the size of the system. It should be evaluated for half of the system but the stability condition, $B > 0$, is not affected by the size. The derivative may thus be evaluated for a system of any given size in the formulation of the stability condition. It can be written as

$$\left(\frac{\partial^2 U}{(\partial X^b)_2}\right)_{X^c} > 0 \text{ or } \left(\frac{\partial Y^b}{\partial X^b}\right)_{X^c} > 0 \text{ or } U_{X^b X^b} > 0$$

The last form uses the short-hand notation for derivatives of characteristic state functions, introduced in Section 2.5.

From the result we may conclude that in order for a substance to be stable it is necessary that it has such properties that any pair of conjugate variables must change in the same direction if all the other extensive variables are kept constant. So far, we have proved this for the basic energy scheme with the conjugate pairs (T,S), $(-P,V)$ and (μ_i, N_i).

$$U_{SS} > 0 \qquad U_{VV} > 0 \qquad U_{N_i N_i} > 0$$

As an example, in a stable system the chemical potential of a component cannot decrease when the content of the same component increases. As another example, when the temperature of a substance is increased at a constant volume, the entropy must also increase in order for the system to be stable.

$$U_{SS} \equiv \left(\frac{\partial T}{\partial S}\right)_{V,N_i} > 0$$

In combination with the fact that the absolute temperature T is always positive, this implies that the heat capacity under constant volume, C_V, must be positive since

$$C_V = T(\partial S/\partial T)_{V,N_i} = T/(\partial T/\partial S)_{V,N_i} > 0$$

but in order to indicate where the limit of stability is, one should rather write this condition as follows

$$\left(\frac{\partial T}{\partial S}\right)_{V,N_i} = \frac{T}{C_V} > 0$$

The limit of stability occurs as C_V goes to infinity, not zero.

Similar considerations can be based upon the entropy scheme, where we have

$$- dS = (-1/T)dU + (-P/T)dV + \Sigma(\mu_i/T)dN_i - (D/T)d\xi$$

and it yields at equilibrium under constant U, V and N_i,

$$- dS = 0.5\frac{B}{T}(d\xi)^2$$

$$\frac{B}{T} = \frac{\partial^2(-S)}{\partial\xi^2} = \left(\frac{\partial^2(-S)}{(\partial X^b)^2}\right)'_{X^c} + \left(\frac{\partial^2(-S)}{(\partial X^b)^2}\right)''_{X^c} = 2\left(\frac{\partial Y^b}{\partial X^b}\right)'_{X^c}$$

Since T is never negative, we find

$$\left(\frac{\partial Y^b}{\partial X^b}\right)_{X^c} > 0$$

where X^b, Y^b is any pair of conjugate variables appearing in the entropy scheme, i.e. $(-1/T,U)$, $(-P/T,V)$ and $(\mu_i/T,N_i)$.

Similar considerations can also be based on the volume scheme introduced in Section 3.4,

$$dV = (T/P)dS - (1/P)dU + + \Sigma(\mu_i/P)dN_i - (D/P)d\xi$$

At equilibrium under constant S, U and N_i it yields

$$\frac{B}{P} = \left(\frac{\partial Y^b}{\partial X^b}\right)_{X^c}$$

The conjugate pairs of variables are here $(T/P,S)$, $(-1/P,U)$ and $(\mu_i/P,N_i)$.

Exercise 5.2.1

In Section 2.7 we saw that Grüneisen's constant can be evaluated from $\gamma = V(\partial P/\partial U)_V$ and it often has a value of about 2. Is this a quantity that is always positive for a stable system?

Solution

γ concerns the variation of P with U but they are not conjugate variables in any of the schemes presented in Section 3.4. Thus, we cannot prove that γ is always positive. On the contrary, it may be negative because α may be negative in rare cases and $\gamma = V\alpha/\kappa_T C_V$.

5.3   Sufficient conditions of stability

So far we have discussed stability with respect to one internal process at a time. However, if there are several internal processes we should write the combined law in the following form,

$$dU = TdS - PdV + \mu_2 dN_2 + \cdots + \mu_c dN_c + \mu_1 dN_1 - \Sigma D^i d\xi^i$$

and at equilibrium we should replace $-\Sigma D^i d\xi^i$ by $+0.5\Sigma\Sigma B^{ij}d\xi^i d\xi^j$. The existence of cross terms $B^{ij}d\xi^i d\xi^i$ necessitates a closer examination. The conditions of stability derived in the preceding section are necessary conditions but we do not yet know if they are sufficient for a test of whether a system is stable. We shall now again start with the combined law in the energy scheme, but decide on the order of the terms, e.g. the one given in the dU expression above. For the whole system we shall again keep all the extensive variables constant and consider a state of equilibrium, where all $D^i_{eq.}$ are zero. Then,

$$dU = 0.5\Sigma\Sigma B^{ij}d\xi^i d\xi^j$$

We shall first focus our attention on one internal process and keep the others constant. Thus, there will be no cross terms. According to the order in which we have arranged the terms in dU we shall first consider the transfer of dS from one half of the system to the other, keeping all the other extensive variables constant in each half. With the method described in the preceding section we obtain the following condition of stability

$$U_{SS} = \left(\frac{\partial T}{\partial S}\right)_{V,N_2,\ldots,N_c,N_1} > 0$$

Next we shall vary V but allow S to vary simultaneously. However, we can eliminate all the cross terms containing dS by first modifying the combined law through a subtraction of $d(TS)$. It is then convenient to introduce the Helmholtz energy F. At equilibrium we obtain

$$\begin{aligned}dF &= d(U - TS)\\ &= -SdT - PdV + \mu_2 dN_2 + \cdots + \mu_c dN_c + \mu_1 dN_1 + 0.5\Sigma\Sigma B^{ij}d\xi^i d\xi^j\end{aligned}$$

Since the internal processes to be considered are those by which an amount of an extensive quantity is exchanged between the two halves of the system, and S is no longer included among those variables, its cross terms have also been eliminated. Instead, T is included which is a potential. One cannot transfer a part of a potential from one half of the system to the other. At equilibrium it will have the same value in the whole system. See Section 1.10.

By considering a system at equilibrium under constant T,V,N_i we have

$$dF = 0.5\Sigma\Sigma B^{ij}d\xi^i d\xi^j$$

For the transfer of dV from one half to the other under constant T,N_i we now obtain a second condition of stability

$$F_{VV} = \left(\frac{\partial[-P]}{\partial V}\right)_{T,N_2,\ldots,N_c,N_1} > 0$$

Before proceeding we should next eliminate all the cross terms involving dV by further modifying the combined law through an addition of $d(PV)$. We have thus included two potentials, T and P, among the variables. For convenience, we shall now introduce the Gibbs energy G and for equilibrium we can write

$$dG = d(F + PV)$$
$$= -SdT + VdP + \mu_2 dN_2 + \cdots + \mu_c dN_c + \mu_1 dN_1 + 0.5\Sigma\Sigma B^{ij}d\xi^i d\xi^j$$

By considering a system at equilibrium under constant T,P,N_i we now have

$$dG = 0.5\Sigma\Sigma B^{ij}d\xi^i d\xi^j$$

The double summation now covers all the extensive variables except for S and V. For the transfer of dN_2 from one half to the other under constant $T,P,N_3,\ldots,N_c,N_1$ we now obtain a third condition of stability

$$G_{N_2N_2} = \left(\frac{\partial\mu_2}{\partial N_2}\right)_{T,P,N_3,\ldots,N_c,N_1} > 0$$

since $\mu_2 = (\partial G/\partial N_2)_{T,P,N_3,\ldots,N_c,N_1}$.

By proceeding in the same way we obtain conditions involving all the components from 2 to c and each time with one more potential among the variables. Finally we obtain

$$\left(\frac{\partial\mu_c}{\partial N_c}\right)_{T,P,\mu_2,\ldots,\mu_{c-1},N_1} > 0$$

It would seem that there is one more derivative in this series, $(\partial\mu_1/\partial N_1)_{T,P,\mu_2,\ldots,\mu_c}$, where all the variables to be kept constant are potentials. However, that derivative is always equal to zero in view of the Gibbs–Duhem relation between the potentials. It says that μ_1 cannot vary if all the other potentials are constant. The final derivative thus yields a trivial condition which will not be included in the set of stability conditions.

In deriving the set of stability conditions we have not neglected any cross term. The result is thus a **sufficient set of stability conditions** with respect to the external variables.

Other sufficient sets of stability conditions can be derived by arranging the terms in the first fundamental equation in a different order or by using another scheme. It may be mentioned again that we have chosen to use N_1 to express the size of the system whereas Gibbs used the volume V. Of course, any extensive variable could be used for that purpose.

Exercise 5.3.1

What would be the last stability condition if we use the combined law written according to the basic entropy scheme?

Consult Table 3.1 in Section 3.4. Select the content of component 1 to define the size of the system.

Solution

We get $(\partial[\mu_c/T]/\partial N_c)_{1/T,P/T,\mu_2/T,\ldots,\mu_{c-1}/T,N_1} > 0$ but since $1/T$ is kept constant, T is also kept constant and because T is always positive we could just as well write this condition as $(\partial\mu_c/\partial N_c)_{T,P,\mu_2,\ldots,\mu_{c-1},N_1} > 0$, which we recognize.

5.4   *Summary of stability conditions*

We have seen that stability conditions are defined through the derivative of a potential with respect to the conjugate extensive variable. In Section 5.2, all the remaining extensive variables in the same set of conjugate pairs were kept constant. In Section 5.3 it was shown that a stability condition is also obtained if one or more of the potentials are kept constant instead, i.e.

$$\left(\frac{\partial Y^b}{\partial X^b}\right)_{Y^c X^d} > 0$$

However, it must be emphasized that all the independent variables appearing in a stability condition must come one from each pair in a set of conjugate pairs. One cannot use a mixture of variables from different sets. Nine possible sets were listed in Section 3.4 and they can all be used for this purpose. Each one yields its own form of the Gibbs–Duhem relation and eight stability conditions, not counting those where a mixture of μ_i and N_i are used. This makes 72 stability conditions, but few of them are really useful.

Exercise 5.4.1

The slope of the G versus T curve for a given substance at constant P is obtained as $(\partial G/\partial T)_P = -S$. If S is always positive, then this slope must always be negative. Examine if there is a similar rule for the curvature, $(\partial^2 G/\partial T^2)_P$.

Hint

Find a relevant stability condition, assuming that the system is stable.

Solution

$(\partial^2 G/\partial T^2)_P = -(\partial S/\partial T)_P$ and in the ordinary energy scheme the conjugate pairs are (T,S), $(-P,V)$ and (μ_i,N_i). One of the stability conditions is thus $(\partial T/\partial S)_{P,N_i} > 0$. We can apply this because N_i is constant for a given substance. Thus, $(\partial^2 G/\partial T^2)_P = -(\partial S/\partial T)_P = -1/(\partial T/\partial S)_P < 0$. It should be noted that this result is quite general whereas the sign of S in $(\partial G/\partial T)_P = -S$ depends upon the choice of zero point for S.

Exercise 5.4.2

Find a stability condition concerned with C_P.

Hint

Remember that $C_P = T(\partial S/\partial T)_{P,N_i}$.

Solution

It is evident that we should look for a stability condition involving $(\partial T/\partial S)_{P,N_i}$. We find the combination of variables in line 1 of Table 3.1 in Section 3.4 listing pairs of conjugate variables. The combined law with S, P and N_i as independent variables is obtained as $dU - d(-PV) = dH = TdS - Vd(-P) + \Sigma\mu_i dN_i$ where $T = (\partial H/\partial S)_{P,N_i}$. We get the stability condition $H_{SS} \equiv (\partial^2 H/\partial S^2)_{P,N_i} = (\partial T/\partial S)_{P,N_i} > 0$ and thus $C_P = T/(\partial T/\partial S)_{P,N_i} > 0$.

Exercise 5.4.3

Formulate as many stability conditions as possible governing the change of U.

Hint

Consult Table 3.1 in Section 3.4. The important point is to define what variables should be kept constant.

Solution

The first set from the entropy scheme yields a number of conditions $\partial(-1/T)/\partial U > 0$ which can be written as $\partial T/\partial U > 0$ since $d(-1/T) = (1/T^2)dT$ and $T > 0$. Thus, $(\partial T/\partial U)_{P/T,\mu_i/T} > 0$; $(\partial T/\partial U)_{V,\mu_i/T} > 0$; $(\partial T/\partial U)_{P/T,N_i} > 0$; $(\partial T/\partial U)_{V,N_i} > 0$. The first set from the volume scheme yields $(\partial P/\partial U)_{T/P,\mu_i/P} > 0$; $(\partial P/\partial U)_{S,\mu_i/P} > 0$; $(\partial P/\partial U)_{T/P,N_i} > 0$; $(\partial P/\partial U)_{S,N_i} > 0$.
In addition, one can let some μ_i/T and some N_i stay constant.

5.5   *Limit of stability*

Let us now compare the stability conditions occurring in a given set of necessary conditions. Suppose we are inside a stable region and want to know which one will first turn negative as we move into a region of instability. We can find this by first examining which derivative is the smallest one inside the stable region. Let us start by comparing any two conditions which differ only by the choice of variable in a conjugate pair to be kept constant, the extensive variable or the potential. By the use of Jacobians we find

$$\left(\frac{\partial Y^b}{\partial X^b}\right)_{Y^c} = \left(\frac{\partial Y^b}{\partial X^b}\right)_{X^c} - \left(\frac{\partial Y^b}{\partial X^c}\right)_{X^b}\left(\frac{\partial Y^c}{\partial X^b}\right)_{X^c} \Big/ \left(\frac{\partial Y^c}{\partial X^c}\right)_{X^b}$$

In view of a Maxwell relation, $(\partial Y^b/\partial X^c)_{X^b}$ and $(\partial Y^c/\partial X^b)_{X^c}$ are equal and $(\partial Y^c/\partial X^c)_{X^b}$ cannot be negative for a stable system. Thus, the last term with its minus sign cannot be positive and we find

$$\left(\frac{\partial Y^b}{\partial X^b}\right)_{Y^c} \le \left(\frac{\partial Y^b}{\partial X^b}\right)_{X^c}$$

It is evident that each time a potential is introduced among the variables to be kept constant, the stability condition gets more restrictive. The most severe condition is the one where only one extensive variable is kept constant, the one chosen to represent the size of the system. Consequently, this derivative must be the first one to go to zero and that happens on the limit of stability. Of course, it is possible that one or several of the other derivatives also go to zero at the same time. However, we can always find the limit of stability by considering the last condition in the set if we know that we *start the search from inside a stable region.*

Let us now consider what happens to the last derivative in a different set of necessary stability conditions. We can write the condition for the limit of stability according to the first set of necessary conditions in the following general form

$$\left(\frac{\partial Y^b}{\partial X^b}\right)_{Y^c, N_1} = 0$$

where Y^c indicates that all potentials except for Y^b and Y^1 are kept constant during the derivation. However, in this situation where the derivative is zero, Y^b is also constant and, according to the Gibbs–Duhem relation, the only remaining potential, Y^1, must also be constant. We thus find that, in this situation, it is possible to change the value of an extensive variable, X^b, without affecting any potential, nor the value of the extensive variable chosen to express the size of the system. However, the other extensive variables will change with X^b, because they are dependent variables, and it would be possible to accomplish the same change of the system by prescribing how anyone of them should change. The above relation thus holds for any conjugate pair of variables. We thus find that the last stability condition, obtained in each set of necessary conditions, are all zero at the same time. Anyone of them could be used to find the limit of stability if one starts from inside a stable region.

It should be emphasized that inside a region of instability the above conditions may again turn positive when other conditions are negative. In the general case it is thus necessary to apply a whole set of stability conditions. It is only when one is able to start from a point inside a stable region that one can identify the limit by applying a single condition.

Exercise 5.5.1

We have seen that $1/C_V > 0$ is a stability condition. Another one is $1/\kappa_T > 0$ because $\kappa_T = -1/V(\partial P/\partial V)_T$ and we know that $F_{VV} \equiv (\partial[-P]/\partial V)_T > 0$ is a stability condition. Show which one of the above conditions is more restrictive. Consider a substance with fixed composition.

Hint

Remember $C_V = T/(\partial T/\partial S)_V = T/U_{SS}$. In order to compare the values of $(\partial[-P]/\partial V)_T$ and $(\partial T/\partial S)_V$ the variables of one must be changed to those of the other. Change $(\partial[-P]/\partial V)_T$ to S and V using Jacobians.

Solution

The result will be $(\partial[-P]/\partial V)_T = U_{VV} - [(U_{SV})^2/U_{SS}]$ but $(\partial T/\partial S)_V = U_{SS}$. In a stable region $U_{VV} > 0$ and $U_{SS} > 0$ and now we see that it is necessary to have $U_{VV} > (U_{SV})^2/U_{SS}$ in order to satisfy the condition $(\partial[-P]/\partial V)_T > 0$. When coming closer to the stability limit, U_{SS} will decrease but, before it reaches zero, it will come to a value $U_{VV} = (U_{SV})^2/U_{SS}$ where $(\partial[-P]/\partial V)_T = 0$. We conclude that $1/\kappa_T > 0$ is more restrictive than $1/C_V > 0$.

Exercise 5.5.2

Show for a unary system that $(\partial[-P]/\partial V)_T$ and $(\partial T/\partial S)_P$ go to zero at the same time, as they should because only one extensive variable, N, is kept constant (and it is omitted from the notation in the case of a substance with fixed composition).

Hint

In order to compare them, they must be expressed in the same set of independent variables, which can be done using Jacobians. Take S and V, for instance.

Solution

We obtain $(\partial[-P]/\partial V)_T = U_{VV} - [(U_{SV})^2/U_{SS}]$ and $(\partial T/\partial S)_P = U_{SS} - [(U_{SV})^2/U_{VV}] = (\partial[-P]/\partial V)_T \cdot U_{SS}/U_{VV}$. If one term goes to zero when U_{SS} and U_{VV} are still > 0, then the other expression also does. The two quantities can be expressed as $1/V\kappa_T$ and T/C_P. It is interesting to note that κ_T and C_P both go to infinity at the limit of stability.

Exercise 5.5.3

We have found that the limit of stability of a unary system can be written as $F_{VV} = (\partial[-P]/\partial V)_{T,N} = 0$. Try to express this criterion in terms of derivatives of G.

Hint

Change variables from V and T to P and T by inverting the derivative.

Solution

$F_{VV} = -(\partial P/\partial V)_T = -1/(\partial V/\partial P)_T = -1/G_{PP} = 0$ because $G_P \equiv (\partial G/\partial P)_V = V$. This gives $G_{PP} = -\infty$.

Exercise 5.5.4

Suppose that for some reason one would like to evaluate the limit of stability for a pure substance by studying the variation of H. Exactly what derivative should one use?

Hint

Using Table 3.1 in Section 3.4, find a form of the combined law having H as one of the variables and only one more extensive variable.

Solution

Line 5 in Table 3.1 is the only one listing H. We may take N ($= N_i$ for a pure substance) as the other extensive variable and should introduce $(-P)$ instead of V/T.
Then,

$- \mathrm{d}[TS + PV)/T] - [(-P)(V/T)] = -\mathrm{d}S = (-1/T)\mathrm{d}H - (V/T)\mathrm{d}(-P) + (\mu/T)\mathrm{d}N$, and thus $1/T = (\partial S/\partial H)_{P,N}$. The criterion for the stability limit is $S_{HH} = (\partial[1/T]/\partial H)_{P,N} = 0$, i.e. $(\partial T/\partial H)_{P,N} = 0$.

Exercise 5.5.5

We have seen that the most severe stability conditions are those with only one extensive variable among those kept constant, the one that defines the size, usually N for a pure substance, and all those conditions are equivalent. Try to define the same stability condition using only extensive variables.

Hint

Change variables using Jacobians.

Solution

One may, for instance, start from $(\partial T/\partial S)_P > 0$ and change variables to S and V. Exercise 5.5.2 shows that the result is $(\partial T/\partial S)_P = U_{SS} - (U_{SV})^2/U_{VV}$. The stability condition may thus be written

$$\begin{vmatrix} U_{SS} & U_{VS} \\ U_{SV} & U_{VV} \end{vmatrix} > 0$$

This matrix is actually called the stiffness matrix.

5.6
Limit of stability of alloys

For an alloy system with c components the limit of stability is given by, for instance,

$$\left(\frac{\partial \mu_c}{\partial N_c}\right)_{T,P,\mu_2,\ldots,\mu_{c-1},N_1} = 0$$

where a set of chemical potentials are kept constant together with T and P. In practice, the Gibbs energy is usually the characteristic state function which is available but it is given as a function of the content of matter rather than chemical potentials. It is thus interesting to change the variables in the derivative. This can be done by the use of Jacobians of a higher order than those discussed before. The result is conveniently written with the notation G_{kl} for $(\partial \mu_k/\partial N_l)_{T,P,N_j}$ which is also equal to $(\partial^2 G/\partial N_k \partial N_l)_{T,P,N_j}$. One has thus obtained the following (see Callen 1988).

$$\left(\frac{\partial \mu_c}{\partial N_c}\right)_{T,P,\mu_2,\ldots,\mu_{c-1},N_1} = \begin{vmatrix} G_{22} & \cdots & G_{2c} \\ \vdots & \ddots & \vdots \\ G_{c2} & \cdots & G_{cc} \end{vmatrix} \Bigg/ \begin{vmatrix} G_{22} & \cdots & G_{2,c-1} \\ \vdots & \ddots & \vdots \\ G_{c-1,2} & \cdots & G_{c-1,c-1} \end{vmatrix}$$

The second determinant can be related to the derivative for the preceding component,

$$\left(\frac{\partial \mu_{c-1}}{\partial N_{c-1}}\right)_{T,P,\mu_2,\ldots,\mu_{c-2},N_cN_1} = \begin{vmatrix} G_{22} & \cdots & G_{2,c-1} \\ \vdots & \ddots & \vdots \\ G_{c-1,2} & \cdots & G_{c-1,c-1} \end{vmatrix} \Bigg/ \begin{vmatrix} G_{22} & \cdots & G_{2,c-2} \\ \vdots & \ddots & \vdots \\ G_{c-2,2} & \cdots & G_{c-2,c-2} \end{vmatrix}$$

Again, the second determinant can be related to the derivative for the preceding component, etc. We thus obtain

$$\left(\frac{\partial \mu_c}{\partial N_c}\right)\left(\frac{\partial \mu_{c-1}}{\partial N_{c-1}}\right)\cdots\left(\frac{\partial \mu_2}{\partial N_2}\right) = \begin{vmatrix} G_{22} & \cdots & G_{2c} \\ \vdots & \ddots & \vdots \\ G_{c2} & \cdots & G_{cc} \end{vmatrix}$$

For convenience, we have here omitted the indices for the derivatives. In a stable region all these derivatives are positive. For the limit of stability we thus obtain simply

$$\begin{vmatrix} G_{22} & \cdots & G_{2c} \\ \vdots & \ddots & \vdots \\ G_{c2} & \cdots & G_{cc} \end{vmatrix} = 0$$

However, this is still not the most practical way of writing the criterion because the Gibbs energy is usually given as a function of the composition, $x_2, x_3, \ldots$ and the size of the system is expressed by the total number of atoms, N, rather than N_1. Thus,

$$G = N \cdot G_m(x_2, x_3, \ldots)$$

It should thus be most practical to express the criterion for the limit of stability in terms of the derivatives of G_m. We should introduce $dx_2, dx_3, \ldots$ and dN in the expression for dG. Using $dx_1 = -\sum_2 dx_i$ we find

$$N_i = Nx_i$$
$$dN_i = Ndx_i + x_i dN$$

$$\sum \mu_i dN_i = N\sum_{i=1} \mu_i dx_i + dN\sum_{i=1} \mu_i x_i = N\sum_{i=2} (\mu_i - \mu_1)dx_i + G_m dN$$

$$dG = -SdT + VdP + N\sum_{i=2}(\mu_i - \mu_1)dx_i + G_m dN - \sum D^i d\xi^i$$

We may proceed as before because we can keep x_j of a whole system constant and exchange dx_j between two halves. The limit of stability will now be given by

$$\left(\frac{\partial(\mu_c - \mu_1)}{\partial x_c}\right)_{T,P,(\mu_2 - \mu_1),\ldots,(\mu_{c-1} - \mu_1),N} = 0$$

Again we shall change the variables to be kept constant by using Jacobians. The final expression will then contain derivatives of the type $(\partial[\mu_k - \mu_1]/\partial x_l)_{T,P,x_j,N}$ and they can be expressed as $(\partial^2 G_m/\partial x_k \partial x_l)_{T,P,x_j}$ or g_{kl} since

$$\mu_k - \mu_1 = \left(\frac{\partial G_m}{\partial x_k}\right)_{T,P,x_j} \equiv g_k$$

This was shown in Section 3.5 where it was mentioned that $\mu_j - \mu_k$ is the diffusion potential between j and k. The molar Gibbs energy G_m is here treated as a function of T, P and all x_i except for x_1 which is chosen as a dependent variable. Thus $dx_1 = - dx_k$. By introducing the notation g_k for first-order derivatives of G_m and g_{kl} for second-order derivatives, we obtain the following convenient form of the limit of stability

$$\begin{vmatrix} g_{22} & \cdots & g_{2c} \\ \vdots & \ddots & \vdots \\ g_{c2} & \cdots & g_{cc} \end{vmatrix} = 0$$

and, before the change of variables, the same stability condition can be written as

$$\left(\frac{\partial g_c}{\partial x_c}\right)_{T,P,g_2,\ldots,g_{c-1},N} = 0$$

It should again be emphasized that g_k and g_{kl} are defined with x_1 as dependent variable.

For a binary system, the condition for the limit of stability reduces to $g_{22} = 0$. Although the limit of stability of a solution is exactly defined by the condition just given, there have been attempts to modify this expression in order to get a function which is more suitable for representing the properties of a solution in its stable range as well. In particular, the determinant goes to infinity at the sides of an alloy system, an effect which can be removed by multiplication with $x_1 x_2 \ldots x_c$. One may further make the expression dimensionless by dividing with RT to the proper power. For a binary system one thus obtains the **stability function**

$$x_1 x_2 g_{22}/RT$$

which is unity over the whole range of composition for an ideal solution and goes to zero at the limit of stability.

Exercise 5.6.1

Show that the stability function, just defined, is unity over the whole system for an ideal A–B–C solution.

Hint

An ideal solution has $G_m = \Sigma x_i (^{\circ}G_i + RT\ln x_i)$. Take the derivatives of G_m remembering that $x_1 = 1 - x_2 - x_3$.

Solution

$g_2 = dG_m/dx_2 = {}^{\circ}G_2 - {}^{\circ}G_1 + RT(\ln x_2 - \ln x_1)$; $g_{23} = RT(1/x_1)$; $g_{22} = RT(1/x_2 + 1/x_1)$ and in the same way $g_{32} = RT(1/x_1)$; $g_{33} = RT(1/x_3 + 1/x_1)$. We thus get: $x_1 x_2 x_3 \begin{vmatrix} g_{22} & g_{23} \\ g_{32} & g_{33} \end{vmatrix}/(RT)^2 = x_1 x_2 x_3 [1/x_2 x_3 + 1/x_2 x_1 + 1/x_1 x_3 + 1/x_1 x_1 - (1/x_1)^2] = x_1 + x_2 + x_3 = 1$.

Exercise 5.6.2

Darken and Gurry (1953) defined an excess stability for binary solutions, $\alpha = {}^E G_2/x_1{}^2$. Darken (1967) instead proposed $\alpha = g_{22} - RT/x_1 x_2$. Show how this second α is related to the excess Gibbs energy.

Hint

Start from $G_m = x_1 {}^{\circ}G_1 + x_2 {}^{\circ}G_2 + RT(x_1 \ln x_1 + x_2 \ln x_2) + {}^E G_m$.

Solution

$g_2 = -{}^{\circ}G_1 + {}^{\circ}G_2 + RT(-1 - \ln x_1 + 1 + \ln x_2) + d^E G_m/dx_2$;
$g_{22} = RT(1/x_1 + 1/x_2) + d^{2E} G_m/dx_2^2$;
$\alpha = [RT(x_2 + x_1)/x_1 x_2] + d^{2E} G_m/dx_2^2 - RT/x_1 x_2 = d^{2E} G_m/dx_2^2$

Exercise 5.6.3

Use a Jacobian transformation to show that the limit of stability in a ternary system is $\begin{vmatrix} g_{22} & g_{23} \\ g_{32} & g_{33} \end{vmatrix} = 0$.

Hint

By omitting the constant variables that will not be exchanged, we can write the stability condition as $\left(\dfrac{\partial(\mu_3 - \mu_1)}{\partial x_3} \right)_{(\mu_2 - \mu_1)} = 0$.

Solution

$$\left(\frac{\partial(\mu_3 - \mu_1)}{\partial x_3}\right)_{(\mu_2 - \mu_1)} = \begin{vmatrix} \dfrac{\partial(\mu_3 - \mu_1)}{\partial x_3} & \dfrac{\partial(\mu_3 - \mu_1)}{\partial x_2} \\ \dfrac{\partial(\mu_2 - \mu_1)}{\partial x_3} & \dfrac{\partial(\mu_2 - \mu_1)}{\partial x_2} \end{vmatrix} \Bigg/ \begin{vmatrix} \dfrac{\partial x_3}{\partial x_3} & \dfrac{\partial x_3}{\partial x_2} \\ \dfrac{\partial(\mu_2 - \mu_1)}{\partial x_3} & \dfrac{\partial(\mu_2 - \mu_1)}{\partial x_2} \end{vmatrix}$$

$$= \begin{vmatrix} \dfrac{\partial^2 G_m}{\partial x_3 \partial x_3} & \dfrac{\partial^2 G_m}{\partial x_3 \partial x_2} \\ \dfrac{\partial^2 G_m}{\partial x_2 \partial x_3} & \dfrac{\partial^2 G_m}{\partial x_2 \partial x_2} \end{vmatrix} \Bigg/ \dfrac{\partial^2 G_m}{\partial x_2 \partial x_2} = \begin{vmatrix} g_{22} & g_{23} \\ g_{32} & g_{33} \end{vmatrix} \Bigg/ g_{22} = 0$$

But $g_{22} > 0$ in the stable region and it does not reach $g_{22} = 0$ before our condition is satisfied. Our condition can thus be written as $\begin{vmatrix} g_{22} & g_{23} \\ g_{32} & g_{33} \end{vmatrix} = 0.$

5.7   *Chemical capacitance*

The diagonal elements in the G_{cc} determinant can be written as $(\partial \mu_j / \partial N_j)_{T,P,N_k}$ and they must all be positive because they are stability conditions according to Section 5.2. In addition, the inverse quantities are sometimes regarded as the chemical capacitance of the component j (see Pelton 1992),

$$\Omega_{jj} = \left(\frac{\partial N_j}{\partial \mu_j}\right)_{T,P,N_k} = 1/G_{jj}$$

This quantity may be of practical importance because it is often of considerable interest to be able to increase the amount of a component j in a system without increasing the chemical potential of the same component too much. A system with a high capacity is said to be well buffered.

An off-diagonal term in the G_{cc} determinant cannot by itself form a stability condition because it concerns variables that do not make a conjugate pair. It may thus be positive or negative in the stable region. Nevertheless, its inverse quantity may also be used as a kind of chemical capacitance. The following relation holds between them,

$$1/\Omega_{jk} = \frac{\partial \mu_j}{\partial N_k} = \frac{\partial^2 G}{\partial N_j \partial N_k} = \frac{\partial \mu_k}{\partial N_j} = 1/\Omega_{kj}$$

Exercise 5.7.1

What gas mixture is best buffered for oxygen: (a) 1 mol of Ar and 10^{-6} mol of O_2 at 1 bar and 1550 K; or (b) 0.99 mol of CO_2 and 0.01 mol of CO at 1 bar and 1550 K ?

Hint

The conditions were chosen in such a way that the equilibrium partial pressure of oxygen is very close to 10^{-6} in case (b) as well as in case (a). Accept this information.

Solution

(a) $P_{O_2} \cong 1 \cdot N_{O_2}/N_{Ar}$; $\ln P_{O_2} = \ln N_{O_2} - \ln N_{Ar}$; $\mu_{O_2} = {}^o\mu_{O_2} + RT\ln P_{O_2} = {}^o\mu_{O_2} + RT(\ln N_{O_2} - \ln N_{Ar})$; $1/\Omega_{O_2O_2} = \partial\mu_{O_2}/\partial N_{O_2} = RT/N_{O_2} = RT/10^{-6}$.

(b) If we add N_{O_2}, most of it will react by $O_2 + 2CO \to 2CO_2$ yielding $N_{CO} = 0.01 - 2N_{O_2}$ and $N_{O_2} = 0.99 + 2N_{O_2}$. We get, using the equilibrium constant K:

$P_{O_2} = K(P_{CO_2}/P_{CO})^2 \cong K(0.99 + 2N_{O_2})^2/(0.01 - 2N_{O_2})^2$;

$1/\Omega_{O_2O_2} = \partial\mu_{O_2}/\partial N_{O_2} = 2RT[2/(0.99 + 2N_{O_2}) - (-2)/(0.01 - 2N_{O_2})] \cong 4RT/0.01$. Thus, $(\Omega_{O_2O_2})^b \gg (\Omega_{O_2O_2})^a$.

5.8   *Limit of stability in phases with sublattices*

The constitution of a phase with more than one sublattice is often defined by the use of site fractions.

$$y_j^s = N_j^s \Big/ \sum_i N_i^s$$

The superscript s identifies a particular sublattice. If N is now the number of moles of formula units, we have

$$G = N G_m(y_j^s)$$

where G_m is the Gibbs energy for one mole of formula units. We now want to express dG in terms of all the dy_j^s and dN. We obtain

$$dG = -SdT + VdP + \sum_s \sum_j \phi_j^s dy_j^s + G_m dN - \Sigma D^i d\xi^i$$

where Σ for each sublattice starts from the second constituent present in that sublattice, the first constituent being the dependent one. The choice of dependent constituent is made separately for each sublattice. ϕ_j^s is the conjugate variable to y_j^s,

$$\phi_j^s = \left(\frac{\partial G}{\partial y_j^s}\right)_{T,P,y_k^s,N} = N\left(\frac{\partial G_m}{\partial y_j^s}\right)_{T,P,y_i^t,y_k^s}$$

y_k^s denotes the site fractions of all the other independent constituents on the same sublattice and y_i^t denotes the site fractions of all the independent constituents on other sublattices. By proceeding as before we obtain for the limit of stability

$$\left(\frac{\partial \phi_c^r}{\partial y_c^r}\right)_{T,P,\varphi_2^l,\ldots,\varphi_c^l,\varphi_2^r,\ldots,\varphi_{c-1}^r,N} = 0$$

and after changing the variables to be kept constant using Jacobians,

$$\begin{vmatrix} g_{11} & \cdots & g_{1k} \\ \vdots & \ddots & \vdots \\ g_{k1} & \cdots & g_{kk} \end{vmatrix} = 0$$

As before, g_{ij} denotes the partial derivatives of G_m but, for convenience, we have now numbered all the independent constituents in all the sublattices from 1 to k. It should be noted that k could be equal to, smaller than or larger than c, the number of components in the system. Some examples may be mentioned.

(i) In simple ionic compounds there is one sublattice for the cations and another one for the anions. The solid solution between NaCl and KCl may thus be represented by the formula $(Na,K)_1 Cl_1$. There is only one independent external variable in addition to T, P and N, for instance x_K. Instead of N and x_K one may choose N_{Na} and N_K, whereas the third one, N_{Cl}, is a dependent variable, being given by the condition of electroneutrality (or stoichiometry). y_K is directly related to x_K by

$$y_K = 2x_K$$

(ii) The β-brass phase has an ideal composition of $Cu_{0.5}Zn_{0.5}$ and, at high temperatures, it has a body-centred cubic (bcc) structure designated A2. As the temperature is lowered, it may start to order by Cu segregating to one family of lattice sites and Zn segregating to another family of lattice sites. One can then define two sublattices and the completely ordered phase, which is stable at zero kelvin, can be represented by the formula $(Cu)_1(Zn)_1$, for which a formula unit contains two moles of atoms. However, at non-zero temperatures some Cu atoms may go into the Zn sublattice and some Zn atoms into the Cu sublattice. That situation can be represented by the formula $(Cu,Zn)_1(Zn,Cu)_1$ where the main constituent in each sublattice is given first. The situation in this phase can be represented by y'_{Zn} and y''_{Zn}, where the superscripts $'$ and $''$ identify the sublattices. These two variables, or x_{Zn} and y'_{Zn}, may be regarded as the independent variables.

In example (ii) there is an internal degree of freedom, which may be represented by y'_{Zn} or y''_{Zn}, or by some other parameter expressing the degree of order more directly. We have thus extended the discussion of the limit of stability to include an internal variable. In fact, any internal variable, ξ^i, can be included in the k variables if it is an extensive quantity divided by the size of the system. In a ferromagnetic alloy it could, for instance, be the number of atoms per mole with spins in a certain direction. A particularly simple case is obtained in a pure element if there is only one internal variable. The stability condition is then

$$B = \frac{\partial^2 G_m}{(\partial \xi^i)^2} > 0$$

However, it should be remembered that a criterion of stability can only be applied to a state of equilibrium and all the elements of the determinant, being partial derivatives of G_m, must be evaluated for that state before the value of the determinant can be calculated. It is thus necessary first to calculate the equilibrium values of all the internal variables.

Exercise 5.8.1

A simple model for ordering in β-brass gives the following expression for one mole of atoms: $G_m = x_A{}^\circ G_A + x_B{}^\circ G_B + 0.5RT[y'_A \ln y'_A + y'_B \ln y'_B + y''_A \ln y''_A + y''_B \ln y''_B] + K(y'_A y''_B + y'_B y''_A)$, where we may regard K as a constant. For ordering alloys, K is negative. $^\circ G_A$ and $^\circ G_B$ are the Gibbs energies of pure A or B in the same structure. At high temperature there is complete disorder and $y'_A = y''_A = x_A$; $y'_B = y''_B = x_B$. Calculate the critical temperature below which the disordered state is no longer stable.

Hint

There are two independent variables in addition to T and P, – the alloy composition and the degree of order. To simplify the calculations it may be convenient to treat y'_A and y''_A as the independent variables. Then $y'_B = 1 - y'_A$; $y''_B = 1 - y''_A$; $x_A = (y'_A + y''_A)/2$; $x_B = (2 - y'_A - y''_A)/2$. Treat T and P as constants.

Solution

Let y'_A be variable 1 and y''_A be variable 2. We find $g_1 = {}^\circ G_A/2 - {}^\circ G_B/2 + 0.5RT(\ln y'_A - \ln y'_B) + K(y''_B - y''_A)$; $g_2 = {}^\circ G_A/2 + {}^\circ G_B/2 + 0.5RT(\ln y''_A - \ln y''_B) + K(-y'_A + y'_B)$; $g_{11} = 0.5RT(1/y'_A + 1/y'_B) = 0.5RT/y'_A y'_B$; $g_{12} = K(-1-1) = g_{21}$; $g_{22} = 0.5RT(1/y''_A + 1/y''_B) = 0.5RT/y''_A y''_B$. The criterion for the limit of stability gives $g_{11}g_{22} - g_{12}g_{21} = 0$; $(0.5RT)^2/y'_A y'_B y''_A y''_B = (-2K)^2$. The critical temperature for ordering in a disordered alloy of composition x_A, x_B is thus $T = 4(-K)x_A x_B/R$.

Exercise 5.8.2

A simple model of a ternary solution gives the following expression: $G_m = x_A{}^\circ G_A + x_B{}^\circ G_B + x_C{}^\circ G_C + RT(x_A \ln x_A + x_B \ln x_B + x_C \ln x_C) + x_A x_B L_{AB} + x_B x_C L_{BC} + x_A x_C L_{AC}$, where L_{AB}, L_{BC} and L_{AC} are regular solution parameters for the three binary systems.

Positive L values result in miscibility gaps on the binary sides which extend into the ternary system. When a homogeneous alloy is cooled inside a miscibility gap it can remain metastable but below a sufficiently low temperature it becomes unstable. Derive an equation for the calculation of that temperature limit for a given alloy composition.

Hint

For the composition there are two independent variables, e.g. x_B and x_C, and x_A must then be regarded as $1 - x_B - x_C$.

Solution

Let x_B be variable 1 and x_C be variable 2. We find $g_1 = - {}^\circ G_A + {}^\circ G_B + RT(- \ln x_A + \ln x_B) + (- x_B + x_A)L_{AB} + x_C L_{BC} - x_C L_{AC}$; $g_2 = - {}^\circ G_A + {}^\circ G_C + RT(- \ln x_A + \ln x_C) - x_B L_{AB} + x_B L_{BC} + (- x_C + x_A)L_{AC}$; $g_{11} = RT(1/x_A + 1/x_B) + (- 1 - 1)L_{AB}$; $g_{12} = RT(1/x_A) - L_{AB} + L_{BC} - L_{AC} = g_{21}$; $g_{22} = RT(1/x_A + 1/x_C) + (- 1 - 1)L_{AC}$. The criterion for the limit of stability gives $g_{11}g_{22} - g_{12}g_{21} = 0$; $(RT)^2(1/x_A^2 + 1/x_A x_B + 1/x_A x_C + 1/x_B x_C) - 2L_{AC}RT(1/x_A + 1/x_B) - 2L_{AB}RT(1/x_A + 1/x_C) + 4L_{AB}L_{AC} = (RT)^2/x_A^2 + 2(- L_{AB} + L_{BC} - L_{AC})RT/x_A + L_{AB}^2 + L_{BC}^2 + L_{AC}^2 - 2L_{AB}L_{BC} - 2L_{BC}L_{AC} + 2L_{AB}L_{AC}$; $(RT)^2(1/x_A x_B + 1/x_A x_C + 1/x_B x_C) - 2L_{AC}RT/x_B - 2L_{AB}RT/x_C - 2L_{BC}RT/x_A = L_{AB}^2 + L_{BC}^2 + L_{AC}^2 - 2L_{AB}L_{BC} - 2L_{BC}L_{AC} - 2L_{AB}L_{AC}$. It is reassuring to see that this equation is symmetric with respect to all three components. It does not matter which x was chosen to be the dependent variable. The numerical value of the temperature limit can easily be evaluated by inserting the alloy composition.

Le Chatelier's principle

When discussing the limit of stability we compared the values of two derivatives which differed only by one of the variables to be kept constant. Using the same method of calculation we can also compare the effect of changing an external variable under a frozen-in internal variable ξ and under a gradual adjustment of ξ according to equilibrium, i.e. $D = 0$. It should be remembered that ξ may be treated as an extensive variable and $- D$ as a potential. We obtain

$$\left(\frac{\partial Y^b}{\partial X^b}\right)_{D=0} = \left(\frac{\partial Y^b}{\partial X^b}\right)_\xi - \left(\frac{\partial Y^b}{\partial \xi}\right)_{X^b}\left(\frac{\partial[-D]}{\partial X^b}\right)_\xi \Big/ \left(\frac{\partial[-D]}{\partial \xi}\right)_{X^b}$$

For simplicity, the variables that have been kept constant in all the derivatives have been omitted from the subscripts but any set of potentials and extensive variables presented in Table 3.1 in Section 3.4 can be used. $(\partial Y^b/\partial \xi)_{X^b}$ and $(\partial[-D]/\partial X^b)_\xi$ are equal due to a

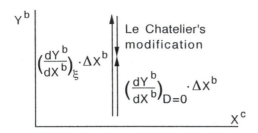

Figure 5.4 Illustration of Le Chatelier's principle. The extensive variable X^b is changed by an amount ΔX^b by an external action. An internal process is first frozen in, $d\xi = 0$, but then proceeds to a new equilibrium, $D = 0$. The initial effect on Y^b is thus partly reversed. During the whole process either the potential or the extensive variable of all other pairs of conjugate variables is kept constant (here represented by X^c on the abscissa.

Maxwell relation and $(\partial[-D]/\partial\xi)_{X^b}$ is equal to the stability B at equilibrium. For a stable system, B is positive and the second term on the right-hand side with its minus sign cannot be positive. We thus obtain

$$0 \le \left(\frac{\partial Y^b}{\partial X^b}\right)_{D=0} \le \left(\frac{\partial Y^b}{\partial X^b}\right)_{\xi}$$

This relation is quite general. It has here derived using the energy scheme. It can also be derived using the other schemes.

Suppose the equilibrium inside a system is disturbed by an action from the outside. For instance, X^b is changed quickly by an amount ΔX^b and there is not enough time for the internal reaction, i.e. ξ is kept constant. Thus, the potential Y^b is changed according to the term appearing on the right-hand side of the inequality (see the left-hand arrow in Fig. 5.4). After a sufficiently long time the internal reaction will occur and ξ will change to a new state of equilibrium and the net change of the two stages may thus be calculated from the term appearing in the middle part of the inequality (see the right-hand arrow pointing upward in the figure). It represents the change of Y^b due to a slow change ΔX^b. The difference between the two changes of Y^b represents the change due to the internal reaction, the so-called Le Chatelier modification. The inequality shows that the change in Y^b will thus be partly reversed during the second stage (see the arrow pointing downward in the figure). This principle was formulated by Le Chatelier but in a less exact manner. It should be emphasized that it concerns two conjugate variables, X^b and Y^b. It should further be emphasized that the extensive variable must be regarded as the primary variable. If, instead, the potential variable is regarded as the primary one, then the opposite result is obtained

$$\left(\frac{\partial X^b}{\partial Y^b}\right)_{D=0} \ge \left(\frac{\partial X^b}{\partial Y^b}\right)_{\xi} \ge 0$$

The derivation of Le Chatelier's principle is based on derivatives and it has thus been proved only for infinitesimal disturbances. The principle may fail for finite changes.

Exercise 5.9.1

Test Le Chatelier's principle on the change of temperature and pressure when an amount of heat is added to a two-phase system of water vapour and water under constant volume. Suppose that evaporation is initially very slow due to a thin film of oil.

Hint

Remember that $dU = dQ - PdV$. At constant V we thus have $dU = dQ$ and could choose U as the variable that is changed by an action from the outside. Its conjugate variable is $-1/T$. The internal variable ξ may be identified with the amount of vapour.

Solution

Identify U with X^b and $-1/T$ with Y^b. For a stable system we get, at constant ξ, i.e. before any change of the amount of vapour, $0 < (\partial[-1/T]/\partial U)_\xi$. This means that $-1/T$, and thus also T, has increased due to increased U. At the same time, the pressure has increased.

   At the higher temperature the equilibrium vapour pressure will be higher. In a second stage of the process there will thus be evaporation and the temperature will decrease in agreement with Le Chatelier's principle, $0 < (\partial[-1/T]/\partial U)_{D=0} < (\partial[-1/T]/\partial U)_\xi$. On the other hand, the pressure has increased during the first stage due to the heating of the vapour present from the beginning. During the second stage, the pressure will increase further, in apparent contradiction to Le Chatelier's principle. However, pressure is not conjugate to the variable U which was changed to a new value in the experiment.

   The result is far from trivial because there would be further evaporation during the second stage only if the increase in pressure of the initial vapour due to its heating is smaller than the increase of the equilibrium vapour pressure due to the heating of the water. By relying upon Le Chatelier's principle we may thus conclude that the heating of the vapour gives a smaller increase of the pressure than the heating of the water would increase the equilibrium vapour pressure.

Exercise 5.9.2

Test Le Chatelier's principle on the change of pressure during a similar experiment where the volume is increased isothermally and then kept constant for a long time.

Hint

At equilibrium at a given temperature water has a certain vapour pressure, the equilibrium pressure.

Solution

The volume increase will result in an instantaneous decrease of the pressure. The equilibrium is thus disturbed and during a second stage there will be further evaporation and the pressure will increase in agreement with Le Chatelier's principle. If the second stage also occurs under isothermal conditions, the pressure will return to its initial value. Evidently, this is a case where $(\partial Y^b/\partial X^b)_{D=0} = 0$.

Exercise 5.9.3

Test Le Chatelier's principle on the change of volume during a similar experiment where the pressure is decreased isothermally and then kept constant for a long time.

Solution

The pressure decrease will result in an initial increase of the volume. Then there will be a further increase during a second stage due to further evaporation. This is in apparent contradiction to Le Chatelier's principle but in full agreement with the inverse form of the principle. It may be interesting to note that, if the external temperature and pressure are kept constant in this experiment, the process will not stop until all the water has evaporated.

Exercise 5.9.4

Somebody has objected to Le Chatelier's principle by referring to the following case. Consider a two-phase system of liquid water–water vapour at constant temperature. It is an experimental fact that an increase in pressure will give an immediate compression and then there will be a further compression due to an internal process, – spontaneous condensation at the increased pressure. Thus, the change due to the internal process will go in the same direction as the primary change, but would not Le Chatelier's principle require that it goes in the opposite direction?

Hint

Identify what is a potential and what is an extensive variable.

Solution

The experimental fact concerns the change of an extensive quantity, volume, caused by change of a potential, pressure. Le Chatelier considered the change of a potential, caused by a change in an extensive quantity. Thus we should expect the opposite result.

6

Applications of molar Gibbs energy diagrams

6.1   Molar Gibbs energy diagrams for binary systems

In this chapter we shall derive some useful thermodynamic relations relating to phase equilibria under constant temperature and pressure, sometimes in exact form but sometimes using approximations in order to bring the final expressions into a suitable form. We shall see how property diagrams for the molar Gibbs energy can be used in such derivations. Most of the applications will make use of the tie-line rule (see Section 3.9).

As an introduction, some basic properties of solutions must be discussed and, in the present section, a simple solution model will be described. A more thorough discussion will be given in Chapter 18.

Let us first consider a case where a solution phase α can vary in composition over a whole binary system from pure A to pure B. It is then convenient to compare the G_m value at any composition with the value one can read on the straight line between the two end-points, usually called the **end-members** of the solution. The difference is often called the **Gibbs energy of mixing** and is denoted with a superscript M. It is illustrated in Fig. 6.1. It is defined by the following equation

$$G_m^\alpha = x_A^o G_A^\alpha + x_B^o G_B^\alpha + {}^M G_m^\alpha$$

A warning should be issued regarding the interpretation of ${}^M G_m$. Usually it is defined with reference to the straight line between points representing the pure components in the same state as the phase under consideration, i.e. the end-members of the solution. However, sometimes it is defined with reference to a different state for one of the components, for instance the state which is most stable at the temperature under consideration. This is illustrated in Fig. 6.1(b) where pure B is more stable as β than as α.

The Gibbs energy diagram gives information on the partial molar Gibbs energies for the two components, i.e. the chemical potentials. For a single phase one can use the construction explained for V_m in Fig. 3.7 and illustrated for G_m in Fig. 3.8. It is again demonstrated in Fig. 6.2(a) and is in agreement with the following relations which

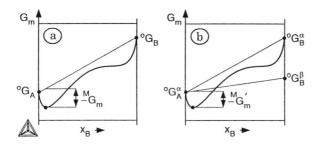

Figure 6.1 Molar Gibbs energy diagram for a binary system illustrating the definition of the Gibbs energy of mixing. The end-members of the solution are used as references in (a) but a more stable state of B is used in (b).

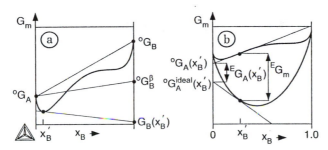

Figure 6.2 (a) Tangent construction to obtain chemical potential. (b) Definitions of excess quantities for an alloy x'_B.

are examples of the more general expression for all partial quantities, derived in Section 3.5.

$$G^\alpha_B = G^\alpha_m + (1 - x_B)dG^\alpha_m/dx_B = G^\alpha_m - x_A dG^\alpha_m/dx_A$$
$$G^\alpha_B - G^\alpha_A = dG^\alpha_m/dx_B$$

An important contribution to the Gibbs energy of mixing comes from the entropy of mixing of the two kinds of atoms. In Section 16.9 we shall consider the case where they are distributed at random and shall find that the entropy of mixing of one mole of atoms will then be $- R(x_A \ln x_A + x_B \ln x_B)$. A solution with only this contribution to the Gibbs energy of mixing is regarded as an ideal solution

$$G^{ideal\ \alpha}_m = x^\circ_A G^\alpha_B + x^\circ_B G^\alpha_B + RT(x_A \ln_A + x_B \ln x_B)$$

Using the equation relating G_B to G_m we obtain

$$G^{ideal\ \alpha}_B = {}^\circ G^\beta_B + RT \ln x_B$$

The ideal solution is illustrated in Fig. 6.2(b). The curve for G^{ideal}_m shows that the term $RT(x_A \ln x_A + x_B \ln x_B)$ is negative and makes the G_m curve look like a hanging rope. It is the main cause of the stability of solutions.

It is common to summarize all other contributions to the Gibbs energy with a term called the **excess Gibbs energy** and denoted by $^E G^\alpha_m$

$$G_m^\alpha = G_m^{\text{ideal }\alpha} + {}^E G_m^\alpha$$
$$= x_A^\circ G_A^\alpha + x_B^\circ G_B^\alpha + RT(x_A \ln x_A + x_B \ln x_B) + {}^E G_m^\alpha$$
$${}^M G_m^\alpha = RT(x_A \ln_A + x_B \ln x_B) + {}^E G_m^\alpha$$

This is also illustrated in Fig. 6.2(b). In the same way we may define the mixing and excess quantities for the partial Gibbs energies,

$$G_B^\alpha = G_B^{\text{ideal }\alpha} + {}^E G_B^\alpha = {}^\circ G_B^\alpha + RT \ln x_B + {}^E G_B^\alpha$$
$${}^M G_B^\alpha = RT \ln x_B + {}^E G_B^\alpha$$

It is evident that the mixing and excess quantities can be calculated directly in the same way as G_B^α,

$${}^M G_B^\alpha = {}^M G_m^\alpha + (1 - x_B) d^M G_m^\alpha / dx_B = {}^M G_m^\alpha - x_A d^M G_m^\alpha / dx_A$$
$${}^E G_B^\alpha = {}^E G_m^\alpha + (1 - x_B) d^E G_m^\alpha / dx_B = {}^E G_m^\alpha - x_A d^E G_m^\alpha / dx_A$$

It should be emphasized that one cannot give an absolute numerical value to the partial Gibbs energies, G_B^α or G_A^α, because there is no natural zero point for Gibbs energy. Numerical values can be given only to differences in Gibbs energy between two states. Thus we can give a value to $G_B^\alpha - {}^\circ G_B^\alpha$, and another value to $G_B^\alpha - {}^\circ G_B^\beta$, where pure α-B or pure β-B are regarded as two different choices of **standard states** for B. Such a value gives the vertical distance between two points (see the B-axis in Fig. 6.2(a)). For alloys, a numerical value can be given only to differences between two states of the *same composition*. The two lines representing ${}^M G_m$ and ${}^M G_m'$ in Fig. 6.1(a) and (b), respectively, are thus vertical. This stems from the fact that one cannot compare the Gibbs energies for A and B. When starting to construct such a diagram one can give the ${}^\circ G_A^\alpha - {}^\circ G_B^\beta$ line any convenient slope.

If one has chosen the end-members of an A–B solution to define the standard states, ${}^\circ G_A$ and ${}^\circ G_B$, then it is evident that the excess Gibbs energy is zero at the two sides of the system where G_m is equal to ${}^\circ G_A$ or ${}^\circ G_B$. For a dilute solution of B in A we may thus try to approximate ${}^E G_m$ as $L x_A x_B$, an expression that goes to zero on both sides. This is the regular solution approximation, and using the equation relating G_B to G_m we find

$$G_B^\alpha = {}^\circ G_B^\alpha + RT \ln x_B + {}^E G_B^\alpha = {}^\circ G_B^\alpha + RT \ln x_B + L^\alpha x_A^2$$

It is common to introduce the **chemical activity**, a_B, through the expression

$$G_B^\alpha = {}^\circ G_B^\alpha + RT \ln a_B = {}^\circ G_B^\alpha + RT \ln x_B + RT \ln f_B$$

where

$$a_B = f_B x_B$$

and f_B is called the **activity coefficient** of B. The activity is thus defined as

$$a_B = \exp[(G_B^\alpha - {}^\circ G_B^\alpha)/RT]$$

and the activity coefficient is obtained through

$$RT \ln_B = {}^E G_B^\alpha$$

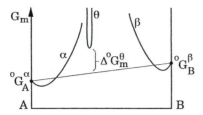

Figure 6.3 Molar Gibbs energy diagram illustrating the definition of the standard Gibbs energy of formation, $\Delta^{o}G_{m}^{\theta}$, of a compound θ.

For a dilute solution, i.e. low x_{B}, we may thus write

$$RT\ln f_{B} = {}^{E}G_{B}^{\alpha} = L^{\alpha}x_{A}^{2} \cong L^{\alpha} \text{ or } f_{B} \cong \exp(L^{\alpha}/RT)$$

With this approximation, the activity is proportional to the content. This is called Henry's law. We also find for low x_{B},

$$RT\ln f_{A} = {}^{E}G_{A}^{\alpha} = {}^{E}G_{m}^{\alpha} + (1 - x_{A})d^{E}G_{m}^{\alpha}/dx_{A} = Lx_{B}^{2} \cong 0$$

The value of f_{A} is thus unity close to the A side. This is called Raoult's law. When Henry's law holds for B and Raoult's law holds for A, then we obtain

$$dG_{m}^{\alpha}/dx_{B} = G_{B}^{\alpha} - G_{A}^{\alpha} = {}^{o}G_{B}^{\alpha} + RT\ln x_{B} + L^{\alpha} - {}^{o}G_{A}^{\alpha} - RT\ln x_{A}$$
$$d^{2}G_{m}^{\alpha}/dx_{B}^{2} = RT/x_{B} + RT/x_{A} = RT(x_{A} + x_{B})/x_{A}x_{B} = RT/x_{A}x_{B} \cong RT/x_{B}$$

For an intermediary phase, which does not extend to the pure components, one must always refer the Gibbs energy to the values of the components in selected states, usually their stable states. For a phase with a well-defined composition one often talks about the **standard Gibbs energy of formation** (see Fig. 6.3). From that diagram we obtain

$$\Delta^{o}G_{m}^{\theta} = {}^{o}G_{m}^{\theta} - x_{A}^{o}G_{A}^{\alpha} - x_{B}^{o}G_{B}^{\beta}$$

This quantity is often denoted by $\Delta_{f}G_{m}^{\theta}$ or $\Delta_{f}^{o}G_{m}^{\theta}$. It is important to mention the reference states to which it refers and also the amount of material being considered, for instance 1 mole of atoms or 1 mole of formula units (like, e.g., $Cr_{23}C_{6}$).

Exercise 6.1.1

An α solution in the A–B system has $a_{B} = 0.9$ at 1000 K when pure α-B is used as standard state. Calculate a_{B} referred to another state of B, called β-B, which is more stable than α-B by 1200 J/mol. Illustrate with a G_{m} diagram.

Hint

The position of the point on the B-axis representing G_{B}^{α} does not depend upon the choice of standard state. We can thus equate any two expressions for G_{B}^{α}.

Solution

$a_B^{ref\ \alpha} = 0.9$; $°G_B^\alpha + RT\ln(a_B^{ref\ \alpha}) = G_B^\alpha = °G_B^\beta + RT\ln(a_B^{ref\ \beta})$; $a_B^{ref\ \beta} = a_B^{ref\ \alpha}\exp[(°G_B^\alpha$
$- °G_B^\beta)/RT] = 0.9\cdot\exp(1200/8.3145\cdot1000) = 1.04$. Since this is > 1, the α solution is supersaturated with B in comparison with the stable β state of B (see diagram).

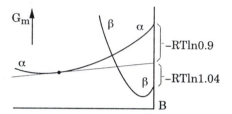

Solution 6.1.1

Exercise 6.1.2

Fe_3C is metastable at all temperatures and could thus decompose into an Fe-rich phase and graphite. At 1169 K the stable Fe phase (γ) dissolves about 1.24 mass% C. Measurements have shown that the Gibbs energy of formation of Fe_3C at 1169 K is negative ($- 1620$ J/mole of formula units). Explain how this can be reconciled with the fact that Fe_3C is not stable.

Hint

Sketch a G_m, x_C diagram.

Solution

Fe_3C falls 1620 J/mol below the line of reference between pure Fe and pure graphite (dashed line in the diagram) but it falls above the common tangent representing the γ + graphite equilibrium. Thus, Fe_3C is more stable than a mixture of the pure elements but less stable than a mixture of an Fe–C solution and pure C.

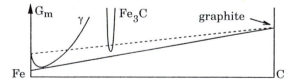

Solution 6.1.2

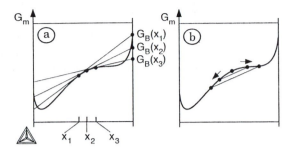

Figure 6.4 (a) Construction showing that a negative curvature results in a decrease of G_B when x_B is increased. (b) Demonstration that a system between the points of inflexion is unstable. A difference in composition will increase.

6.2   *Instability of binary solutions*

In Figs. 6.1 and 6.2 we have sketched molar Gibbs energy curves, each with two minima and a central region where $\partial^2 G_m/\partial x_B^2$ is negative. This region falls between two points of inflexion and according to Section 5.6 they should define the limit of stability. In Fig. 6.4(a) tangents have been drawn at some compositions between the points of inflexion and it can be seen that G_B decreases when x_B increases. As we have seen in Section 5.3, this is also a violation of the condition of stability.

The change in the total Gibbs energy of the system, when one half of the system grows richer in A and the other one in B, is illustrated in Fig. 6.4(b). The tie-line rule requires that the total Gibbs energy is represented by a point on the line connecting the points representing the two parts. Since the overall composition is not changed, the total Gibbs energy moves down along a vertical line at the alloy composition. Such a system is thus unstable against fluctuations in composition.

Exercise 6.2.1

In a binary solution one usually discusses two activities, a_A and a_B, and for a stable system each of them increases monotonously with the content of the same component. However, one may define the activity of an intermediary species, e.g. A_2B. Prove that a_{A_2B} also increases as the content of A_2B increases in a stable system and thus has a maximum at the very composition of A_2B.

Hint

Instead of the chemical activity, a_{A_2B}, let us consider the chemical potential μ_{A_2B}, which is equal to $2G_A + G_B$, or, even better, $\mu_{A_{2/3}B_{1/3}}$, i.e. $(2G_A + G_B)/3$. It may be studied in a G_m,x_B diagram.

Solution

For any alloy the tangent in the G_m, x_B diagram gives μ_A and μ_B on the two sides and the intersection of the tangent with a vertical line at $x_B = 1/3$ thus gives $(2G_A + G_B)/3$. By inspection it is evident that the intersection has its highest position for the alloy $x_B = 1/3$. Otherwise, the G_m curve must have a negative curvature somewhere (see diagram).

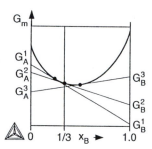

Solution 6.2.1

6.3   *Illustration of the Gibbs–Duhem relation*

The molar Gibbs energy diagram in Fig. 6.5(a) shows a stoichiometric compound, θ, with a well-defined composition A_aB_b, possibly because it is a crystalline phase with two sublattices. Often, the composition of such a phase cannot vary appreciably without a very steep increase of the Gibbs energy. It is thus practically impossible to vary N_A and keep N_B constant and the definition of partial Gibbs energy, given in Section 3.5, cannot be used. However, the tangent construction in Fig. 6.2 can still be used. For this case we shall prefer to talk about chemical potentials and use the notations μ_A and μ_B. The situation is not drastically different for a phase with variable composition (see Fig. 6.5(b)). For both types of phase one may select the value of μ_A and the value of μ_B will then be fixed. One could also talk about the chemical potential of the compound, $\mu_{A_aB_b}$. From Section 3.12 we get

$$G_{A_aB_b} = \mu_{A_aB_b} = \Sigma a_i^s \mu_i = a\mu_A + b\mu_B; \quad \mu_B = (\mu_{A_aB_b} - a\mu_A)/b$$

By comparing the values defined by two different tangents we find, because $\mu_{A_aB_b}$ is fixed,

$$x_A(\mu_A'' - \mu_A') + x_B(\mu_B'' - \mu_B') = 0$$

where $a = x_A$ and $b = x_B$ if $a + b$ is chosen as 1. This may be regarded as the Gibbs–Duhem relation integrated for constant composition.

For a phase with variable composition one may also select the value for μ_A, and

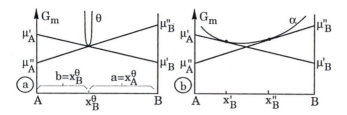

Figure 6.5 Molar Gibbs energy diagrams for (a) a binary stoichiometric phase θ and (b) a binary solution phase α. If μ_A is controlled by some method, then the value of μ_B is given in both cases.

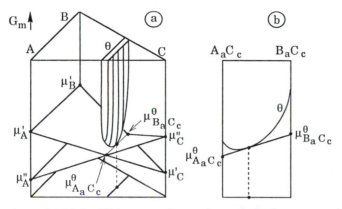

Figure 6.6 Molar Gibbs energy diagram for a solution between two binary stoichiometric phases, a so-called line compound.

the value for μ_B will then be fixed by the expression for the molar Gibbs energy,

$$G_m = x_A\mu_A + x_B\mu_B; \quad \mu_B = (G_m - x_A\mu_A)/x_B$$

but here G_m varies with the composition and the composition varies with the choice of μ_A, as demonstrated for a solution phase α in Fig. 6.5(b). The Gibbs–Duhem relation holds

$$x_A d\mu_A + x_B d\mu_B = 0$$

but not in the integrated form given for the stoichiometric compound, because all the tangents do not intersect in the same point.

In a ternary system one may have a solution between two binary stoichiometric phases if they are isomorphic (have the same structure). It may be called a **line phase** or **line compound** because it appears as a line in the composition triangle. Figure 6.6 shows a Gibbs energy diagram for such a case. The composition can now be varied with one degree of freedom and we may consider two components A_aC_c and B_aC_c.

The line compound may be represented by the formula $(A,B)_aC_c$ and the composition may be represented by molar contents defined as

$$x_{B_aC_c} = N_{B_aC_c}/(N_{A_aC_c} + N_{B_aC_c})$$

where the two Ns represent the moles of formula units. Another method of representation is based on the molar contents evaluated for each sublattice, the so-called site fractions which were discussed in Section 3.11.

From Fig. 6.6 it is evident that the molar Gibbs energy for the line compound can be expressed in the following ways if $a + c = 1$.

$$G_m = x_A\mu_A + x_B\mu_B + x_C\mu_C = y_A\mu_{A_aC_c} + y_B\mu_{B_aC_c}$$

The diagram in Fig. 6.6(a) with two possible tangent planes shows that the values of $\mu_{A_aC_c}$ and $\mu_{B_aC_c}$ are well defined by the composition of the line compound but μ_A, μ_B and μ_C are not.

Exercise 6.3.1

Consider the chemical potential of Fe_3C in a solution phase of Fe, C and Mn, using basic principles. Show that it is actually equal to $3\mu_{Fe} + \mu_C$ by making a calculation using Section 3.10.

Hint

In the Fe–C–Mn system we usually use Fe, C and Mn as the components but now we should change to a new set of components. Fe_3C is one component and the others could be C and Mn. Notice that $dG = \Sigma\mu_i dN_i$ cannot change its value just because we change the set of components to be considered.

Solution

Let the amounts of the components be N_{Fe}, N_C and N_{Mn} in the old description and N'_{Fe_3C}, N'_C and N'_{Mn} in the new one. Denote the chemical potentials in the new description with μ'_i. The mass balance for each element gives $N_{Mn} = N'_{Mn}$, $N_{Fe} = 3N'_{Fe_3C}$ and $N_C = N'_C + N'_{Fe_3C}$. We thus get $\Sigma\mu'_i dN'_i = \Sigma\mu_i dN_i = \mu_{Fe}dN_{Fe} + \mu_{Mn}dN_{Mn} + \mu_C dN_C = \mu_{Fe}\cdot3dN'_{Fe_3C} + \mu_{Mn}dN'_{Mn} + \mu_C dN'_C + \mu_C dN'_{Fe_3C} = (3\mu_{Fe} + \mu_C)dN'_{Fe_3C} + \mu_C dN'_C + \mu_{Mn}dN'_{Mn}$. Thus, $\mu'_C = \mu_C$; $\mu'_{Mn} = \mu_{Mn}$; and $\mu'_{Fe_3C} = 3\mu_{Fe} + \mu_C$.

Exercise 6.3.2

Prove analytically that $x_A\mu_A + x_B\mu_B + x_C\mu_C = y_A\mu_{A_aC_c} + y_B\mu_{B_aC_c}$ when $a + c = 1$ and the composition falls on the line between A_aC_c and B_aC_c.

Hint

$x_A = ay_A/(a + c) = ay_A$; $x_B = ay_B$; $y_A + y_B = 1$. The mole fraction of C is constant, $x_C = c$.

Solution

$$y_A\mu_{A_aC_c} + y_B\mu_{B_aC_c} = y_A(a\mu_A + c\mu_C) + y_B(a\mu_B + c\mu_C) = y_A a\mu_A + (y_A + y_B)c\mu_C +$$
$$y_B a\mu_B = x_A\mu_A + x_B\mu_B + x_C\mu_C$$

Exercise 6.3.3

Consider a solution phase in a binary A–B system. Define A and AB as the two components. Where in the system will μ_{AB} have its highest value?

Hint

In a preceding section we treated this question graphically but now we should do it analytically. Start with the Gibbs–Duhem relation at constant T and P and replace the old set of components, A and B, with the new set, A and AB.

Solution

Let N_A and N_B be the contents according to the old set and N'_A and N'_{AB} according to the new set. Then $N_A = N'_A + N'_{AB}$; $N_B = N'_{AB}$; $\mu_{AB} = \mu_A + \mu_B$. The Gibbs–Duhem relation gives: $0 = N_A d\mu_A + N_B d\mu_B = (N'_A + N'_{AB})d\mu_A + N'_{AB}(d\mu_{AB} - d\mu_A) = N'_A d\mu_A + N'_{AB}d\mu_{AB}$; $d\mu_{AB}/d\mu_A = -N'_A/N'_{AB}$. When $N'_A = 0$, i.e. when $N_A = N'_{AB} = N_B$, we get $d\mu_{AB}/d\mu_A = 0$ and there is a maximum in μ_{AB} at the composition of AB.

6.4

Two-phase equilibria in binary systems

In a two-phase equilibrium we have the following two conditions at constant T and P because the chemical potential for each component must be the same in the two phases.

$$G_B^\alpha = \mu_B = G_B^\beta$$
$$G_A^\alpha = \mu_A = G_A^\beta$$

It is evident that these conditions can be satisfied only by a **common tangent**, as illustrated in Fig. 6.7(a). The lowest possible G_m for each composition is shown in Fig. 6.7(b) and it is evident that some mixture of $\alpha + \beta$ represents the stable state for an alloy between the two tangent points.

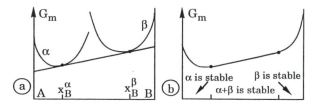

Figure 6.7 The common-tangent construction for finding the equilibrium compositions of two phases at given T and P.

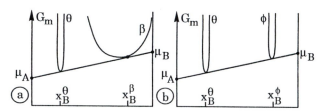

Figure 6.8 Molar Gibbs energy diagram for the equilibrium between a stoichiometric phase and a solution phase or another stoichiometric phase.

It was illustrated in Fig. 6.5 that, for a stoichiometric phase with its well-defined composition, it is not possible to define the chemical potentials since one can draw different tangents without changing the composition of the point of tangency markedly. On the other hand, the chemical potentials of a phase can be defined by equilibrium with a second phase. Figure 6.8(a) illustrates this case when the second phase is a solution phase or another stoichiometric phase. When it is a solution phase, we obtain

$$G_m^\theta = x_A^\theta \cdot G_A^\beta(x_B^\beta) + x_B^\theta \cdot G_B^\beta(x_B^\beta)$$

By solving this equation, one can determine the composition of the solution phase, β, and from the known properties of β one can then calculate μ_A and μ_B. In Fig. 6.8(b) both phases, θ and ϕ, are stoichiometric phases, and we obtain the following relations

$$G_m^\theta = x_A^\theta \mu_A + x_B^\theta \mu_B \qquad \mu_A = \frac{x_B^\theta G_m^\phi - x_B^\phi G_m^\theta}{x_B^\theta - x_B^\phi}$$

$$G_m^\theta = x_A^\phi \mu_A + x_B^\phi \mu_B \qquad \mu_B = \frac{x_A^\theta G_m^\phi - x_A^\phi G_m^\theta}{x_A^\theta - x_A^\phi}$$

Exercise 6.4.1

In a binary system, where the mutual solubilities are very small, there are two stable stoichiometric phases $\alpha(A_3B_2)$ and $\beta(AB_3)$. Calculate the chemical potential of B in a 50:50 alloy in terms of the quantities G_m^α and G_m^β. Base the calculation on a construction in the G_m diagram.

Hint

Remember that the Gibbs energy of a two-phase state falls on the common tangent. Start by drawing a solid line representing all stable states. It should show that both stoichiometric phases are stable.

Solution

(See diagram.) Evidently, the alloy is $\alpha + \beta$. With $x_A^\alpha = 0.6$ and $x_A^\beta = 0.25$ the construction gives $\mu_B = G_B^{\alpha+\beta} = G_m^\alpha + (G_m^\beta - G_m^\alpha)(x_A^\alpha - 0)/(x_A^\alpha - x_A^\beta) = G_m^\alpha + (G_m^\beta - G_m^\alpha)60/(60 - 25) = (60/35)G_m^\beta - (25/35)G_m^\alpha$.

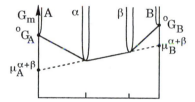

Solution 6.4.1

Exercise 6.4.2

Suppose a solution phase β has a higher A content than required by equilibrium with an A-rich stoichiometric phase θ. When it comes into contact with θ, its surface layer will lose A and grow richer in B. Suppose that the lattice parameter varies strongly with the composition. The surface layer will then be under coherency stresses as long as it is coherent with the bulk of α. Examine if this phenomenon will increase or decrease the chemical potentials for the two-phase equilibrium.

Hint

The stressed α material has an additional energy and should thus have a new G_m curve which is higher than the ordinary G_m curve but tangent to it at the initial α composition where there should be no stresses.

Solution

The common tangent construction shows that μ_B is higher but μ_A is lower. See diagram.

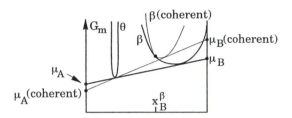

Solution 6.4.2

6.5   *Allotropic phase boundaries*

One can sometimes draw a line inside a two-phase region to show where the two phases would have the same Gibbs energy value if they had the same composition. It is the critical limit for a hypothetical diffusionless phase transformation. Such a transformation is very similar to an allotropic transformation in a pure element and the line, sometimes called the **allotropic phase boundary**, is often denoted by T_o. This name is derived from the word 'allotropy', i.e. the property of a substance of being found in two or more forms. Figure 6.9 illustrates the relation between the allotropic phase boundary and the molar Gibbs energy diagram.

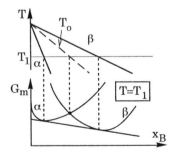

Figure 6.9 The relation between Gibbs energy curves of two phases and their two-phase field in the phase diagram. The dashed line is the so-called T_o line or allotropic phase boundary.

Exercise 6.5.1

Consider a binary system with three phases of variable compositions and in a eutectic equilibrium (see Section 11.5) with each other. Show reasonable positions of the three allotropic phase boundaries. Extrapolate all of them below the eutectic temperature.

Hint

Consider in particular how the three allotropic phase boundaries intersect each other when extrapolated. Will there be one or three points of intersection? It may be instructive to sketch a molar Gibbs energy diagram.

Solution

It is evident that $T_o^{\alpha+L}$ and $T_o^{L+\beta}$ will intersect inside the $\alpha + \beta$ phase field. Consider a G_m diagram at the eutectic temperature, showing one G_m curve for each phase (full lines in the lower portion of the diagram). Each one of the three intersections is a point on a T_o line. When the temperature is decreased, the L curve is lifted relative to the other two unit (see thin line) all three curves finally intersect in one point. There the three T_o lines will intersect. In the phase diagram (upper portion) we should thus draw the $T_o^{\alpha+\beta}$ line through the intersection of the other two.

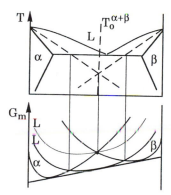

Solution 6.5.1

Exercise 6.5.2

Consider an isobarothermal section of a ternary system with three phases taking part in a three-phase equilibrium. Each two-phase field has an allotropic phase boundary. Show reasonable positions of the three lines. Is it necessary or possible that they intersect inside or outside the three-phase triangle? Will there be one or three points of intersection?

Hint

In the ternary case, an allotropic phase boundary is the line of intersection between two G_m surfaces. The two phases must have the same G_m value along that line.

Solution

Consider the point where the $\alpha + \beta$ and $\alpha + \gamma$ lines intersect. There, α has the same G_m as β and γ. Consequently, β and γ also have the same G_m there and the $\beta + \gamma$ line must intersect in the same point. By simple constructions one may show that the intersection falls inside the three-phase triangle in a regular triangle but not in a very thin triangle (see diagram).

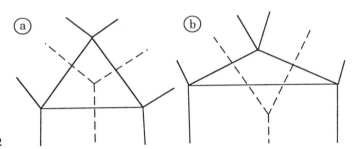

Solution 6.5.2

6.6   *Effect of a pressure difference on a two-phase equilibrium*

In order to illustrate the effect of pressure, we shall consider an incompressible phase, β. The application of a hydrostatic pressure will lift up its Gibbs energy curve by the amount $P^\beta V_m^\beta$. This is illustrated in Fig. 6.10(a). Since V_m^β is usually dependent on the composition, the curve will be somewhat deformed. The construction with a tangent will yield $P^\beta V_A^\beta$ and $P^\beta V_B^\beta$ where V_A^β and V_B^β are defined from V_m^β in the same way as G_A^β and G_B^β are defined from G_m^β (see Fig. 3.7).

The relative position of the Gibbs energy curves for different phases can change with pressure due to differences in the molar volumes. The equilibrium conditions can thus be modified by the application of a hydrostatic pressure. This effect is even stronger if the pressure is applied to one of the phases only, which may happen due to the effect of interfacial energy in a curved interface. In Fig. 6.10(b) the phase under pressure is a stoichiometric phase and its molar Gibbs energy is increased by $P^\theta V_m^\theta$. We may, for instance, imagine that the phases are contained in a cylinder where the balance of surface tensions, σ, at the wall of the cylinder gives a constant radius of curvature $\rho = 2\sigma/P^\theta$. The diagram illustrates that the solubility of θ in α is increased due to the pressure in θ, assuming ordinary pressure in α. It should be emphasized that this case must be treated with care because the two phases are under different pressures and the law of additivity does not apply to the Gibbs energy unless special precautions are taken. This will be explained in Section 14.7. However, in applying the common-tangent

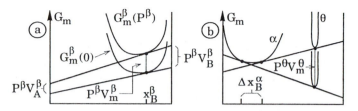

Figure 6.10 (a) Effect of a hydrostatic pressure on the molar Gibbs energy of single phase. The phase is assumed to be incompressible. (b) The effect of pressure in a stoichiometric phase on the equilibrium composition of a coexisting phase not under pressure.

construction we have only made use of the definition of the chemical potentials.

The effect on the solubility can be estimated if one knows the curvature of the G_m^{α} curve. The difference in slope between the two tangents can be estimated as $[x_B^{\alpha/\theta}(P^{\theta}) - x_B^{\alpha/\theta}(0)] \cdot d^2 G_m^{\alpha}/dx_B^2$ if the change in composition is small. If the distance between the two phases is reasonably constant, we obtain, by multiplying with that distance,

$$P^{\theta} V_m^{\theta} = [x_B^{\alpha/\theta}(P^{\theta}) - x_B^{\alpha/\theta}(0)] \cdot d^2 G_m^{\alpha}/dx_B^2 \cdot (x_B^{\theta} - x_B^{\alpha})$$
$$x_B^{\alpha/\theta}(P^{\theta}) - x_B^{\alpha/\theta}(0) = P^{\theta} V_m^{\theta}/(x_B^{\theta} - x_B^{\alpha})d^2 G_m^{\theta})dx_B^2$$

When α is a dilute solution we may approximate $d^2 G_m^{\alpha}/dx_B^2$ with $RT/x_B^{\alpha/\theta}$ (see the end of Section 6.1) and obtain

$$x_B^{\alpha/\theta}(P^{\theta}) - x_B^{\alpha/\theta}(0) = P^{\theta} V_m^{\theta} x_B^{\alpha/\theta}/RT(x_B^{\theta} - x_B^{\alpha})$$

This equation is often applied to a spherical interface and $2\sigma/r$ is then substituted for P^{θ}. In that form it is known as the Gibbs–Thomson equation. For large changes in solubility one should make an integration over the pressure increase and allow $x_B^{\alpha/\theta}$ on the right-hand side to vary during the integration. For an infinitesimal increase in P^{θ} we get

$$dx_B^{\alpha/\theta}/x_B^{\alpha/\theta} = [V_m^{\theta}/RT(x_B^{\theta} - x_B^{\alpha})] \cdot dP^{\theta}$$

For the case where $(x_B^{\theta} - x_B^{\alpha})$ is reasonably constant integration yields

$$\ln[x_B^{\alpha/\theta}(P^{\theta})/x_B^{\alpha/\theta}(0)] = P^{\theta} V_m^{\theta}/RT(x_B^{\theta} - x_B^{\alpha})$$

If the phase under pressure can also vary in composition, its equilibrium composition will also change. This case is illustrated in Fig. 6.11. The change in composition of the phase under pressure can be evaluated from Fig. 6.11(b) where a tangent to the initial G_m^{β} curve has been drawn for the β composition of the new equilibrium. The diagram defines a quantity ΔG_m which is given as

$$\Delta G_m = P^{\beta}(x_A^{\alpha} V_A^{\beta} + x_B^{\alpha} V_B^{\beta})$$

but also as

$$\Delta G_m = (x_B^{\beta} - x_B^{\alpha}) \cdot d^2 G_m^{\beta}/d_B^2 \cdot [x_B^{\beta/\alpha}(P^{\beta}) - x_B^{\beta/\alpha}(0)]$$

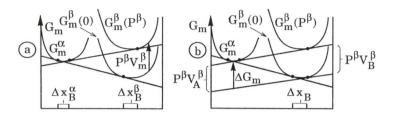

Figure 6.11 Molar Gibbs energy diagram illustrating the change in composition of a phase, β, under pressure when in equilibrium with another phase, α.

if $d^2G_m^\beta/dx_B^2$ is reasonably constant. By equating the two expressions we obtain

$$x_B^{\beta/\alpha}(P^\beta) - x_B^{\beta/\alpha}(0) = P^\beta(x_A^\alpha V_A^\beta + x_B^\alpha V_B^\beta)/[(x_B^\beta - x_B^\alpha)\cdot d^2 G_m^\beta/dx_B^2]$$

If the β phase is a dilute solution of B in A, then $x_A^\alpha V_A^\beta + x_B^\alpha V_B^\beta$ is approximately equal to the molar volume for pure A in the β state, which we shall simply denote by V_m^β, and $d^2G_m^\beta/dx_B^2$ can be approximated by RT/x_B^β yielding

$$dx_B^{\beta/\alpha}/x_B^{\beta/\alpha} = [V_m^\beta/RT(x_B^\beta - x_B^\alpha)]\cdot dP^\beta$$

We can take into account the effect of P^β on both phases if they are both dilute solutions of B in A but then we cannot treat $x_B^\beta - x_B^\alpha$ as a constant. However, the expressions for $dx_B^{\alpha/\beta}/x_B^{\alpha/\beta}$ and $dx_B^{\beta/\alpha}/x_B^{\beta/\alpha}$ are identical,

$$dx_B^{\alpha/\beta}/x_B^{\alpha/\beta} = dx_B^{\beta/\alpha}/x_B^{\beta/\alpha}$$

Integration yields

$$x_B^{\alpha/\beta}(P^\beta)/x_B^{\alpha/\beta}(0) = x_B^{\beta/\alpha}(P^\beta)/x_B^{\beta/\alpha}(0)$$

$$\frac{x_B^{\beta/\alpha}(P^\beta) - x_B^{\alpha/\beta}(P^\beta)}{x_B^{\beta/\alpha}(P^\beta)} = \frac{x_B^{\beta/\alpha}(0) - x_B^{\alpha/\beta}(0)}{x_B^{\beta/\alpha}(0)}$$

The quantity on the left-hand side appears in the equation for $dx_B^{\beta/\alpha}$ with a slightly different notation, $(x_B^\beta - x_B^\alpha)/x_B^\beta$. This ratio can thus be treated as a constant during the integration of $dx_B^{\beta/\alpha}$, yielding

$$x_B^{\beta/\alpha}(P^\beta) - x_B^{\beta/\alpha}(0) = (P^\beta V_m^\beta/RT)\cdot x_B^{\beta/\alpha}/(x_B^\beta - x_B^\alpha)$$
$$x_B^{\alpha/\beta}(P^\beta) - x_B^{\alpha/\beta}(0) = (P^\beta V_m^\beta/RT)\cdot x_B^{\alpha/\beta}/(x_B^\beta - x_B^\alpha)$$

It should again be emphasized that the equations in the present section were derived only for an incompressible phase.

Exercise 6.6.1

The precipitation of Sn from a supersaturated solid solution of Sn in Pb sometimes results in a lamellar aggregate of a Sn phase with very little Pb and a Pb phase will less Sn. The aggregate, comprising alternate layers, grows into a Pb-rich matrix of the original composition. Experimental studies have been made of the coarseness of such a structure in terms of a quantity w, the combined width of one lamella of each phase. When discussing theoretical predictions the measured w is compared with the critical value w^* which would completely stop the growth of the Sn phase because, due to the effect of surface tension, it would put this phase under such a high pressure that it would be in equilibrium with the original Pb matrix in spite of its supersaturation. This pressure can be calculated from the effect of surface tension. In one study an alloy with $x_{Sn} = 0.112$ was investigated at a temperature where the equilibrium value was 0.06. The investigators assumed that the new phase had the equilibrium composition, $x_{Sn} = 0.06$, and using the Gibbs–Thomson equation for large changes they calculated w^* from $w^* = 2\sigma V_m/RT\ln(0.112/0.06)$. They found that the observed w was about 100 times larger than w^* instead of twice according to a simple theory. Check their calculation.

Hint

Make a careful analysis of what pressure the surface tension will impose on the Sn phase under the simplifying assumption that the Pb lamellae are not under an increased pressure.

Solution

The pressure in the Sn phase will balance the surface tension, which gives a force of $2\sigma L$ if L is the length of the lamella. The area of the edge is fw^*L if f is the fraction of the Sn phase. Thus, $P^*fw^*L = 2\sigma L$. The lever rule gives $f = (0.112 - 0.06)/(1 - 0.06) = 0.055$. Now relate the pressure to the change in solubility, $\ln(0.112/0.06) = P^*V_m/RT(1 - 0.112) \cong P^*V_m/RT$. Combining these equations yields $w^* = 2\sigma/P^*f = 2\sigma V_m/fRT\ln(0.112/0.06)$. The investigators missed the factor $f(= 0.055)$ which explains most of the discrepancy.

Driving force for the formation of a new phase

In Section 5.3 we considered the evaluation of the driving force in some simple cases. Now we will study more complicated cases involving changes in composition.

When we take some A and B away from a large quantity of a solution phase, α, it is like taking them from one reservoir each, with the chemical potentials equal to G_A^α and G_B^α, respectively. As long as the amount of the α phase is large, we can take A and B in any proportion without changing the values of G_A^α and G_B^α. We can thus form a small amount of a new phase, θ, of any composition without changing the Gibbs energy of the

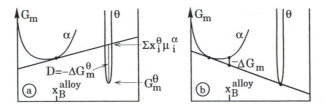

Figure 6.12 (a) Molar Gibbs energy diagram. (b) Method for evaluation of the driving force for the formation of a new phase θ from a supersaturated β solution.

whole system, provided that the new phase falls on the α tangent. If the new phase falls below the tangent, the decrease counted per mole of atoms in the new phase is obtained as

$$- \Delta G_m^\theta = x_A^\theta \cdot G_A^\alpha(x_B^\alpha) + x_B^\theta \cdot G_B^\alpha(x_B^\alpha) - G_m^\theta(x_B^\theta)$$

This is illustrated in the molar Gibbs energy diagram in Fig. 6.12(a). By convention, the change of Gibbs energy accompanying a reaction is defined as $\Delta G_m = G_m^{product} - G_m^{reactant}$. It is evident that the decrease in Gibbs energy, $- \Delta G_m$, is equal to the driving force for the precipitation of the θ phase from a supersaturated β solution, counted per mole of θ, if the extent of the reaction, ξ, is expressed as the number of moles of θ, N^θ,

$$D = - \left(\frac{\partial G}{\partial \xi} \right)_{T,P,N_i} = - \left(\frac{\partial G}{\partial N^\theta} \right)_{T,P,N_i} = - \Delta G_m^\theta$$

The magnitude of the driving force for the precipitation of θ from a supersaturated α solution, counted per mole of θ, can be estimated from the supersaturation Δx_B^α in almost the same way as the effect of pressure on solubility was evaluated. By comparing Fig. 6.12(a) with Fig. 6.10(b) we obtain

$$D = - \Delta G_m^\theta = P^\theta V_m^\theta = \Delta x_B^\alpha \cdot d^2 G_m^\alpha / dx_B^2 \cdot (x_B^\theta - x_B^\alpha)$$

As the process continues, the supersaturation will decrease gradually and so will the driving force. It may thus be interesting to evaluate the integrated driving force which should represent an average value for the whole process. The method of evaluation is illustrated in Fig. 6.12(b). One usually evaluates the integrated driving force for the transformation of the whole system, i.e. the difference in Gibbs energy between the final $\alpha + \theta$ mixture and the initial supersaturated α. It is simply given by the short vertical line in Fig. 6.12(b).

Exercise 6.7.1

Consider the formation of a small amount of β from a large reservoir of α under conditions such that the reservoir has the potentials G_A^α and G_B^α and the new phase has G_A^β and G_B^β (accepting that such conditions can somehow be realized). (a) Construct a reasonable molar Gibbs energy diagram and use it for deriving an expression for the driving force per mole of β phase. Express the result in terms of the potentials and the compositions of the two phases. (b) Suppose the composition of α has been decided. How should one choose the composition of β in order to get the largest driving force?

Hint

(a) Using the given potentials one can draw the tangents to the two Gibbs energy curves. Evaluate the distance between them at the proper composition. (b) In this exercise, the tangent to the α curve is given. The question is how we can find the point on the β curve which lies as low as possible relative to the α tangent. In principle, it can be found without drawing the corresponding β tangent but it would be most helpful to do so, so long as one draws that tangent correctly.

Solution

(a) See left-hand diagram. (b) One should choose the composition obtained from a parallel tangent construction (see right-hand diagram).

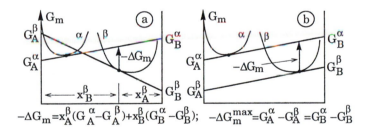

$$-\Delta G_m = x_A^\beta (G_A^\alpha - G_A^\beta) + x_B^\beta (G_B^\alpha - G_B^\beta); \quad -\Delta G_m^{max} = G_A^\alpha - G_A^\beta = G_B^\alpha - G_B^\beta$$

Solution 6.7.1

Exercise 6.7.2

On solidification an Fe–C melt normally first precipitates γ (fcc-Fe with dissolved C) and then either graphite (in grey cast iron), which gives a stable state, or cementite, Fe_3C, (in white cast iron), which gives a metastable state. Compare the driving forces for the nucleation of graphite and cementite from the melt at the temperature where the extrapolated lines for $L/L +$ graphite and $L/L +$ cementite intersect (see diagram).

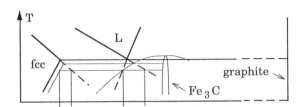

Exercise 6.7.2

Hint

Make the comparison by means of a schematic molar Gibbs energy diagram and assume that the melt is in equilibrium with γ which has precipitated first. Start by drawing curves for L, cementite and graphite with a common tangent. Then draw a curve for γ showing that L is not stable. Finally, draw the common tangent to γ and L and evaluate by how much the curves for graphite and cementite fall below it. The eutectic liquid has $x_C^L = 0.17$.

Solution

$\Delta G_m^{gr}/\Delta G_m^{cem} = (1 - 0.17)/(0.25 - 0.17) = 10$ (see diagram). The larger composition difference is favoured during nucleation but, of course, not during growth.

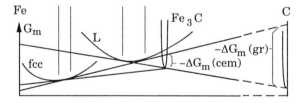

Solution 6.7.2

Exercise 6.7.3

Show with the construction in Fig. 6.12(b) the magnitude of the integrated driving force counted per mole of the θ phase formed.

Hint

The magnitude is $-\Delta G_m/f^\theta$ if f^θ is the final fraction of θ in the alloy. The question is how to find this by construction. Notice that f^θ can be found graphically using the lever rule.

Solution

Draw a straight line joining the final α and the initial α. Extend the line to the composition of θ. Also use the tangent to the final α point shown in Fig. 6.12(b). Read the distance between intersections on the θ composition.

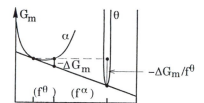

Solution 6.7.3

Partitionless transformation under local equilibrium

So far, we have mainly considered stationary states and for a state with more than one phase we have assumed equilibrium between the phases which is a reasonable approximation after a long enough time at a high enough temperature. The situation is quite different during a phase transformation but it is still common to assume that full equilibrium is established locally at the phase interface even when it is migrating through the material. The local-equilibrium approximation was introduced in Section 4.8 and will now be our starting point for an examination of **partitionless transformations**. The local conditions at migrating interfaces will be further discussed in Chapter 13.

When a $\beta \rightarrow \alpha$ transformation occurs in an alloy without any difference in composition between the reactant phase (also called parent phase) and the product phase (also called daughter phase or growing phase), it is called a partitionless transformation. The two phases will fall on the same vertical line in the molar Gibbs energy diagram. Figure 6.13(a) shows the construction for a binary system. Under constant T and P, the driving force is given by the vertical distance between two points representing the initial β and the growing α

$$D = G_m^\beta - G_m^\alpha = - \Delta G_m$$

It is evident that the partitionless transformation cannot possibly occur unless the composition of the phases falls on the left-hand side of the point of intersection between the two G_m curves.

Whether or not a transformation can actually occur under the conditions illustrated in Fig. 6.13(a) will be discussed in Section 13.4. An attractive possibility is

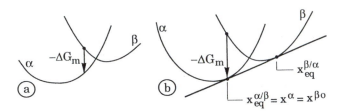

Figure 6.13 (a) Change in Gibbs energy for a partitionless $\beta \to \alpha$ transformation. (b) A partitionless transformation under local equilibrium. Here the whole decrease in Gibbs energy drives the diffusion in the parent phase β. The quantity $x^{\beta o}$ is the initial composition of the β alloy and also the equilibrium composition of α.

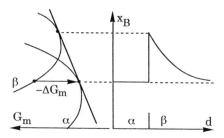

Figure 6.14 Use of the common-tangent construction to find the boundary conditions for diffusion.

illustrated in Fig. 6.13(b). It is based on the assumption of local equilibrium at the interface and that is why the common-tangent construction is used here. This illustration presumes that the parent phase is so supersaturated that its composition falls on the equilibrium composition of the growing phase. It should be realized that the local-equilibrium assumption implies that there is a gradient within the parent phase, as illustrated in Fig. 6.14. There the composition axis has been turned vertically in order to demonstrate how the molar Gibbs energy diagram can yield the boundary conditions for diffusion.

Figure 6.14 demonstrates that the local-equilibrium assumption implies that there is a pile-up (spike) of one of the components in front of the migrating interface. After an induction period during which this spike is being built up, one could expect a steady-state process where the rate of migration and the composition profile stay constant. As the interface migrates through the system and pushes the spike forward, it makes material of the initial alloy composition move up on the spike and on the top it will be deposited on the growing phase which is here assumed to have the same composition. During this process the material passes through regions of higher and higher alloy content. In each such region the chemical potentials can be described by the end-points of the tangent to the G_m curve at the local composition. The value of G_m for the material we consider will be found on that tangent and at the initial composition. It is thus evident that the material will gradually decrease its Gibbs energy by an amount

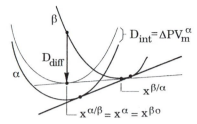

Figure 6.15 Partitionless $\beta \rightarrow \alpha$ transformation under local equilibrium and a pressure difference.

$-\Delta G_m$ corresponding to the arrow in the G_m diagram. The length of the arrow represents the integrated driving force dissipated by diffusion in the spike. Under our assumptions, all the driving force is used to drive the diffusion and the transformation is completely diffusion controlled.

In rapid transformations an appreciable driving force may be required in order to make the phase interface move with the high speed. A driving force may also be required in order to balance the pressure difference across a curved phase interface, caused by its interfacial energy, $2\sigma/\rho$. The total driving force on the interface, D_{int}, may actually be regarded as a pressure difference $\Delta P = D_{int}/V_m$. In a very crude but useful approach it is assumed that the rate of migration, v, of an interface is proportional to the net pressure difference,

$$v = M \cdot \Delta P_{net} = M \cdot (D_{int}/V_m - 2\sigma/\rho)$$

where M is the mobility of the interface, σ is the specific interfacial energy and ρ is the radius of curvature , assuming a spherical shape.

The driving force on the interface has an effect on the local equilibrium between the two phases, as illustrated in Fig. 6.15. The G_m curve for the growing phase is lifted by an amount D_{int} relative to the curve for the parent phase as if there actually were a pressure difference D_{int}/V_m. Due to this construction, the equilibrium composition of the growing phase is displaced and the local-equilibrium assumption now requires that the parent phase is initially even more supersaturated and falls on the other side of the equilibrium composition of the growing phase, i.e. inside its one-phase field. The amount of driving force dissipated by diffusion will in general be higher than before.

Exercise 6.8.1

Consider the partitionless growth of α into a spherical β particle of radius ρ in a binary alloy. Suppose there is local equilibrium at the interface and no driving force is required in order to make the interface move at a velocity v. Make a reasonable construction in a G_m diagram illustrating that this could occur at an alloy composition inside the $\alpha + \beta$ two-phase region.

Hint

From α's point of view, the radius of curvature is $-\rho$.

Solution

The negative pressure $2\sigma/(-\rho)$ will displace the G_m curve for α downward relative to the curve for β. We could just as well lift the curve for β (see diagram).

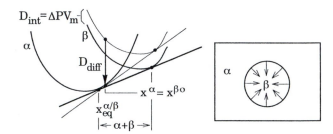

Solution 6.8.1

Exercise 6.8.2

In Section 3.5 it was mentioned that $\mu_1 - \mu_2$ can be regarded as the diffusion potential and in Section 6.7 we used $\Delta(\mu_2 - \mu_1)$ as the integrated driving force for diffusion. From Fig. 6.15 we can now find that the integrated driving force is $-\Delta G_m$ but ΔG_m may be expressed as $x_A \Delta \mu_A + x_B \Delta \mu_B$. Explain how the results can be so different.

Solution

Here we have considered the transport of both components in the same direction but in different amounts. Previously, we considered diffusion in opposite directions but in equal amounts.

6.9   *Activation energy for a fluctuation*

Sometimes one is interested in the formation of a fluctuation for which the driving force is negative. In such cases one instead talks about the **activation energy**. For the moment, we shall make two assumptions: (i) the fluctuation is only in composition, not in structure; and (ii) the size will not be prescribed. We have already demonstrated that a

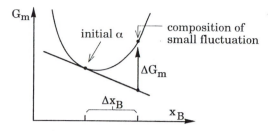

Figure 6.16 Molar Gibbs energy diagram for a fluctuation in composition.

system is not stable against fluctuations in composition if d^2G_m/dx^2 is negative. We shall now consider the case of a positive curvature, Fig. 6.16. The activation energy per mole of atoms in a fluctuation Δx_B is represented by ΔG_m in the diagram. By introducing the curvature of the G_m^α curve we directly obtain an approximate expression if both Henry's and Raoult's laws hold,

$$\Delta G_m \cong \tfrac{1}{2}(\Delta x_B^\alpha)^2 \cdot d^2G_m^\alpha/dx_B^2 \cong \tfrac{1}{2}(\Delta x_B^\alpha)^2 RT/x_A^\alpha x_B^\alpha \cong \tfrac{1}{2}(\Delta x_B^\alpha)^2 RT/x_B^\alpha$$

However, in this case we should examine the validity of the approximation by also carrying out an exact calculation. By comparing with the evaluation of the driving force for the precipitation of a new phase we find without any approximation

$$\Delta G_m = G_m^f - x_A^f G_A^\alpha - x_B^f G_B^\alpha = x_A^f(G_A^f - G_A^\alpha) + x_B^f(G_B^f - G_B^\alpha)$$

where the superscript 'f' denotes the fluctuation. Henry's and Raoult's laws yield

$$\Delta G_m \cong RT[x_A^f \ln(x_A^f/x_A^\alpha) + x_B^f \ln(x_B^f/x_B^\alpha)]$$

For $|x_B^f - x_B^\alpha| \ll x_B \ll 1$ we obtain approximately

$$\Delta G_m = \tfrac{1}{2}RT(x_B^\alpha - x_B^f)^2/x_B^f$$

This is in agreement with the previous result.

Exercise 6.9.1

Consider a binary liquid with 0.1% of B in A at 1273 K. Evaluate the activation energy for the formation of fluctuations with 0.05 and 0.15% of B, respectively. Express the results as joule per mole of atoms in the fluctuations.

Hint

It might be justified to use a dilute solution approximation but not the special approximation for $|x_B^f - x_B^\alpha| \ll x_B^f$.

Solution

(a) $\Delta G_m = RT[0.9995\ln(0.9995/0.9990) + (0.0005\ln(0.0005/0.0010)] = 1.625RT$;

(b) $\Delta G_m = RT[0.9985\ln(0.9985/0.9990) + 0.0015\ln(0.0015/0.0010)] = 1.147RT$.

Notice that the approximate equation would have given:

(a) $\Delta G_m = 0.5RT(0.0010 - 0.0005)^2/0.0005 = 2.646RT$;

(b) $\Delta G_m = 0.5RT(0.0010 - 0.0015)^2/0.0015 = 0.882RT$.

6.10   *Ternary systems*

The property diagram for G_m at constant T and P as function of the molar content in a ternary phase gives three-dimensional diagrams with a surface like a canopy. It can be shown that for a stable phase it is everywhere convex downwards and Fig. 3.11 was drawn in accordance with that fact. In that diagram the tangent plane to an alloy was also drawn, the intersections of which give the partial Gibbs energies in the alloy, i.e. the chemical potentials. We shall now apply such diagrams to various cases of phase equilibria.

Equilibrium between two phases requires that they have the same value for the chemical potential of each component. In a binary system this leads to the common-tangent construction where the intersections with the sides represent the chemical potentials. In a ternary system it leads to a common tangent-plane construction where the intersections with the three edges represent the chemical potentials. With the two Gibbs energy surfaces given, one can allow this tangent plane to roll under them and thus describe a series of possible equilibrium situations, each one represented by a tie-line between the two tangent points in the plane. The result will be a two-phase field, formed by projection on the compositional triangle (see Fig. 6.17 where two tie-lines are projected).

The general equilibrium condition in a ternary system is of course $G_A^\alpha = \mu_A = G_A^\beta$, $G_B^\alpha = \mu_B = G_B^\beta$ and $G_C^\alpha = \mu_C = G_C^\beta$. These three equations leave one degree of freedom for the two-phase equilibrium since each phase can vary its composition by two degrees of freedom. The two-phase region in a ternary phase diagram will thus be an area covered by tie-lines. Each tie-line connects two points, representing the coexisting phases in a possible state of equilibrium. This conclusion still holds even if there is a restriction to the variation in composition of one of the phases but the equilibrium equations will then be modified, as we shall now see.

Let us first consider the equilibrium between a line compound and a solution phase. It can be illustrated with the molar Gibbs energy diagram in Fig. 6.18. It should be noticed that here $a + c = 1$ because the diagram is for one mole of atoms. The construction shows that the equilibrium condition can be derived by considering two of the sides of the triangular tangent plane.

$$G_{A_aC_c}^\theta = a\mu_A + c\mu_C = aG_A^\alpha + cG_C^\alpha$$

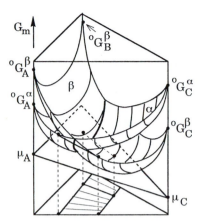

Figure 6.17 Molar Gibbs energy diagram for a two-phase equilibrium in a ternary system. The two-phase field is created by the common tangent-plane rolling under the two surfaces.

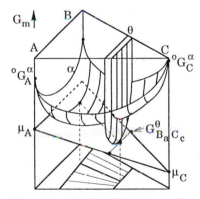

Figure 6.18 Molar Gibbs energy diagram for a ternary system with an ordinary solution phase, α, and a line compound, θ. In this diagram $a + c = 1$ because the whole diagram is for one mole of atoms.

$$G^{\theta}_{B_aC_c} = a\mu_B + c\mu_C = aG^{\alpha}_B + cG^{\alpha}_C$$

These equilibrium conditions leave one degree of freedom because there are two equations and three possible variations in composition, one for the line compound and two for the solution phase. By taking the difference between the equations we find that

$$(G^{\theta}_{A_aC_c} - G^{\theta}_{B_aC_c})/a = G^{\alpha}_A - G^{\alpha}_B$$

Let us next consider the equilibrium between two parallel line compounds θ and ϕ with the formulas $(A,B)_aC_c$ and $(A,B)_bC_d$ (see Fig. 6.19). The previous type of equation applies to each one of these phases although the chemical potentials on the

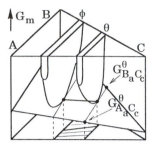

Figure 6.19 Molar Gibbs energy diagram for a ternary system with two line compounds.

right-hand side cannot be referred to any one of the phases but are simply the chemical potentials of the two-phase equilibrium.

$$G^\theta_{A_aC_c} = a\mu_A + c\mu_C$$
$$G^\theta_{B_aC_c} = a\mu_B + c\mu_C$$
$$G^\phi_{A_bC_d} = b\mu_A + d\mu_C$$
$$G^\phi_{B_bC_d} = b\mu_B + d\mu_C$$

where $a + c = b + d = 1$. By eliminating the unknown potentials we find a single equilibrium condition

$$(G^\theta_{A_aC_c} - G^\theta_{B_aC_c})/a = (G^\phi_{A_bC_d} - G^\phi_{B_bC_d})/b$$

It is interesting to see that this relation follows directly from Fig. 6.19 if it is realized that the two straight lines drawn in the tangent plane are parallel and the horizontal lengths of their projections are b and a, respectively.

We have thus found that there will again be one degree of freedom because now there are two possible variations in composition, one for each line compound. If one selects a composition for one phase, the composition of the other one is given by this equation. The result will be similar but mathematically more complicated if the two line compounds are not parallel.

Exercise 6.10.1

Use a geometrical interpretation of the equation for calculating a partial molar quantity in order to prove that the intercepts of a tangent plane in a ternary G_m diagram represent the partial Gibbs energies, as already indicated by Fig. 3.11.

Hint

Use a construction similar to the one shown in Fig. 3.8 but applied to G_C in Fig. 3.11. Choose x_B and x_C as the independent composition variables.

Solution

$$G_C = G_m + \partial G_m/\partial x_C - x_B \partial G_m/\partial x_B - x_C \partial G_m/\partial x_C$$
$$= G_m + (1 - x_C)\partial G_m/\partial x_C - x_B \partial G_m/\partial x_B.$$

Exercise 6.10.2

Consider the equilibrium between a line compound $(A,B)_a C_c$ and a stoichiometric compound $A_l B_m C_n$ in a ternary system. Show how the chemical potential of the element C can be calculated.

Hint

For the stoichiometric compound, ϕ, there is only one relation. For the line compound, θ, there are two. Find a combination of Gs that eliminates μ_A and μ_B.

Solution

$^\circ G_m^\phi = l\mu_A + m\mu_B + n\mu_C;\ \ G_{A_aC_c}^\theta = a\mu_A + c\mu_C;\ \ G_{B_aC_c}^\theta = a\mu_B + c\mu_C.$ Eliminate μ_A and μ_B by taking $a\,{}^\circ G_m^\phi - lG_{A_aC_c}^\theta - mG_{B_aC_c}^\theta$ which is found to be equal to $(an - cl - cm)\mu_C$. We obtain $\mu_C = (a\,{}^\circ G_m^\phi - lG_{A_aC_c}^\theta - mG_{B_aC_c}^\theta)/(an - cl - cm)$.

6.11   *Solubility product*

According to Section 3.3 the Gibbs energy of a phase ϕ is always related to the chemical potentials μ_i by the following relation

$$G_m^\phi = \Sigma x_i^\phi \mu_i$$

where x_i^ϕ represents the composition of the phase. When the phase is a compound, the composition is constant and it is described by the indices in the formula, e.g. l,m,n in $A_l B_m C_n$. For one mole of formula units we have, if $l + m + n = 1$,

$$^\circ G_m^\phi = l\mu_A + m\mu_B + n\mu_C$$

The superscript $^\circ$ is used in order to indicate that the value refers to the compound itself, the 'pure compound', and not to a compound phase, diluted by other components being dissolved in it.

Figure 6.20 illustrates the equilibrium between a compound ϕ and a solution phase, α. There is only one equilibrium condition and it is obtained by inserting the partial Gibbs energies of the solution phase instead of the chemical potentials in the last equation. So,

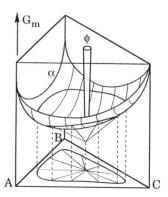

Figure 6.20 Molar Gibbs energy diagram for a ternary system with an ordinary solution phase α, and a ternary, stoichiometric phase, ϕ.

$$°G_m^\phi = lG_A^\alpha + mG_B^\alpha + nG_C^\alpha$$

Let us consider the solubility curve of ϕ in α close to the A corner and introduce activities instead of chemical potentials. The activity a_i is defined through the equation

$$G_i = °G_i + RT\ln a_i$$

where $°G_i$ is the partial Gibbs energy of some state of reference for i. We can thus obtain

$$(°G_m^\phi - l°G_A - m°G_B - n°G_C)/RT = l\cdot\ln a_A + m\cdot\ln a_B + n\cdot\ln a_C$$

Using the standard Gibbs energy of formation of the ϕ phase from the pure components in their states of reference, which is equal to the expression in parentheses, we get

$$\exp(\Delta°G_m^\phi/RT) = (a_A)^l(a_B)^m(a_C)^n$$

In a dilute solution the activity of minor components is approximately proportional to the content expressed, for instance, as the molar content. The activity of the major component is approximately unity and can thus be omitted from the equations. Thus,

$$\exp(\Delta°G_m^\phi/RT)/(f_B)^m(f_C)^n = (x_B)^m(x_C)^n$$

The left-hand side is often denoted by K and is regarded as the solubility product. The solubility curve for a compound phase in a terminal solution may thus be approximated by a hyperbolic curve in a linear phase diagram and with a straight line in a logarithmic phase diagram. As an example, in Fig. 6.21 an isobarothermal section of the phase diagram Fe–Cr–C is presented. The diagram shows the solubilities of three carbides in γ. In the logarithmic diagram the solubility lines are almost straight although the compositions of the carbides are not quite constant.

Exercise 6.11.1

Consider the equilibrium between two ternary stoichiometric phases. Even though

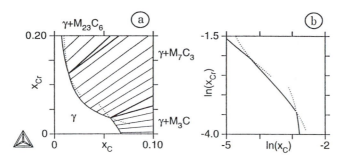

Figure 6.21 Isothermal and isobaric section of the Fe–Cr–C phase diagram near the corner of Fe. The solubility curves for strictly stoichiometric compounds would have been straight lines in the logarithmic diagram (b) and hyperbolic in the linear diagram (a).

the compositions are fixed, there is a degree of freedom from a thermodynamic point of view because there are three chemical potentials. After a value has been chosen for one of them, the other two are fixed. Derive equations for their calculation.

Hint

There are only two equations relating the three potentials, one for each phase. Choose one of the potentials as the independent one and eliminate one of the other two.

Solution

The two conditions are $°G_m^\phi = a\mu_A + b\mu_B + c\mu_C$ and $°G_m^\theta = l\mu_A + m\mu_B + n\mu_C$. A Gibbs energy diagram demonstrates that there is indeed one degree of freedom. We can thus take any value of μ_C, for instance, and then calculate the other two. By eliminating μ_B by multiplying the first equation with m and the other with b and subtracting, we get $\mu_A = [m°G_m^\phi - b°G_m^\theta + (bn - cm)\mu_C]/(am - bl)$ and in the same way $\mu_B = [l°G_m^\phi - a°G_m^\theta + (an - cl)\mu_C]/(bl - am)$.

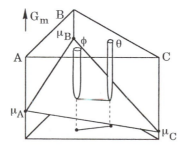

Solution 6.11.1

Exercise 6.11.2

Sketch the whole $\gamma + M_7C_3$ two-phase field in Fig. 6.21 and include a series of tie-lines.

Hint

Tie-lines are straight lines in diagrams with linear scales. When the scales are changed to logarithmic, only those tie-lines remain straight that are horizontal or vertical or have a slope of unity.

Solution

See diagram.

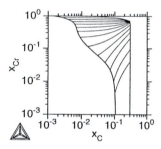

Solution 6.11.2

Phase equilibria and potential phase diagrams

Gibbs' phase rule

We saw in Section 3.1 that the internal energy of a c-component system is a function of $c + 2$ independent, extensive state variables, with the possible addition of internal variables, and the fact is that the equilibrium state of the system is completely determined by the $c + 2$ variables. Consequently, if the state of a system is known, one may calculate the change of internal energy U by specifying the change of these variables, assuming that there is no entropy-producing process inside the system. So, for a reversible change we have

$$dU = TdS - PdV + \Sigma \mu_i dN_i$$

where T, $-P$ and μ_i are potentials. We have also seen that one can instead introduce other independent variables, for instance the potentials T and $-P$,

$$dG = d(U - ST + VP) = -SdT - Vd(-P) + \Sigma \mu_i dN_i$$

The state of the system is still determined by $c + 2$ independent variables. However, when we further introduced all the chemical potentials μ_i as variables in Section 3.3, we obtained a relation between the $c + 2$ variables which did not involve any other state function,

$$0 = d(G - \Sigma N_i \mu_i) = -SdT + VdP - \Sigma N_i d\mu_i$$

Instead, this equation gave a relation between the $c + 2$ potential variables. As mentioned in Section 3.3 it is usually called the Gibbs–Duhem relation. As a consequence, only $c + 1$ of the potentials T, $-P$ and μ_i are independent and anyone of them may be regarded as the dependent potential. In order to define the state of a system completely it is thus necessary to use at least one extensive variable and that is for the purpose of defining the size of the system. It is convenient to use the total content of matter, N, for this purpose or the content of one of the components, N_j. If one is only interested in the properties of a substance, one may disregard the size of the system and regard the state as completely defined by $c + 1$ potentials. In order to represent all the states we then need a diagram with $r = c + 1$ axes, a state diagram according to Section 1.1. We shall call r the **dimensionality** of that diagram. In the following, when we talk about the properties of a system, we shall disregard its size.

If μ_1 is chosen as the dependent potential, then it is convenient to divide by N_1 and thus introduce molar quantities per mole of component 1.

$$\mathrm{d}\mu_1 = -S_{m1}\mathrm{d}T - V_{m1}\mathrm{d}(-P) - \sum_2^c z_i\mathrm{d}\mu_i$$

In this connection it may again be emphasized that one should always specify how the formula unit is defined for molar quantities like S_m and V_m. In Section 3.7 the molar quantities, obtained by dividing by N_1, were identified with the subscript 'm1' and N_i/N_1 was denoted by z_i. For clarity this notation is adopted in the present discussion.

When considering more than one phase in mutual equilibrium, one has a relation of the above type for each phase and every such relation should be obeyed simultaneously if the phases are to stay in equilibrium during the change. Of course, T must have the same value in all the phases and the same holds for all μ_i. Neglecting the effect of surface energy, the same holds for P. On the other hand, the molar quantities have different values in the various phases. We should thus write the Gibbs–Duhem relation for any phase α as follows

$$\mathrm{d}\mu_1 = -S_{m1}^\alpha\mathrm{d}T - V_{m1}^\alpha\mathrm{d}(-P) - \sum_2^c z_i^\alpha\mathrm{d}\mu_i$$

For each new phase, added to the equilibrium, there will thus be one more relation between the potentials and the number of independent variables will decrease by one. This is expressed by Gibbs' phase rule

$$v = c + 2 - p$$

and v is called the **variance** or the **number of degrees of freedom** for the equilibrium with p phases.

The independent variables in Gibbs' phase rule are primarily the potentials because the derivation of the expression for the variance is based upon the Gibbs–Duhem relation which concerns the change of potentials. An extensive quantity must be included in the set of independent variables in order to define the size of the system but that feature is not covered by Gibbs' phase rule and will not be further discussed here. On the other hand, instead of a potential one may alternatively use one of the molar quantities S_{m1}^α, V_{m1}^α, and z_i^α for any phase α, because they are intensive variables and are strictly related to T, $-P$ and μ_i. However, one cannot use the average of a molar quantity of a system, averaged over all the phases present, because then one will also introduce internal variables representing the fractions of various phases. It may again be emphasized that the molar quantities are not potentials like T, $-P$ and μ_i although they are intensive quantities. They will generally have different values in the individual phases.

It should be emphasized that c is the number of independent components. In an alloy system it is usually the number of elements but in a system with molecules it may not be immediately evident how many species should be included in the set of independent components because it is affected by stoichiometric constraints. In a complicated system it may be difficult to identify the number of stoichiometric constraints. We shall return to this problem in Chapter 12.

We may encounter even more complicated cases in systems with molecules of restricted capability to react with each other. In order to describe such cases with Gibbs' phase rule one sometimes includes all molecules or 'chemical substances' and then subtracts a term for the number of 'independent reactions' in order to obtain the number of components. However, the problem remains and is now focused on defining the number of independent reactions. As a consequence, we shall not modify Gibbs' phase rule in this way. By components we shall understand a set of chemical substances necessary and sufficient to define the over-all composition of every phase, taking due notice of all the chemical reactions which can occur and also of all stoichiometric constraints.

In the remainder of the present chapter we shall discuss the consequences of Gibbs' phase rule for a kind of diagram which will be introduced soon, the so-called phase diagrams.

Exercise 7.1.1

Consider as a system the content of a vessel. In the vessel one has enclosed a certain amount of water. Then one varies T and P and studies what happens to V in an attempt to decide whether the system behaves as a unary system. Due to its larger volume, it is easy to see when a gas phase forms. Discuss what one would expect to happen.

Hint

The discussion should be based upon Gibbs' phase rule. Remember that it was derived by first considering each phase separately. It is thus necessary to define a set of independent components which is capable of giving each phase its composition. The problem is trivial if one does not need to consider the dissociation of H_2O into H_2 and O_2, because all phases then have the same composition. There is only one independent component, H_2O. In order to make this exercise more interesting, suppose instead that an efficient catalyst for the dissociation is present.

Solution

The gas phase will contain H_2O, H_2 and O_2. If the dissolution of H_2 and O_2 in liquid H_2O is negligible, then the gas phase must still have the stoichiometric composition H_2O (unless some H_2 or O_2 leaks out of the vessel) and the system behaves as a unary system. If the solubilities cannot be neglected and if the solubility of H_2 is not exactly twice that of O_2, then the system behaves as a binary system. For the water + vapour equilibrium we get $v = c + 2 - p = 2 + 2 - 2 = 2$ and for each given T both phases can exist over a range of P values (but of course a very small range).

Exercise 7.1.1

Consider as a system the content of a vessel. In the vessel one has enclosed a certain amount of solid $CaCO_3$. On heating there is a dissociation into a new solid phase CaO and gaseous CO_2. Discuss whether this system behaves as a unary, binary or ternary system.

Hint

Examine how many independent components are required in order to form the compositions of all the phases.

Solution

It is most natural to select CaO and CO_2 as the independent components. $CaCO_3$ can form by combining them. Gibbs' phase rule thus yields for the three-phase equilibrium $v = c + 2 - p = 2 + 2 - 3 = 1$. We may conclude that at any given P all three phases may coexist only at a single T. Thus, on heating under atmospheric pressure $CaCO_3$ will suddenly dissociate into solid CaO and gaseous CO_2 when a certain T is reached.

7.2   *Fundamental property diagram*

Let us first discuss a T,P diagram for a substance with one component, A, and one phase, α. According to Gibbs' phase rule the state is completely determined by giving the values of T and P, i.e. by giving a point in the T,P diagram. In this sense we may thus regard the T,P diagram as a state diagram according to Section 1.1. The value of μ_A for a particular substance can be calculated and plotted as a surface above the T,P state diagram, yielding a three-dimensional diagram, see Fig. 7.1. This type of diagram we may regard as a property diagram for the particular substance under consideration. Actually, this diagram can be looked at from any direction and any one of T, P and μ_A may be regarded as the dependent variable. The state may be defined by a point on any side of the property diagram. As a state diagram one may thus use a diagram formed by any two of the potentials.

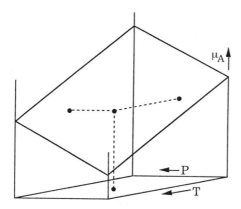

Figure 7.1 Fundamental property diagram for a unary system with one phase. Any one of the three potentials can be chosen as the dependent variable (property). The potential P has been plotted in the negative direction because $-P$ appears naturally in thermodynamic equations.

This kind of property diagram is of special interest because it is composed of a complete set of potentials. We shall call it the **fundamental property diagram** and it has the axes T, P and one μ_i for each component. In a unary system μ_A is identical to G_m and the surface in the diagram thus represents a fundamental equation, $G_m(T,P)$.

For a higher-order system it represents a fundamental equation $\mu_1(T,P,\mu_2,\mu_3,\ldots)$ which is of a type we have not defined before. In principle, we could calculate a point on the surface from any one of the fundamental equations, if it is available. One can then follow the surface by applying the Gibbs–Duhem relation. For a unary system we get

$$S dT - V dP + N_A d\mu_A = 0$$

The direction of the surface is given by the relations

$$\left(\frac{\partial \mu_A}{\partial(-P)}\right)_T = -\frac{V}{N_A} < 0; \quad \left(\frac{\partial \mu_A}{\partial T}\right)_P = -\frac{S}{N_A}; \quad \left(\frac{\partial(-P)}{\partial T}\right)_{\mu_A} = -\frac{S}{V}$$

As many times before, we take $-P$ as a potential rather than $+P$. The numerical values of the last two ratios depend on what reference we choose for the entropy. If we accept the common choice of $S = 0$ at $T = 0\,\mathrm{K}$, then all the ratios are positive at $T > 0$ and all the derivatives are negative. Figure 7.1 was constructed accordingly. Similar expressions can be derived for a system with several components and we can summarize all the expressions in a general form

$$\left(\frac{\partial Y^a}{\partial Y^b}\right)_{Y^c} = -\frac{X^b}{X^a} < 0$$

where Y^c represents all the potentials except Y^a and Y^b.

Since a point on any side of the T,P,μ_A diagram defines the state, we can use the third axis for the representation of some other property. We may, for instance, represent the refractive index r as a function of T and P but that would not be a fundamental property diagram. However, knowing a point on the surface, we may follow the surface by applying an equation similar to the Gibbs–Duhem relation.

$$\mathrm{d}r = \left(\frac{\partial r}{\partial T}\right)_P \mathrm{d}T + \left(\frac{\partial r}{\partial P}\right)_T \mathrm{d}P$$

The surface in Fig. 7.1 was given as a plane for the sake of simplicity. That would require that V/N_A and S/N_A are constant for the α phase, independent of T and P. In reality, they are not constant and the surface would be curved. We shall now examine in what direction it will be curved. The fundamental property diagram is independent of the size of the system since only potentials are concerned. However, we have the right to consider a system of a constant size and to define that size by any extensive variable. If we take X^a as that variable, we obtain for the curvature in a section of constant Y^c

$$\left(\frac{\partial Y^a}{\partial Y^b}\right)_{Y^c} = -\frac{X^b}{X^a}$$

$$\left(\frac{\partial^2 Y^a}{\partial(Y^b)^2}\right)_{Y^c} = \left(\frac{\partial(-X^b/X^a)}{\partial Y^b}\right)_{Y^c} = \left(\frac{\partial(-X^b/X^a)}{\partial Y^b}\right)_{Y^c,X^a}$$

$$= -\frac{1}{X^a}\left(\frac{\partial X^b}{\partial Y^b}\right)_{Y^c,X^a} = -\frac{1}{X^a}\bigg/\left(\frac{\partial Y^b}{\partial X^b}\right)_{Y^c,X^a} < 0$$

in view of the stability conditions presented in Section 5.4. The result is illustrated in Fig. 7.2 for an element A with the choice of $\mu_A > 0$ at $T = 0\,\mathrm{K}$ and $P = 0$. The surface looks like part of a dome and is everywhere convex, as seen from the origin. A different choice of reference for μ_A will simply displace the whole surface vertically.

Let us return to the simple picture in Fig. 7.1. Suppose that we make a similar diagram for the same substance in another possible structure (phase), β, and plot the two surfaces in the same coordinate frame. We can then compare the two phases at the same T and P, for instance, and evaluate the difference in μ_A, see Fig. 7.3.

Let us consider a possible transition from phase β to phase α at the fixed values of T and P. We cannot evaluate the driving force for that transition without knowing the detailed mechanism, i.e. the reaction path. However, we can evaluate the integrated driving force for the transition, $\int D\mathrm{d}\xi$. We should then use the form of the combined law having $T, -P$ and μ_A as the variables:

$$0 = -S\mathrm{d}T - V\mathrm{d}(-P) - N_A\mathrm{d}\mu_A - D\mathrm{d}\xi$$

T and P must be regarded as independent variables if they are kept constant. The third potential, μ_A, must then be regarded as a dependent variable. In addition, we may choose one of the extensive variables as independent in order to define the size of the

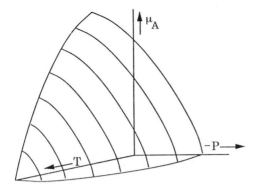

Figure 7.2 Fundamental property diagram for a unary phase. The surface is everywhere convex. The property surface has here been extended to negative pressures which is not unrealistic for solid substances.

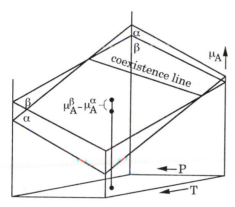

Figure 7.3 Fundamental property diagram for a unary system with two phases. One may regard $\mu_A^\beta - \mu_A^\alpha$ as the integrated driving force for transformation from β to α at given values of T and P.

system and it must come from the conjugate pair which has not yet been used. It must thus be N_A. For a system at constant T, $-P$ and N_A we obtain

$$\int D\mathrm{d}\xi = -\int N_i \mathrm{d}\mu_i = -N_A(\mu_A^\alpha - \mu_A^\beta) = N_A(\mu_A^\beta - \mu_A^\alpha)$$

It is evident that the phase with the lower μ_A value will be the more stable phase. At the combination of T and P, marked in Fig. 7.3, α is thus the more stable phase. Furthermore, the line of intersection of the two surfaces must be a line of coexistence because on that line there is no driving force for a change. This line is shown in Fig. 7.3. In the figure the α phase is stable in front of the coexistence line and the β phase behind it. It is evident that the coexistence line represents a ridge on the composite surface representing the stable states. We may generalize this observation and conclude that the

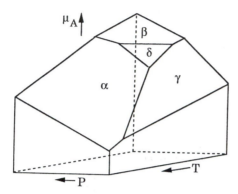

Figure 7.4 Fundamental property diagram for a substance having four different structures (phases). The two-phase lines are all ridges. There are no re-entrant angles.

surface representing stable states in a system with several phases is composed of pieces, one for each stable phase, and joined by coexistence lines which are all ridges. An example with four phases is shown in Fig. 7.4. If we combine this with the previous conclusion that the surface for each single phase is convex, we may conclude that the composite surface is also convex.

Exercise 7.2.1

Show the derivation of an expression for the curvature $(\partial^2 T/\partial P^2)_{\mu_A}$ and find its sign.

Hint

For a stable phase $(\partial P/\partial V) < 0$, whether T or S and N_A or μ_A are kept constant.

Solution

$$SdT - VdP + N_A d\mu_A = 0; \quad (\partial T/\partial P)_{\mu_A} = V/S.$$

$$(\partial^2 T/\partial P^2)_{\mu_A} = [\partial(V/S)/\partial P]_{\mu_A} = [\partial(V/S)/\partial P]_{\mu_A,S} = (1/S)(\partial V/\partial P)_{\mu_A,S} < 0.$$

Exercise 7.2.2

The following diagram is a typical property diagram for a substance with three phases, solid(S), liquid(L) and gas. It has a critical point where the difference between liquid and gas vanishes. The composite shape is not everywhere convex. Should it not be?

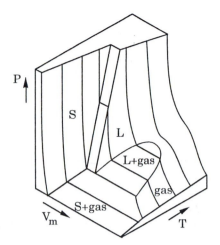

Exercise 7.2.2

Solution

It is a property diagram but it is not a fundamental property diagram which should have only potential axes. V_m is an intensive quantity but not a potential.

Exercise 7.2.3

Using the criterion that the more stable phase in a unary system under constant T and P has the lower chemical potential, it is possible to obtain a so-called phase diagram from Fig. 7.3. (This will be demonstrated in Fig. 7.5.) Suppose that one would instead like to choose μ_A and T as the independent variables and construct a phase diagram with these axes. What criterion could then be used for deciding where each phase is stable?

Hint

The answer can be found by again considering the same form of the combined law. What extensive variable should be regarded as independent when μ_A and T are chosen as independent potentials?

Solution

In this case, the conjugate pairs are (T,S), $(-P,V)$ and (μ_A, N_A). We must choose V as the independent extensive variable. By keeping V constant together with T and μ_A we obtain $\int D d\xi = \int V dP = V(P^\alpha - P^\beta) > 0$ if α is the more stable phase. It is evident that the phase with the highest P will be the more stable phase. Recognizing that it is $- P$ that can be regarded as a potential, one should write the result as $\int Dd\xi = V(- P^\beta + P^\alpha) > 0$ in conformity with the previous result $\int Dd\xi = N_A(\mu_A^\beta - \mu_A^\alpha) > 0$.

Topology of potential phase diagrams

A coexistence line in the fundamental property diagram can be projected onto any side of the diagram, for instance the T,P side (see Fig. 7.5). In that T,P diagram (Fig. 7.5(b)) we may indicate on which side of the line each phase is stable, i.e. has a lower μ_A value than the other phase. We may further indicate that the coexistence line represents the $\alpha + \beta$ equilibrium. Such a diagram is called a phase diagram and it is actually a state diagram used for plotting coexistence lines. In this chapter we shall mainly be concerned with phase diagrams. In order to emphasize the character of the axis variables we may call the present diagram a **potential phase diagram**. It is worth remembering that it is actually a projection of the fundamental property diagram. When $T, - P,\mu_A$ is used as the complete set of potentials, one usually projects in the direction of μ_A and presents the T,P phase diagram. However, it should be remembered that in Section 3.4 it was shown that there are at least nine ways of writing the Gibbs–Duhem relation and there are thus at least nine sets of potentials that can be used in the construction of potential phase diagrams.

   Knowing one point on the coexistence line in the fundamental property diagram we can determine the line by applying the Gibbs–Duhem relation to both phases using the fact that dT, dP and $d\mu_A$ must be the same in both phases:

$$d\mu_A = - S_m^\alpha dT + V_m^\alpha dP$$
$$d\mu_A = - S_m^\beta dT + V_m^\beta dP$$

This system of equations defines the direction of the $\alpha + \beta$ coexistence line in the fundamental property diagram. The direction of the projected line in the T,P phase diagram, i.e. the $\alpha + \beta$ phase field, is obtained by eliminating $d\mu_A$ from the Gibbs–Duhem relations

$$\frac{dP}{dT} = \frac{S_m^\alpha - S_m^\beta}{V_m^\alpha - V_m^\beta}$$

As an example Fig. 7.6 shows the equilibrium between graphite and diamond in a T,P phase diagram for carbon. Except for low temperatures the equilibrium line is almost a straight line because the differences in S_m and V_m stay rather constant. At low temperature the line becomes parallel to the T axis because the difference in S_m goes to zero at absolute zero in agreement with the third law of thermodynamics.

   Using the alternative form of the Gibbs–Duhem relation, obtained from line 5 in Table 3.1 (Section 3.4), we may introduce $(H_m^\alpha - H_m^\beta)/T$ instead of $(S_m^\alpha - S_m^\beta)$:

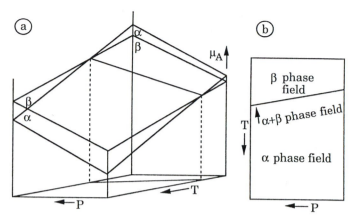

Figure 7.5 Projection of the fundamental property diagram onto the T,P state diagram, yielding a potential phase diagram.

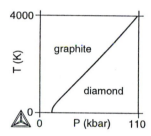

Figure 7.6 The T,P phase diagram for carbon, according to a thermodynamic assessment.

$$\frac{d(-P)}{d(-1/T)} = -\frac{(H^\alpha - H^\beta)}{(V^\alpha - V^\beta)/T}$$

$$\frac{dP}{dT} = \frac{(H_m^\alpha - H_m^\beta)/T}{V_m^\alpha - V_m^\beta}$$

This is known as Clapeyron's relation. It should be realized that the molar volumes of condensed phases are so small that pressures of about 1 bar have an effect on the equilibrium temperature which is negligible for many purposes.

Suppose there is a third possible phase. We shall then have a third surface in the property diagram. There will be three coexistence lines and one point of intersection, a triple point, and by projection they will all show up on the phase diagram (see Fig. 7.7).

It is immediately evident that all the angles between the three intersecting lines in the phase diagram are less than 180°. We have thus found the 180° rule which says that the corners of a one-phase field must have angles less than 180°. The dashed lines in Fig. 7.7 represent metastable extrapolations of the two-phase coexistence lines and they

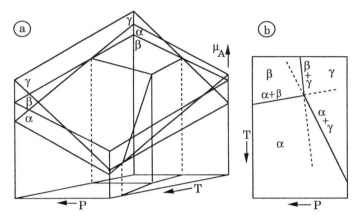

Figure 7.7 Projection of a property diagram onto the T,P state diagram, yielding a phase diagram with three univariant and one invarian phase equilibria. Metastable equilibria are shown with dashed lines.

fall inside the one-phase field of the third phase. The geometrical elements of the potential phase diagram are called **phase fields** and they are:

(a) Points where three phases are in equilibrium. We cannot change any variable without changing the kind of equilibrium. We call this an invariant equilibrium or a zero-dimensional phase field.

(b) Lines where two phases are in equilibrium. We can change only one variable independently without leaving the line. We call this a univariant equilibrium or a one-dimensional phase field.

(c) Surfaces where a single phase exists. We can change two independent variables without leaving this kind of phase field. We call this a divariant equilibrium or a two-dimensional phase field.

The dimensionality of a phase field in the potential phase diagram is thus equal to the variance of the corresponding phase equilibrium. We shall denote the dimensionality by d and can calculate it from Gibbs' phase rule. With one component it yields

$$d = v = c + 2 - p = 3 - p$$

A three-phase equilibrium thus has a variance of 0 and appears as a point ($d = 0$). A single phase has a variance of 2 and it thus requires a surface ($d = 2$) to be represented.

More phases can be added but there will be no new kind of geometrical element. We should never expect more than three surfaces to meet in a point in the property diagram. As an example of a more complex phase diagram, Fig. 7.8 reproduces the Fe phase diagram.

Most of the lines are here fairly straight similar to the curve in Fig. 7.6. An exception is the two branches of the bcc + fcc line and together they look like a

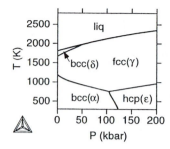

Figure 7.8 The T,P phase diagram of Fe according to an assessment of experimental information.

parabola. Clapeyron's relation shows that the heat of transformation has different signs for the two branches and must go through zero at some intermediate temperature. That point would be found at a negative pressure. The reason is a magnetic transition in the bcc phase.

Exercise 7.3.1

Derive an equation for the $\alpha + \beta$ line in a unary T,P phase diagram under the conditions that ΔH_m and ΔV_m can be regarded as constants.

Hint

Start with Clapeyron's relation.

Solution

$dP = (\Delta H_m / \Delta V_m)(dT/T)$ and $P - P_o = (\Delta H_m / \Delta V_m)\ln(T/T_o)$ under constant ΔH_m and ΔV_m. In addition, a point on the line, T_o, P_o, must be known. It should be noticed that it is sometimes more convenient to approximate ΔS_m as constant than ΔH_m. The result is then a straight line in a linear T,P phase diagram. When one of the phases is a gas, one may approximate ΔV_m by RT/P and integration yields, if ΔH_m is constant, $\ln P = K\exp(-\Delta H_m/T)$.

Exercise 7.3.2

Pure Fe at 1 bar transforms from γ to δ at 1667 K and from δ to L at 1811 K. It has been reported that in the T,P phase diagram the corresponding one-dimensional phase fields approach each other and intersect at $T = 1991$ K and $P = 5.2$ GPa.

The following table was presented, giving information on these two phase fields at the intersection. The intersection is, of course, a triple point and the third phase field is $L + \gamma$. Calculate its slope.

Reaction	$\Delta V (cm^3/mol)$	$dT/dP (K/GPa)$
$\gamma \rightarrow \delta$	0.074	62
$\delta \rightarrow L$	0.278	35

Hint

At a triple point $\Sigma \Delta V = 0$ and $\Sigma \Delta H = 0$. Insert ΔH from $dP/dT = \Delta H/T\Delta V$ in the second relation. Consider 1 mole.

Solution

$V^L - V^\gamma = V^L - V^\delta + V^\delta - V^\gamma = 0.278 + 0.074 = 0.352;$ $H^L - H^\gamma = H^L - H^\delta + H^\delta - H^\gamma$ gives $T(V^L - V^\gamma)(dP/dT)^{L+\gamma} = T(V^L - V^\delta)(dP/dT)^{L+\delta} + T(V^\delta - V^\gamma)(dP/dT)^{\delta+\gamma}$ and $(dT/dP)^{L+\gamma} = 0.352/(0.278/35 + 0.074/62) = 38.5.$

Exercise 7.3.3

A T,P phase diagram for a unary system (pure A) is given. It shows four phases. Construct a reasonable T,μ_A property diagram at P_1. It should show all the stable and metastable two-phase equilibria at P_1

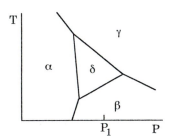

Exercise 7.3.3

Hint

The T values for all the two-phase equilibria at P_1 are easily found by extrapolation. Approximate all the T,μ_A lines by straight lines, intersecting at the three-phase equilibria.

Solution

See diagram.

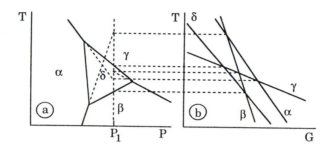

Solution 7.3.3

Exercise 7.3.4

When studying a pure liquid substance one finds that it solidifies to β on slow cooling but to α on rapid cooling. Sketch T,P phase diagrams for two cases: (a) where α is a stable low-temperature phase; and (b) where it is not stable at any temperature.

Hint

In order to compare the stabilities of α and β at low temperature it may be convenient to compare their vapour pressures. The gas phase should thus be included in the diagrams. The invariant melting points of β and α are found on the same L/gas phase boundary.

Solution

See diagram.

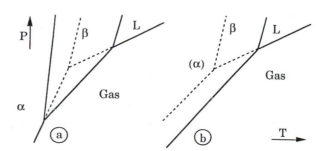

Solution 7.3.4

Exercise 7.3.5

Consider a unary T,P phase diagram with a triple point $\alpha + \beta + \lambda$. For the three lines meeting there, we have $(dP/dT)^{\alpha+\beta} = (H_m^\alpha - H_m^\beta)/[T(V_m^\alpha - V_m^\beta)]$, etc. Then $H_m^\alpha - H_m^\beta + H_m^\beta - H_m^\gamma + H_m^\gamma - H_m^\alpha = 0$ gives $(V_m^\alpha - V_m^\beta)(dP/dT)^{\alpha+\beta} + (V_m^\beta - V_m^\gamma)(dP/dT)^{\beta+\gamma} + (V_m^\gamma - V_m^\alpha)(dP/dT)^{\gamma+\alpha} = 0$. It is thus possible to calculate the slope of the third phase boundary from the other two if all the molar volumes are known. Suppose one only knows that $V_m^\beta - V_m^\gamma = 0$. From the above equation it would then seem that $(dP/dT)^{\alpha+\beta} = (dP/dT)^{\gamma+\alpha}$ and $\alpha + \beta$ and $\gamma + \alpha$ should form a common line. Examine this conclusion by drawing a schematic phase diagram.

Hint

With the given condition we know the slope of the $\beta + \gamma$ phase boundary.

Solution

To make $\alpha + \beta$ and $\gamma + \alpha$ coincide would be to make the angle of the α phase 180° in violation of the 180° rule. The mistake was to omit a term which has an indeterminate value $0 \cdot \infty$, since the $\beta + \gamma$ line is vertical if $V_m^\beta - V_m^\gamma = 0$. The diagram shows a reasonable construction.

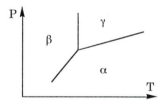

Solution 7.3.5

7.4   *Potential phase diagrams in binary and multinary systems*

So far we have discussed a system with one component, a unary system. In a binary system we have two components and four potentials, $T, -P, \mu_A$ and μ_B. The fundamental property diagram will be four-dimensional and cannot be visualized. The phase diagram will be three-dimensional and it will be composed of four geometrical elements as illustrated in Fig. 7.9. They are all phase fields:

(a) Points where four phases are in equilibrium. We cannot change any variable without changing the kind of equilibrium.

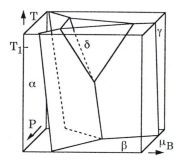

Figure 7.9 The T,P,μ_B phase diagram for a binary system with four phases.

(b) Lines where three phases are in equilibrium. We can change only one variable independently without leaving the line.

(c) Surfaces where two phases are in equilibrium. We can change two independent variables without leaving this phase field.

(d) Volumes where a single phase exists. We can change three independent variables without leaving this kind of phase field. Its equilibrium is trivariant.

For higher-order systems, ternary, quaternary, quinary, etc., the principles will be the same. The phase diagram will have $c + 1$ axes, where c is the number of components. The geometrical elements will be points, lines, surfaces, volumes, hyper-volumes, etc., and they will represent phase equilibria which have a variance of zero, one, two, three, four, etc., in accordance with Gibbs' phase rule.

Suppose one wants to calculate a state of equilibrium under the requirement that it must consist of p specified phases. Then one must, in addition, specify the values of v independent variables, where v is given by Gibbs' phase rule, $v = c + 2 - p$. On the other hand, suppose one wants to calculate a state of equilibrium without specifying any phase. Then one must specify the values of v independent variables, where v is equal to $c + 1$ because the phase diagram will have $c + 1$ axes. That corresponds to the case of one specified phase. This does not violate Gibbs' phase rule because one will always fall inside a one-phase field, $p = 1$. One will never be able to hit exactly on the other types of geometrical elements.

Figure 7.3 illustrated the integrated driving force for a transition from β to α. The same situation cannot be illustrated for a higher-order system but the integrated driving force can be derived in the same way under conditions where T,P and all the chemical potentials except for μ_1 are kept constant. The combined law yields

$$N_1 d\mu_1 = -SdT + VdP - \sum_2^c N_k d\mu_k - Dd\xi = -Dd\xi$$

$$\int Dd\xi = -N_1(\mu_1^\alpha - \mu_1^\beta) = N_1(\mu_1^\beta - \mu_1^\alpha)$$

It is thus necessary that μ_1 is lowest in the stable phase if all the other potentials are kept constant.

In the above integration it was assumed that N_1 is kept constant which was the way to define the size of the system. However, it must be noted that the content of all the other components will most probably change during a transition carried out under the conditions considered here. It may be of more practical interest to derive the integrated driving force for a transition under constant T, P and composition. It can be obtained from the combined law expressed in terms of G

$$dG = -SdT + VdP + \Sigma\mu_i dN_i - Dd\xi = -Dd\xi$$

$$\int Dd\xi = G^\beta - G^\alpha$$

Exercise 7.4.1

Try to formulate the equivalence of the 180° rule for a point where four phases coexist in a binary three-dimensional phase diagram.

Solution

The phase field for each single phase must have a pointed tip. Also, in any projection the four three-phase lines cannot form any angle equal to or larger than 180°.

Exercise 7.4.2

Suppose we want to calculate the equilibrium between three particular phases in a ternary system. For how many variables should we then specify values in order to get a unique answer.

Hint

With those variables we must specify a point in the v-dimensional space.

Solution

$v = c + 2 - p = 3 + 2 - 3 = 2$, e.g. T and P.

Sections of potential phase diagrams

In order to visualize a higher-order potential phase diagram one may decrease the number of dimensions by making a **section** at a constant value of some potential, an equipotential section. It will show exactly the same geometrical elements as a potential phase diagram for a system with one component less. One may section several times and thus decrease the dimensions of a higher-order phase diagram until it can be plotted as a two-dimensional diagram. It is common first to keep P constant and then T. One may then continue and keep the chemical potential of some component constant.

At each sectioning one will lose the geometrical element of the lowest dimensionality. This is demonstrated in Fig. 7.10 which was obtained by taking a horizontal $(T = T_1)$ section through the potential phase diagram in Fig. 7.9. The chance of hitting the four-phase point is negligible and no four-phase point should be included in this type of diagram. The topology of a diagram will thus be the same whether the number of dimensions is decreased by sectioning at a constant value of a potential or by reducing the number of components by one. In order to distinguish the two cases, one may call the diagram with axes for all the independent potentials a **complete potential phase diagram**. It has $c + 1$ axes.

The dimensionality of a phase field with p phases in a section of a potential phase diagram may be derived from Gibbs' phase rule,

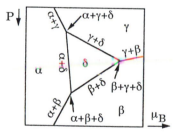

Figure 7.10 Equipotential (isothermal) section of the potential phase diagram in Fig. 7.9 at $T = T_1$.

$$d = c + 2 - p - n_s$$

where n_s is the number of sectionings. In order to avoid confusion with the variance of a phase equilibrium, which is given by Gibbs' phase rule and is independent of what kind of diagram is used, this will be called the **phase field rule**. The number of axes in the diagram, r, will also decrease by sectioning, $r = c + 1 - n_s$, and we can thus write the phase field rule as

$$d = r + 1 - p$$

Phase fields for which $d < 0$ will normally not show up in the final diagram, as demonstrated by the negligible chance of hitting the four-phase point in the above case.

It is evident from the second form of the phase field rule that a diagram with r axes has the same topology independent of how many sectionings have been used to obtain it. By inspecting a diagram without knowing the number of components, it is thus impossible to tell if it is a section or not.

Exercise 7.5.1

Consider the equilibrium Fe + S(gas) $\leftrightarrow$ FeS under a constant pressure. Can it exist in a range of T?

Hint

Fe and FeS are two different solid phases.

Solution

We have two components, Fe and S, i.e. $c = 2$, and three phases, Fe, gas and FeS, i.e. $p = 3$. If we section at some pressure, we have $n_s = 1$. Thus $d = c + 2 - p - n_s = 2 + 2 - 3 - 1 = 0$. Under these conditions the equilibrium can exist only at a particular T.

7.6   *Binary systems*

As an example of a sectioned phase diagram, Fig. 7.11 shows the bcc-W and WC phases in the W–C phase diagram at 1 bar. Two different sets of axes are used. Since a chemical potential has no natural zero point, a reference must be chosen. In this case graphite at 1 bar and the actual temperature was chosen for carbon.

It is interesting to note that the univariant two-phase field approximates very well to a straight line in Fig. 7.11(b). Its slope is obtained from the Gibbs–Duhem relation for constant P, applied to each one of the phases. In order to calculate the slope of the line in Fig. 7.11(b) we first apply the Gibbs–Duhem relation in its alternative form to each one of the phases. By choosing W as component 1 we get

$$d(\mu_W/T) = H_{m1}^W d(1/T) - z_C^W d(\mu_C/T)$$
$$d(\mu_W/T) = H_{m1}^{WC} d(1/T) - z_C^{WC} d(\mu_C/T)$$

On the line of coexistence the change of each one of the potentials must be the same in both phases. We may thus eliminate $d(\mu_W/T)$ by subtracting one equation from the other, to obtain

$$\frac{\partial(\mu_C/T)}{\partial(1/T)} = \frac{H_{m1}^{WC} - H_{m1}^W}{z_C^{WC} - z_C^W}$$

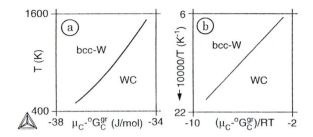

Figure 7.11 Isobaric section at 1 bar of the W–C phase diagram with two potential axes, drawn in two alternative ways. $1/T$ has been plotted in the negative direction because $-1/T$ appears naturally as a potential in thermodynamic equations.

Here, $z_C^W = 0$ and $z_C^{WC} = 1$. Since the solubility of carbon in bcc-W is very low, we can approximate H_{m1}^W with the enthalpy of pure bcc-W, ${}^\circ H_W^{bcc}$, to give

$$\frac{\partial(\mu_C/T)}{\partial(1/T)} = H_{m1}^{WC} - {}^\circ H_W^{bcc}$$

However, in order to define a numerical value for the right-hand side, it is necessary to choose a state of reference for carbon. By introducing graphite as the state of reference for carbon, we obtain

$$\frac{\partial([\mu_C - {}^\circ G_C^{gr}]/T)}{\partial(1/T)} = H_{m1}^{WC} - {}^\circ H_W^{bcc} - {}^\circ H_C^{gr}$$

because $\partial({}^\circ G_C^{gr}/T)/\partial(1/T) = {}^\circ H_C^{gr}$. The right-hand side is the heat of formation of one mole of WC units from the pure elements, a quantity we may denote by $\Delta_f^\circ H_{WC}$. The fact that the curve in Fig. 7.11(b) is almost straight, indicates that the heat of formation is approximately constant. By definition $\mu_C - {}^\circ G_C^{gr}$ is equal to $RT\ln a_C$ where a_C is the carbon activity, referred to graphite. We may thus write the slope as

$$\frac{R\partial\ln a_C}{\partial(1/T)} = \Delta_f^\circ H_{WC}$$

and we could have plotted $R\ln a_C$ as the ordinate axis and still have the almost straight line.

In Fig. 7.11(a) the potentials T and $\mu_C - {}^\circ G_C^{gr}$ have been used on the axes. The slope can now be derived from the usual form of the Gibbs–Duhem relation. It is equal to $-S_{m1}^{WC} + {}^\circ S_W^{bcc} + {}^\circ S_C^{gr}$ and the ordinate axis could have been interpreted as $RT\ln a_C$. From the fact that the slope is reasonably constant we may conclude that the entropy of formation of WC is approximately constant, but not as constant as the heat of formation.

The situation will be more complicated if one or both of the phases can vary in composition. As an example, a complete Fe–C phase diagram at a constant pressure is presented in Fig. 7.12, using the axes $1/T$ and $(\mu_C - {}^\circ G_C^{gr})/T$. The strong curvatures are

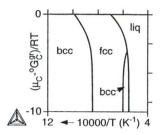

Figure 7.12 The Fe–C phase diagram at 1 bar, plotted with two potential axes.

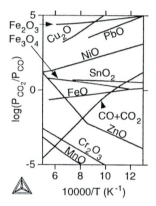

Figure 7.13 Combination of isobaric phase diagrams for many M–O systems at 1 bar. The oxygen potential is represented by P_{CO_2}/P_{CO} in a hypothetical gas which is not present, except for the line $CO + CO_2(g)$.

caused by the strong variation in composition of the fcc phase and liquid. All the lines turn vertical at low values of μ_C. That is where the C content goes to zero and all phases become pure Fe. The difference in composition thus goes to zero.

In the case of reactions involving oxygen it is natural to use an O_2 gas of 1 bar as reference. However, we may also express the oxygen potential by the ratio of the partial pressures of CO_2 and CO in an ideal gas and use as a reference a gas where these partial pressures are equal. Figure 7.13 gives an example of such a diagram with information from a large number of M–O systems. An oxide is stable above each line. Below the line the stable state is either the pure metal or a lower oxide. The diagram is calculated for 1 bar and the state for pure Zn above the boiling point is thus Zn gas of 1 bar because the O_2 pressure is low enough to be neglected. This diagram is often called the Ellingham diagram. It should be emphasized that the effect of pressure is so small that this diagram could be used for any pressure down to zero and up to many bars, except for (i) the line $CO + CO_2$ which holds only for $P_{CO} + P_{CO_2} = 1$ bar and (ii) the line for gaseous Zn.

Exercise 7.6.1

Find an explanation why the line in Fig. 7.11(a) is less curved than the line in Fig. 7.11(b).

Hint

Consider the values at low temperatures of the quantities represented by the slopes.

Solution

The slopes of the two lines represent $\Delta_f^\circ S_{WC}$ and $\Delta_f^\circ H_{WC}$, respectively. The first one starts from zero at $0\,\mathrm{K}$ but the second one starts from a considerable value. Any change in slope will thus be much more visible in the first one.

Exercise 7.6.2

From the information given in Fig. 7.13 construct an Fe–O potential phase diagram at a constant pressure of 1 bar.

Hint

It is not necessary to change the axes. The liquid phase cannot be included due to lack of information.

Solution

See diagram.

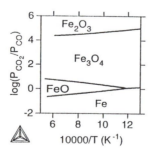

Solution 7.6.2

Exercise 7.6.3

Consider a system with graphite in a vessel under a pressure of 1 bar and a temperature of $1000\,^\circ\mathrm{C}$. The vessel can expand and accommodate a gas. What would be the partial pressures in the gas if a small amount of oxygen is introduced?

Hint

In this case the ordinate axis in Fig. 7.13 expresses not only the oxygen potential but also gives the actual value of P_{CO_2}/P_{CO}.

Solution

The system would place itself on the $CO + CO_2$ line and from Fig. 7.13 we read for $1000\,°C$: $\log(P_{CO_2}/P_{CO}) = -2$ which together with $P_{CO_2} + P_{CO} = 1$ bar yields $P_{CO_2} = 0.01$ bar and $P_{CO} = 0.99$ bar.

7.7   *Ternary systems*

For a ternary system one may obtain a two-dimensional phase diagram by sectioning at constant T and P. Figure 7.14 shows such a diagram for the Ti–O–Cl system and the axes represent μ_O/RT and μ_{Cl}/RT, expressed by the logarithm of the partial pressures of O_2 and Cl_2 in an imagined ideal gas that would be in equilibrium with the system.

Again we find that the univariant phase equilibria are represented by lines which look straight, a fact that can again be illustrated by application of the Gibbs–Duhem relation. For constant T and P we get, from the alternative form,

$$d(\mu_{Ti}/T) = -z_O^\alpha d(\mu_O/T) - z_{Cl}^\alpha d(\mu_{Cl}/T)$$
$$d(\mu_{Ti}/T) = -z_O^\beta d(\mu_O/T) - z_{Cl}^\beta d(\mu_{Cl}/T)$$

from which

$$\frac{\partial \mu_{Cl}}{\partial \mu_O} = \frac{\partial(\mu_{Cl}/T)}{\partial(\mu_O/T)} = -\frac{z_O^\alpha - z_O^\beta}{z_{Cl}^\alpha - z_{Cl}^\beta}$$

Here all z_i quantities are defined as N_i/N_{Ti}. It is interesting to note that the slope can be calculated directly from the compositions involved.

A sectioned potential diagram like Fig. 7.14 is sometimes called a Kellogg diagram. It must be emphasized that here the gas phase is not considered in the phase equilibria. The partial pressure is simply a popular means of expressing the chemical potential of volatile elements. It may be expressed in bar and the reference states are chosen as an ideal gas with a partial pressure for O_2 or Cl_2 of 1 bar. Thus we have, for instance,

$$2(\mu_O - °G_O^{ref})/RT = \ln P_{O_2}$$

Alternatively, one may express chemical potentials through the content in any other phase that happens to be present or could be present. As an example, Fig. 7.15

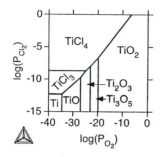

Figure 7.14 The Ti–O–Cl phase diagram at 1 bar and 1273 K, plotted with two potential axes. The potentials are expressed in terms of the partial pressures (in bar) in an ideal gas which is not present.

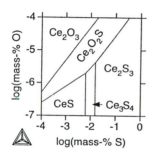

Figure 7.15 The Ce–O–S phase diagram at 1 bar and 1873 K, plotted with two potential axes. The potentials are expressed in terms of the contents in liquid iron which is not present.

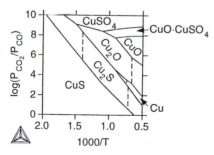

Figure 7.16 The Cu–O–S phase diagram sectioned at 1 bar and a potential of SO_2 equal to the potential of pure SO_2 gas of 1 bar.

shows a case where the logarithm of the contents of O and S in liquid iron are used for representing the Ce–O–S phase diagram at constant temperature and pressure. With these axes one can directly see what cerium compound should form first from liquid iron if the cerium content is gradually increased. However, the diagram does not reveal what cerium contents are required in the liquid iron phase.

One may also section a ternary phase diagram at some value of a chemical potential and keep the temperature as an axis. Figure 7.16 shows such a case sectioned at a constant value of $\mu_S + 2\mu_O$ and plotted with μ_O/RT versus $1/T$. Here μ_O/RT is expressed by the ratio of the partial pressures of CO_2 and CO in an ideal gas.

Exercise 7.7.1

The diagram shows at what O_2 and N_2 pressures three nitrides can form from pure Si at 1840 K. (a) Use the slopes in order to evaluate the O content in α and β, both of which are usually considered to be Si_3N_4. (b) Their coexistence lines are missing in the phase diagram. Calculate their slopes.

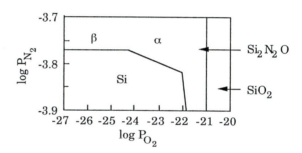

Exercise 7.7.1

Hint

The axes may be regarded as potential axes for O and N because T is constant. We can thus apply the relation $\partial\mu_O/\partial\mu_N = -(z_N^\gamma - z_N^\delta)/(z_O^\gamma - z_O^\delta)$. The Si phase does not dissolve noticeable amounts of O or N. In all the oxides and oxynitrides we can assume the following valencies: $+4$ for Si, -2 for O, -3 for N.

Solution

(a) z_i is here defined as N_i/N_{Si}. We thus get:

For α/Si: $-45 = -(z_N^\alpha - 0)/(z_O^\alpha - 0) = -z_N^\alpha/z_O^\alpha$. Applying electroneutrality, $4 = 2z_O^\alpha + 3z_N^\alpha = 2z_O^\alpha + 3\cdot45z_O^\alpha = 137z_O^\alpha$; $z_O^\alpha = 0.0292$; $z_N^\alpha = 45\cdot0.0292 = 1.3139$. The diagram thus predicts that the formula for the α phase is $Si_1N_{1.3139}O_{0.0292}$ or $Si_{2.978}Va_{0.022}N_{3.013}O_{0.087}$.

For β/Si: $\infty = -(z_N^\beta - 0)/(z_O^\beta - 0)$; $z_O^\beta = 0$. The formula for β is Si_3N_4.

(b) For α/β we get: $\partial\mu_N/\partial\mu_O = -(0.0292 - 0)/(1.3139 - 1.3333) = 1.505$.

For α/Si_2N_2O we get: $\partial\mu_N/\partial\mu_O = -(0.0292 - 0.5)/(1.3139 - 1) = 1.500$. The two new coexistence lines will thus be parallel and almost vertical in our diagram because of the very enlarged scale for $\log P_{N_2}$.

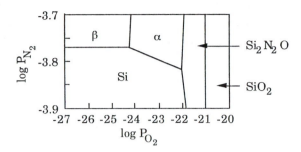

Solution 7.7.1

Exercise 7.7.2

For the invariant equilibrium $TiCl_4 + TiO_2 + Ti_3O_5$ in Fig. 7.14 it has been found that the partial pressure of Ti is $5 \cdot 10^{-22}$ bar. Construct a reasonable $\log P_{O_2}, \log a_{Ti}$ diagram for these three phases at the constant values of T and P.

Hint

Evidently, the potential diagram in Fig. 7.14 was obtained from the fundamental property diagram by first sectioning twice (at constant T and P) and then projecting in the μ_{Ti} direction. Now we are asked to project in the μ_{Cl} direction. Start by plotting the point for the three-phase equilibrium at $\log P_{Ti} = -21.3$ and a value of $\log P_{O_2}$ obtained from Fig. 7.14. Then we can calculate the slopes of invariant equilibria in terms of the compositions involved. When one obtains an indeterminate value one should go back to the derivation of the equation used.

Solution

At constant T,P we have a three-dimensional property diagram looking like Fig. 7.7(a) but with μ_{Ti}, μ_O and μ_{Cl} on the axes. Figure 7.14 is the projection on the μ_O, μ_{Cl} side. Now we want the projection on the μ_O, μ_{Ti} side. Then we must project in the μ_{Cl} direction and define z_i as N_i/N_{Cl}. For TiO_2 and Ti_3O_5 we get z_O and z_{Ti} equal to infinity. We should thus go back to the Gibbs–Duhem relation for two phases, α and β, and get $x_{Ti}^\alpha d\mu_{Ti} + x_O^\alpha d\mu_O + x_{Cl}^\alpha d\mu_{Cl} = 0$; $x_{Ti}^\beta d\mu_{Ti} + x_O^\beta d\mu_O + x_{Cl}^\beta d\mu_{Cl} = 0$.

For $TiO_2/TiCl_4 : x_{Cl}^\alpha = 0$ and already the first equation yields
$d\mu_{Ti}/d\mu_O = -x_O^\alpha/x_{Ti}^\alpha = -2/1$ and $d\ln a_{Ti}/d\ln P_{O_2} = 0.5 \cdot d\mu_{Ti}/d\mu_O = -1$.

For $Ti_3O_5/TiCl_4 : x_{Cl}^\gamma = 0$ and already the first equation yields
$d\mu_{Ti}/d\mu_O = -x_O^\alpha/x_{Ti}^\alpha = -5/3$ and $d\ln a_{Ti}/d\ln P_{O_2} = 0.5 \cdot d\mu_{Ti}/d\mu_O = -5/6$.

For TiO_2/Ti_3O_5: $x_{Cl}^{\alpha} = x_{Cl}^{\beta} = 0$ and the only solution to the two equations is $d\mu_{Ti} = 0$ and $d\mu_O = 0$. This two-phase equilibrium will thus occur in one point only (see diagram). The reason is that we have projected the property diagram in the direction of the $TiO_2 + Ti_3O_5$ coexistence line.

In a two-dimensional potential phase diagram we normally expect to see two-dimensional phase fields for single phases and one-dimensional phase fields for two phases in equilibrium. As expected, the phase field for $TiCl_4$ is two-dimensional but not the one for TiO_2 or Ti_3O_5. However, since TiO_2 and Ti_3O_5 do not dissolve any Cl, their properties are not affected by μ_{Cl}. The μ_{Cl} axis in the fundamental property diagram is thus parallel to the property surface of both phases and hence parallel to the line representing their intersection. In the μ_{Cl} projection these surfaces will become lines and their intersection, representing a two-phase equilibrium, will become a point (compare Fig. 7.7).

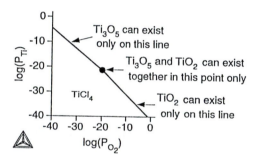

Solution 7.7.2

Exercise 7.7.3

In principle, one could plot potential phase diagrams at constant T and P for various M–O–S systems with the axes for μ_O and μ_S. However, it is more practical to use axes for $\log P_{SO_2}$ and $\log P_{O_2}$ where the pressures refer to an imagined atmosphere, in equilibrium with the system. Start with such a diagram and let it have a line for a constant value of $\log P_{SO_2}$. Show how one can find the slope of the corresponding line in the μ_S, μ_O diagram.

Hint

It is usually assumed that the imagined atmosphere is ideal and thus $\mu_{O_2} = {}^{\circ}\mu_{O_2} + RT\ln P_{O_2}$ and $\mu_{SO_2} = {}^{\circ}\mu_{SO_2} + RT\ln P_{SO_2}$. Remember that $\mu_{SO_2} = \mu_S + 2\mu_O = \mu_S + \mu_{O_2}$.

Solution

$2\mu_O = \mu_{O_2} = {}^{\circ}\mu_{O_2} + RT\ln P_{O_2}$; $(2\mu_O - {}^{\circ}\mu_{O_2})/RT\ln 10 = \log P_{O_2}$ and $\mu_S = \mu_{SO_2} - $

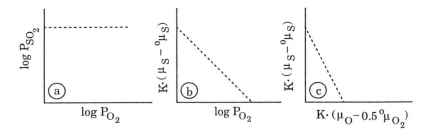

Solution 7.7.3

$\mu_{O_2} = {}^{\circ}\mu_{SO_2} - {}^{\circ}\mu_{O_2} + RT\ln P_{SO_2} - RT\ln P_{O_2}$; $(\mu_S - {}^{\circ}\mu_S)/RT\ln10 = \log P_{SO_2} - \log P_{O_2}$ where ${}^{\circ}\mu_S$ has been defined as ${}^{\circ}\mu_{SO_2} - {}^{\circ}\mu_{O_2}$. Denote $1/RT\ln10$ by K (see diagram).

7.8 *Direction of phase fields in potential phase diagrams*

In the discussions of two-dimensional phase diagrams we have several times derived equations for the slope of two-dimensional phase fields. We shall now give a more general treatment. The direction of phase fields is governed by the Gibbs–Duhem relation, which applies to each one of the p phases in an equilibrium, e.g. for the phase α:

$$- S_m^\alpha dT + V_m^\alpha dP - \Sigma x_i^\alpha d\mu_i = 0$$

If all the phases stay in equilibrium with each other when some variation is made, each of $d\mu_i$, dT and dP must have the same value for all phases. By combining the Gibbs–Duhem relation for all phases one obtains a system of equations for the coexistence of the phases in the fundamental property diagrams. With p phases we have p Gibbs–Duhem relations and can thus eliminate $p - 1$ of the variables. If we would choose to eliminate μ_i for i from 1 to $p - 1$, we should multiply each equation by a factor that we can represent by a determinant. For example, the equation for the α phase should be multiplied by a factor

$$\begin{vmatrix} x_1^\beta & x_2^\beta & \cdots & x_{p-1}^\beta \\ x_1^\gamma & x_2^\gamma & \cdots & x_{p-1}^\gamma \\ \vdots & \vdots & \cdots & \vdots \\ x_1^\varepsilon & x_2^\varepsilon & \cdots & x_{p-1}^\varepsilon \end{vmatrix}$$

As a short-hand notation such a determinant will be written by giving the diagonal elements $|x_1^\beta \quad x_2^\gamma \quad \cdots \quad x_{p-1}^\varepsilon|$. By adding the equations for all the phases, we obtain

$$- |S_m^\alpha x_1^\beta x_2^\gamma \ldots x_{p-1}^\varepsilon| \, dT + |V_m^\alpha x_1^\beta x_2^\gamma \ldots x_{p-1}^\varepsilon| \, dP - \Sigma |x_i^\alpha x_1^\beta x_2^\gamma \ldots x_{p-1}^\varepsilon| \, d\mu_i = 0$$

Using an alternative form of the Gibbs–Duhem relation from Section 3.4 we instead obtain

$$|H_m^\alpha x_1^\beta x_2^\gamma \ldots x_{p-1}^\varepsilon| \, d(1/T) + |V_m^\alpha x_1^\beta x_2^\gamma \ldots x_{p-1}^\varepsilon| \, dP/T$$
$$- \Sigma |x_i^\alpha x_1^\beta x_2^\gamma \ldots x_{p-1}^\varepsilon| \, d(\mu_i/T) = 0$$

The factors in front of $d\mu_i$ or $d(\mu_i/T)$ for i from 1 to $p-1$ are zero because two columns have the same elements. For instance, with $i = p-1$ the first and last columns in the last determinant are identical. It should be emphasized that the equation can be formulated in many ways by including different μ_i in the set of eliminated variables. All such alternative equations apply simultaneously and together they give the direction of the phase field. We shall now consider various cases by considering different values of $p - c$ and in some cases different values of p.

   For $p = c + 1$ we have a univariant equilibrium according to Gibbs' phase rule, $v = 1$, and shall thus obtain a linear phase field in the potential phase diagram. All the $d\mu_i$ or $d(\mu_i/T)$ can be eliminated mathematically because $c = p - 1$. Furthermore, in this case each column in each determinant contains all the x_i in a phase and we can make use of $\Sigma x_i = 1$ in the remaining terms,

$$|H_m^\alpha \, 1 \, x_2^\gamma \ldots x_c^\varepsilon| \, d(1/T) + |V_m^\alpha \, 1 \, x_2^\gamma \ldots x_c^\varepsilon| \, dP/T = 0$$

This gives the direction of the phase field in a $(1/T),P$ phase diagram produced by projection of the complete phase diagram. As an example, for a binary system with three phases the equation gives

$$|H_m^\alpha \, 1 \, x_2^\gamma| \, d(1/T) + |V_m^\alpha \, 1 \, x_2^\gamma| \, dP/T = 0$$

This gives the slope of the phase field for a univariant phase equilibrium in a projection onto the $(1/T),P$ side of the phase diagram. It can also be written as follows,

$$\frac{dP}{dT} = \frac{(x_2^\gamma - x_2^\beta)H_m^\alpha + (x_2^\alpha - x_2^\gamma)H_m^\beta + (x_2^\beta - x_2^\alpha)H_m^\gamma}{(x_2^\gamma - x_2^\beta)V_m^\alpha + (x_2^\alpha - x_2^\gamma)V_m^\beta + (x_2^\beta - x_2^\alpha)V_m^\gamma} \cdot \frac{1}{T}$$

   In Section 12.2 we shall see that the equation can be reduced to a much simpler form. In fact, the numerator is equal to the heat of the three-phase reaction between α, β and γ and is thus independent of the choice of reference states chosen for the H values. The denominator is equal to the change in volume. Thus, the value of dP/dT is independent of the choice of references, as it should be.

   If we had eliminated $d(\mu_1/T)$ and dP instead of $d(\mu_1/T)$ and $d(\mu_2/T)$ we would have obtained

$$\frac{d(\mu_2/T)}{d(1/T)} = \frac{(V_m^\gamma - V_m^\beta)H_m^\alpha + (V_m^\alpha - V_m^\gamma)H_m^\beta + (V_m^\beta - V_m^\alpha)H_m^\gamma}{(V_m^\gamma - V_m^\beta)x_2^\alpha + (V_m^\alpha - V_m^\gamma)x_2^\beta + (V_m^\beta - V_m^\alpha)x_2^\gamma}$$

which would be the slope of the phase field for a univariant phase equilibrium in a projection onto the $(\mu_2/T),(1/T)$ side of the complete phase diagram. The value of the numerator here depends upon the choice of reference states for the H values and that choice will thus affect the value of $d(\mu_2/T)/d(1/T)$.

For $p = c$ we have a divariant equilibrium, $v = 2$, and the corresponding phase field will form a surface in the phase diagram. We can, for instance, eliminate all $d(\mu_i/T)$ terms except for $d(\mu_c/T)$ and obtain a relation between $d(1/T)$, dP and $d(\mu_c/T)$, representing the direction of the two-dimensional phase field in a three-dimensional projection of the complete phase diagram:

$$|H^\alpha_m x^\beta_1 x^\gamma_2 \ldots x^\varepsilon_{c-1}| \, d(1/T) + |V^\alpha_m x^\beta_1 x^\gamma_2 \ldots x^\varepsilon_{c-1}| \, dP/T = |x^\alpha_c x^\beta_1 x^\gamma_2 \ldots x^\varepsilon_{c-1}| \, d(\mu_c/T)$$

Under isobaric conditions we obtain a one-dimensional phase field, the slope of which is given by

$$|x^\alpha_1 x^\beta_2 \ldots x^\delta_{c-1} H^\varepsilon_m| \, d(1/T) = |x^\alpha_1 x^\beta_2 \ldots x^\delta_{c-1} x^\varepsilon_c| \, d(\mu_c/T)$$

For $p = c - 1$ we obtain a similar equation which has terms for $d(\mu_c/T)$ as well as $d(\mu_{c-1}/T)$. Under isobarothermal conditions it simplifies to

$$|x^\alpha_c x^\beta_1 \ldots x^\varepsilon_{c-2}| \, d\mu_c + |x^\alpha_{c-1} x^\beta_1 \ldots x^\varepsilon_{c-2}| \, d\mu_{c-1} = 0$$

and we may thus evaluate the slope $d\mu_c/d\mu_{c-1}$ for the one-dimensional phase field in the constant T and P section of the phase diagram. We can see that it is completely defined by the ratio of two subdeterminants of the complete composition determinant.

For a two-phase equilibrium in a ternary system at constant T and P, the equation reduces to

$$\frac{\partial \mu_2}{\partial \mu_3} = -\frac{x^\alpha_3 x^\varepsilon_1 - x^\alpha_1 x^\varepsilon_3}{x^\alpha_2 x^\varepsilon_1 - x^\alpha_1 x^\varepsilon_2} = -\frac{z^\alpha_3 - z^\varepsilon_3}{z^\alpha_2 - z^\varepsilon_2}$$

This is an example where the final result is simplified by introducing the z variables defined as $z_i = x_i/x_1$. This equation was derived in a more direct way when ternary systems were discussed in Section 7.7. We could apply the present method to two-phase equilibria in general, obtaining

$$\sum_2^c (z^\alpha_i - z^\beta_i) d\mu_i = -(S^\alpha_{m1} - S^\beta_{m1})dT + (V^\alpha_{m1} - V^\beta_{m1})dP$$

$$\sum_2^c (z^\alpha_i - z^\beta_i) d(\mu_i/T) = (H^\alpha_{m1} - H^\beta_{m1})d(1/T) + (V^\alpha_{m1} - V^\beta_{m1})dP/T$$

Exercise 7.8.1

Calculate the change of μ_O for the $Al + Al_2O_3$ two-phase equilibrium when the pressure is increased. The densities of the phases are 2.7 and 3.5 g/cm^3, respectively.

Hint

Since $p = 2$ and also $c = 2$, we have the case $p = c$ and there is a relation between

$d(\mu_c/T)$, $d(1/T)$ and dP. It is necessary to define the problem better. Let us assume that the intention was to keep T constant.

Solution

Let Al be α: $V_m^\alpha = (1/2.7)\cdot 27 = 10\,\mathrm{cm}^3/\mathrm{mole}$ of atoms. Let Al_2O_3 be β: $V_m^\beta = (1/3.5)\cdot(102/5) = 5.8\,\mathrm{cm}^3$ mole of atoms. $(\partial[\mu_0/T]/\partial P)_T = (x_1^\alpha V_m^\beta - x_1^\beta V_m^\alpha)/T(x_1^\alpha x_2^\beta - x_1^\beta x_2^\alpha) = (1\cdot 5.8 - 0.4\cdot 10)/T(1\cdot 0.6 - 0) = 3/T\,\mathrm{cm}^3/\mathrm{mol\,K}$. Since $T = \mathrm{constant}$, $J = Nm$ and $Pa = N/m^2$ we get $(\partial\mu_0/\partial P)_T = 3\cdot 10^{-6}\,\mathrm{J/mol\,Pa}$.

Exercise 7.8.2

The direction of a one-dimensional phase field for a case with $p = c + 1$ has been derived. Try to apply the same method to the phase field for an invariant equilibrium.

Hint

According to Gibbs' phase rule we now have $p = c + 2$. We thus have one equation more and could now eliminate all μ_i and also another variable, say T.

Solution

Instead of multiplying the expression for the α phase with $|x_1^\beta\, x_2^\gamma \ldots x_{p-1}^\varepsilon|$, which is no longer a determinant because it has one row more than columns, we could now multiply with the determinant $|x_1^\beta\, x_2^\gamma \ldots x_{p-1}^\varepsilon H_m^\phi|$. Using similar determinants for the other phases and adding all the equations we obtain $|H_m^\alpha\, x_1^\beta\, x_2^\gamma \ldots x_{p-1}^\varepsilon\, H_m^\phi|\,d(1/T) + |V_m^\alpha\, x_1^\beta\, x_2^\gamma \ldots x_{p-1}^\varepsilon\, H_m^\phi|\,dP/T - \Sigma\,|x_i^\alpha\, x_1^\beta\, x_2^\gamma \ldots x_{p-1}^\varepsilon\, H_m^\phi|\,d(\mu_c/T) = 0$. In this case the determinants in front of $d(1/T)$ as well as $d(\mu_c/T)$ are zero and only the dP term remains. One thus obtains $dP = 0$ which confirms that no potential can vary in the phase field for an invariant equilibrium.

Exercise 7.8.3

The equation for dP/dT of a three-phase equilibrium in a binary system contains ordinary molar quantities and mole fractions. Show how the equation is modified if z fractions are introduced instead of x fractions.

Hint

Remember that $z_i = x_i/x_1$. We should thus divide all terms by $x_1^\alpha x_1^\beta x_1^\gamma$.

Solution

If we also remember that for any extensive quantity A one has $A_{m1} = A_m/x_1$, then we obtain

$$\frac{dP}{dT} = \frac{(z_2^\gamma - z_2^\beta)H_{m1}^\alpha + (z_2^\alpha - z_2^\gamma)H_{m1}^\beta + (z_2^\beta - z_2^\alpha)H_{m1}^\gamma}{(z_2^\gamma - z_2^\beta)V_{m1}^\alpha + (z_2^\alpha - z_2^\gamma)V_{m1}^\beta + (z_2^\beta - z_2^\alpha)V_{m1}^\gamma} \cdot \frac{1}{T}$$

7.9   *Extremum in temperature and pressure*

For convenience we shall now use the relation derived from the Gibbs–Duhem relation in its ordinary form, i.e. we shall use S instead of H.

For $p = c$ we obtain, by rearranging the terms in the determinants,

$$-|x_1^\alpha x_2^\beta \ldots x_{c-1}^\delta S_m^\varepsilon| dT + |x_1^\alpha x_2^\beta \ldots x_{c-1}^\delta V_m^\varepsilon| dP = |x_1^\alpha x_2^\beta \ldots x_{c-1}^\delta x_c^\varepsilon| d\mu_c$$

Suppose the composition determinant on the right-hand side is zero,

$$|x_1^\alpha x_2^\beta \ldots x_c^\varepsilon| = 0$$

Under isobaric conditions this would yield $dT/d\mu_c = 0$ for the linear phase field obtained in the μ_c, T phase diagram and the phase field must go through a temperature extremum. The requirement for this to occur can also be written in the following form because $\Sigma x_i = 1$ in each phase,

$$|1\, x_2^\beta \ldots x_c^\varepsilon| = 0$$

This is a well-known equation from the theory of determinants and shows that the phases fall on the same point (i.e. have the same composition) for $c = p = 2$, they fall on a straight line for $c = p = 3$, on a plane surface for $c = p = 4$, etc. The first two cases are descrbed by Konovalov's and von Alkemade's rule, respectively (see Sections 9.8 and 9.9). Furthermore, if one knows that there is such a temperature extremum under isobaric conditions, then one can conclude that the composition determinant must be zero and the equation shows that there will also be a pressure extremum under isothermal conditions. For a binary case, $c = p = 2$, this is illustrated in Fig. 7.17.

For $p = c - 1$ we obtain, under isobaric conditions,

$$-|x_1^\alpha x_2^\beta \ldots x_{c-2}^\delta S_m^\varepsilon| dT + |x_1^\alpha x_2^\beta \ldots x_{c-2}^\delta V_m^\varepsilon| dP$$
$$= |x_1^\alpha x_2^\beta \ldots x_{c-2}^\delta x_c^\varepsilon{}_{-1}| d\mu_{c-1} + |x_1^\alpha x_2^\beta \ldots x_{c-2}^\delta x_c^\varepsilon| d\mu_c$$

In order to obtain an extremum in T at constant P (and thus in P at constant T), it is now necessary that two determinants are zero,

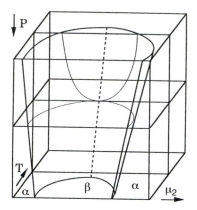

Figure 7.17 Potential phase diagram for a binary system showing a divariant phase field having a T extremum in an isobaric section (see thin horizontal curve). It follows that an isothermal section will show a P extremum (see thin vertical curve).

$$|x_1^\alpha \, x_2^\beta \ldots x_{c-2}^\delta \, x_c^\varepsilon| = 0$$
$$|x_1^\alpha \, x_2^\beta \ldots x_{c-2}^\delta \, x_{c-1}^\varepsilon| = 0$$

For a binary system this condition has no meaning because $p = 1$. For $p = 2$ and $c = 3$ it implies that the two phases fall on the same point in the composition plane (in agreement with a generalization of Konovalov's rule), for $p = 3$ and $c = 4$ it implies that the three phases fall on a straight line in the composition volume (in agreement with a generalization of Alkemade's rule), etc. For a ternary system this can be demonstrated easily

$$|x_1^\alpha \, x_3^\beta| = 0$$
$$|x_1^\alpha \, x_2^\beta| = 0$$

By adding the two equations we get

$$0 = |x_1^\alpha (x_2^\beta + x_3^\beta)| = |x_1^\alpha \, 1| = x_1^\alpha - x_1^\beta$$

or $x_1^\alpha = x_1^\beta$. By inserting this in the two primary conditions we get $x_2^\alpha = x_2^\beta$ and $x_3^\alpha = x_3^\beta$. This case is illustrated in Fig. 7.18 which may be regarded as a diagram corresponding to the P section through the diagram in Fig. 7.17 but with one more axis due to the third element. It follows from the initial equation that here will also be an extremum in P under isothermal conditions but we would need four dimensions to show a diagram corresponding to the whole diagram in Fig. 7.17.

For $p = c - 2$ the conditions for an extremum in T at constant P (and thus in P for constant T) is obtained as a set of three determinants equal to zero and this means that the compositions of the phases fall on the same point for $p = 2$ and $c = 4$, same line for $p = 3$ and $c = 5$, same plane for $p = 4$ and $c = 6$, etc.

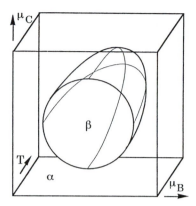

Figure 7.18 Potential phase diagram showing a divariant phase field with a T extremum at a certain combination μ_B, μ_C. The complete phase diagram has been sectioned at some constant P.

Exercise 7.9.1

Consider a three-phase equilibrium at 1 bar in a ternary system between pure A, a compound B_1C_1 and a third phase with variable composition. Can this equilibrium go through a T maximum? Under what conditions?

Hint

Notice that $p = 3$ and $c = 3$ and thus $c - p$.

Solution

$$0 = \begin{vmatrix} x_1^\alpha & x_2^\alpha & x_3^\alpha \\ x_1^\beta & x_2^\beta & x_3^\beta \\ x_1^\gamma & x_2^\gamma & x_3^\gamma \end{vmatrix} = \begin{vmatrix} 1 & x_2^\alpha & x_3^\alpha \\ 1 & x_2^\beta & x_3^\beta \\ 1 & x_2^\gamma & x_3^\gamma \end{vmatrix} = \begin{vmatrix} 1 & 0 & 0 \\ 1 & 0.5 & 0.5 \\ 1 & x_B^\beta & x_C^\beta \end{vmatrix} = 0.5x_C^\beta - 0.5x_B^\beta$$

or $x_B^\beta = x_C^\beta$. The variable phase must fall on the straight line between A and B_1C_1 in order for a T maximum to occur. However, if this is fulfilled, we cannot tell if it will be a T maximum or minimum.

Exercise 7.9.2

Consider a ternary system where the potential of the third component is kept constant (by means of a high diffusivity and equilibrium with an external reservoir). The pressure is also kept constant. Suppose one will thus find that there is a maximum temperature for a certain $\alpha + \beta$ equilibrium. What conclusion can be drawn regarding the compositions of the two phases? What would be the most convenient composition variable to use in such a case?

Hint

$p = c - 1$. An equation has already been derived for that case. Under conditions of constant P and μ_c it yields

$$|x_1^\alpha \, x_2^\beta \ldots x_{c-2}^\delta \, x_{c-1}^\varepsilon| \, d\mu_{c-1} = -|x_1^\alpha \, x_2^\beta \ldots x_{c-2}^\delta \, S_m^\varepsilon| \, dT$$

Solution

At the maximum, T is constant and the equation yields $|x_1^\alpha \, x_2^\beta \ldots x_{c-2}^\delta \, x_{c-1}^\varepsilon| = 0$ and for a ternary system $|x_1^\alpha \, x_2^\beta| = 0$ or $x_1^\alpha/x_2^\alpha = x_1^\beta/x_2^\beta$. The ratio of components 1 and 2 is thus the same in the two phases. The most convenient composition variable in this case is $u_i = x_i/(1 - x_c)$ since $u_1 + u_2 + \ldots + u_{c-1} = 1$ and from $|x_1^\alpha \, x_2^\beta \ldots x_{c-2}^\delta \, x_{c-1}^\varepsilon| = 0$ we get $|u_1^\alpha \, u_2^\beta \ldots u_{c-2}^\delta \, u_{c-1}^\varepsilon| = |1 \, u_2^\beta \ldots u_{c-2}^\delta \, u_{c-1}^\varepsilon| = 0$. For a ternary system we get $|1 \, u_2^\beta| = u_2^\alpha - u_2^\beta = 0$.

Molar phase diagrams

Molar axes

If one starts from a potential phase diagram, one may decide to replace one of the potentials by its conjugate variable. However, the potential phase diagram has no information on the size of the system and it is thus logical to introduce a molar quantity rather than the extensive variable. By replacing all the potentials with their conjugate molar variables, one gets a molar diagram. One would like to retain the diagram's character of a true phase diagram, which means that there should be a unique answer as to which phase or phases are stable at each location. In this chapter we shall examine the properties of molar diagrams and we shall find under what conditions they are true phase diagrams. Only then may they be called **molar phase diagrams**. However, we shall start with a simple demonstration of how a diagram changes when molar axes are introduced.

Figure 8.1(a)–(d) demonstrates what happens to a part of the T,P potential phase diagram for Fe when S_m and V_m axes are introduced. Initially the P axis is plotted in the negative direction because V is conjugate to $-P$. It can be seen that the one-phase fields separate and leave room for a two-phase field. It can be filled with tie-lines connecting the points representing the individual phases in the two-phase equilibrium. It is self-evident how to draw them when one axis is still a potential but they yield additional information when all axes are molar (Fig. 8.1(d)).

Figure 8.2(a)–(d) is a similar demonstration using a part of the Fe phase diagram with a three-phase equilibrium, a triple point. It forms a tie-triangle when both potentials have been replaced (Fig. 8.2(d)). All the phase fields are then two-dimensional. One may also notice that each one-phase field from the potential diagram maintains its general shape. Their corners still have angles less than $180°$ (see the $180°$ rule formulated in Section 7.3).

It should be emphasized that the phase fields never overlap in these diagrams. They may all be classified as true phase diagrams because each point represents one and only one phase equilibrium. Three requirements must be fulfilled in order for this to happen. Firstly, the two one-phase fields meeting at a two-phase line in a potential phase diagram must move away from each other and leave room for an extended two-phase

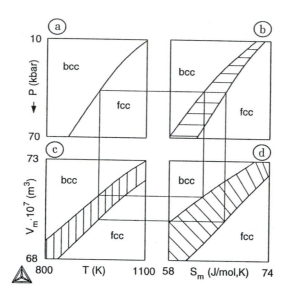

Figure 8.1 Introduction of molar axes instead of potential axes in a part of the unary phase diagram for Fe with two phases.

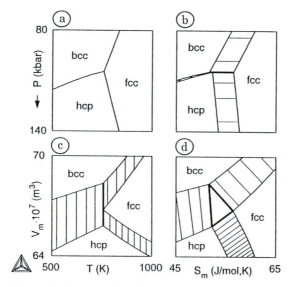

Figure 8.2 Introduction of molar axes instead of potential axes in a part of the unary phase diagram for Fe with three phases. All phase fields become two-dimensional.

field, when a molar axis is introduced. Secondly, the one-phase field extending from the two-phase field in the direction of increasing values of a potential must also extend to increasing values of the conjugate molar variable that is introduced. If it goes the other way, it would overlap the two-phase region. The other one-phase field must extend in

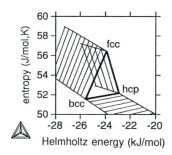

Figure 8.3 S,H diagram for Fe. This is not a true phase diagram because S and H never appear in the same set of conjugate variables.

the other direction, before as well as after replacing the potential with the conjugate molar variable. Thirdly, a one-phase field is nowhere allowed to fold over itself.

The last two requirements are fulfilled if the system is everywhere stable because of the stability condition in Section 5.4,

$$\left(\frac{\partial Y^b}{\partial X^b}\right)_{Y^c X^d} > 0$$

The potential Y^b and its conjugate variable X^b thus increase in the same direction. However, as already emphasized, this stability condition requires that all the variables to be kept constant, Y^c, X^d, come from the same set of conjugate pairs as Y^b and X^b. Nine such sets were presented in Section 3.4 but it is necessary to examine what happens to them when the size of the system is measured in different ways. This will be discussed in the next section. Figure 8.3 is an example of what can happen if one uses two molar variables which do not appear in the same set of conjugate variables, S and H. Figure 8.3 is not a true phase diagram according to the definition given at the very beginning of this section.

The first requirement can be tested as follows, using the form of the Gibbs–Duhem relation with molar quantities introduced in Section 7.1,

$$d\mu_1 = -S_{m1}dT + V_{m1}dP - \sum_2^c z_i d\mu_i = -\sum_2^{c+2} X_{m1}^j dY^j$$

Consider two phases, α and β, which are initially in equilibrium with each other. The system is then moved away from equilibrium by changing the value of one potential, Y^j, keeping the other independent potentials in the summation constant. Applying the Gibbs–Duhem relation to each of the two phases and taking the difference, we obtain

$$d(\mu_1^\beta - \mu_1^\alpha) = (X_{m1}^{j\alpha} - X_{m1}^{j\beta})dY^j$$

Suppose α is the phase favoured by the increased Y^j value. Then μ_1^α must be smaller than μ_1^β as demonstrated by Fig. 7.3. We thus obtain

$$X_{m1}^{j\alpha} - X_{m1}^{j\beta} = \frac{d(\mu_1^\beta - \mu_1^\alpha)}{dY^j} > 0$$

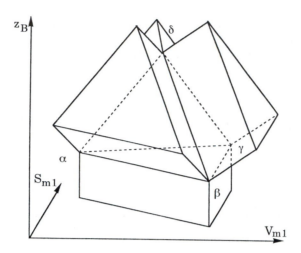

Figure 8.4 Four-phase equilibrium in a phase diagram with three molar axes. The four-phase field is tetrahedral and is covered by triangular prisms representing three-phase equilibria. The two- and one-phase fields are not outlined but they are also three-dimensional.

It is thus evident that the two one-phase fields will move apart by a positive distance $X_{m1}^{j\alpha} - X_{m1}^{j\beta}$ when X_{m1}^{j} is introduced as an axis instead of Y^{j}. The one-phase fields will separate and give room for the two-phase field in between, $X_{m1}^{j\alpha} - X_{m1}^{j\beta}$ being the length of the tie-line.

In a binary system there are three independent potential axes. If they are all replaced by molar axes, all the phase fields become three-dimensional and the invariant four-phase equilibrium expands into a tetrahedron. This is demonstrated by Fig. 8.4 which corresponds to the central region of Fig. 7.9.

It was emphasized that the topology of potential phase diagrams is very simple and each geometrical element is a phase field. A phase diagram with only molar axes has a relatively simple topology. All the phase fields have the same dimensionality as the diagram itself. For the unary system in Fig. 8.2 all the phase fields have two dimensions and for the binary system in Fig. 8.4 they have three dimensions.

Exercise 8.1.1

Suppose one studies the total vapour pressure of a liquid mixture of two metals, A and B, at a constant temperature. One finds that the total vapour pressure increases if more B is added to the mixture. Show whether the vapour or the liquid is richer in B.

Hint

At constant T, the P, μ_B potential phase diagram will be two-dimensional. Sketch it using μ_A as the dependent potential variable. Remember that the conjugate composition variable to μ_B would then be $z_B = N_B/N_A$. High pressure should favour the liquid, being much denser than the vapour.

Solution

The construction (see diagram) shows that the vapour would be richer in B than the liquid if measured relative to A.

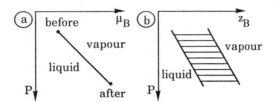

Solution 8.1.1

Exercise 8.1.2

Experiments carried out with a pure element at constant T and V after rapid cooling from a higher T showed that an α phase was present initially. Then, a β phase nucleated and grew and consumed all α. Then, a γ phase nucleated and grew but some β remained. Finally the first phase, α, reappeared and consumed the remaining β and some of the γ. Sketch a simple T,V phase diagram with three phases and indicate a possible choice of T and V which could yield such a result.

Hint

Choose a simple three-phase equilibrium and draw the adjoining two-phases fields. Extrapolate the two-phase fields below the three-phase line.

Solution

The point representing the system must fall inside the α-phase field at the higher T.

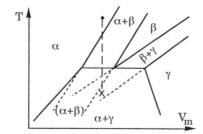

Solution 8.1.2

At the new T the point must fall on the β side of the $\alpha + \beta$ two-phase field but inside the $\beta + \gamma$ two-phase field. Then it automatically falls inside the $\alpha + \gamma$ two-phase field if it is below the three-phase line (see diagram).

8.2　*Sets of conjugate pairs containing molar variables*

A molar variable can easily be introduced in a stability condition by dividing X^b with the quantity used to define the size of the system because that quantity is kept constant. Expressing the size by N, we get for instance,

$$\left(\frac{\partial Y^b}{\partial X^b_m}\right)_{Y^c X^d} = N \cdot \left(\frac{\partial Y^b}{\partial X^b}\right)_{Y^c X^d} > 0$$

However, with this measure of size the Gibbs–Duhem relation gives

$$S_m dT - V_m dT + \Sigma x_i d\mu_i = 0$$

where one of the x_i is dependent on the others because $\Sigma x_i = 1$. Choosing x_1 as the dependent one, we obtain $x_1 = 1 - \sum_2 x_i$

$$-d\mu_1 = S_m dT - V_m dP + \sum_2 x_i d(\mu_i - \mu_1)$$

Then it is logical to regard μ_1 as the dependent potential but the consequence is that the conjugate variable to x_i is no longer μ_i but $(\mu_i - \mu_1)$.

If we instead measure the size with the amount of a certain component, N_1, then we obtain the form introduced in Section 7.1,

$$-d\mu_1 = S_{m1} dT - V_{m1} dP + \sum_2 z_i d\mu_i$$

In this way one may keep μ_i but its conjugate variable is $z_i = N_i/N_1$ and S_{m1} and V_{m1} are also defined by dividing with N_1.

Sometimes it is convenient to measure the size as the total amount of more than one component, e.g. of those which do not easily evaporate. Suppose they are the first k components. Using $u_1 = 1 - \sum_2^k u_i$ we obtain

$$S_{m(1\ldots k)} dT - V_{m(1\ldots k)} dP + \sum_1^c u_{i(1\ldots k)} d\mu_i = 0$$

$$-d\mu_1 = S_{m(1\ldots k)} dT - V_{m(1\ldots k)} dP + \sum_2^k u_{i(1\ldots k)} d(\mu_i - \mu_1) + \sum_{k+1}^c u_{i(1\ldots k)} d\mu_i$$

where the S, V and u variables are defined in Section 3.7.

Table 8.1 *Sets of conjugate pairs of state variables using molar quantities defined by dividing with ΣN_i.*

T, S_m	$-P, V_m$	$\sum_2 (\mu_i - \mu_1), x_i$
$T, (TS_m - PV_m)/T$	$-P/T, TV_m$	$\sum_2 (\mu_i - \mu_1), x_i$
$T/P, PS_m$	$P, (TS_m - PV_m)/P$	$\sum_2 (\mu_i - \mu_1), x_i$
$-1/T, U_m$	$-P/T, V_m$	$\sum_2 (\mu_i - \mu_1)/T, x_i$
$-1/T, H_m$	$-P, V_m/T$	$\sum_2 (\mu_i - \mu_1)/T, x_i$
$-P/T, H_m/P$	$-1/P, PU_m/T$	$\sum_2 (\mu_i - \mu_1)/T, x_i$
$T/P, S_m$	$-1/P, U_m$	$\sum_2 (\mu_i - \mu_1)/P, x_i$
$-1/T, TU_m/P$	$-T/P, F_m/T$	$\sum_2 (\mu_i - \mu_1)/P, x_i$
$T, S_m/P$	$-1/P, F_m$	$\sum_2 (\mu_i - \mu_1)/P, x_i$

These three methods of measuring the size of the system can be applied to all the rows in Table 3.1, Section 3.4. We may thus construct Tables 8.1, 8.2 and 8.3 for the sets of conjugate potentials and molar variables.

Each row defines a set of conjugate variables and each pair can be used to construct a stability condition if the variables to be kept constant are taken from the same set.

So far, we have always chosen to regard μ_1 as the dependent potential. However, we are presented with an infinite number of choices. For example, it may be convenient to regard $(\mu_1 + \mu_2)/2$ as the dependent variable when the size is measured through $N_1 + N_2$.

Exercise 8.2.1

At the end of Section 5.6 we found that the stability limit in a binary solution is $g_{22} = 0$. Show how this condition can be obtained from the list of conjugate variables presented in Table 8.1.

Table 8.2 *Sets of conjugate pairs of state variables using molar quantities defined by dividing with* N_i.

T, S_{m1}	$-P, V_{m1}$	$\sum_2 \mu_i z_i$
$T, (TS_{m1} - PV_{m1})/T$	$-P/T, TV_{m1}$	$\sum_2 \mu_i z_i$
$T/P, PS_{m1}$	$P, (TS_{m1} - PV_{m1})/P$	$\sum_2 \mu_i z_i$
$-1/T, U_{m1}$	$-P/T, V_{m1}$	$\sum_2 \mu_i/T, z_i$
$-1/T, H_{m1}$	$-P, V_{m1}/T$	$\sum_2 \mu_i/T, z_i$
$-P/T, H_{m1}/P$	$-1/P, PU_{m1}/T$	$\sum_2 \mu_i/T, z_i$
$T/P, S_{m1}$	$-1/P, U_{m1}$	$\sum_2 \mu_i/P, z_i$
$-1/T, TU_{m1}/P$	$-T/P, F_{m1}/T$	$\sum_2 \mu_i/P, z_i$
$T, S_{m1}/P$	$-1/P, F_{m1}$	$\sum_2 \mu_i/P, z_i$

Hint

The index 2 in g_{22} indicates a derivative with respect to x_2, with x_1 as a dependent variable. Thus, one should use a set of conjugate variables containing x_2.

Solution

From the first row of Table 8.1 we can formulate the condition $(\partial[\mu_2 - \mu_1]/\partial x_2)_{T,P,N} = 0$. However, x_1 is a dependent variable and then $\mu_2 - \mu_1 = \partial G_m/\partial x_2$ and our stability condition can be expressed as $\partial^2 G_m/\partial x_2^2 = 0$ and g_{22} is the notation for that derivative.

Exercise 8.2.2

Two diagrams of the Mo–N system are presented. How would you interpret them?

Hint

In diagram (a) notice that the phase field for the gas is not included but isobars for

Table 8.3 *Sets of conjugate pairs of state variables using molar quantities defined by dividing with* $N_1 + N_2$.

T,S_{m12}	$-P,V_{m12}$	$(\mu_2-\mu_1),u_{i(12)}$	$\sum_3 \mu_i,u_{i(12)}$
$T,(TS_{m12}-PV_{m12})/T$	$-P/T,TV_{m12}$	$(\mu_2-\mu_1),u_{i(12)}$	$\sum_3 \mu_i,u_{i(12)}$
$T/P,PS_{m12}$	$P,(TS_{m12}-PV_{m12})/P$	$(\mu_2-\mu_1),u_{i(12)}$	$\sum_3 \mu_i,u_{i(12)}$
$-1/T,U_{m12}$	$-P/T,V_{m12}$	$(\mu_2-\mu_1)/T,u_{i(12)}$	$\sum_3 \mu_i/T,u_{i(12)}$
$-1/T,H_{m12}$	$-P,V_{m12}/T$	$(\mu_2-\mu_1)/T,u_{i(12)}$	$\sum_3 \mu_i/T,u_{i(12)}$
$-P/T,H_{m12}/P$	$-1/P,PU_{m12}/T$	$(\mu_2-\mu_1)/T,u_{i(12)}$	$\sum_3 \mu_i/T,u_{i(12)}$
$T/P,S_{m12}$	$-1/P,U_{m12}$	$(\mu_2-\mu_1)/P,u_{i(12)}$	$\sum_3 \mu_i/P,u_{i(12)}$
$-1/T,TU_{m12}/P$	$-T/P,F_{m12}/T$	$(\mu_2-\mu_1)/P,u_{i(12)}$	$\sum_3 \mu_i/P,u_{i(12)}$
$T,S_{m12}/P$	$-1/P,F_{m12}$	$(\mu_2-\mu_1)/P,u_{i(12)}$	$\sum_3 \mu_i/P,u_{i(12)}$

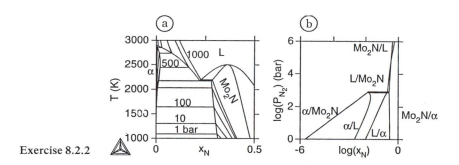

Exercise 8.2.2

N_2 are given. In order to interpret diagram (b) it is helpful first to construct a $T,\log P_{N_2}$ diagram and then replace the T axis with a $\log x_N$ axis.

Solution

Diagram (a) above is basically a T,x_N diagram at 1 bar. The lines for various N_2 pressures should be understood as isoactivity lines for N expressed as P_{N_2} of a gas which is not present. Using these values of P_{N_2} it is easy to construct a $T, \log P_{N_2}$ diagram (see (a) below). For convenience, we shall make T the abscissa. Next we shall introduce x_N (with a logarithmic scale) instead of T, i.e. a molar quantity instead of a potential. The two-phase fields will open up but there may be overlapping because the new variable, x_N, is not conjugate to the old one, T. As an example, the $\alpha + L$ field falls inside the $\alpha + Mo_2N$ field.

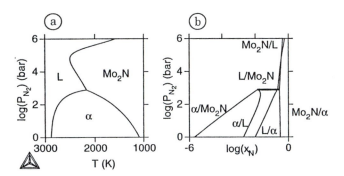

Solution 8.2.2

8.3   *Phase boundaries*

Since all the phase fields in a molar diagram have the same dimensionality as the diagram has axes, it is evident that all other geometrical elements, surfaces, line and points in a three-dimensional diagram, are not phase fields. They separate phase fields and may be called **phase boundaries**. When discussing the topology of a molar phase diagram in terms of the phase boundaries, it is possible and convenient to choose a smaller elementary unit than a phase field. A smaller unit is shown in Fig. 8.5(a) and it is composed of four linear phase boundaries meeting at a point. Topologically it may be represented by two intersecting lines as shown in Fig. 8.5(b). Any complicated two-dimensional phase diagram with molar axes is composed of such units.

It can be seen by inspection of the three-dimensional diagram in Fig. 8.4, that it is possible to divide it into four topologically identical, elementary units, each one composed of a point where eight phase fields meet, although only four of them are shown. Six linear phase boundaries radiate from these points. They are all shown for the β and δ points. Topologically, this elementary unit can be represented by three intersecting planes as shown in Fig. 8.6. Evidently, the topology of a complicated three-dimensional molar diagram can be represented by a system of intersecting surfaces.

When studying two-dimensional molar diagrams, Masing (1949) observed that the number of phases in the phase fields changes by one unit when one crosses a linear phase boundary. This is easily verified by inspection of Fig. 8.2(d). Masing's rule was later generalized by Palatnik and Landau (1964) who gave it the following form

$$D^+ + D^- = r - b$$

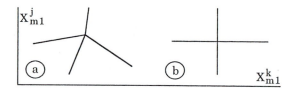

Figure 8.5 (a) Elementary unit of a phase diagram with two molar axes. (b) Topological equivalence.

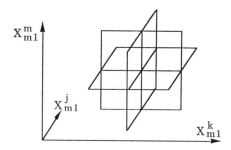

Figure 8.6 Elementary unit of a phase diagram with three molar axes.

where D^+ and D^- are the number of phases that appear and disappear, respectively, as one crosses a phase boundary of dimensionality b, and r is the number of axes in the molar diagram. This rule may be referred to as the **MPL boundary rule**, after Masing, Palatnik and Landau.

It may be added that phase boundaries sometimes have special names. The boundary between a liquid phase and a liquid + solid phase field is called the **liquidus** and the corresponding boundary for the coexisting solid phase is called the **solidus**. The boundary between a solid and the two-phase field with another solid is sometimes called the **solvus**.

Exercise 8.3.1

In the central region of Fig. 8.4 there is a tetrahedron, representing a four-phase field. Apply the MPL rule in order to find how many phases there are outside the α–β line and outside the β point.

Hint

There are only four phases in the system and D^+ must be zero when we move out from the four-phase field because there is no new phase that can be added.

Solution

This is a three-dimensional diagram, $r = 3$, and the dimensionality of the line is one,

$b = 1$. We get $D^- = r - b = 3 - 1 = 2$. The number of phases has decreased from 4 to 2. We have moved into the $\alpha + \beta$ two-phase region by crossing the α–β line. The dimensionality of a point is zero, $b = 0$, and we get $D^- = r - b = 3 - 0 = 3$. The number of phases has decreased from 4 to 1. We have moved into the β one-phase region by crossing the β point.

8.4   *Sections of molar phase diagrams*

A diagram with a full set of molar axes may be called a **complete molar phase diagram**. For practical reasons one often likes to reduce the number of axes. A popular method is to section at a constant value of a potential, e.g. P or T. The resulting diagram looks just like a complete molar phase diagram for a system with one component less. Another method is to section at a constant value of a molar variable, a so-called *isoplethal section* or an *isopleth*.

Since all phase fields in a molar phase diagram have the same dimensionality as the phase diagram has axes, all kinds of phase fields may show up in that kind of section whereas a phase field with the maximum number of phases (i.e. for an invariant equilibrium) will disappear in an equipotential section because the section cannot be expected to go exactly through the point. The topology of a molar section is simplified if it is again assumed that it will not be possible to place a section exactly through a point. All two-dimensional sections with molar axes will be composed of the elementary unit shown in Fig. 8.5 and all three-dimensional sections will be composed of the elementary unit shown in Fig. 8.6 independent of how many potential or molar axes have been sectioned. Of course, if one adds a component, one must section once more in order to keep the number of dimensions. As an example, two sections through Fig. 8.4 are indicated in Fig. 8.7. In each case, the section gives the same arrangement of lines as in Fig. 8.5(a). Furthermore, the MPL boundary rule applies to the sections, since the value of $r - b$ does not change by sectioning.

Inspection of the two sections in Fig. 8.7 reveals that one shows an intersection between phase fields of 2, 3, 3 and 4 phases and the other 1, 2, 2 and 3 phases. We may thus give the general picture shown in Fig. 8.8. For the sections shown in Fig. 8.7 we have $e = 3$ and 4, respectively, where e is the highest number of phases in any of the two adjoining phase fields. In fact, the maximum value of e in a section, which is also the maximum number of phases in a phase field, depends upon the number of sectioned molar axes, n_{ms},

$$e_{max} = 3 + n_{ms}$$

Exercise 8.4.1

On the right-hand side of the tetrahedron in Fig. 8.7 there is a triangular prism.

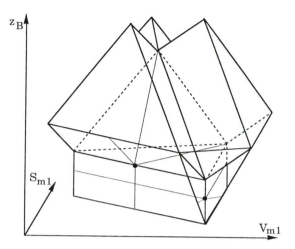

Figure 8.7 Two sections through the molar phase diagram of Fig. 8.4. The sections are shown with thin lines.

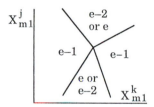

Figure 8.8 Elementary unit of a molar phase diagram sectioned a sufficient number of times to make it two-dimensional. The diagram may have units with different e values from 3 up to a maximum, determined by the number of sectionings.

Make a section through that prism parallel to the side of the tetrahedron. Make a sketch of the intersection obtained at the front edge of the prism. Indicate the number of phases in the four adjoining phase fields.

Hint

It may be useful to go back to the Exercise 8.3.1.

Solution

See diagram.

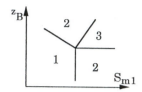

Solution 8.4.1

Schreinemakers' rule

When studying isobarothermal sections of ternary diagrams Schreinemakers (1912) found that the extrapolations of the boundaries of the one-phase field in the elementary unit must either both fall inside the three-phase fields or one inside each of the two two-phase fields. This is illustrated in Fig. 8.9 and is called Schreinemakers' rule. It can be generalized in the following way (Hillert, 1985).

Let us examine if Schreinemakers' rule applies to different e values and start by considering a complete phase diagram constructed with molar axes only. A discussion of thermodynamic properties should then be based upon the internal energy. For reversible changes we obtain

$$dU = TdS - PdV + \sum_1^c \mu_i dN_i$$

In Section 3.10 we saw that it is always possible to introduce a new set of components instead of the old one by combining the components in a new way as long as we get a complete set of independent components and do not change the value of the sum, $\Sigma \mu_i dN_i$. We can do so by selecting c points in the compositional space and make sure that they can be used to define a new set of independent components by checking that three of them never fall on a line, four of them never fall on a plane, etc. We shall use this method of changing to a new set of components but we shall then consider entropy and volume as components, whose amounts are expressed by S and V, and whose chemical potentials are T and $-P$, respectively. The introduction of $c+2$ new components instead of the old ones will now be effected by selecting $c+2$ points in the state diagram. They will each be identified by an index d.

We can follow the procedure outlined in Section 3.10 and obtain

$$dU = \sum_1^{c+2} \mu_i dN_i = \sum_1^{c+2} \mu_d dN_d$$

where $\mu_d = \sum_i a_i^d \mu_i$ and $N_i = \sum_d a_i^d N_d$. For these generalized chemical potentials, the following Maxwell relation is obtained

$$\left(\frac{\partial \mu_j}{\partial N_k}\right)_{N_j} = \frac{\partial^2 U}{\partial N_k \partial N_j} = \left(\frac{\partial \mu_k}{\partial N_j}\right)_{N_k}$$

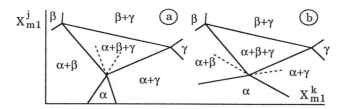

Figure 8.9 Elementary unit of a phase diagram with two molar axes. Two of the phase boundaries of the one-phase field are shown.

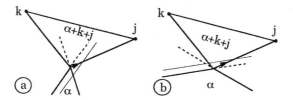

Figure 8.10 General proof of Schreinemakers' rule.

When considering the cases in Fig. 8.9 with a tie-triangle in the section, we shall include the β and γ corners in the set of new components. In a more general case we shall denote them by k and j (see Fig. 8.10(a) and (b)).

At the point under consideration, one of the two boundaries, the extrapolations of which we discuss, represents equilibrium with k, and is thus an equipotential line for k. If it extrapolates outside the α–k–j triangle, the potential of k must increase on moving closer to the point j, because this path intersects equipotential lines for k situated closer to the point k, i.e.

$$\frac{\partial \mu_k}{\partial N_j} > 0$$

(see thin line in Fig. 8.10(b)). Then, from the Maxwell relation,

$$\frac{\partial \mu_j}{\partial N_k} > 0$$

from which it follows that the second boundary must also extrapolate outside the α–k–j triangle. On the other hand, if the k boundary extrapolates into the triangle, a movement towards the point j will intersect equipotential lines for k further away from the point k (see thin line in Fig. 10(a)). Both derivatives must then be negative, and both boundaries must extrapolate into the triangle. It has thus been shown that the extrapolations of both phase boundaries under consideration must fall either outside the highest-order phase field or inside it, in agreement with Schreinemakers' rule. It may be emphasized that the rule also holds for equipotential sections. In order to prove it in such a case, one must use a Maxwell relation based on a thermodynamic function which allows the

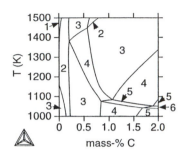

Figure 8.11 Calculated phase diagram for system with seven components. The complete phase diagram has two potential axes and six molar axes, and has been sectioned at one constant potential (P) and five constant molar quantities (x_i). Schreinemakers' rule holds at all intersections. Numbers given are number of phases in each phase field.

corresponding potentials to be kept constant, for instance G in the case actually considered by Schreinemakers, constant T and P.

In the derivation of Schreinemakers' rule it is essential that the two boundaries of the highest-order phase field of those considered are straight lines. That this happens in the ternary case under isobarothermal conditions is self-evident because then the tie-triangle is situated in the plane of the diagram. In a quaternary system the sides of a four-phase equilibrium will be planar and the intersections shown in a two-dimensional section will be straight lines. The components k and j then represent two-phase mixtures situated in the section. On the other hand, a three-phase equilibrium will not be bounded by planar sides and its boundaries in the two-dimensional section will not be straight lines. Then the boundaries of the one-phase field will not be equipotential lines for any components k and j chosen in the section. It may be concluded that the proof, given above, is not rigorous except when an equilibrium of the highest order allowed in the section is involved. However, experience shows that Schreinemakers' rule is obeyed in most cases, and it may be used as a convenient guide when other information is lacking. As an example, the result of a computer-operated calculation of a section through a seven-component system is presented in Fig. 8.11. The rule is satisfied at all the intersections in this diagram.

Figure 8.12 shows an apparent violation of Schreinemakers' rule at the corner of the bct phase field. However, this is not a true phase diagram because S_m and u_i never appear in the same set of conjugate variables.

Usually, Schreinemakers' rule is used to predict the directions of phase boundaries. On the other hand, if the phase boundaries are given, for instance from calculation or experiment, then the rule can help to give the number of phases in the various phase fields. Suppose the arrangement in Fig. 8.13(a) is given, but the numbers of phases in the four adjoining phase fields are not known. One should then extrapolate all the lines, as shown in Fig. 8.13(b). Two of the phase fields will contain one extrapolation each, and these phase fields will be opposite to one another. According to Schreinemakers' rule, these will be the phase fields with the same number of phases, $e - 1$ in Fig. 13(c). Of the

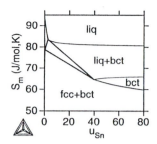

Figure 8.12 S_m, u_{Sn} diagram for Pb–Sn at 1 bar. It shows an apparent violation of Schreinemakers' rule but is not a true phase diagram.

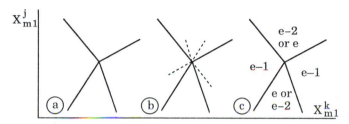

Figure 8.13 Use of Schreinemakers' rule to decide which phase fields have equal numbers of phases.

two remaining phase fields, one will contain two extrapolations and the other none. These phase fields will contain one phase more and one phase less than the others, respectively. However, the rule does not allow us to tell which has more and which less. It would be possible to predict the number of phases in all the phase fields of Fig. 8.11 by this method, if it were known that the phase field in the upper left corner has one phase.

Exercise 8.5.1

Find an error in the following isothermal section of a system with a solid miscibility gap (here denoted by $\alpha + \beta$). The letter 'S' signifies the liquid phase (German 'Schmelze'). (From Masing, 1949.)

Hint

Remember Schreinemakers' rule.

Solution

The $(\alpha + \beta)/\beta$ boundary cuts into the three-phase triangle but the $(S + \beta)/\beta$ boundary cuts into a two-phase field. The diagram should look as the following sketch.

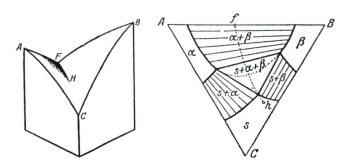

Exercise 8.5.1

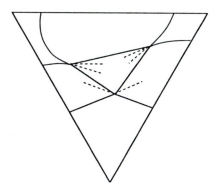

Solution 8.5.1

Exercise 8.5.2

A diagram for a multicomponent system is given but the numbers of phases have been left out except for one phase field. Try to decide the numbers of phases in all the other phase fields.

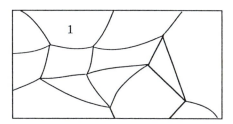

Exercise 8.5.2

Hint

Discuss first what kind of phase diagram it is.

Solution

It looks like a molar diagram because at each point of intersection there are four lines. It may thus be reasonable to use Schreinemakers' rule. The result is shown in the following diagram.

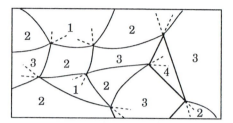

Solution 8.5.2

Exercise 8.5.3

It has been emphasized that Schreinemakers' rule cannot be proved for an H_m, V_m diagram. Show how the method used in this paragraph would fail for that type of diagram.

Hint

The method is based upon a Maxwell relation to be derived from the combined law. In the present case we need a form where H and V are the variables.

Solution

We could try to use the entropy scheme, $dS = (1/T)dU + (P/T)dV - (\mu_i/T)dN_i$ and we get $(\partial[1/T]/\partial V)_{U,N_i} = (\partial[P/T]/\partial U)_{V,N_i}$. The rule could thus be proved for a U_m, V_m diagram. However, if we introduce H through $U = H - PV$ we lose V as a variable, $dS = (1/T)dH - (V/T)dP - (\mu_i/T)dN_i$. We could try to change back to V by adding PV/T: $d(S + PV/T) = (1/T)dH + Pd(V/T) - (\mu_i/T)dN_i$.
Schreinemakers' rule could thus be proved for an $H_m, V_m/T$ diagram but not for an H_m, V_m diagram. This fact is immediately evident from Tables 8.1–8.3 because H_m and V_m do not appear together in any set, i.e. in any one row.

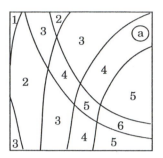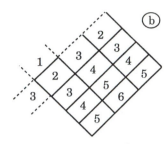

Figure 8.14 Two diagrams topologically equivalent to the sectioned molar phase diagram in Fig. 8.11.

Topology of sectioned molar diagrams

Before leaving the discussion of sections of molar phase diagrams we should further consider the topology of diagrams with several phases. Figure 8.5 showed the elementary unit of a two-dimensional molar diagram. The result of sectioning can vary depending upon the direction of sectioning and the regularity of the diagram before sectioning. However, topologically the whole section can be regarded as composed of intersecting lines, and the elementary unit will be the same as in Fig. 8.5(b). By the same reasoning, a three-dimensional diagram will have elementary units like the one in Fig. 8.6 and will give units like the one in Fig. 8.5(b) after sectioning. A many-dimensional molar phase diagram, after being sectioned a sufficient number of times, may look something like the one illustrated in Fig. 8.14(a). It was constructed to be topologically equivalent to the phase diagram in Fig. 8.11. In Fig. 8.14(b) it has been further simplified but it still has the same topology. This is an unusually simple case. The lines may very well intersect in a more complicated manner, as illustrated in Fig. 8.15.

The observation by Masing can be generalized. For each one of the lines in a two-dimensional section of a molar phase diagram there is a phase which ceases to exist on the line. It is illustrated for a complicated case in Fig. 8.16(a), using the topologically equivalent diagram in Fig. 8.16(b). These lines running through a complicated phase diagram have been called 'zero-phase-fraction' lines by Gupta, Morral and Nowotny (1986) and they can be used as a valuable tool for identifying the phase fields and even for constructing a phase diagram from experimental information. The same principle applies to the surfaces in three-dimensional sections of molar phase diagrams.

Exercise 8.6.1

In order to investigate the phase relations in a quinary system Gupta, Morral and Nowotny (1986) established equilibrium in 21 alloys by isothermal treatment at 1400 K and 1 bar. All alloys had the same molar content of two components. The

(a)

1	2	3	2	1
2	3	4	3	2
3	4	5	4	3
2	3	4	3	2
1	2	3	2	1

(b)

1	2	1/2	2	1
2	3	(4)	3	2
3	4	5	4	3
2	3	4	3	2
1	2	3	2	1

(c)

1	2		1
2	3	4	3	2
3	4	5	4	3
2	3	4	3	2
1	2	3	2	1

Figure 8.15 Some possibilities for the topology of a sectioned molar phase diagram with several phases. Both axes are molar axes.

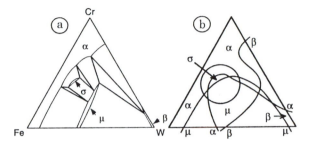

Figure 8.16 (a) The Fe–W–Cr phase diagram at 1 bar and 1673 K. α and β are both bcc but do not mix completely. μ and σ are intermetallic phases. (b) Topologically identical diagram but drawn with lines without any sharp points. These lines represent the limit of existence for one phase each, as given by the letters outside the triangle. The circle is the limit for σ.

phases found in the various alloys could thus be shown in a composition triangle (see the diagram where the compositions are represented by the relative fractions of the three remaining components). Draw a reasonable phase diagram.

Exercise 8.6.1

Hint

The diagram is a molar phase diagram. Start by drawing lines showing the limit of existence of each phase (zero-phase-fraction lines). Improve the diagram by making the various phase boundaries reasonably straight. Phase boundaries for invariant equilibria must be quite straight. Improve the diagram further by applying Schreinemakers' rule.

Solution

At constant T and P the maximum number of phases in a quinary system is five. None of the alloys falls in such a phase field. All the phase boundaries may thus be curved but we may find that it is possible to use straight lines which is preferable when we do not know in which direction a line should be curved. The following diagram shows a possible solution.

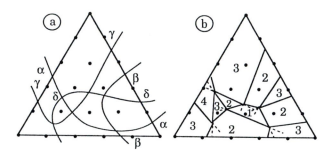

Solution 8.6.1

Exercise 8.6.2

Draw the zero-phase-fraction line for the β phase in the following T,x phase diagram for a binary system at 1 bar.

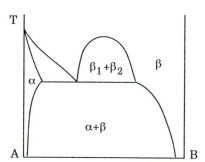

Exercise 8.6.2

Hint

Due to the miscibility gap we can distinguish between β_1 and β_2. Draw the line for each one of them. Then join the two lines. Furthermore, imagine that the three-phase horizontal is a very thin triangle.

Solution

See thick line in diagram. It may be regarded as two lines meeting at the consolute point, one each for β_1 and β_2.

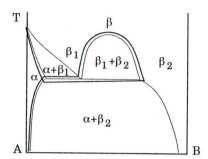

Solution 8.6.2

9 Projected and mixed phase diagrams

9.1 Schreinemakers' projection of potential phase diagrams

Another method of reducing the number of axes is based on projection. By projecting all the features onto one side of the phase diagram, one will retain all the features, but the features of the highest dimensionality will no longer be visible because the dimensionality of a geometrical element will decrease by one unit by projection and they may thus overlap each other and also overlap features of the next-higher dimensionality. As an example, Fig. 9.1(b) shows a P,T diagram obtained by projection of Fig. 7.9 (shown again as Fig. 9.1(a)) in the μ_B direction. Such a P,T diagram is called Schreinemakers' projection (Schreinemakers, 1911). In a system with c components it is obtained by projecting in the directions of $(c - 1)\,\mu_i$ axes. It will show invariant equilibria with $c + 2$ phases as points, univariant equilibria with $c + 1$ phases as lines and in the angles between them there will be surfaces representing divariant equilibria with c phases. Using a short-hand notation developed by Schreinemakers, the coexistence lines for $c - 1$ phases are here identified also by giving in parentheses the phases from the invariant equilibrium which does *not take part*. For example, the (α) curve represents the α-absent equilibrium, i.e. $\beta + \gamma + \delta$. By comparison with Fig. 9.1(a) it can be seen that the angle between (α) and (β) is covered by the $\gamma + \delta$ surface but also by the $\alpha + \delta$ surface which extends to the (γ) line and by the $\beta + \gamma$ surface which extends to the (δ) line. The α one-phase field covers the whole diagram and the other one-phase fields each cover part of it.

Suppose we have a binary system with five phases, denoted 1, 2, 3, 4 and 5. An invariant equilibrium would have four phases. Suppose the system shows two such equilibria and by giving the absent phase they may be denoted (1) and (5). The complete phase diagram would have three dimensions (same as for a one-phase field). Projection would give just one of the diagrams shown in Fig. 9.2 but by presenting two diagrams obtained by projection in slightly different directions as a stereographic pair one can preserve the three-dimensional information. It is thus evident that the apparent intersection between the lines (1,4) and (5,3) is not an intersection in three dimensions. Therefore, it does not represent an invariant equilibrium.

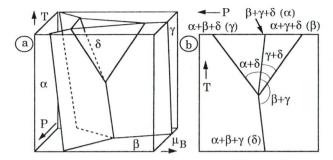

Figure 9.1 (a) The binary potential phase diagram of Fig. 7.9 reproduced to illustrate the projection in the μ_B direction. (b) The diagram obtained by projection. The positions of some of the two-phase surfaces are shown.

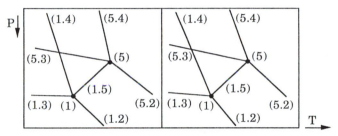

Figure 9.2 Stereographic pair of Schreinemakers' projection of a binary system, showing the three-dimensional shape. The phases not taking part in an equilibrium are given in parentheses. Lines (1,4) and (5,3) do not intersect.

T,P diagrams obtained by projection are particularly useful for multinary systems and are obtained by projecting in the direction of all the independent chemical potentials. We shall return to such diagrams in Section 9.5 but first we shall consider simpler cases.

In a projected diagram one sometimes includes a series of parallel sections drawn with thinner lines. Such lines may be called equipotentials (or isotherms or isobars when appropriate). Such a section was presented in Fig. 7.10. Figure 9.3(b) shows a diagram with several parallel sections. In order to simplify this picture, only the equilibria with the δ phase are shown here. Arrows within the figure show the direction of decreasing temperature.

Sometimes one uses both projecting and equipotential sectioning in order to reduce the number of axes. One may be interested in the changes of various phase equilibria with T and the chemical potential of some volatile component, e.g. oxygen, and one is willing to limit the information by making an equipotential section at $P = 1$ bar. Figure 9.4 gives an illustration from a quaternary system. According to Gibbs' phase rule an invariant equilibrium is obtained with $c + 2 = 6$ phases for $c = 4$, and six invariant equilibria should emanate from it. Let us denote the phases by 1 to 6. A section at $P = P_1$ will cut through the lines (1), (2) and (3). They will thus appear as points in the

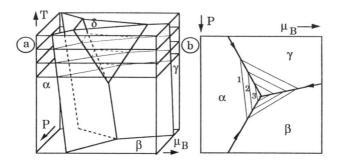

Figure 9.3 (a) Equipotential sections inserted in the potential phase diagram of Fig. 7.9. (b) the Projection of the same potential phase diagram with inserted equipotential sections of two-phase surfaces involving the δ phase.

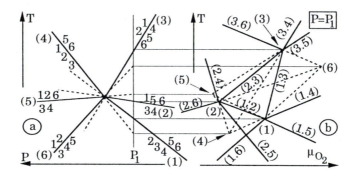

Figure 9.4 *Left*: Schreinemakers' projection of a quaternary system. *Right*: Section at $P = P_1$. The new axis, μ_{O_2}, is one of those projected in the T,P diagram.

right-hand part where one of the projected axes, μ_{O_2}, is now shown. The surfaces extending between the lines in the T,P diagram (see Fig. 9.1(b)) will appear as lines in the section. A major difference between the two diagrams is that to the left, all other potentials were projected but, to the right, one of them, P, was sectioned.

Exercise 9.1.1

Find the section of the $2 + 4 + 5 + 6$ equilibrium in the T,μ_{O_2} diagram of Fig. 9.4. Then, find its surface in the T,P diagram.

Hint

How would that equilibrium be denoted using the absent phases?

Solution

It would be (1,3). In the T,μ_{O_2} diagram it is represented by a line between the points

(1) and (3). In the T,P diagram its surface covers the angle between the (1) and (3) lines.

Exercise 9.1.2

A series of Fe–Cr alloys are sealed in a silica capsule together with chips of an Fe–C and an Fe–N alloy. After heat treatment for a long time at 1273 K, it is found that four phases are present in the capsule, not counting the gas phase. The experiment is repeated several times with different amounts of chips and in some of those experiments four phases are again found. It may be assumed that Cr is not transferred between the alloys but C and N are distributed to equilibrium. Can the potentials of C and N be different in the capsules containing four phases?

Hint

The gas phase allows C and N to be transferred between the specimens but, except for that, the gas phase is omitted from the discussion. We may thus treat P as constant because its value is too low to affect the properties of the solid phases. In addition, T is kept constant. With four components we thus have the same situation as in a binary system at variable P and T. We may use Fig. 9.1(b) and identify μ_B with μ_{Cr}, T with μ_C and $-P$ with μ_N.

Solution

In each capsule we have uniform μ_C and μ_N at the end of the experiment. It is thus useful to look at a μ_N,μ_C diagram obtained by projection in the μ_{Cr} direction. Each experiment should be represented by a point in that diagram but individual specimens would fall on different positions along the projected μ_{Cr} axis. Such a diagram is given in Fig. 9.1(b). We can see that all experiments falling between lines (α) and (β) may cut through three two-phase surfaces, together involving all four phases. With all such combinations of μ_C and μ_N we will cut through all four one-phase fields in Fig. 9.1(a). Four phases can thus be found in several of the experiments with different values of μ_C and μ_N.

9.2

The phase field rule and projected diagrams

In Section 7.5 we derived the phase field rule for equipotential sections. It will now be derived for projections and the two forms will then be combined. We begin by realizing that the dimensionality of a diagram, r, which is $c + 1$ for the complete potential phase diagram, will be one unit less for each projection. Let n_{pr} denote the number of projections, then

$$r = c + 1 - n_{pr}$$

Phase fields with a low dimensionality will not change their dimensionality by the projection. For instance, a point will still be a point. The variance v of a phase equilibrium is given by Gibbs' phase rule and we can use that rule to calculate the dimensionality of the corresponding phase field, d. Let us first consider the case where the dimensionality of a phase field has not changed. By introducing r and eliminating c we get

$$d = v = c + 2 - p = r + 1 - p + n_{pr}$$

However, when the dimensionality of the diagram after repeated projections has decreased to the dimensionality of the phase field under discussion, i.e. $r = d$, which happens when $n_{pr} = p - 1$, then the dimensionality of the phase field will start to decrease by the next projection because it cannot be larger than the dimensionality of the new diagram. As an example, the three-dimensional one-phase fields in Fig. 9.1(a) become two-dimensional in the projection in Fig. 9.1(b). We thus have

$$d = r \text{ for } n_{pr} \geq p - 1$$

By introducing c instead of r, this can be written as

$$d = r = c + 1 - n_{pr} \text{ for } p \leq 1 + n_{pr}$$

Sometimes one uses both projecting and equipotential sectioning in order to reduce the number of axes to two. Since each such sectioning decreases both the dimensionality of all the phase fields and the dimensionality of the diagram by one unit, we can still use the relations between d and r. However, in the relations between d and c we must replace c by $c - n_s$, where n_s is the number of equipotential sections. The phase field rule thus takes the following form,

$$d = c + 2 - p - n_s = r + 1 - p + n_{pr} \text{ for } p \geq 1 + n_{pr}$$
$$d = r = c + 1 - n_{pr} - n_s \qquad\qquad \text{ for } p \leq 1 + n_{pr}$$

A practical example is given in Fig. 9.5. It concerns the Fe–O–S system. Since there are four lines radiating from each point we may conclude that the invariant equilibria concern four phases. The phase field rule gives, for large p,

$$0 = c + 2 - p - n_s = 3 + 2 - 4 - n_s = r + 1 - p + n_{pr} = 2 + 1 - 4 + n_{pr}$$
$$n_s = 1; \ n_{pr} = 1$$

It is evident that one has sectioned at a constant value of some potential, probably P at 1 bar. Then one has projected once, in the direction of μ_S or μ_{Fe}. However, it must be remembered that the complete phase diagram was first obtained from the fundamental property diagram by projecting in the direction of some μ_i, the one which was consider-

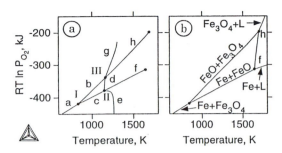

Figure 9.5 Left: Projected and sectioned (P = 1bar) phase diagram for the Fe–O–S system. (a) Fe + Fe$_3$O$_4$ + FeS; (b) FeO + Fe$_3$O$_4$ + FeS; (c) Fe + FeO + FeS; (d) FeO + FeS + L; (e) Fe + FeS + L; (f) Fe + FeO + L; (g) Fe$_3$O$_4$ + FeS + L; (h) FeO + Fe$_3$O$_4$ + L. The invariant four-phase equilibria are (I) Fe + FeO + Fe$_3$O$_4$ + FeS; (II) Fe + FeO + FeS + L; (III) FeO + Fe$_3$O$_4$ + FeS + L. *Right*: The Fe–O bottom plate.

ed as the dependent variable. Figure 9.5 has thus been obtained from the fundamental property diagram by projecting twice, and sectioning once, and it does not matter which projection was made first, μ_S or μ_{Fe}. It should be emphasized that n_{pr} represents the number of projections of the complete phase diagram. The first projection of the fundamental property diagram is not included in n_{pr}.

It is interesting to note that the two three-phase lines h (FeO + Fe$_3$O$_4$ + L) and f (Fe + FeO + L) stop inside the diagram. They stop at invariant three-phase points in the binary Fe–O system which overlaps the diagram. In principle, the whole surface of the diagram is covered by the binary Fe–O diagram which may be regarded as the bottom plate of the three-dimensional phase diagram, where $\mu_S = -\infty$, assuming that the fundamental property diagram was first projected in the μ_{Fe} direction to give a phase diagram. On the bottom plate, there is no S and the number of components c is thus 2 instead of 3. Three-phase equilibria would thus appear as points and two-phase equilibria as lines. That bottom plate is shown to the right but was not included to the left because it would have made the diagram difficult to interpret. Only the binary end-points for FeO + Fe$_3$O$_4$ + L and Fe + FeO + L were marked.

In many cases one should consider the top plate as well as the bottom plate. A $\log P_{SO_2}, \log P_{O_2}$ diagram of the Cu–Mn–O–S system under $P = 1$ bar and $T = 1000$ K would be an example. The top and bottom would represent the Cu–O–S and Mn–O–S systems, respectively, if the projected axis is taken as $(\mu_{Cu} - \mu_{Mn})$. These diagrams are given in Fig. 9.6.

The solubilities of Cu in the Mn phases and of Mn in the Cu phases are all very low. The equilibria between the Cu–O–S phases will not be affected by the presence of Mn, nor the Mn–O–S phases by Cu. Both diagrams can thus be plotted in the same $RT\ln P_{SO_2}, RT\ln P_{O_2}$ coordinate frame to form the Cu–Mn–O–S diagram, Fig. 9.7. The lines in the ternary systems can be copied into the quaternary system and they become surfaces in the projected direction and still appear as lines in the Cu–Mn–O–S diagram.

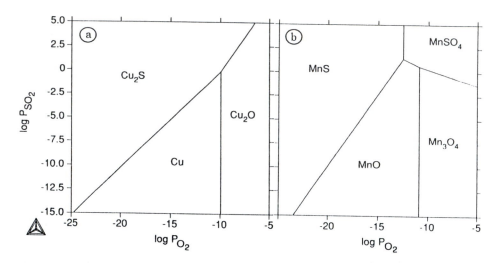

Figure 9.6 Potential phase diagrams for (a) Cu–O–S and (b) Mn–O–S at 1 bar and 1000 K. These phase diagrams are two-dimensional and are not projections.

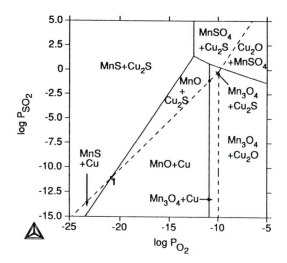

Figure 9.7 Projected potential diagram for Cu–Mn–O–S at 1 bar and 1000 K. For clarity, all the lines from the Cu–O–S side are presented with dashed lines here.

New two-phase surfaces form between the previous one-phase fields and they are identified in the diagram.

An interesting question is the choice of projected axis in Fig. 9.7. In order to treat Cu and Mn in a symmetric way, it is convenient to consider $(\mu_{Cu} + \mu_{Mn})$ as the projected axis to give a phase diagram from the fundamental property diagram and then $(\mu_{Cu} - \mu_{Mn})$ as the axis used for projection of the phase diagram to reduce the number of axes to two.

Exercise 9.2.1

Only three lines intersect at the invariant equilibrium I in Fig. 9.5. What line is not shown and why not?

Hint

The fourth line should be the one without FeS. The reason why it is not shown may be found by comparing with Exercise 7.7.2.

Solution

The $Fe + FeO + Fe_3O_4$ line is not shown because one has projected the diagram exactly in its direction and the line thus appears as a point. The reason is that the projection has been made in the μ_S direction and S does not dissolve in any one of these three phases. Thus, S can have no effect on the equilibrium $Fe + FeO + Fe_3O_4$ and its line goes exactly in the μ_S direction. It exists at a certain T, P_{O_2}, only, and it is shown in binary Fe–O diagram to the right in Fig. 9.5.

Exercise 9.2.2

At 1000 K one measures the emf of an electrolytic cell where one electrode is a mixture of MnS, MnO, Cu_2S and Cu and the other is a mixture of Cu_2O and Cu. The electrolyte is solid zirconia stabilized with calcia. Use Fig. 9.7 to estimate the resulting emf.

Hint

The electrical current can pass through this electrolyte mainly by the diffusion of O^{-2} ions. The emf will thus be an expression of the difference in oxygen potential between the two electrodes and it can be estimated from the difference in $RT \ln P_{O_2}$ for two points in the diagram representing the electrodes.

Solution

The point for $Cu_2O + Cu$ can be taken anywhere on the corresponding line yielding $\log P_{O_2} = -9.8$. The other point is obtained as the intersection between two lines. It is marked as point 1 in the diagram and yields $\log P_{O_2} = -21.2$. We get $\Delta\mu_O = 0.5\Delta\mu_{O_2} = 0.5RT(\ln P'_{O_2} - \ln P''_{O_2}) = 0.5RT\ln 10 \cdot (-9.8 + 21.2) = 13.1RT$. Remembering that the O ion is divalent we get $E \cdot 2\mathscr{F} = \Delta\mu_O$ where $\mathscr{F}$ is Faraday's constant (96 486 coulomb/mol) and thus $E = 0.567$ V.

9.3 *Relation between molar diagrams and Schreinemakers' projected diagrams*

As demonstrated by Figs. 7.7 and 7.9, the elementary units of potential diagrams are very simple from the topological point of view. In this sense, the projections of such diagrams are more interesting. This is evident if one considers the dashed extrapolations shown in the projected diagram in Fig. 9.8(b). Between the lines there are two extrapolations in one case, one extrapolation in two cases, and no extrapolation in one case. In fact, this is the only way to draw four lines in different directions if the 180° rule is to be obeyed. It is evident that, in the projected direction, the four phases are related to each other in a special way. This phenomenon will now be examined. In order to simplify the discussion the method of identifying a univariant line by giving within parentheses the absent phase is used in Fig. 9.8(b).

If potential axes are chosen for plotting the complete, three-dimensional phase diagram of a binary system, the four phases of an invariant equilibrium will fall on one point. If one molar quantity is introduced, say, X_{m1}^j instead of Y^j, then the four phases will fall on a line, just as the three phases in the three-phase equilibrium in Fig. 8.2(a) fall on a line in Fig. 8.2(b). In that case, it is easy to see the order in which the phases are arranged along the line. The hcp phase must be placed between bcc and fcc. Otherwise, there would be some overlapping of the one-phase fields which is not allowed according to Section 8.1. Using the same reasoning, it is easy to see the order in which the four phases of Fig. 9.8 will be arranged when a molar quantity is introduced. One simply looks at the direction of the two-phase surfaces. Each one will turn into a two-phase volume when z_B is introduced instead of the μ_B axis. In Fig. 9.9 these volumes are demonstrated for the three surfaces between α, β and γ. It is evident that γ must fall between α and β along the z_B axis.

The composition is here expressed with the z_B variable because it is the conjugate variable to μ_B according to Section 8.2. The relative positions of all the four phases along a z_B axis, going through the invariant phase equilibrium are demonstrated schematically in Fig. 9.10.

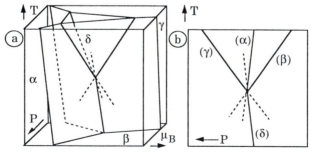

Figure 9.8 (a) Binary diagram and (b) projection in the μ_B direction, taken from Fig. 9.1. The extrapolations of three-phase lines are marked with dashed lines.

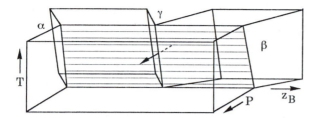

Figure 9.9 Introduction of a molar axis into the potential diagram of Fig. 9.8. Only the lower half of that diagram is used here. The surface marked with horizontal tie-lines represents the $\alpha + \beta + \gamma$ equilibrium.

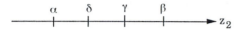

Figure 9.10 One-dimensional molar phase diagram at constant T and P, showing the relative positions of four phases along a z_B axis, introduced instead of a μ_B axis, through the invariant equilibrium in Fig. 9.8(a).

When the phase diagram of Fig. 9.8(a) is projected in the μ_B direction and Fig. 9.8(b) is formed, much information is lost. However, the information regarding the order of arrangement along the projected direction, obtained when the molar quantity is introduced, is not lost. This is because some conclusions can still be drawn regarding the directions of the two-phase surfaces. They are situated between the three-phase lines. This was demonstrated in Fig. 9.1(a). It is thus possible to get an impression of the relative positions of the six surfaces and thus of the relative positions of the phases along the molar axis of the projection.

The simplest method to interpret an experimental diagram like Fig. 9.8(b) is to draw the four extrapolations and then turn the diagram in the same way as Fig. 9.8(b) with respect to the dashed extrapolations. The compositions of the phases will then be arranged in the order given by Fig. 9.10 or in the completely reverse order. A more logical method will be described in the following section.

Exercise 9.3.1

In Exercise 9.1.2 we considered a heat treatment of Fe–Cr specimens under carburizing and nitriding conditions at constant T and P. It has been found that four phases could be present in some experiments, each one made with several alloys. Now try to find what is the maximum number of phases in any one specimen.

Hint

As already explained, we can use Fig. 9.1(b) because our quaternary system at constant T and P behaves like a binary system at variable T and P. In our case the two axes should be μ_C and μ_N and the projection has been made in the z_{Cr} direction.

Solution

If the experiment is carried out under predetermined values of μ_C and μ_N then it is extremely unlikely that one would fall on the four-phase point in the potential diagram and even on any of the three-phase lines. We should then expect no more than two phases in any one of the specimens. On the other hand, if predetermined amounts of C and N are added to the set of specimens in a capsule, then one could not use Fig. 9.1(b) directly. For each specimen one should rather consider a molar diagram like Fig. 8.4 and it is evident that an experiment could hit the four-phase field and four phases could thus occur in the same specimen and this could even happen in more than one specimen in the same experiment.

9.4   *Coincidence of projected surfaces*

The method to determine the relative compositions of phases, now to be described, can be used in higher-order systems as demonstrated in the next section, but a binary system will be considered first.

Suppose one could gradually change the properties of the system in such a way that the $\beta + \gamma$ surface in Fig. 9.8 would rotate around an axis roughly parallel to the T axis. One could thus make the two lines (α) and (δ) in the projection approach each other without changing the topology of the projected diagram. At the moment of coincidence, one has a situation such as that illustrated in Fig. 9.11.

The $\beta + \gamma$ surface is now parallel to the direction of projection, μ_B, and a continued rotation would put the $\beta + \gamma$ surface on the other side of the (α) and (δ) lines. Thus, β and γ would be transposed in Fig. 9.10. It is possible to conclude that the β and γ phases have the same value of z_B if the μ_B projections of the (δ) and (α) lines coincide when they meet at the four-phase point. Evidently, before the gradual rotation the β and γ phases must have been neighbours along the z_B line in Fig. 9.10. When the lines coincide, the phases fall on the same point on the z_B axis. This rule of coincidence is closely related to Konovalov's rule which will be discussed in Section 9.8. The relative positions of the phases for the various cases of coincidence are shown in Fig. 9.12. Other cases of coincidence may occur but not until at least one of these has occurred. Before any rotation, the phases must have been arranged along the z_B axis as shown in Fig. 9.10 or in the completely reverse order.

It is interesting to note from Fig. 9.8 that it should be possible to rotate the $\alpha + \beta$ surface in such a way that the (γ) and (δ) lines approach each other and finally coincide. However, the 180° rule prevents this from happening before there is another coincidence.

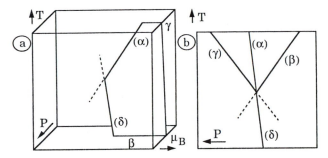

Figure 9.11 Modification of Fig. 9.8 by rotation of the $\beta + \gamma$ surface until it is parallel to the μ_B axis. In the μ_B projection the $\beta + \gamma + \delta(\alpha)$ and $\alpha + \beta + \gamma(\delta)$ lines will now coincide and their extrapolations will not be visible.

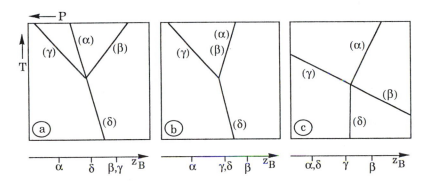

Figure 9.12 Three cases of coincidence of three-phase lines in a projected potential phase diagram obtained by modifying Fig. 9.8.

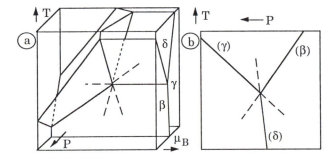

Figure 9.13 Modification of Fig. 9.8 by rotating the entire diagram until the $\beta + \gamma + \delta(\alpha)$ line is parallel to μ_B axis. All three surfaces, $\beta + \gamma$, $\gamma + \delta$ and $\delta + \beta$, are then parallel to μ_B axis. In the μ_B projection to the right, $\beta + \gamma + \delta(\alpha)$ degenerates to a point, and will thus coincide with all the other lines without them coinciding with one another.

What happens if three of the four phases β, γ and δ, have the same z_B value will now be investigated. The three surfaces representing $\beta + \gamma$, $\beta + \delta$ and $\gamma + \delta$ must all be parallel to the μ_B axis and the $\beta + \gamma + \delta(\alpha)$ line must point in the μ_B direction. It thus degenerates to a point. This case is illustrated in Fig. 9.13.

Exercise 9.4.1

Prove mathematically that the compositions of β and γ must coincide when the μ_B axis in a binary system is parallel to the $\beta + \gamma$ surface, as in Fig. 9.11.

Hint

Use one of the last two equations in Section 7.8.

Solution

For a binary system we get

$$(z_B^\beta - z_B^\gamma)\mathrm{d}\mu_B = - (S_{m1}^\beta - S_{m1}^\gamma)\mathrm{d}T + (V_{m1}^\beta - V_{m1}^\gamma)\mathrm{d}P.$$

When the $\beta + \gamma$ surface is parallel to μ_B, then we can change μ_B and stay on the surface without changing T or P, i.e. with $\mathrm{d}T = 0$ and $\mathrm{d}P = 0$. The coefficients of $\mathrm{d}\mu_B$ must thus be zero, i.e. $z_B^\beta - z_B^\gamma$.

Exercise 9.4.2

Suppose one has measured μ_C as function of T at 1 bar for a ternary A–B–C system. What conclusion could be drawn if the diagram looks like Fig. 9.11(b) with μ_C inserted instead of P?

Hint

Compared to Fig. 9.11 we have one component more but the dimensionality has been reduced to the same by keeping P constant. In both cases μ_A and μ_B are the potentials that are not shown, i.e. those used to reduce the number of axes by projection from the four-dimensional fundamental property diagram.

Solution

One of μ_A and μ_B is the dependent variable and the final projection has been made in the direction of the other one. From the coincidence of the (α) and (δ) lines we may conclude that $z_B^\beta = z_B^\gamma$ if μ_A is the dependent variable and $z_A^\beta = z_A^\gamma$ if μ_B is the dependent one. These two results are identical since $z_B = N_B/N_A = 1/z_A$.

Exercise 9.4.3

Darken (1948) has published the following phase diagram for the Fe–Si–O system. We may regard it as an isobaric section at 1 bar and a projection along some chemical-potential axis. Let us concentrate on the invariant equilibrium 3. The lines emanating from that point are as follows: (e) magnetite + fayalite + melt; (h) wüstite + fayalite + melt; (g) magnetite + wüstite + fayalite; (i) magnetite + wüstite + melt.

It is known from experiments that the composition of the melt at point 3 falls at a lower Si content than fayalite ($2FeO \cdot SiO_2$). Check if that is confirmed by the diagram. Knowing that the diagram concerns two more phases, metal and silica, it is possible to identify the other three invariant equilibria. Do that. Examine where the compositions of the melt at points 2, 4 and 5 fall relative to the other phases.

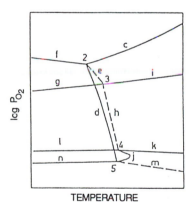

Exercise 9.4.3   TEMPERATURE

Hint

First decide which axis is projected. Then find out on which composition variable the diagram can give information. Remember that we can interpret the $\log P_{O_2}$ axis as a way of expressing the oxygen potential without having a gas present.

Solution

We see no axis for Si or Fe. The diagram is thus projected in the direction of one of them and the other is the dependent variable. We can get information on the ratio

N_{Si}/N_{Fe}. With the method illustrated in Fig. 9.1 we can mark the lines around point 3 in the following way: e-(w); g-(L); h-(mag); i-(fay). Their arrangements can be obtained from their relative compositions (see two alternatives in sketch 3a below). From the information given about the four lines we find that the first alternative applies here. We should thus correct Darken's diagram for point 3 as shown in sketch 3b.

We can go from point 3 to 4 by following line h which represents an equilibrium without magnetite. Point 4 is thus an equilibrium without magnetite which is understandable because we have gone to a lower value of P_{O_2}. Magnetite being the highest oxide has disappeared and the new phase is probably metal (see sketch 4).

As we follow line e from point 3 to 2 we lose wüstite and gain silica (see sketch 2). Finally, we may guess that point 5, falling between points 4 and 2, contains metal and silica in addition to melt and fayalite (see sketch 5).

Accepting Darken's constructions for the points 2, 4 and 5 we find the relative compositions as given in the sketches.

By comparing the position of L in the various diagrams, we find that it is thus necessary that L changes its composition relative to fayalite between 4 and 5 and moves back between 2 and 3. This is why the lines e and j are curved.

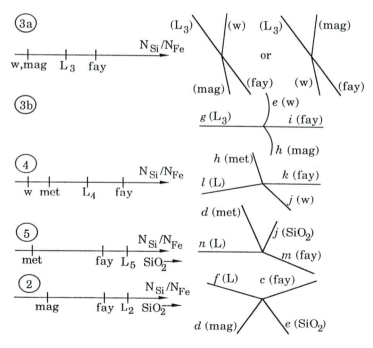

Solution 9.4.3

Exercise 9.4.4

In Exercise 9.2.1 we discussed the reason why only three lines seem to intersect at point I in Fig. 9.5. How could one express the reason in terms of the present discussion on coincidence?

Hint

There are three components in the system. The potential of one is used as an axis. We could apply the arguments to the remaining two components.

Solution

$LogP_{O_2}$ is plotted on one axis and it represents the chemical potential of O. The composition variable which is the same in the three phases, Fe, FeO and Fe_3O_4, is thus the ratio between the amounts of the other components, S and Fe, i.e. $z_S \equiv N_S/N_{Fe}$. It is in fact zero for these three phases because they all have $N_S = 0$.

9.5   *Projection of higher-order invariant equilibria*

The topological examination may be extended to higher-order invariant equilibria and adjoining univariant equilibria, although the visibility is then lost. However, it has been shown by analytical methods (Schreinemakers, 1915; Morey and Williamson, 1918; Morey, 1936) that the same principles, which have been derived here by inspection, apply. Three components will yield a four-dimensional phase diagram and it must be projected twice in order to yield a two-dimensional picture. It may show an invariant five-phase equilibrium and five adjoining four-phase lines. If no lines coincide, they can arrange themselves in three different ways, as illustrated in Fig. 9.14.

For a closer discussion of the compositions of the phases taking part in the five-phase equilibrium, the two potentials on the axes will be kept constant at the values of the invariant equilibrium, while the two projected potentials will be replaced by their conjugate molar quantities. The five phases will fall on different points on the plane formed by the two molar quantities, and Fig. 9.15 illustrates the arrangement of the phases in the three different cases. Three phases may here be regarded as neighbours if their points can be connected to form a triangle with no other point inside and if the triangle can be changed into a line without any one of its points first moving inside any other such triangle. If two lines in the projected potential phase diagram coincide, then the three phases they have in common will fall on a straight line in the molar diagram.

Four components yield a five-dimensional phase diagram and it must be projected three times in order to yield a two-dimensional picture. A six-phase equilib-

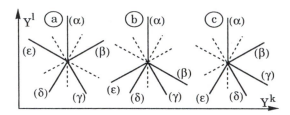

Figure 9.14 Possible Schreinemakers' projections for a ternary system obtained by projecting the Y^j, Y^k, Y^m, Y^n phase diagram in Y^m and Y^n directions. Points represent invariant five-phase equilibria. The five lines emanating from each point represent four-phase equilibria, and are identified by giving the absent phase in parentheses. Dashed lines are metastable extrapolations.

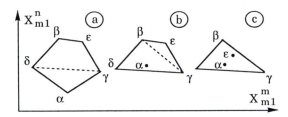

Figure 9.15 Molar phase diagram at constant Y^k and Y^j showing the relative positions of the five phases in the invariant equilibria in Fig. 9.14. The change occurring when lines (ε) and (β) in the left-hand picture are rotated to approach each other can be illustrated by moving point α towards the straight line between δ and γ. The middle picture is obtained by letting (ε) and (β) lines pass one another, thus making point α cross the δ–γ tie-line. The right-hand picture is obtained by letting (α) and (δ) lines rotate and pass one another, whereupon ε will cross the β–γ line.

rium will be invariant and represented by a point from which six lines will emanate, representing univariant equilibria with one phase absent in each. The relative positions of the six lines will reveal how the six phases are arranged in the three-dimensional compositional space formed by three molar quantities. However, this is not easy to visualize.

The rule relating the coincidence of lines in a projected potential phase diagram to the positions of the common phases in the space formed by the molar quantities has been called the 'Coincidence theorem' (Morey, 1936). The theorem can be generalized as demonstrated by the following example. Suppose that two of the points, α and δ, in the left-hand picture of Fig. 9.15 coincide. Then δ, α and γ fall on a line, and lines (β) and (ε) in Fig. 9.14 should coincide. However, β, δ and α would also fall on a line, and lines (γ) and (ε) in Fig. 9.14 should also coincide. As a consequence all three lines, (β), (γ) and (ε), should coincide. It is thus possible to generalize the coincidence theorem as follows. Consider a two-dimensional projection of an r-dimensional potential diagram. It may have an invariant equilibrium involving $r + 1$ phases. From this point, $r + 1$ univariant equilibria, each with r phases, emanate. The theorem concerns the positions of the phases in the $(r - 2)$-dimensional space formed by the molar quantities conjugate to the projected potentials. If t of the phases fall in a $(t-2)$-dimensional section through that space, then all the univariant equilibria, which contain the t phases, coincide in the two-dimensional projection. There would be $r + 1 - t$ such equilibria.

Exercise 9.5.1

In Exercise 9.4.3 we discussed Darken's $\log P_{O_2}, T$ diagram for the Fe–Si–O system. The univariant equilibria around point 4 can be identified as shown in the diagram.

We regarded Darken's diagram as a section at a constant pressure of 1 bar and as a projection in the μ_{Si} direction. However, Darken himself considered the diagram as a projection in the P directions as well as the μ_{Si} direction. Thus, he should have included the gas phase and in the above picture there should have been a line marked (gas) and representing the change of the L + met + fay + w equilibrium with pressure. We can easily add that line because it is well known that the pressure has a very small effect on the stabilities of condensed phases. The line should thus be almost vertical and go upwards in the plane of the picture. Make this modification to the diagram and use it then in order to test what can be said about the molar volume of the liquid phase relative to the others.

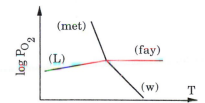

Exercise 9.5.1

Hint

Extend all the lines across point 4 and compare with the pictures in Fig. 9.14. Then look at the corresponding picture in Fig. 9.15. Remember that we have already decided that the value of N_{Si}/N_{Fe} for L in point 4 is lower than the value for fayalite.

Solution

We find that our case resembles the picture in the middle of Figs. 9.14 and 9.15. We can identify the phases as shown to the upper left in the diagram below. From our knowledge of the compositions of the phases and an assumption that the molar volumes are roughly the same in all the condensed phases we can draw the w + met + fay + gas quadrangle and place the point for L as seen to the lower right in the diagram V_m/x_{Fe} of L is thus above the tie-line met + fay.

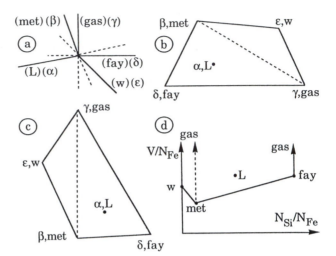

Solution 9.5.1

9.6   # The phase field rule and mixed diagrams

The number of axes in a complete phase diagram, whether a potential one or a molar one, is $r = c + 1$. For a closed system one has fixed the composition and has actually sectioned at $c − 1$ molar axes. The number of remaining axes is $r = c + 1 − (c − 1) = 2$. For a closed system the equilibrium state is thus uniquely defined by choosing values for T and P or their conjugate variables. This is called **Duhem's theorem**.

In the most common type of phase diagram there is a temperature axis and a composition axis. It is thus an example of phase diagrams with a mixture of potential and molar axes. Such diagrams are more complicated and due to the large variety no general description will be given. However, it is worth discussing how the phase field rule can be generalized to such diagrams but first it should be emphasized that the discussion only concerns true phase diagrams, i.e. diagrams obtained from a single set of conjugate variables. The nine possibilities were discussed in Section 3.4 and they resulted in 27 sets when molar variables were introduced in Section 8.2. As a consequence, all the variables in a mixed diagram, including those that have been projected or sectioned, must come from one of the 27 sets.

Figure 8.2 demonstrates that the dimensionality of each phase field increases by one unit for each molar quantity being introduced, until it is equal to the number of axes in the phase diagram. This increase does not disappear on sectioning a molar axis. The phase field rule generalized to a projected and sectioned diagram in Section 9.2 may thus be generalized as follows

$$d = c + 2 − p − n_s + n_m$$
$$= r + 1 − p + n_{pr} + n_m \quad \text{for } p \geq 1 + n_{pr} + n_m \text{ i.e. for } d \leq r$$

where n_{pr} is the number of projections, n_s is the total number of sectioned quantities, potentials as well as molar quantities, and n_m is the total number of molar quantities used, i.e. sectioned molar quantities, n_{ms}, as well as molar axes in the final diagram, n_{ma}. Of course, $n_m = n_{ms} + n_{ma}$. On the other hand, if we project a phase diagram in the direction of an axis, then it does not matter what kind of variable was chosen on that axis, a potential or its conjugate molar quantity. The projected phase diagram will look the same and all the projections will thus have the same effect on the phase field rule. The number of projected molar quantities should not be included in n_m.

A limit of validity of this relation is obtained from the fact that the dimensionality d of a phase field in a diagram can never be larger than the dimensionality r of the diagram itself. The limit of validity is thus $p = 1 + n_{pr} + n_m$. For this and all lower p values we thus have

$$d = r = c + 1 - n_{pr} - n_s \text{ for } p \leq 1 + n_{pr} + n_m$$

A few more considerations of the properties of mixed diagrams should be added. The lowest possible dimensionality of a phase field will occur for the maximum number of phases. In a potential phase diagram that dimensionality will be zero but it is evident from the preceding discussion that it will increase by one unit for each molar axis and the lowest possible dimensionality will thus be equal to the number of molar axes in the final diagram, n_{ma}, but this will occur at the maximum number of phases. By inserting $n_{ms} + n_{ma} = n_m$ we obtain

$$n_{ma} = d_{min} = c + 2 - p_{max} - n_s + n_{ma} + n_{ms}$$
$$= r + 1 - p_{max} + n_{pr} + n_{ma} + n_{ms}$$
$$p_{max} = c + 2 - n_s + n_{ms} = r + 1 + n_{pr} + n_{ms}$$

The MPL boundary rule can be applied to mixed diagrams but only with caution. It is important first to distinguish between phase fields and phase boundaries. The rule concerns two adjoining phase fields separated by a phase boundary. As an example, we may examine the case illustrated in Fig. 8.2(c). It is reproduced in Fig. 9.16 without tie-lines and with the three-phase field bcc + fcc + hcp marked with a thick line. All the other lines are phase boundaries. The MPL rule cannot be applied to the contact between fcc and bcc + hcp phase fields because they are not connected by a phase boundary but separated by the three-phase field bcc + fcc + hcp. For the contact between fcc and bcc + fcc + hcp the rule gives

$$b = r - D^+ - D^- = 2 - 2 - 0 = 0$$

in agreement with the fact that these two phase fields meet at a point, only. For the contact between bcc + hcp and bcc + fcc the rule gives

$$b = r - D^+ - D^- = 2 - 1 - 1 = 0$$

This is also correct because these two phase fields do not make contact along the thick horizontal line, where they are separated by the bcc + fcc + hcp phase field. They only make contact at the upper end-point of the thick line. Cases like this can be easily analyzed by imagining that the one-dimensional phase field is a very thin triangle (Palatnik and Landau, 1964). That method is also helpful if one wants to draw zero-

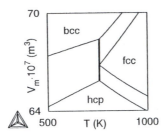

Figure 9.16 Mixed phase diagram from Fig. 8.2(c), reproduced without tie-lines. The thick line represents the three-phase field. All other lines represent phase boundaries.

phase-fraction lines. Each one-dimensional phase field will have one such line on each side.

Exercise 9.6.1

The following T,%C phase diagram is for a quaternary A–B–C–D system at 20% B and 20% D and at 1 bar. Test it with the phase field rule.

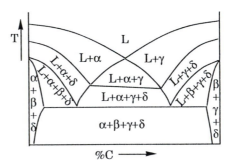

Exercise 9.6.1   %C $\longrightarrow$

Solution

There are four components, $c = 4$. The complete phase diagram has been sectioned three times, $n_s = 3$ but two of the sections were for molar quantities, $n_{ms} = 2$. In the final diagram there is one molar axis, $n_{ma} = 1$. There is no projection, $n_{pr} = 0$. In the diagram we can see a horizontal line. Let us test if it is a phase field or just a boundary between two-dimensional phase fields. A line has the dimensionality 1 and it thus gives $1 = d = c + 2 - p - n_s + n_{ma} + n_{ms} = 4 + 2 - p - 3 + 1 + 2 = 6 - p$; $p = 5$. If the horizontal line is a phase field, it should have five phases. From the neighbouring phase fields we find $\alpha + \beta + \gamma + \delta + L$. We may conclude that this line is a phase field. The diagram is two-dimensional, $r = 2$. Let us now check for what number of phases a field should be two-dimensional. $p \le 1 + n_{pr} + n_{ma} + n_{ms} = 1 + 0 + 1 + 2 = 4$. This is confirmed by the diagram.

Exercise 9.6.2

Test the equations for p_{max} and d by application to the following sections taken through a quaternary phase diagram at 1 bar.

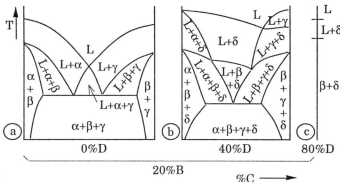

Exercise 9.6.2

Solution

$p_{max} = r + 1 + n_p + n_{ms} = c + 2 - n_s + n_{ms}$ and the smallest $d = c + 2 - p_{max} - n_s + n_{ma} + n_{ms}$ for large p. Diagram (a) has 0%D and thus $c = 3$. It has been sectioned twice (20%B and 1 bar), $n_s = 2$, but one of the sectionings was for a molar quantity (%B) and thus $n_{ms} = 1$. This gives $p_{max} = 2 + 1 + 0 + 1 = 4$ and we can indeed see four phases at the horizontal line, $L + \alpha + \beta + \gamma$. The diagram has one molar axis (molar content of either A or C) and thus $n_{ma} = 1$. We thus get $d = c + 2 - p - n_s + n_{ma} + n_{ms} = 3 + 2 - 4 - 2 + 1 + 1 = 1$ and that is the horizontal line. Diagram (b) has one more component but also one more sectioned molar quantity. Thus, $p_{max} = 2 + 1 + 0 + 2 = 5$ and indeed we see five phases at the horizontal line, $L + \alpha + \beta + \gamma + \delta$. We get the smallest $d = 4 + 2 - 5 - 3 + 1 + 2 = 1$ and this is for the horizontal line. Diagram (c) has 20%B and 80%D and c is thus 2 (only B and D). We get $p_{max} = 1 + 1 + 0 + 1 = 3$ and we see three phases at a point, $L + \alpha + \beta$. The smallest $d = 2 + 2 - 3 - 2 + 0 + 1 = 0$, in agreement with the fact that the three phases occur at a point.

Exercise 9.6.3

Two sections of a system with a solid miscibility gap are here reproduced from Masing (1949). In the so-called vertical section a phase field $S + \alpha + \beta$ appears twice. Check if the author has made an error or not.

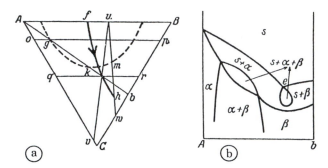

Exercise 9.6.3   (a)   (b)

Hint

The three-phase field in an isothermal section is a triangle.

Solution

In the vertical section the three-phase triangle cannot appear twice at the same temperature. Consequently, the two $S + \alpha + \beta$ regions must be separated in temperature. By sketching a series of triangles in the left-hand diagram, which is a temperature projection, we can see that the $S + \alpha + \beta$ region, situated below the eutectic point e, must fall at lower temperatures than the other one.

9.7   *Selection of axes in mixed diagrams*

For mixed diagrams it is particularly important to pay attention to how the axes are selected. As already emphasized, they must all come from a set of conjugate variables and one from each conjugate pair. The various possibilities are listed in Tables 8.1–8.3, Section 8.2. A number of examples will now be given in order to demonstrate what could otherwise happen.

Figure 9.17 shows part of the T, S_m diagram for Fe. Two two-phase fields overlap which is made possible by the fact that T and S_m do not represent different conjugate pairs in any of the set in the Tables. This is not a true phase diagram.

Figure 9.18 shows the S_m, μ_{Pb} diagram for the Pb–Sn system at 1 bar. Two two-phase fields overlap because S_m and μ_{Pb} do not come from the same set of conjugate pairs. This is not a true phase diagram. One should have combined S_m with $\mu_{Pb} - \mu_{Sn}$ or S_{m1} with μ_{Pb}.

Figure 9.19 shows the x_{Cr}, a_C diagram for Fe–Cr–C at 1 bar and 1200 K. The intersecting phase boundaries in the upper left corner, forming two 'swallow-tails',

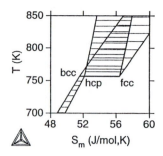

Figure 9.17 T, S_m diagram for Fe. This is not a true phase diagram.

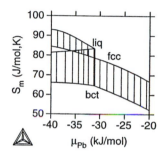

Figure 9.18 S_m, μ_{Pb} diagram for Pb–Sn at 1 bar. This is not a true phase diagram.

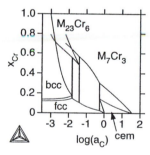

Figure 9.19 x_{Cr}, a_C diagram for Fe–Cr–C at 1 bar and 1200 K. This is not a true phase diagram.

indicate that this is not a true phase diagram. The activity a_C can be regarded as an expression for μ_C/T and should have been combined with u_{Cr} or z_{Cr} but not x_{Cr}.

Figure 9.20 shows the T, x_{Cr} diagram for Fe–Cr–C at 1 bar and $a_C = 0.3$, relative to graphite. The two intersecting phase boundaries on the right-hand side indicate that this is not a true phase diagram. The two axes, T and x_{Cr}, do belong to the same set of conjugate variables but one must also consider the sectioned axes. In this case one has sectioned at constant P and a_C, i.e. μ_C/T. However, μ_C/T and x_{Cr} do not belong to the same set.

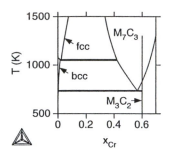

Figure 9.20 T,x_{Cr} diagram for Fe–Cr–C at 1 bar and $a_C = 0.3$. This is not a true phase diagram.

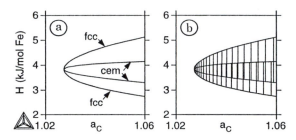

Figure 9.21 H_{m1},a_C diagram for Fe–C at 1 bar. This is not a true phase diagram as revealed by the overlapping two-phase fields, shown when the tie-lines are included in (b).

Figure 9.21(a) shows the H_{m1},a_C diagram for Fe–C at 1 bar. This is not a true phase diagram although H_{m1} and μ_C/T, here represented by a_C, come from the same set of conjugate variables. That is revealed by the tie-lines which are included in Fig. 9.21(b). The reason is that the numerical values used for H_{m1} refer to standard states of Fe and C at 298 K but a_C refers to graphite at the actual temperatures. It is evident that one should also be careful when representing the oxygen potential with P_{O_2}. It is only under isothermal conditions that it should be combined with an axis for a molar quantity given relative to references at 298 K. It should be noted that the diagrams in Figs. 7.11 and 7.12 used $\mu_C - {}^{\circ}G_C^{gr}$ or $(\mu_C - {}^{\circ}G_C^{gr})/T$ as an axis and ${}^{\circ}G_C^{gr}$ was defined at the actual temperature which varied. That did not cause any problem because there was no molar axis in those diagrams.

Exercise 9.7.1

Figure 9.20 showed an incorrect selection of axes. If one really wanted to section at a constant value of a_C, what composition axis should one have used?

Hint

Consult the Tables in Section 8.2.

Solution

a_C represents μ_C/T which may be combined with $-1/T$, $-P$ and z_{Cr} (according to fifth row in Table 8.2) or u_{Cr} (according to fifth row in Table 8.3).

9.8

Konovalov's rule

The rule that two one-phase fields are separated from each other by a positive distance, when the proper molar quantity is introduced instead of a potential, was described in Section 8.1. That rule is not as trivial as it may appear. It was discovered experimentally by Konovalov (1881) when measuring the vapour pressure of liquid solutions of water and various organic substances under isothermal conditions. He established that, compared with the solution, the vapour contains a higher relative content of that component which, when added to the solution, increases the total vapour pressure. In addition, he found two cases with a pressure maximum and realized that the liquid and vapour must have the same composition at such a point. A case of this type is shown in Fig. 9.22(b), and it is evident that it is simply due to the fact that the molar quantity which is used, here z_B, replaces a potential whose axis happens to be parallel to a *line* tangential to the linear two-phase field in the potential diagram. Except for that, the system has no unique properties at this point. The point is sometimes called a **singular point** and the equilibrium under this special condition is called **singular equilibrium**.

Figure 9.23(a) shows a three-dimensional diagram for the same kind of system but including both temperature and pressure axes. It was presented in Fig. 7.17 and it was then concluded that an extremum in P at constant T must lead to an extremum in T at constant P. The corresponding diagram, where z_B has been introduced instead of μ_B, is shown in Fig. 9.23(b) and it confirms that the two phases have the same composition at the point of extremum considered. In fact, there is a whole series of such points, marked as a dotted line. This is the locus of points of tangency for tangents parallel to the μ_B or z_B axis. That line represents a singular equilibrium and could be included in the T,P diagram, obtained by projecting in the μ_B (i.e. z_B) direction, Schreinemakers' projection. The other lines shown in the same diagram represent univariant three-phase equilibria because this is a binary system (see Fig. 9.1(a), for instance). The line representing singular equilibrium is called a **singular curve**. Singular equilibrium will be further discussed in Section 11.6 and 12.7–12.10.

A major difference between univariant lines and singular curves should be noted. A univariant line shows exactly where a particular equilibrium occurs. A singular

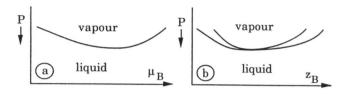

Figure 9.22 An isothermal section of a binary diagram with a singular point for two phases illustrating Konovalov's rule.

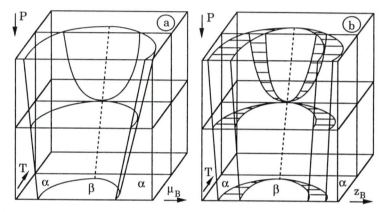

Figure 9.23 (a) A two-phase equilibrium in a binary system illustrated with the complete three-dimensional potential phase diagram. Points of tangency for lines parallel to the μ_B axis are marked with a dotted line. (b) After μ_B has been replaced by its conjugate molar quantity z_B, the phases still coincide along the dotted line. Any point of sectioning through this line will give a point of extremum in the sectioned phase diagram.

curve shows the maximum extension of a divariant equilibrium which is otherwise not shown in the diagram. It would thus be wise to indicate on what side of a singular curve the particular equilibrium exists. This is done in Fig. 9.24 which is a projection of Fig. 9.23.

Points of extremum in P and T were discussed in Section 7.9 and Konovalov's rule was actually derived there in an analytical way, using the ordinary molar quantities, S_m, V_m and x_i. In Chapter 7 and the present one we have mainly used molar quantities defined by dividing the integral quantities with the content of a certain component, N_A for instance. We denote these quantities with S_{m1}, V_{m1} and z_i. However, if all the molar quantities we are interested in are molar contents, then the results look the same in both notations. As an example, the insertion of $x_i = x_1 z_i$ in the result for $p = c$ in Section 7.9 yields

$$0 = |\, 1\ x_2^\beta \ldots x_c^\varepsilon\, | = |\, x_1^\alpha\ x_2^\beta \ldots x_c^\varepsilon\, | = |\, 1\ z_2^\beta \ldots z_c^\varepsilon\, | \cdot (x_1^\alpha x_1^\beta \ldots x_1^\varepsilon)$$

and thus

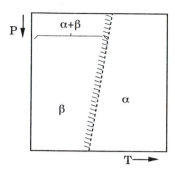

Figure 9.24 Singular curve showing the maximum extension of the $\alpha + \beta$ equilibrium in Fig. 9.23. The $\alpha + \beta$ surface is folded and to the left of the curve one would intersect that surface twice by moving in the projected direction.

$$0 = |\, 1 \; z_2^{\beta} \ldots z_c^{\varepsilon} \,|$$

In spite of the fact that we shall now continue to use z_i it should be remembered that the results hold for x_i as well.

The importance of Konovalov's rule stems from the fact that composition is often used as an experimental variable. A system with a composition at a point of maximum or minimum undergoes an azeotropic or congruent transformation on passing through it and such a point is often given a special name, **azeotropic** (actually meaning 'boiling unchanged') or **congruent**.

Exercise 9.8.1

The following phase diagram is an isobarothermal section at 1273 K and 1 bar of the Fe–Cr–N phase diagram under conditions where N_2 gas does not form. An isoactivity line for N has been drawn in the γ phase field. Show a reasonable continuation of it after first sketching the corresponding μ_{Cr}, a_N phase diagram.

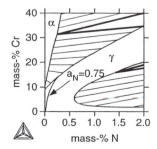

Exercise 9.8.1

Hint

Notice that there is a tie-line for which the α and γ phases have the same Cr content. It should be a point of extremum for the N potential (or N activity).

Solution

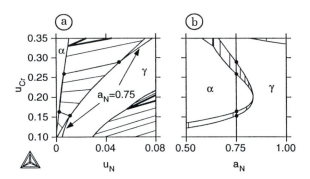

Solution 9.8.1

Exercise 9.8.2

It is known that the three-phase equilibrium $L/fcc + Mg_2Si$ in the Al–Mg–Si phase diagram at 1 bar has a temperature maximum of $868\,K$ at $x_{Mg}^L = 0.110$ and $x_{Si}^L = 0.045$. Estimate the content of Mg in the Al-rich fcc phase at this equilibrium.

Hint

The solubility of Si in the fcc phase is much lower than that of Mg.

Solution

At the point of extremum the three phases must be on a straight line. Neglecting the Si content of fcc completely we find

$$0 = \begin{vmatrix} 1 & x_{Mg}^{fcc} & x_{Si}^{fcc} \\ 1 & x_{Mg}^L & x_{Si}^L \\ 1 & x_{Mg}^{Mg_2Si} & x_{Si}^{Mg_2Si} \end{vmatrix} = \begin{vmatrix} 1 & x_{Mg}^{fcc} & 0 \\ 1 & x_{Mg}^L & x_{Si}^L \\ 1 & 2/3 & 1/3 \end{vmatrix}$$

$$= x_{Mg}^L \cdot 1/3 - x_{Si}^L \cdot 2/3 - x_{Mg}^{fcc} \cdot 1/3 + x_{Mg}^{fcc} \cdot x_{Si}^L = 0,$$

$x_{Mg}^{fcc} = (x_{Mg}^L - 2x_{Si}^L)/(1 - 3x_{Si}^L) = (0.110 - 2 \cdot 0.045)/(1 - 3 \cdot 0.045) = 0.023$. In reality, $x_{Si}^{fcc} = 0.0037$ which is not quite negligible. The more exact result is $x_{Mg}^{fcc} = 0.030$.

9.9   *General rule for singular equilibria*

It is evident that Konovalov's rule does not only apply to composition. It may thus be generalized. Suppose that a linear two-phase field in a Y^k, Y^j diagram, determined at constant values of all the other potentials except Y^1, which is chosen as the dependent potential, shows a Y^j maximum or minimum. At the point of extremum the two phases must have the same value of X^k_{m1}. Furthermore, if Y^j is kept constant and another potential is allowed to vary, it will also have an extremum at the same value of X^k_{m1}.

Let us now consider a two-phase equilibrium in an isobaric potential diagram for a ternary system, which is three-dimensional. Thus, $p = c - 1$. Suppose there is a point of tangency for a *plane* parallel to the μ_B, μ_C plane (i.e. an isothermal plane) as shown in Fig. 9.25(a) which is a reproduction of Fig. 7.18. As demonstrated in Fig. 9.25(b), the two phases thus have the same composition and the point of extremum is a congruent transformation point. This was already proved in Section 7.9 using an analytical method.

The point of extremum in Fig. 9.25 may be characterized as a doubly singular point. It would also appear in a diagram with a P axis under a constant value of T equal to the extreme value. In order to show in one diagram that this point is an extremum for P as well as T, one would need a fourth dimension. It is evident that the doubly singular point in Fig. 9.25 would fall at a different T value if the constant P value was different and in a P, T projection all such points would form a line, a **doubly singular** curve.

From Section 7.9 it is evident that Konovalov's rule is just a special case of a more general rule. In fact, for the ternary case, $p = c = 3$, it was formulated by von Alkemade (1893). His rule was originally formulated for a liquid which solidifies to two solid phases and P is regarded as constant. It may be stated as follows, 'The direction of

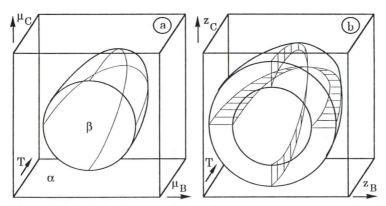

Figure 9.25 (a) Isobaric potential phase diagram of a ternary system with a doubly singular point on a divariant phase field. Thin lines represent points of tangency for lines parallel to the μ_B or μ_C axis. Their intersection is a point of tangency on a μ_B, μ_C plane. It gives an extremum for T. (b) By replacing μ_B and μ_C with their conjugate molar quantities, z_B and z_C, it is shown that the two phases in the point of T extremum must have the same z_B and the same z_C value. The point of extremum thus defines the composition of an alloy which can transform congruently between the two phases.

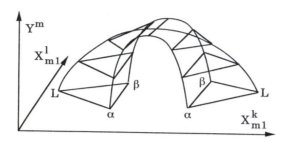

Figure 9.26 Ternary phase diagram at constant P and with two molar axes showing a three-phase equilibrium with an extremum for T (here represented by Y^m), illustrating von Alkemade's rule. The triangles are parallel to the X^k_{m1}, X^l_{m1} plane.

falling temperature of the liquid in equilibrium with two solid phases is always away from the tie-line between the solid phases. If the liquid falls on the tie-line, then the three-phase equilibrium is at a temperature maximum.' Figure 9.26 illustrates von Alkemade's rule. It is evident that Alkemade neglected the possibility of having a temperature minimum.

The reasoning applied to Konovalov's rule can also be applied to von Alkemade's rule. If T is kept constant at the extreme value and P is varied with the three phases present, then one will find that P also has an extremum. At a different constant value of P, the T extremum would occur at a different value. The locus of these three-phase equilibria would also give a line in Schreinemakers' projection, a singular curve.

From the mathematical study of conditions of extrema given in Section 7.9 it is evident that Konovalov's rule can be applied to two-phase equilibria and von Alkemade's rule to three-phase equilibria in systems with $c > p$, although they were originally formulated for $c = p$. Konovalov's rule: T at constant P and P at constant T have extreme values for a two-phase equilibrium if the two phases have the same composition, i.e. fall on the same point; von Alkemade's rule: T at constant P and P at constant T have extreme values for a three-phase equilibrium if the compositions of the three phases fall on a straight line. We can combine these cases into a general **rule for singular equilibria**: T at constant P and P at constant T have extreme values for an equilibrium between p phases if their compositions fall on a point for $p = 2$ (Konovalov's rule), on a line for $p = 3$ (Alkemade's rule), on a plane for $p = 4$, etc. In all these cases a curve representing the locus of these equilibria can be plotted in the T,P diagram obtained by Schreinemakers' projection. For $p = c$ such a line is called a singular line, for $p = c - 1$ a doubly singular line, etc. The connection between such lines will be demonstrated in Fig. 11.13.

Finally, it may be instructive to apply the phase field rule to the diagram in Fig. 9.22(a). For the two-phase field liquid + vapour we get

$$d = c + 2 - p - n_s + n_m = 2 + 2 - 2 - 1 + 1 = 2$$

because we have sectioned once, $n_s = 1$, by keeping temperature, which is a potential, constant. There is one molar variable, used as axis in the P,z_B diagram, $n_m = 1$. The result agrees with the diagram because it shows a two-dimensional phase field for the two phases. However, if we section once more, at a constant value of z_B, then $n_s = 2$ and we get $d = 2 + 2 - 2 - 2 + 1 = 1$. The phase diagram is now just a vertical line and in general it will show that the two-phase field extends over a range of P values in agreement with the calculated $d = 1$. However, the special section, going through the point of extremum (the singular point), will show the two phases coexisting at a point, and one should thus have expected to obtain $d = 0$. It is evident that one should exercise special care when applying the phase field rule to systems with singular points. This problem will be discussed further in Chapter 12.

Exercise 9.9.1

Try to construct a diagram similar to Fig. 9.25 for a case where α falls between L and β at the maximum.

Solution

See diagram.

Solution 9.9.1

Direction of phase boundaries

Use of distribution coefficient

In this chapter we shall examine in more detail the direction of phase boundaries in molar and mixed phase diagrams. As an introduction we shall first discuss some approximate calculations based upon the use of the distribution coefficient of a component between two phases but later we shall use a more general method.

In multinary systems one is often interested in the distribution of a particular component between two phases. One may for instance define a distribution coefficient (also called partition coefficient) which can be used to represent experimental data and to carry out calculations of phase boundaries and changes in chemical potentials.

Let us consider the equilibrium between two solution phases, α and β, which exist already without an element B. On adding B one finds that it partitions between the two phases in a characteristic manner, which can be derived from the equilibrium condition $G_B^\alpha = G_B^\beta$. By applying a general model for a solution phase we obtain

$$^\circ G_B^\alpha + RT\ln x_B^\alpha + {}^E G_B^\alpha = {}^\circ G_B^\beta + RT\ln x_B^\beta + {}^E G_B^\beta$$

We may now define a distribution coefficient $K_B^{\alpha/\beta}$ as

$$K_B^{\alpha/\beta} = x_B^\alpha/x_B^\beta = \exp\left\{\frac{1}{RT}[{}^\circ G_B^\beta - {}^\circ G_B^\alpha + {}^E G_B^\beta - {}^E G_B^\alpha]\right\}$$

In many cases the distribution coefficient is relatively independent of composition. This occurs when the composition dependence of the partial Gibbs energy of each phase is mainly given by the $RT\ln x$ term. In such cases the distribution coefficient may be a useful tool. As an example we may consider the case where both phases are dilute solutions in a major component A. The excess Gibbs energy terms may then be approximated by a regular solution parameter L and we find, for low B contents,

$$K_B^{\alpha/\beta} = \exp(\Delta G_B/RT) \quad \text{where} \quad \Delta G_B = {}^\circ G_B^\beta - {}^\circ G_B^\alpha + L^\beta - L^\alpha$$

It should be emphasized that ΔG_B, being a Gibbs energy, may be represented as $\Delta H_B - T\Delta S_B$ and we thus obtain

$$K^{\alpha/\beta}_B = K_o \exp(\Delta H_B / RT)$$

K_o and ΔH_B may often be approximated as constant.

When there are several minor components, we can define a distribution coefficient for each one

$$x^\alpha_j / x^\beta_j = K^{\alpha/\beta}_j$$

For the major component we obtain, from $G^\alpha_A = G^\beta_A$,

$$^oG^\alpha_A + RT\ln x^\alpha_A + {}^EG^\alpha_A = {}^oG^\beta_A + RT\ln x^\beta_A + {}^EG^\beta_A$$

but it is not useful to define a distribution coefficient for this component. Instead we can apply another approximation if the total content of alloying elements is small,

$$\ln x_A = \ln(1 - \Sigma x_j) \cong -\Sigma x_j$$

and neglecting the excess Gibbs energy for this component we obtain

$$\Sigma x^\beta_j - \Sigma x^\alpha_j = (^oG^\beta_A - {}^oG^\alpha_A)/RT$$

For a binary system we thus have two equations derived from the equilibrium conditions for the two components. For any temperature and pressure we can calculate two unknown quantities, i.e. the compositions of the two phases. The temperature dependence of the various parameters will give the directions of the two phase boundaries in a T,x diagram. In an isobarothermal section of a ternary system there will be three equations and each of the two lines will be represented by a line. With the approximations used here we have been able to simplify all the equilibrium equations to linear equations and the phase boundaries will thus be approximately straight lines as far as the dilute solution approximation is valid. It is thus possible to construct the A-rich corner of a ternary diagram from the binary diagrams by simply using a ruler. Two examples are given in Fig. 10.1 and it should be noticed that the construction of the second one is based upon an extrapolation of the phase boundaries in one of the binary systems to negative compositions. This is non-physical but in accordance with the form of the mathematical equations.

Exercise 10.1.1

Fe has two allotropic modifications, γ(fcc) and α(bcc). At 1423 K γ is more stable by 71 J/mol but α can be stabilized by alloying with 5 atom.% Si. Estimate how much Si is required if the alloy also contains 0.5 atom.% Ni, which has a distribution coefficient between γ and α of 1.3.

Hint

First evaluate the distribution coefficient for Si from the information.

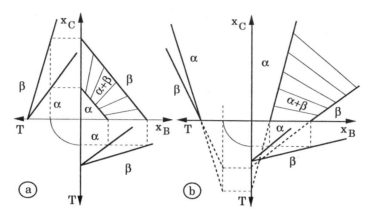

Figure 10.1 Isobarothermal sections of ternary phase diagrams showing equilibrium between two phases, both with the same major component.

Solution

Using $\Sigma x_j^\gamma - \Sigma x_j^\alpha = ({}^\circ G_{Fe}^\gamma - {}^\circ G_{Fe}^\alpha)/RT$ we find for the binary Fe–Si alloy:
$x_{Si}^\gamma - x_{Si}^\alpha = x_{Si}^\alpha (K_{Si}^{\gamma/\alpha} - 1) = 0.05(K_{Si}^{\eta/\alpha} - 1) = -71/8.314 \cdot 1423 = -0.006;$
$K_{Si}^{\eta/\alpha} = 1 - 0.12 = 0.88$, and for the ternary Fe–Si–Ni alloy: $x_{Si}^\gamma + x_{Ni}^\gamma - x_{Si}^\alpha -$
$x_{Ni}^\alpha = x_{Si}^\alpha (K_{Si}^{\gamma/\alpha} - 1) + x_{Ni}^\alpha (K_{Ni}^{\gamma/\alpha} - 1) = x_{Si}^\alpha (0.88 - 1) + 0.005(1.3 - 1) = -0.006;$
$x_{Si}^\alpha = 0.0075/0.12 = 0.0625.$

10.2   *Calculation of allotropic phase boundaries*

On an allotropic phase boundary the two phases have the same composition (see Section 6.5). When comparing two phases we get the following expression by definition if we apply the regular solution model to both phases (${}^E G_m = x_A x_B L$, see Section 6.1) because the ideal entropy term will be the same for the two phases and will drop out.

$$G_m^\beta - G_m^\alpha = x_A({}^\circ G_A^\beta - {}^\circ G_A^\alpha) + x_B({}^\circ G_B^\beta - {}^\circ G_B^\alpha) + x_A x_B(L^\beta - L^\alpha)$$

For low B contents it may be convenient to rearrange the equation as

$$G_m^\beta - G_m^\alpha = {}^\circ G_A^\beta - {}^\circ G_A^\alpha + x_B[({}^\circ G_B^\beta - {}^\circ G_B^\alpha) + (L^\beta - L^\alpha) - ({}^\circ G_A^\beta - {}^\circ G_A^\alpha)] - x_B^2(L^\beta - L^\alpha)$$

At sufficiently low B contents we can neglect the last square term. Close to the temperature of the allotropic phase transformation for pure A we can neglect the term $({}^\circ G_A^\beta - {}^\circ G_A^\alpha)$ in the bracket which is there close to zero and we thus get

$$G_m^\beta - G_m^\alpha = {}^\circ G_A^\beta - {}^\circ G_A^\alpha + x_B \cdot \Delta^\circ G_B^{\alpha \to \beta A}$$

where we have introduced the following notation

$$\Delta^\circ G_B^{\alpha \to \beta A} = {}^\circ G_B^\beta - {}^\circ G_B^\alpha + L^\beta - L^\alpha$$

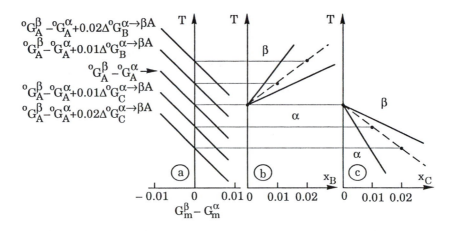

Figure 10.2 The effect of two different types of alloying element on the allotropic phase boundary. The equilibrium phase boundaries (solid lines) fall one on each side of the allotropic phase boundaries (dashed lines). The diagrams are calculated with $\Delta^\circ G_B^{\alpha\to\beta A} = RT\ln 2$ and $\Delta^\circ G_C^{\alpha\to\beta A} = - RT\ln 2$.

We have already seen that the distribution coefficient of B between α and β can be approximated by an expression for low B contents

$$K_B^{\alpha/\beta} = \exp\left\{\frac{1}{RT}[^\circ G_B^\beta - {}^\circ G_B^\alpha + L^\beta - L^\alpha]\right\}$$

We thus find the following relation between the parameters used in the calculation of allotropic boundaries as well as ordinary phase boundaries

$$\Delta^\circ G_B^{\alpha\to\beta A} = RT\ln K_B^{\alpha/\beta}$$

Within a narrow range of temperature and composition, it is reasonable to assume that $\Delta^\circ G_B^{\alpha\to\beta A}$ is constant and we can then describe the effect of the alloying element as a parallel displacement of the curve for $^\circ G_A^\beta - {}^\circ G_A^\alpha$ by the amount $x_A\cdot\Delta^\circ G_B^{\alpha\to\beta A}$. We shall thus get two types of alloying effect, which are demonstrated by B and C in Fig. 10.2. There it is assumed that $^\circ G_A^\beta - {}^\circ G_A^\alpha$ varies linearly with temperature.

We can obtain an equation for the allotropic phase boundary by inserting $G_m^\beta - G_m^\alpha = 0$ in the previous equation,

$$x_B^{\text{allot.}} = - (^\circ G_A^\beta - {}^\circ G_A^\alpha)/\Delta^\circ G_B^{\alpha\to\beta A} = - (^\circ G_A^\beta - {}^\circ G_A^\alpha)/RT\ln K_B^{\alpha/\beta}$$

Close to the transition point T_o for pure A we obtain

$$x_B^{\text{allot.}} = (T - T_o)(^\circ H_A^\beta - {}^\circ H_A^\alpha)/T_o\Delta^\circ G_B^{\alpha\to\beta A}$$

This type of construction is especially interesting for iron because its high-temperature phase δ is identical to its low-temperature phase α. As a consequence, the allotropic phase boundary must be strongly curved as demonstrated in Fig. 10.3. It should be noticed that one can extrapolate all phase boundaries mathematically, even to negative

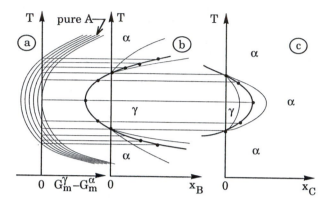

Figure 10.3 The effect of two types of alloying element on the allotropic phase boundaries (thick lines) when the low-temperature phase comes back at high temperature. The phase boundaries are here given with thin lines.

alloy contents if one avoids the use of mathematical expressions containing $\ln x_B$. The two types of alloying effects on iron, the stabilization of austenite (γ) by element B and ferrite (α) by element C, thus look like each other's mirror images.

It should finally be emphasized that the approximate equations derived in this section are valid only up to a few atomic percent of the alloying element.

Exercise 10.2.1

Suppose pure A has an α/β transition at 1000 K. An alloying element B, which itself has the α structure at all temperatures, has been found first to expand the temperature range of the β phase but at higher B contents β will exist at high temperatures, only, and then disappear completely. Find the congruent point for the α/β equilibrium from the following kind of expression for both phases: $G_m = x_A{}^\circ G_A + x_B{}^\circ G_B + RT(x_A \ln x_A + x_B \ln x_B) + L x_A x_B$, where ${}^\circ G_A^\alpha - {}^\circ G_A^\beta = R(T - 1000)$; ${}^\circ G_B^\alpha - {}^\circ G_B^\beta = -RT$); $L^\alpha = 200R$ and $L^\beta = -1000R$.

Hint

At a point of extremum, where the ordinary phase boundaries are horizontal, the allotropic phase boundary coincides with them and is also horizontal. It is much easier to calculate this point from the allotropic phase boundary than from the ordinary ones.

Solution

The allotropic phase boundary is given by $G_m^\alpha - G_m^\beta$:
$$x_A({}^\circ G_A^\alpha - {}^\circ G_A^\beta) + x_B({}^\circ G_B^\alpha - {}^\circ G_B^\beta) + (L^\alpha - L^\beta)x_A x_B = 0; \quad x_A R(T - 1000) + x_B(-$$

$$RT) + (200 + 1000)Rx_Ax_B = 0; \quad RT(x_A - x_B) - 1000Rx_A + 1200Rx_Ax_B = 0;$$
$$T = (1000x_A - 1200x_Ax_B)/(x_A - x_B) = 1000(1 - 2.2x_B + 1.2x_B^2)/(1 - 2x_B);$$
$$dT/dx_B = 1000[(1 - 2x_B)(-2.2 + 2.4x_B) - (1 - 2.2x_B + 1.2x_B^2)(-2)]/$$
$$(1 - 2x_B)^2 = 0; \quad x_B = 0.092; \quad T = 990 \text{ K}.$$

10.3

Variation of a chemical potential in a two-phase field

We shall now consider the effect of a ternary alloying addition on a two-phase equilibrium which exists already in a binary system. The effect of the minor binary component on the chemical potential can be estimated rather accurately from the distribution coefficient of the alloying element between the two phases without using any information on the direction of the phase boundaries in the ternary system. In Section 7.8 we considered the effect of any small change in composition of phases in a ternary system by combining two Gibbs–Duhem relations at constant T and P. We can easily introduce the distribution coefficient

$$d\mu_C = \frac{x_B^\alpha x_A^\beta - x_B^\beta x_A^\alpha}{x_C^\beta x_A^\alpha - x_C^\alpha x_A^\beta} \cdot d\mu_B = x_B^\alpha x_A^\beta \frac{1 - K_{BA}^{\beta/\alpha}}{x_C^\beta x_A^\alpha - x_C^\alpha x_A^\beta} \cdot d\mu_B$$

By dividing through with $(x_A^\alpha + x_B^\alpha)(x_A^\beta + x_B^\beta)$ we can change from the x composition to u (see Section 3.7). The distribution coefficient for B and A between the two phases can be defined with both types of variable

$$K_{BA}^{\beta/\alpha} = x_B^\beta x_A^\alpha / x_A^\beta x_B^\alpha = u_B^\beta u_A^\alpha / u_A^\beta u_B^\alpha$$

At low contents of B in both phases we can approximate u_A^α and u_A^β with unity and we can apply Henry's law to B in the α phase in the following form if the C content in α is also low,

$$\mu_B = G_B^\alpha = {}^\circ G_B^\alpha + RT\ln f_B^\alpha + RT\ln u_B^\alpha$$
$$d\mu_B = (RT/u_B^\alpha) \cdot du_B^\alpha$$

The equation is thus simplified to

$$\frac{d\mu_C}{du_B^\alpha} = RT \cdot \frac{1 - K_{BA}^{\beta/\alpha}}{u_C^\beta - u_C^\alpha}$$

By approximating the right-hand side with its value close to the binary A–C side of the system, we can easily integrate and obtain

$$\mu_C^{\text{ternary}} - \mu_C^{\text{binary}} = RT \cdot \frac{1 - K_{BA}^{\beta/\alpha}}{u_C^\beta - u_C^\alpha} \cdot u_B^\alpha$$

where u_B^α is the B content of α in the ternary alloy. By introducing the activity for C we instead obtain

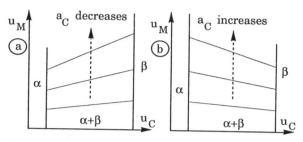

Figure 10.4 The effect of the slope of tie-lines on the activity of a component in a two-phase field.

$$\ln \frac{a_C^{\text{ternary}}}{a_C^{\text{binary}}} = \frac{1 - K_{BA}^{\beta/\alpha}}{u_C^\beta - u_C^\alpha} \cdot u_B^\alpha$$

This is a useful equation for rough calculations. It demonstrates that an alloying element which concentrates to the phase which is richest in C, i.e. which has $K_{BA}^{\beta/\alpha} > 1$ if β is the C-rich phase, will decrease the C activity for the two-phase equilibrium $\alpha + \beta$. An alloying element that concentrates to the C-poor phase will increase the C activity. From the derivation it is evident that this effect is additive for several alloying elements if evaluated for μ_C or $\ln a_C$.

The value of $K_{BA}^{\beta/\alpha}$ is directly related to the slope of the tie-lines in the u_C, u_M phase diagram. We can thus illustrate the two cases with the phase diagrams in Fig. 10.4 where the u parameters are used. The alloying element will have no effect on the C activity of the two-phase equilibrium if the tie-lines are horizontal, i.e. if they are directed towards the C corner which is situated infinitely far away in a diagram with the u variable.

The equation shows that μ_C does not change in a two-phase field where $K = 1$, i.e. where the two phases have the same content of B relative to A. This is thus a point of extremum and the present result is in complete agreement with Konovalov's rule. An illustrative example was given in Exercise 9.8.2.

The chemical potential of a two-phase equilibrium can also be strongly affected by a difference in pressure, caused by the surface energy in a curved phase interface. The complete form of the Gibbs–Duhem equation is the following:

$$x_A d\mu_A + x_B d\mu_B + x_C d\mu_C = V_m dP - S_m dT$$

Now we shall let the pressure vary in the β phase but keep the temperature constant. The expression for $d\mu_C$ derived earlier in this section, will thus have one more term which can be written as

$$\frac{x_A^\alpha V_m^\beta dP^\beta}{x_C^\beta x_A^\alpha - x_C^\alpha x_A^\beta} \quad \text{or} \quad \frac{u_A^\alpha dP^\beta}{u_C^\beta u_A^\alpha - u_C^\alpha u_A^\beta} \cdot \frac{V_m^\beta}{(1 - x_C^\beta)}$$

For low B contents we can thus write

$$d\mu_C = k \cdot d\mu_B^\alpha + 1 \cdot dP^\beta$$

where $k = RT(1 - K_{BA}^{\beta/\alpha})/(u_C^\beta - u_C^\alpha)$ and $l = V_m^\beta/(1 - x_C^\beta)(u_C^\beta - u_C^\alpha)$.

Exercise 10.3.1

Low-carbon steels are sometimes carburized in order to increase the surface hardness. This is done at a temperature where γ(fcc) is the stable phase. A hard and brittle carbide called cementite, Fe_3C, may form if one uses a high carbon activity in the gas. In the binary system it has a carbon activity of 1.04 when in equilibrium with γ at 1173 K. What would be the highest carbon activity to be used if one wants to avoid cementite for a steel with 1.5 atom.% Cr and 3 atom.% Ni. They can both replace Fe in cementite and the distribution coefficient $K_{M/Fe}^{cementite/\gamma}$ is 6 for Cr and 0.1 for Ni.

Hint

The effects of two alloying elements on μ_C or $\ln a_C$ are additive. The alloy contents given are for the initial low-carbon steel and we should evaluate the u variable because it does not change when C is added due to its definition. We obtain $u_{Cr}^\gamma = 0.015$ and $u_{Ni}^\gamma = 0.03$. For cementite $u_C = 1/3$ and for γ in equilibrium with cementite at 1173 K we have 1.23 mass% C which gives $u_C = 0.059$.

Solution

$\ln(a_C^{alloy}/a_C^{binary}) = (1 - 6) \cdot 0.015/[(1/3) - 0.059] + (1 - 0.1) \cdot 0.03/[(1/3) - 0.059] = -0.27 + 0.10 = -0.17$; $a_C^{alloy} = a_C^{binary} \cdot \exp(-0.17) = 1.04 \cdot 0.84 = 0.88$. This is the highest value one should use.

10.4   *Direction of phase boundaries*

So far, we have discussed the direction of phase boundaries in some simple cases. For the general case we need a more powerful method and we should then turn to the Gibbs–Duhem relation. In fact, we have already calculated the directions of phase fields in potential phase diagrams by the application of the Gibbs–Duhem relation. However, in order to calculate the directions of the phase boundaries in molar phase diagrams we must introduce the molar quantities instead of the potentials as variables in the Gibbs–Duhem relation. No general treatment can be given here in view of the large variety that can occur in mixed phase diagrams. Only the special case will be treated where T and P are retained but all the chemical potentials are replaced by molar contents.

The fact that the molar quantities of two phases in equilibrium are generally different, although the potentials are equal, makes it necessary to choose a special phase, for instance α, if one wants to express the potentials through molar quantities. If T and P are retained, then it is convenient to express the changes of the chemical potentials μ_i through the composition dependence of the partial Gibbs energies in the α phase, G_i^α. In order to make the enthalpy appear in the final expression instead of the entropy we shall use the potentials occurring in the special form of the Gibbs–Duhem relation containing enthalpy:

$$H_m^\beta d(1/T) + (V_m^\beta/T)dP - \Sigma x_i^\beta d(\mu_i/T) = 0$$

We shall now introduce the properties of the special phase α by using $\mu_i = G_i^\alpha$ and with $1/T$, P and x_j^α for $j > 1$ as the independent variables, treating x_1^α as the dependent composition variable. We can then eliminate $d(\mu_i/T)$ using

$$d(\mu_i/T) = \frac{\partial(G_i^\alpha/T)}{\partial(1/T)} d(1/T) + \frac{\partial G_i^\alpha}{\partial P} dP/T + \sum_{j=2}^{c} \frac{\partial G_i^\alpha}{\partial x_j^\alpha} dx_j^\alpha/T$$

where we can insert

$$\partial(G_i^\alpha/T)/\partial(1/T) = H_i^\alpha$$
$$\partial G_i^\alpha/\partial P = V_i^\alpha$$

$$\partial G_i^\alpha/\partial x_j^\alpha = g_{ij}^\alpha - \sum_{l=2}^{c} x_l^\alpha g_{jl}^\alpha$$

since

$$G_1^\alpha = G_m^\alpha + \partial G_m^\alpha/\partial x_i^\alpha - \sum_{l=2}^{c} x_l^\alpha \partial G_m^\alpha/\partial x_i^\alpha$$

The notation g_{ij} is here used for $\partial^2 G_m/\partial x_i \partial x_j$ treating all x_i except for x_1 as independent variables. The same definition was used in Section 6.6. When inserting these expressions we shall also replace H_m^β and V_m^β by $\Sigma x_i^\beta H_i^\beta$ and $\Sigma x_i^\beta V_i^\beta$.

$$\Sigma x_i^\beta H_i^\beta d(1/T) + \Sigma x_i^\beta V_i^\beta dP/T - \Sigma x_i^\beta \left[H_i^\alpha d(1/T) + V_i^\alpha dP/T + \right.$$

$$\left. \sum_{j=2}^{c} \left(g_{ij}^\alpha - \sum_{l=2}^{c} x_l^\alpha g_{jl}^\alpha \right) \right] dx_j^\alpha/T = 0$$

However,

$$\sum_{i=1}^{c} x_i^\beta \sum_{j=2}^{c} g_{ij}^\alpha - \sum_{i=1}^{c} x_i^\beta \sum_{j=2}^{c} \sum_{l=2}^{c} x_l^\alpha g_{jl}^\alpha = \sum_{i=2}^{c} \sum_{j=2}^{c} x_i^\beta g_{ij}^\alpha - \sum_{j=2}^{c} \sum_{l=2}^{c} x_l^\alpha g_{jl}^\alpha =$$

$$\sum_{i=2}^{c} \sum_{j=2}^{c} (x_i^\beta - x_i^\alpha) g_{ij}^\alpha$$

since $\Sigma x_i^\beta = 1$ and $g_{ij} = 0$ for $i = 1$ because the fraction of component 1 is treated as a dependent variable in the definition of g_{ij}. We obtain, because $d(1/T) = -dT/T^2$

$$\sum_{i=2}^{c}\sum_{j=2}^{c}(x_i^\beta - x_i^\alpha)g_{ij}^\alpha dx_j^\alpha + \sum_{i=1}^{c}x_i^\beta(H_i^\beta - H_i^\alpha)dT/T - \sum_{i=1}^{c}x_i^\beta(V_i^\beta - V_i^\alpha)dP = 0$$

Contrary to the Gibbs–Duhem relation this equation always concerns two phases, and all the terms become zero when applied to the phase which was chosen for expressing the chemical potentials. When applied to more than two phases it yields a system of equations and some variables can then be eliminated with the method used for calculating the direction of phase fields from the Gibbs–Duhem relation. The elements of the determinants will then be $\Sigma(x_i^\beta - x_i^\alpha)g_{ij}^\alpha$ instead of x_i^β. However, we shall apply the equation to equilibria concerning two coexisting phases and the equation can then be applied directly.

For a binary system under isobaric conditions we get for the phase boundaries

$$\left(\frac{dx_2^\alpha}{dT}\right)_{coex} = \frac{x_1^\beta(H_1^\alpha - H_1^\beta) + x_2^\beta(H_2^\alpha - H_2^\beta)}{(x_2^\beta - x_2^\alpha)g_{22}^\alpha T} = \frac{\Delta H_m^{\beta\,in\,\alpha}}{(x_2^\beta - x_2^\alpha)g_{22}^\alpha T}$$

$$\left(\frac{dx_2^\beta}{dT}\right)_{coex} = \frac{x_1^\alpha(H_1^\beta - H_1^\alpha) + x_2^\alpha(H_2^\beta - H_2^\alpha)}{(x_2^\alpha - x_2^\beta)g_{22}^\beta T} = \frac{\Delta H_m^{\alpha\,in\,\beta}}{(x_2^\alpha - x_2^\beta)g_{22}^\beta T}$$

The numerator is equal to the heat of solution of the other phase (α or β) in the phase being considered (β or α). Either of these two equations can be used to evaluate the slope of a phase boundary but also to calculate the width of a two-phase field if the slope is known.

$$x_2^\beta - x_2^\alpha = \frac{\Delta H_m^{\beta\,in\,\alpha}}{g_{22}^\alpha T(dx_2^\alpha/dT)_{coex}} = \frac{\Delta H_m^{\alpha\,in\,\beta}}{g_{22}^\beta T(dx_2^\beta/dT)_{coex}}$$

Exercise 10.4.1

Derive an equation for the solubility of pure component 2 in α which is almost pure component 1.

Hint

Section 6.1 gives $g_{22}^\alpha \equiv d^2 G_m^\alpha/d(x_2^\alpha)^2 \cong RT/x_2^\alpha$ if x_2^α is low. Also use $x_2^\beta - x_2^\alpha \cong 1$.

Solution

$dx_2^\alpha/dT = \Delta H_m/(RT^2/x_2^\alpha)$; $d(\ln x_2^\alpha)/d(1/T) = -\Delta H_m/R$; $\ln x_2^\alpha = K\cdot\exp(-\Delta H_m/RT)$.

Exercise 10.4.2

Take the equation for the direction of phase boundaries in the T,P,x_i space. Apply it to a binary case at constant P. Then consider the $\alpha/\alpha + \beta$ phase boundary in the T,x phase diagram in a system where α is almost pure A and β is a bcc phase close to the 50–50 composition. Suppose β has a sharp transformation at T_0 from a perfectly ordered to a perfectly disordered state (which would never happen). Calculate the angle of the $\alpha/\alpha + \beta$ phase boundary at T_0 (or, more precisely, the difference in direction, dx/dT, of this phase boundary just below and just above T_0).

Hint

At constant P: $[x_1^\beta(H_1^\alpha - H_1^\beta) + x_2^\beta(H_2^\alpha - H_2^\beta)]dT = (x_2^\beta - x_2^\alpha)g_{22}^\alpha T dx_2^\alpha$. Notice that $x_1^\beta H_1^\beta + x_2^\beta H_2^\beta = H_m^\beta$ and that the entropy of disordering is $-R(x_1 \ln x_1 + x_2 \ln x_2) = R\ln 2$ for $x_1 = x_2 = 0.5$. For the dilute solution of component 2 in α we may use $g_{22}^\alpha = RT/x_1^\alpha x_2^\alpha \cong RT/x_1^\alpha$.

Solution

$x_1^\beta = x_2^\beta = 0.5$ gives $dx_2^\alpha/dT = 0.5(H_1^\alpha - H_1^\beta + H_2^\alpha - H_2^\beta)/(0.5 - 0)g_{22}^\alpha T$. By taking the difference between just below and just above the transition, we eliminate H_1^α and H_2^α and thus get $\Delta[dx_2^\alpha dT] = (H_m^\beta - H_m^{\beta'})/0.5g_{22}^\alpha T_0$. But $H_m^\beta - H_m^{\beta'} = \Delta H_m^{\text{ordering}}$ and at the transition point the two states have the same Gibbs energy and thus $\Delta H_m^{\text{ord}} - T_0 \Delta S_m^{\text{ord}} = 0$; $\Delta[dx_2^\alpha/dT] = T_0 \Delta S_m^{\text{ord}}/0.5g_{22}^\alpha T_0 = R\ln 2/(0.5RT_0/x_2^\alpha) = 2x_2^\alpha \ln 2/T_0$.

Exercise 10.4.3

The direction of an α/β phase field in a T,P diagram for a pure element is given by the Clapeyron equation. Derive an expression for the direction of a corresponding phase boundary in an H,S diagram.

Hint

When the axes are not both potentials, the two phase boundaries α/β and β/α will be separated. One may consider either one. Let us take α. The final expression then contains properties of the α phase. Let us express them in terms of G^α and its derivatives G_T^α, G_P^α, etc. It may then be convenient to start by considering H^α and S^α as functions of T and P. However, T and P are not independent because we want to follow the $\alpha + \beta$ two-phase equilibrium. Therefore, we know from the Clapeyron equation that $dT = dP(V_m^\alpha - V_m^\beta)/(S_m^\alpha - S_m^\beta)$. We may thus evaluate $(dH^\alpha/dP)^{\alpha+\beta}$ and $(dS^\alpha/dP)^{\alpha+\beta}$ and take their ratio and insert the expression for dT/dP.

Solution

$H^\alpha = G^\alpha + TS^\alpha = G^\alpha - TG^\alpha_T;\ (\partial H^\alpha/\partial T)_P = G^\alpha_T - G^\alpha_T - TG^\alpha_{TT} = -TG^\alpha_{TT};$
$(\partial H^\alpha/\partial P)_T = G^\alpha_P - TG^\alpha_{TP};\ \ dH^\alpha = (\partial H^\alpha/\partial P)_T dP + (\partial H^\alpha/\partial T)_P dT =$
$[G^\alpha_P - TG^\alpha_{TP} - TG^\alpha_{TT} dT/dP]dP.$ We also get $dS^\alpha = (\partial S^\alpha/\partial P)_T dP + (\partial S^\alpha/\partial T)_P dT =$
$[-G^\alpha_{TP} - G^\alpha_{TT} dT/dP]dP.$ Finally, by dividing the two expressions, we get
$dH^\alpha/dS^\alpha = T + G^\alpha_P/[-G^\alpha_{TP} - G^\alpha_{TT} dT/dP].$ By inserting the value for dT/dP along
the $\alpha + \beta$ equilibrium we find $(dH^\alpha/dS^\alpha)^{\alpha+\beta} = T - G^\alpha_P/[G^\alpha_{TP} + G^\alpha_{TT}(V^\alpha_m -$
$V^\beta_m)/(S^\alpha_m - S^\beta_m)].$

Exercise 10.4.4

A published Ti–Mo phase diagram shows the liquidus as a dashed line, indicating
insufficient experimental information. Try to predict its position from the solidus
by evaluating the width of the two-phase field.

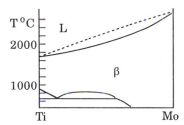

Exercise 10.4.4

Hint

Consider the centre of the system. Lacking detailed information, we may assume
that the heat of solution is equal to the average of the heat of solidification of the
elements, -15.45 and -32.54 kJ/mol. Since the temperature is so high, it may be
justified to neglect the contribution from the excess Gibbs energy in g_{22} which is
thus approximated by $RT/x_1 x_2$.

Solution

$x^L_2 - x^\beta_2 = -\Delta H_m (dT/dx^\beta_2)_{coex}/(RT/x_1 x_2)T = (-24000) \cdot 1200/4 \cdot 8.314 \cdot$
$(2150)^2 = -0.19.$ It seems that the authors have made the same calculation.

10.5   *Congruent melting points*

It is immediately clear from our equations for dx/dT that for a congruent transformation point in a binary system, i.e. for $x^\beta_2 = x^\alpha_2$, the phase boundaries must be horizontal
and such a point must be a point of temperature extremum. This is also in agreement

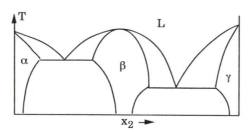

Figure 10.5 Binary T,x phase diagram at 1 bar. The point of congruent melting of β must have horizontal phase boundaries. At the melting points of the two components the phase boundaries are not horizontal.

with Konovalov's rule (see also Goodman and Cahn, 1981). However, at the side of the system where x_2^β and x_2^α both approach zero, g_{22}^α approaches infinity as RT/x_2^α and the whole denominator approaches $RT^2(x_2^\beta/x_2^\alpha - 1)$ which is not zero. Thus, the phase boundaries do not turn horizontal on the sides of the system. The two cases are demonstrated in Fig. 10.5.

The slopes of the phase boundaries at the left-hand side of the binary system in Fig. 10.5 can be evaluated from the limiting value of $g_{22}^\alpha = RT/x_2^\alpha$ when $x_2^\alpha \to 0$ (see Section 6.1).

$$\left(\frac{dx_2^\alpha}{dT}\right)_{coex} = \frac{1}{x_2^\beta - x_2^\alpha}\cdot\frac{{}^\circ H_1^\alpha - {}^\circ H_1^\beta}{(RT/x_2^\alpha)T} = \frac{{}^\circ H_1^\alpha - {}^\circ H_1^\beta}{RT^2}\cdot\frac{K_2^{\alpha/\beta}}{1 - K_2^{\alpha/\beta}}$$

$$\left(\frac{dx_2^\beta}{dT}\right)_{coex} = \frac{1}{x_2^\alpha - x_2^\beta}\cdot\frac{{}^\circ H_1^\beta - {}^\circ H_1^\alpha}{(RT/x_2^\beta)T} = \frac{{}^\circ H_1^\alpha - {}^\circ H_1^\beta}{RT^2}\cdot\frac{1}{1 - K_2^{\alpha/\beta}}$$

and the width of the two-phase field at some temperature T below the transformation point T_o for pure component 1 is obtained from the difference,

$$\frac{x_2^\beta - x_2^\alpha}{T_o - T} = \frac{{}^\circ H_1^\beta - {}^\circ H_1^\alpha}{RT^2}$$

If one of the phases is liquid, one can often neglect the solubility in the solid phase and one thus obtains a simple expression for the freezing-point depression,

$$T_o - T = x_2^L\cdot RT^2/({}^\circ H_m^L - {}^\circ H_m^\alpha).$$

It should be emphasized that it would be difficult to see the horizontal part of a phase boundary at a congruent transformation point if the properties of the phase change so rapidly that g_{22} is very large. An obvious case is the $\beta/(\beta + L)$ boundary when β is almost stoichiometric, i.e. the composition of β does not vary noticeably. The phase boundary of the surrounding phase, in our case $L/(L + \beta)$ can also be very sharp if the properties of the liquid changes rapidly with composition at the particular composition of the congruent transformation. For such cases it may be interesting to evaluate the curvatures of the two phase boundaries. At the congruent point we have $x_2^\alpha = x_2^\beta$ and the heat of solution of each phase in the other one is simply the heat of transformation of the other phase into the phase under consideration. For this special case we can write $\Delta H_m^{\alpha \text{ in } \beta} = H_m^\beta - H_m^\alpha = -\Delta H_m^{\beta \text{ in } \alpha}$ and thus

$$\frac{dx_2^\beta}{dx_2^\alpha} = \frac{dx_2^\beta/dT}{dx_2^\alpha/dT} = \frac{(H_m^\beta - H_m^\alpha)(x_2^\beta - x_2^\alpha)g_{22}^\alpha T}{(x_2^\alpha - x_2^\beta)g_{22}^\beta T(H_m^\alpha - H_m^\beta)} = \frac{g_{22}^\alpha}{g_{22}^\beta}$$

$$\frac{d^2 T}{d(x_2^\alpha)^2} = \frac{g_{22}^\alpha T}{H_m^\alpha - H_m^\beta}\left(\frac{dx_2^\beta}{dx_2^\alpha} - 1\right) = \frac{g_{22}^\alpha T}{H_m^\alpha - H_m^\beta}\left(\frac{g_{22}^\alpha}{g_{22}^\beta} - 1\right)$$

$$\frac{d^2 T}{d(x_2^\beta)^2} = \frac{g_{22}^\beta T}{H_m^\beta - H_m^\alpha}\left(\frac{dx_2^\alpha}{dx_2^\beta} - 1\right) = \frac{g_{22}^\beta T}{H_m^\alpha - H_m^\beta}\left(1 - \frac{g_{22}^\beta}{g_{22}^\alpha}\right) = \frac{d^2 T}{d(x_2^\alpha)^2}\left(\frac{g_{22}^\alpha}{g_{22}^\beta}\right)^2$$

For a stoichiometric phase, g_{22}^β would be very large and for the liquid at a congruent melting point we then get

$$\frac{d^2 T}{d(x_2^L)^2} = \frac{g_{22}^L T}{H_m^\beta - H_m^L}$$

It should be noted that $H_m^\beta - H_m^L$ is negative and so is $d^2 T/d(x_2^L)^2$.

It should be emphasized that another possibility of finding a horizontal phase boundary is by having g_{22}^α approach zero, i.e. a limit of stability.

Exercise 10.5.1

The following diagram for the Zr–Th system has been proposed. The liquidus was not known experimentally and has been drawn tentatively with a thin line. Criticize it.

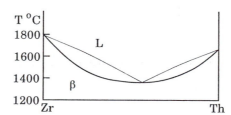

Exercise 10.5.1

Hint

The line has not been carefully drawn at the melting minimum.

Solution

Both the solidus and the liquidus must be horizontal at the point of minimum according to Konovalov's rule. It is not very probable that g_{22} in a liquid would be very large, especially if g_{22} in the solid is not, and the present diagram does not indicate that it is.

Exercise 10.5.2

In elementary textbooks one can sometimes see a series of sketched phase diagrams like the following. Criticize it.

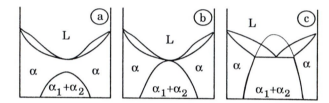

Exercise 10.5.2

Hint

The author may not have remembered that there are two different effects which can make a phase boundary horizontal.

Solution

In diagram (a) the two phase boundaries at the minimum are horizontal because it is a congruent transformation point. It is an effect of the combined properties of the two phases. The top of the miscibility gap, $\alpha_1 + \alpha_2$, is horizontal because $g_{22}^{\alpha} = 0$ and that is a property of the α phase alone. It would be highly unlikely that these two phenomena should occur at the same composition, as indicated in diagram (b).

Exercise 10.5.3

The T,x phase diagram of Al–Zn shows an unusual feature. The solidus line turns almost horizontal in the centre of the system but the liquidus does not. It thus seems to be due to some property of the solid phase rather than the interaction between the two phases. Examine the possible explanation by inspecting the equation for the slope of a phase boundary. If a conclusion is reached, try to test it by examining other features of the diagram.

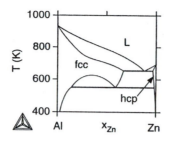

Exercise 10.5.3

Hint

If the explanation is to be found in the G_m function of the solid, then the same factor may have consequences for other phase equilibria with the solid.

Solution

The equation suggests that g^α_{22} is very small at the centre of the system. We may thus be close to a limit of stability where g^α_{22} goes through zero to turn negative. Indeed, at lower temperatures one can see the top of a miscibility gap in the α phase where a homogeneous α alloy starts to decompose in regions of two different compositions.

Exercise 10.5.4

The liquidus in the Fe–S system becomes almost horizontal at 20 atom.% S. From the previous exercise one could perhaps expect to see a miscibility gap in the liquid phase just above the most horizontal point and, indeed, it was proposed long ago. However, it has not been observed experimentally. Try to find an explanation.

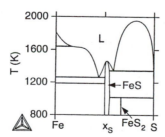

Exercise 10.5.4

Hint

It is more common that a solution grows more ideal at higher temperatures. If there is a tendency to form a miscibility gap it is normally expected to grow stronger at lower temperatures.

Solution

g_{22}^L is close to zero on the liquidus and may go through zero below the liquidus. It seems most probable that a miscibility gap would be seen below the liquidus if solidification could be prevented in some way, i.e. there would be a metastable miscibility gap in the undercooled liquid phase.

Exercise 10.5.5

Calculate the value of g_{22}^L to give the melting point of a stoichiometric phase such a strong curvature that it looks sharp. Compare with the value for an ideal solution.

Hint

Suppose Richard's rule can be applied, $\Delta H_m = H_m^\alpha - H_m^L = -RT$. The maximum may look sharp if the radius of curvature $-1/(d^2T/dx^2)$ is less than $0.005/T$.

Solution

$-d^2T/dx^2 = -g_{22}T/(-RT) > T/0.005$; $g_{22} > 200RT$. If the solution is ideal we have $g_{22} = RT/x_1 x_2$ which is very much lower.

Exercise 10.5.6

Use Richard's rule to estimate the slope of the liquidus from the melting point of A in a binary A–B system where the solubility of B in solid A is very low.

Hint

Richard's rule says that the heat of melting of a metal is approximately $RT_{m.p.}$.

Solution

$K_B^{\alpha/L} = 0$ gives $(dT/dx_B^L)_{coex.} = RT^2/\Delta H_m = RT_{m.p.}^2/(-RT_{m.p.}) = -T_{m.p.}$.

Exercise 10.5.7

A pure element has a sharp melting temperature at a given pressure. The solid and liquid phases can be in equilibrium with each other at a different temperature if a second element is added and also if the pressure is changed. Consider how much should be added to the liquid phase if the temperature is changed by ΔT and the pressure by ΔP. Do this by deriving an expression for the value of $x_2^L - x_2^\alpha$.

Hint

Go back to the general equation for the direction of phase boundaries and apply it to a binary system.

Solution

This time we keep the dP term, $(x_2^\alpha - x_2^L)g_{22}^L dx_2^L = -[x_1^\alpha(H_1^\alpha - H_1^L) + x_2^\alpha(H_2^\alpha - H_2^L)]dT/T + [x_1^\alpha(V_1^\alpha - V_1^L) + x_2^\alpha(V_2^\alpha - V_2^L)]dP$. For small x_2: $RT(x_2^\alpha - x_2^L)dx_2^L/x_2^L = -(^\circ H_1^\alpha - ^\circ H_1^L)dT/T + (^\circ V_1^\alpha - ^\circ V_1^L)dP$. We can easily integrate if ΔT and ΔP are small and $(x_2^\alpha - x_2^L)/x_2^L$ is a constant. By integrating from 0 to x_2 we get $x_2^\alpha - x_2^L = [-(^\circ H_1^\alpha - ^\circ H_1^L)\Delta T/T + (^\circ V_1^\alpha - ^\circ V_1^L)\Delta P]/RT$. We may thus evaluate x_2^L if x_2^α is small or if $K_B^{\alpha/L}$ is known.

Exercise 10.5.8

When adding a third component C to a certain binary system A–B, one found that the depression of the freezing point of a stoichiometric phase $A_a B_b$ only depended upon the molar content x_C and was independent of whether one kept x_A, x_B or x_A/x_B constant. Examine if this result can be expected in general. Suppose the pressure is constant.

Hint

Apply the general equation for the direction of phase boundaries to the ternary case, making C the component 1. Remember that g_2 is the derivative of G_m with respect to x_2, keeping x_3 constant, i.e. with $dx_1 = -dx_2$. Writing G_m as $\Sigma x_i(^\circ G_i + RT\ln x_i) + {}^E G_m(x_2,x_3)$ we get: $g_2 = {}^\circ G_2 - {}^\circ G_1 + RT(\ln x_2 - \ln x_1) + \partial^E G_m/\partial x_2$; and $g_{22} = RT(1/x_2 + 1/x_1) + \partial^2 {}^E G_m/\partial x_2^2$, etc. Look for the predominating term when x_1 is small. Furthermore, the liquid composition is close to that of $A_a B_b$.

Solution

For $x_C \equiv x_1 \to 0$ we get $[(x_2^\beta - x_2^L)g_{22}^L + (x_3^\beta - x_3^L)g_{23}^L]dx_2^L + [(x_2^\beta - x_2^L)g_{32}^L + (x_3^\beta - x_3^L)g_{33}^L]dx_3^L = -[x_2^L(H_2^\beta - H_2^L) + x_3^L(H_3^\beta - H_3^L)] \cdot dT/T$ under constant P. The predominating term in g_{22} is RT/x_1 and all the other second derivatives of g

have the same predominating term. By neglecting other terms we get

$$[(x_2^\beta - x_2^L + x_3^\beta - x_3^L) \cdot RT/x_1^L] dx_2^L + [(x_2^\beta - x_2^L + x_3^\beta - x_3^L) RT/x_1^L] dx_3^L =$$
$$[a(H_A^\beta - H_A^L) + b(H_B^\beta - H_B^L)] dT/T \text{ with } a + b = 1.$$ Suppose component C (i.e. component 1) does not dissolve in the stoichiometric phase, β, then $x_2^\beta + x_3^\beta = 1$ and $(x_2^\beta - x_2^L + x_3^\beta - x_3^L) RT/x_1^L = x_1^L RT/x_1^L = RT.$ The left-hand side of the equation is reduced to $RT(dx_2^L + dx_3^L)$ which is equal to $-RTdx_C^L.$ Furthermore, $x_2^\beta = a$ and $x_3^\beta = b$ and on the right-hand side we find that $aH_A^\beta + bH_B^\beta = H_m^\beta$ and the bracket can be expressed as $\Delta^\circ H_m^\beta$, the heat of formation of β. We can thus obtain $dT/dx_C^L = -\Delta^\circ H_m^\beta/RT^2.$ The depression of the freezing point is thus independent of whether one keeps x_A, x_B or x_A/x_B constant. For small additions of C it only depends on x_C.

10.6   *Vertical phase boundaries*

It is also interesting to discuss the possibility of finding a vertical phase boundary. This requires that the numerator is zero, i.e. that the heat of reaction, when β is dissolved in α, is zero. An example is given in Fig. 10.6 showing a so-called retrograde solidus line.

As another example we may take the well-known case of the so-called γ loop in binary iron diagrams with α-stabilizing alloying elements (see Fig. 10.3). Here both phases are rich in iron and we can approximate the numerator in the expression for dx_2^β/dT with $H_{Fe}^\alpha - H_{Fe}^\gamma$ and for dx_2^γ/dT with $H_{Fe}^\gamma - H_{Fe}^\alpha$. The characteristic γ loop thus depends upon the fact that the enthalpy difference between α- and γ-Fe changes sign and goes through zero in this range of temperature.

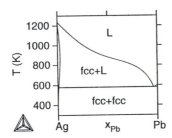

Figure 10.6 The T,x phase diagram for Ag–Pb. The solidus of the Ag phase is retrograde.

Exercise 10.6.1

From the detail of the Fe–O phase diagram, what can be said about the heat of solution of γ-Fe in the wüstite phase?

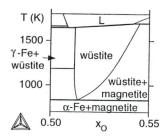

Exercise 10.6.1

Hint

Examine the boundary representing the solubility of γ-Fe in wüstite (w).

Solution

Since the γ phase is almost pure Fe, the numerator in the expression for dx_O^w/dT is $x_{Fe}^\gamma(H_{Fe}^w - H_{Fe}^\gamma) + x_O^\gamma(H_O^w - H_O^\gamma) \cong H_{Fe}^w - H_{Fe}^\gamma$, i.e. the heat of solution of γ-Fe in wüstite. This quantity is thus close to zero over a wide range of temperature because the boundary is almost vertical.

10.7 *Slope of phase boundaries in isothermal sections*

For a ternary system under isobarothermal conditions we get

$$[(x_2^\beta - x_2^\alpha)g_{22}^\alpha + (x_3^\beta - x_3^\alpha)g_{32}^\alpha]dx_2^\alpha + [(x_2^\beta - x_2^\alpha)g_{23}^\alpha + (x_3^\beta - x_3^\alpha)g_{33}^\alpha]dx_3^\alpha = 0$$

We can here introduce the slope of the α + β tie-line,

$$n = (x_3^\beta - x_3^\alpha)/(x_2^\beta - x_2^\alpha)$$

$$\frac{dx_3^\alpha}{dx_2^\alpha} = -\frac{g_{22}^\alpha + ng_{32}^\alpha}{g_{23}^\alpha + ng_{33}^\alpha}$$

As an application we shall examine when the $\alpha/(\alpha + \beta)$ phase boundary is parallel to the x_2 axis, i.e. when $dx_3^\alpha/dx_2^\alpha = 0$. We find the condition

$$g_{22}^\alpha/g_{32}^\alpha = -n$$

When the α phase is a dilute solution of components 2 and 3 in 1, the leading term in g_{22}^α/RT is $1/x_2^\alpha$ and it may be more convenient to recast the result into one of the following forms by inserting $g_{22}^\alpha/RT - 1/x_2^\alpha + 1/x_2^\alpha$ instead of g_{22}^α/RT

$$x_2^\alpha = -\frac{1}{g_{22}^\alpha/RT - 1/x_2^\alpha + ng_{32}^\alpha/RT}$$

$$x_2^\alpha = -\frac{x_2^\beta}{(x_2^\beta - x_2^\alpha)(g_{22}^\alpha/RT - 1/x_2^\alpha) + (x_3^\beta - x_3^\alpha)g_{32}^\alpha/RT - 1}$$

The latter equation can be rearranged into a form which is even more convenient because the ideal entropy of mixing gives a contribution of RT/x_1^α to both g_{22} and g_{32},

$$x_2^\alpha = -\frac{x_2^\beta}{(x_2^\beta - x_2^\alpha)(1/x_1^\alpha + 1/x_2^\alpha - g_{22}^\alpha/RT) + (x_3^\beta - x_3^\alpha)(1/x_1^\alpha - g_{32}^\alpha/RT) + x_1^\beta/x_1^\alpha}$$

We have thus made the first term in the denominator so small that it can often be neglected. One could then write

$$x_2^\alpha \cong \frac{x_2^\beta}{(x_3^\beta - x_3^\alpha)(1/x_1^\alpha - g_{32}^\alpha/RT) + x_1^\beta x_1^\alpha}$$

It is common to introduce Wagner's interaction parameter ε_2^3 which will be discussed in Section 17.7. It yields

$$x_2^\alpha \cong -\frac{x_2^\beta}{\varepsilon_2^3(x_3^\beta - x_3^\alpha) - x_1^\beta/x_1^\alpha} \cong -\frac{x_2^\beta}{\varepsilon_2^3(x_3^\beta - x_3^\alpha)}$$

Exercise 10.7.1

According to Schreinemakers' rule the phase boundary $\alpha/\alpha + \gamma$ in an isobarothermal section of a ternary phase diagram must be directed towards the β point if $\alpha/\alpha + \beta$ is directed towards the γ point. Prove this using the equation for the direction of phase boundaries.

Hint

Denote the slope of the $\alpha + \gamma$ tie-line by $n^{\alpha/\gamma}$ and the slope of the $\alpha + \beta$ tie-line by $n^{\alpha/\beta}$.

Solution

$n^{\alpha/\gamma} = (dx_3/dx_2)^{\alpha/\beta} = -(g_{22}^\alpha + n^{\alpha/\beta}g_{32}^\alpha)/(g_{23}^\alpha + n^{\alpha/\beta}g_{33}^\alpha)$. Thus, $-g_{22}^\alpha - n^{\alpha/\beta}g_{32}^\alpha = n^{\alpha/\gamma}g_{23}^\alpha + n^{\alpha/\gamma}n^{\alpha/\beta}g_{33}^\alpha$. By rearranging the terms we get $-g_{22}^\alpha - n^{\alpha/\gamma}g_{23}^\alpha = n^{\alpha/\beta}g_{32}^\alpha + n^{\alpha/\beta}n^{\alpha/\gamma}g_{33}^\alpha$ and thus we can form $n^{\alpha/\beta} = -(g_{22}^\alpha + n^{\alpha/\gamma}g_{32}^\alpha)/(g_{23}^\alpha + n^{\alpha/\gamma}g_{33}^\alpha)$ which is equal to $(dx_3/dx_2)^{\alpha/\gamma}$.

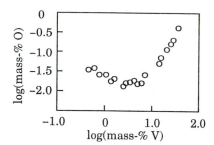

Exercise 10.7.2

Exercise 10.7.2

The diagram shows the solubility of V_2O_3 in liquid Fe at 1873 K. It is evident that the solubility of O goes through a minimum. Apply the regular solution model and show the requirement for the minimum to occur. What parameter or combination of parameters can be evaluated from the position of the minimum?

Hint

Apply the regular solution model in order to express the derivatives of G_m. Insert the expressions in the final equation obtained without any approximations. Parameters will occur in two places. Neglect one combination of parameters at a time and examine if the result is realistic.

Solution

For a very dilute solution we expect the solubility curve for V_2O_3 to give a solubility product $(x_V)^2(x_O)^3 = constant$ which yields $2\log(\text{mass}\%V) + 3\log(\text{mass}\%O) = constant$. That line can be fitted to the data points at low V contents but not at higher. There we must take into account the deviation from ideal behavior and shall try the regular solution model and identify the components as Fe $= 1$; V $= 2$; and O $= 3$. The regular solution model gives $G_m = \Sigma x_i(^\circ G_i + RT\ln x_i) + x_1x_2L_{12} + x_2x_3L_{23} + x_3x_1L_{31}$. When evaluating g_{22} and g_{23} we must remember that x_1 is a dependent variable. $g_2 = {}^\circ G_2 - {}^\circ G_1 + RT(x_2/x_2 + \ln x_2 - x_1/x_1 - \ln x_1) + (x_1 - x_2)L_{12} + x_3L_{23} - x_3L_{31}$; $g_{22} = RT(1/x_2 + 1/x_1) - 2L_{12}$; $g_{22}/RT - 1/x_1 - 1/x_2 = -2L_{12}/RT$; $g_{23} = RT(1/x_1) - L_{12} + L_{23} - L_{31}$; $g_{23}/RT - 1/x_1 = (L_{23} - L_{12} - L_{31})/RT$. With $\beta = V_2O_3$ we have $x_2^\beta = 0.4$; $x_3^\beta = 0.6$; $x_1^\beta = 0$. The diagram shows that the minimum is at $\log(\text{mass}\%V) = 0.5$ which gives mass$\%V = 3.2$; $x_2^\alpha = 0.035$. We get $0.035 = 0.4/[0.4 - 0.035)(2L_{12}/RT) + (0.6 - 0)(-L_{23} + L_{12} + L_{31}) + 0]$ from the equation for the minimum. There are two simple alternatives.

Alt.1: If $L_{23} - L_{12} - L_{31}$ can be neglected then $2L_{12}/RT = 31$. This would give a very large miscibility gap in the Fe–V system which has not been observed.

Alt.2: If $2L_{12}/RT$ is much less than 31, then its term can be neglected and we get $(L_{23} - L_{12} - L_{31})RT = -19$ (or $\varepsilon_V^O = \varepsilon_O^V = -19$ if we use the last equation). This is regarded as a reasonable value because O and V have a strong affinity to each other.

Exercise 10.7.3

The following diagram shows the solubilities of the three oxides in liquid Fe at 1823 K according to an experimental study. All curves show minima. Use this information in order to estimate the Cr content of the two spinels.

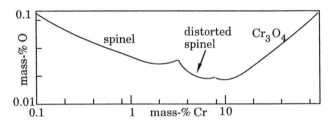

Exercise 10.7.3

Hint

Start by evaluating ε_{Cr}^O from the minimum for the phase with a known composition, Cr_3O_4. Knowing ε_{Cr}^O one can then calculate the Cr content for another oxide from its minimum. Both spinels can be represented by the general formula $(Fe,Cr)_3O_4$. Considering the limited accuracy of the data it is justified to approximate mass fraction Cr as molar content Cr.

Solution

Let β = oxide; α = liquid; 2 = Cr; 3 = O. Then $x_3^\beta = x_O^{oxide} = 4/7$ for all the oxides and $x_3^\alpha = x_O^L \cong 0$. From the known composition of Cr_3O_4: $0.1 = x_{Cr}^L = x_2^\alpha = -x_2^\beta/\varepsilon(x_3^\beta - x_3^\alpha) = -(3/7)/\varepsilon(4/7)$; $\varepsilon = -3/4 \cdot 0.1 = -7.5$. Using this value we find:

For undistorted spinel: $0.02 = -x_{Cr}^{oxide}/(-7.5)(4/7)$; $x_{Cr}^{oxide} = 0.6/7$. The formula is $Fe_{2.4}Cr_{0.6}O_4$.

For distorted spinel: $0.06 = -x_{Cr}^{oxide}/(-7.5)(4/7)$; $x_{Cr}^{oxide} = 1.8/7$. The formula is $Fe_{1.2}Cr_{1.8}O_4$.

10.8

The effect of a pressure difference between two phases

In Section 10.4 we derived an expression for the change in composition of an α phase in equilibrium with a β phase caused by changes in T and P. It was then assumed that T and P had always the same values in both phases. The same derivation can be carried out even if P changes in different ways in the two phases. This will occur when they are separated by a curved interface. In Section 14.6 we shall find the equilibrium condition $P^\beta = P^\alpha + 2\sigma/r$. Now we shall simply assume that α and β can be in equilibrium even at a difference in pressure. The result will then be

$$\sum_{i=2}\sum_{j=2}(x_i^\beta - x_i^\alpha)g_{ij}^\alpha dx_j^\alpha = \sum_{i=1}^{s} x_i^\beta(V_i^\beta dP^\beta - V_i^\alpha dP^\alpha) - \sum_{i=1} x_i^\beta(H_i^\beta - H_i^\alpha)dT/T$$

Let us now apply this equation to a binary case where $dP^\alpha = 0$ and $dT = 0$. Using $V_m^\beta = x_1^\beta V_1^\beta + x_2^\beta V_2^\beta$ we get

$$dx_2^\alpha = \frac{V_m^\beta dP^\beta}{(x_2^\beta - x_2^\alpha)g_{22}^\alpha}$$

An expression for the simultaneous change in the β phase can be obtained by first exchanging α and β in the general equation and then applying it to the case $dP^\alpha = 0$ and $dT = 0$,

$$(x_2^\beta - x_2^\alpha)g_{22}^\beta dx_2^\beta = (x_1^\alpha V_1^\beta + x_2^\alpha V_2^\beta)dP^\beta$$

$$dx_2^\beta = \frac{(x_1^\alpha V_1^\beta + x_2^\alpha V_2^\beta)dP^\beta}{(x_2^\beta - x_2^\alpha)g_{22}^\beta}$$

It is interesting to see that α and β change their composition in the same direction.

It should be noted that these equations were actually derived graphically by means of molar Gibbs energy diagrams in Figs. 6.10 and 6.11.

Exercise 10.8.1

For the α/β equilibrium in a ternary system at a constant T and P^α one obtains $(V_m^\beta/RT)dP^\beta = h^\alpha dx_2^\alpha + k^\alpha dx_3^\alpha$. Show that $h^\alpha = x_2^\beta/x_2^\alpha - x_1^\beta/x_1^\alpha$ and $k^\alpha = x_3^\beta/x_3^\alpha - x_1^\beta/x_1^\alpha$ if α and β are ideal solutions.

Hint

For an ideal solution $g_{22}/RT = 1/x_1 + 1/x_2$; $g_{23} = g_{32} = RT/x_1$; $g_{33}/RT = 1/x_1 + 1/x_3$.

Solution

The dx_2^α coefficient becomes

$$(x_2^\beta - x_2^\alpha)g_{22}^\alpha + (x_3^\beta - x_3^\alpha)g_{32}^\alpha = (x_2^\beta - x_2^\alpha)(x_1^\alpha + x_2^\alpha)/x_1^\alpha x_2^\alpha + (x_3^\beta - x_3^\alpha)/x_1^\alpha =$$
$$(x_2^\beta - x_2^\alpha)(x_1^\alpha + x_2^\alpha)/x_1^\alpha x_2^\alpha + (1 - x_1^\beta - x_2^\beta - 1 + x_1^\alpha + x_2^\alpha)/x_1^\alpha =$$
$$(x_2^\beta x_1^\alpha + x_2^\beta x_2^\alpha - x_2^\alpha x_1^\alpha - x_2^\alpha x_2^\alpha - x_2^\alpha x_1^\beta - x_2^\alpha x_2^\beta + x_2^\alpha x_1^\alpha + x_2^\alpha x_2^\alpha)/x_1^\alpha x_2^\alpha =$$
$$(x_2^\beta x_1^\alpha - x_2^\alpha x_1^\beta)/x_1^\alpha x_2^\alpha = x_2^\beta/x_2^\alpha - x_1^\beta/x_1^\alpha.$$

Sharp and gradual phase transformations

Experimental conditions

There will be a driving force for a phase transformation if the conditions of a system are changed in such a way that the system moves from one phase field into another in the phase diagram. In this chapter we shall examine the character of such phase transformations and we shall find that they depend upon the experimental method of controlling and changing the conditions. It is important first to realize that the possibility of efficiently controlling the various state variables is very different. For gaseous and liquid phases it is comparatively easy to control the pressure. It can be kept constant or it can be changed gradually according to an experimental program. At any moment it is very uniform in the system apart from effects due to the surface tension of curved phase interfaces. For solid systems it is more difficult to control the pressure, in particular during a phase transformation resulting in a volume change. This may give rise to local deformation and internal stresses. On the other hand, solid phases are so dense that the thermodynamic effect of pressure differences and stresses can often be neglected. From a practical point of view we may often regard the pressure as an experimental variable which can be reasonably well maintained at a low enough level to have a negligible effect.

The temperature can often be kept relatively constant but in a large piece of material it may be difficult to change the temperature according to an experimental programme. This is due to the limited rate of heat conduction. As a consequence, in a well-controlled experiment the required change of temperature must be slow enough. Another way to change the temperature is to control the flow of heat. If the pressure is kept constant we have

$$dH = dU + d(PV) = dU + PdV + VdP = dQ + VdP = dQ$$

and this is thus a way of controlling the enthalpy rather than the temperature. Again, the rate of heat conduction may be a limiting factor and in order for an experiment to be well controlled it can only involve slow internal changes or small specimens. The heat content is changed locally if there is a spontaneous phase transformation. Only slow phase transformations or small specimens can thus be studied if one wants to have at least approximately isothermal conditions.

If the chemical potential of an element is changed gradually in a system by changing its value in the surroundings, considerable potential differences within the system will normally prevail for a long time unless the change is extremely slow. This is because equilibration of the chemical potential requires a change of the composition which can only be accomplished by diffusion or convection. Diffusion is usually many orders of magnitude slower than heat conduction. As a consequence, it is much more common to keep the composition constant during an experiment than to keep the chemical potentials constant.

There are cases where a particular component is much more mobile than the other components. This may happen for elements with small atoms when dissolved interstitially in solid phases. An example of some practical importance is carbon in steel. An even better example is hydrogen in most metals and alloys. In such cases one may have some success in controlling the chemical potential of that particular component.

A phase transformation may itself give rise to severe difficulties in the control of the experimental conditions. Under the given values of the potential variables the new phase will most probably have different values for all the molar quantities and there will be a tendency for their conjugate potential variables to change during a phase transformation, independent of what potential is being changed experimentally. In practice, the difficulties in carrying out a well-controlled experiment may be the same whatever potential one has decided to change. As an example, if the changed conditions give rise to a phase transformation, then the transformation may in turn give rise to a redistribution of the components by diffusion, heat flow by conduction and material transport by plastic and elastic deformation.

Due to the complications caused by a phase transformation in a solid material it may be somewhat easier to carry out a well-controlled experiment under constant values of some extensive variables rather than potentials. However, that will affect the character of the phase transformation. This will be evident from the discussion in this chapter.

Exercise 11.1.1

A solid substance is kept at its melting point T_1 under a certain high pressure P_1. Discuss what happens if the pressure is suddenly released. Suppose that the liquid form of the substance is less dense.

Hint

The solid phase with its higher density was favoured by the high pressure. T_1 being the melting point at P_1 is thus above the melting point at $P = 0$.

Solution

Melting will most probably start somewhere. The melt will instantaneously be at the new melting point which is lower than T_1. Heat will thus start to flow into the melted region from the remaining solid which may thus cool down to the new melting point. Thus, a mixture of the two phases may be established and its temperature will be at the new melting point. However, this may cause heat flow into the system from the surroundings if they are kept at T_1. The whole system will thus melt eventually. On the other hand, if the new melting point is very low compared to T_1, then the whole system may melt without any additional heat (see Sections 13.2 and 13.3).

11.2

Characterization of phase transformations

In this section we shall neglect the difficulties mentioned in Section 11.1 in connection with controlling the variables. We shall limit the discussion to cases where we have selected one variable to be varied. From the phase diagram point of view this means that c of the $c + 1$ independent variables in a set of external state variables will be kept constant by sectioning, $n_s = c$, and the selected variable can be represented on the resulting one-dimensional phase diagram, $r = 1$. It should be noticed that we cannot have any projected variable when discussing a phase transformation. All variables must then have a value.

When the selected variable is changed gradually, the system may move from one phase field into another and a phase transformation may thus occur. It can be represented by a reaction formula obtained by combining the names of the phase fields. For instance, when moving from an α phase field into a γ phase field we expect the transformation $\alpha \rightarrow \gamma$. In doing so we must pass an $\alpha + \gamma$ phase field and one may characterize the transformation as a **sharp** one if the $\alpha + \gamma$ phase field has no extension in the one-dimensional phase diagram, $d = 0$. Otherwise, it may be characterized as a **gradual** transformation and has $d = 1$. These cases may be illustrated by starting with two-dimensional phase diagrams. The main part of Fig. 11.1 is a two-dimensional diagram obtained by starting with only potential variables, $c - 1$ of which have then been sectioned, $n_s = c - 1$, (e.g. an isobaric section of a binary system). A further sectioning (making $n_s = c$) could be made at $T = T_1$, giving the one-dimensional phase diagram in the lower part of the figure. The phase field rule yields for $\alpha + \gamma$: $d = c + 2 - p - n_s + n_m = c + 2 - 2 - c + 0 = 0$ and confirms that the phase transformation $\alpha \rightarrow \gamma$ should be a sharp one if μ_B is increased gradually. A similar result would be obtained if one could keep μ_B constant at μ_{B_1} and gradually increase T (see the right-hand part of Fig. 11.1).

Figure 11.2 shows the diagram for the same system when the chemical potential has been replaced by the conjugate molar content, x_B. By sectioning at $T = T_1$ one

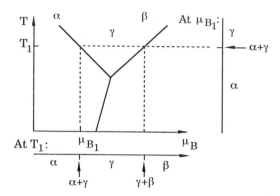

Figure 11.1 Illustration of the conditions for a sharp phase transformation in a simple case where all the external variables to be kept constant are potentials.

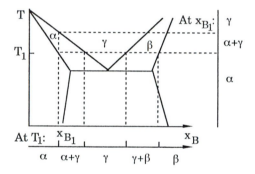

Figure 11.2 Illustration of the conditions for a gradual phase transformation in the simple case where only one molar variable is used.

now obtains the one-dimensional diagram in the lower part. The two-phase fields have opened up and the $\alpha \rightarrow \gamma$ transformation will be gradual if x_B is increased gradually. It is evident that a transformation can never be sharp if it occurs under a gradual increase of a molar quantity because all phase fields have some extension in the direction of a molar quantity.

A section at $x_B = x_{B_1}$ is shown to the right of the figure. It also shows a gradual transformation when T is changed gradually but that result cannot be predicted without inspecting the phase diagram or applying the phase field rule. In this case $c = 2$, $p = 2$, $n_s = 2$ (P and x_B) and $n_m = 1$ (x_B), yielding $d = c + 2 - p - n_s + n_m = c + 2 - 2 - c + 1 = 1$.

Let us now return to the case of a sharp transformation in Fig. 11.1. If the gradual change of μ_B at $T = T_1$ is continued, then the sharp transformation $\alpha \rightarrow \gamma$ will be followed by another sharp transformation $\gamma \rightarrow \beta$ at a higher value of μ_B. It is then interesting to discuss what would happen if the section $T = T_1$ were made right through the three-phase point. The lower part of the figure would show a point for the $\alpha + \beta + \gamma$ equilibrium instead of two points for $\alpha + \gamma$ and $\gamma + \beta$. However, with $p = 3$ the phase field rule would yield $d = c + 2 - p - n_s + n_m = c + 2 - 3 - 0 + 0 = -1$. Since $d = 0$ represents a point, one may conclude that $d = -1$ represents a phase field which should not show up at all in the one-dimensional diagram, the reason being that it is practically impossible to place the section exactly through the three-phase equilibrium. Let us for a while neglect that practical difficulty and suppose that the section actually goes right through the three-phase equilibrium. What phase transformation would one then observe on gradually increasing μ_B? One could expect to observe the sharp transformation $\alpha \to \gamma$, followed by $\gamma \to \beta$ but also a direct transformation $\alpha \to \beta$. We may regard this as a case of **overlapping sharp transformations**.

Let us next replace μ_B by x_B and still assume that $T = T_1$ can be chosen and controlled in such a way that the section goes right through the three-phase equilibrium, in this case right through the three-phase horizontal in Fig. 11.2. The lower part of that figure would then show an $\alpha + \beta + \gamma$ region instead of the three regions, $\alpha + \gamma$, γ and $\gamma + \beta$. For three phases the phase field rule would now give $d = c + 2 - p - n_s + n_m = c + 2 - 3 - c + 1 = 0$ yielding the incorrect prediction of a three-phase point instead of a line, the horizontal.

In order to understand this puzzling result one should remember that a transformation can never be sharp when taking place under a gradual change of a molar quantity. If the phase field rule gives $d = 0$ for a molar axis, the interpretation must be that it is practically impossible to carry out such an experiment. It thus corresponds to the improbable case of $d = -1$ for a potential axis. We may conclude that, if a molar quantity is varied, $d = 0$ predicts **overlapping gradual transformations** (in the present case $\alpha \to \beta$ or $\alpha \to \gamma$ followed by $\gamma \to \beta$). However, it is as unlikely as the case of overlapping sharp transformations for $d = -1$.

It is evident that the only way to get a sharp transformation is to vary a potential. Usually this is T and one keeps P and the composition constant, $n_s = c$ and $n_m = c - 1$. The sharp transformation will then occur when

$$0 = d = c + 2 - p - n_s + n_m = c + 2 - p - c + (c - 1) = c + 1 - p$$
$$p = c + 1$$

Under the same conditions, $p = c + 2$ will yield $d = -1$, i.e. overlapping sharp transformations. The present discussion results in two schemes for the character of phase transformations. When a potential is varied gradually, we obtain

for $d = +1$:   Gradual transformation

for $d = 0$:   Sharp transformation

for $d = -1$:   Overlapping sharp transformations

When a molar quantity is varied gradually, we obtain

for $d = +1$:   Gradual transformation

for $d = 0$:   Overlapping gradual transformations

In a sharp transformation ($d = 0$) the fractions of the phases (i.e. the extent of the transformation) are not fixed by the value of the changing variable. This is why the corresponding state of phase equilibrium is sometimes called 'indifferent' (Prigogine and Defay, 1958). On the other hand, the compositions of all the phases are fixed. This is why any sharp transformation is sometimes called 'azeotropic' although that term is usually reserved for the case with an extremum discussed in connection with Konovalov's rule in Section 9.8. Cases with an extremum have been neglected in the present discussion but will be further discussed in Sections 12.7–12.10.

Also, overlapping sharp transformations $(d = -1)$ are sometimes called 'indifferent', because the extent of transformation is not fixed. In that case, however, there is more than one transformation and their relative progress is also not fixed.

Before leaving this topic, it should be emphasized that the present discussion is based on considerations of equilibrium. In practice, there are many kinetic obstacles and it is not impossible to observe overlapping transformations (often regarded as competing reactions) if the experimental conditions come close to the improbable ones, for which the phase field rule predicts overlapping transformations.

Exercise 11.2.1

What kind of transformation involving p phases will occur if $p - 3$ molar quantities and an appropriate number of potentials are kept constant and a potential is varied?

Hint

In order to get a one-dimensional diagram, the total number of constant (i.e. sectioned) variables must be $n_s = c$ if there are no projections.

Solution

There are $p - 3$ molar variables, $n_m = p - 3$, and we get $d = c + 2 - p - n_s + n_m = c + 2 - p - c + (p - 3) = -1$. There is thus a negligible chance to get a transformation involving p phases. We may only succeed by choosing a particular value for one of the potentials. We would then obtain a case of overlapping sharp transformations.

Exercise 11.2.2

Consider the equilibrium: solid Cu + solid Ag + liquid + vapour. What type of transformation should one expect between these phases if T is changed gradually for a system with constant composition and pressure?

Hint

There is some low mutual solubility between solid Cu and solid Ag but that has no effect on the present problem. The vapour pressure of Ag is higher than that of Cu.

Solution

$c = 2$, $p = 4$ and Gibbs' phase rule yields $v = c + 2 - p = 2 + 2 - 4 = 0$. This equilibrium would thus show up as a point in the complete potential phase diagram. Under the present experimental conditions, $n_m = c - 1$ (all composition variables), and $n_s = c$ (constant composition and pressure), we predict $d = c + 2 - p - n_s + n_m = 2 + 2 - 4 - c + (c - 1) = - 1$. The chance of observing the corresponding phase transformation would be negligible since it would require that a particular value of P could be chosen and kept constant. If we were to succeed in doing this, the system could transform with all four phases present but it would be a case of overlapping sharp transformations. They would be Cu + Ag → liq., Cu + Ag → vapour, Ag + liq. → vapour, liq. → Cu + vapour.

Exercise 11.2.3

Consider the same system as in Exercise 11.2.2 but suppose that the system is heated gradually.

Hint

As before, $n_s = c$, but instead of a gradually changing temperature, we should now consider a gradually changing enthalpy. Thus $n_m = c - 1 + 1 = c$.

Solution

Under the new experimental conditions $d = c + 2 - p - n_s + n_m = 2 + 2 - 4 - c + c = 0$. We would observe the same overlapping transformations but they would now be gradual.

Exercise 11.2.4

A pure element under a constant pressure melts at a given temperature T if the temperature is increased gradually. Discuss how the melting will be affected if one were to keep the volume constant instead of the pressure.

Hint

In both cases $n_s = 1$. In the first instance of melting there can be no difference because the system does not 'feel' which variable is kept constant.

Solution

Melting starts at the same temperature but is no longer a sharp transformation. $d = c + 2 - p - n_s + n_m = 1 + 2 - 2 - 1 + 1 = 1$ shows that with a molar variable ($n_m = 1$) the melting will be gradual. It will thus proceed gradually as the temperature increases.

11.3   *Microstructural character*

We shall now discuss how the phases will be distributed within a system as a result of a phase transformation. With most materials one needs a microscope in order to study the distribution of the phases, thus the term **microstructure**.

During a gradual transformation the new product will only occupy some fraction of the volume. It there are many nuclei, the result may be an intimate mixture of the old and new phases, with a gradual change of the fractions and of the compositions of the phases. Such a transformation may be regarded as **microstructurally gradual**. On the other hand, in a sharp phase transformation the new phase or phases will completely replace the old ones but it may still be interesting to discuss the microstructural appearance during the transformation because it is never instantaneous, due to kinetic restrictions. Thus, let us first consider the effect of the limited rate of heat conduction with reference to Fig. 11.1.

Whether one regards T or H_m (enthalpy per mole of the system) as the controlling variable, heat supplied from the surroundings must normally flow into the system through the surface layer. If there is no other kinetic restriction, then the phase transformation should start at the surface where the temperature must be at least slightly higher than in the interior. After some time there will be a massive surface layer of the new β phase. It will have a sharp interface to the old α phase in the interior. Thus, the phase transformation will be **microstructurally sharp** in both cases. Figure 11.3 illustrates the variation of the local value of the molar enthalpy as a function of the distance from the surface, assuming that the whole system was initially in a state of α at the temperature of equilibrium with β. The P axis has been added to this diagram in order to illustrate that P is kept constant. The difference between the two cases is that with T as the controlling variable the process will not stop until the microstructurally sharp transformation has proceeded through the entire system. With H_m as the controlling variable, the process will stop when the average value of H_m has reached the prescribed value.

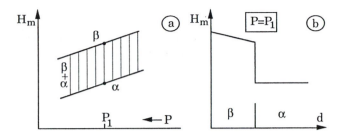

Figure 11.3 Phase transformation $\alpha \to \beta$ during heat conduction from the surroundings. It will be microscopically sharp if Gibbs' phase rule predicts that it should be sharp. d is the distance from the surface.

On the other hand, suppose that the phase transformation is so slow due to kinetic reasons that it would be possible to increase H_m of the initial α phase to a value falling inside the $\alpha + \beta$ phase field and, thus, to increase T to a value falling inside the β phase field before the phase transformation starts. Wherever there are β nuclei, they could then grow and an $\alpha + \beta$ mixture would develop, the final fractions of α and β being controlled by the lever rule applied to the prescribed H_m value. The kinetics may be so slow that α and β are not in equilibrium, not even at the α/β interfaces, until the fraction of β has approached its final value. In that case the progressing phase transformation may look similar to the microstructurally gradual transformation.

We have here considered a phase transformation which is sharp when a potential is varied. We have found that in order to predict its microscopic appearance one must first examine if the transformation is slow due to other kinetic restrictions. If that is the case, the transformation may be microstructurally gradual. If the transformation is fast enough to follow the changes of the controlling variable, then it may be microstructurally sharp. In both cases the result will be the same whether one varies the potential or its conjugate molar quantity. However, in the remainder of the present chapter we shall always use a potential as the variable.

Exercise 11.3.1

Is it possible to solidify a pure liquid substance by increasing P if the solid form is denser? If so, will the solidification be complete or only partial? Will it be microscopically sharp or not?

Hint

It all depends upon what other variable is controlled. One will probably try to keep some variable constant. Consider two conditions, isothermal (very slow) and adiabatic (very rapid). It may be helpful to sketch the appropriate phase diagrams.

Solution

We get complete solidification if T is kept constant. Adiabatic conditions are more difficult to discuss, because they give, according to the first law,: $dQ = dU + PdV = 0$ and $dQ = dH - VdP = 0$. Neither U nor H is thus constant when P is changed. In order to find a state function which is constant, we must assume reversible conditions, and, using the second law, we then get

$dQ = TdS - Dd\xi = TdS = 0$. For this case we should thus use an S_m,P diagram. We can then see that the solidification can be partial or complete depending upon how large the P change is. This conclusion may not change if there is some internal entropy production due to the transformation.

When considering the microstructural character we may first examine the adiabatic case and accept that the change of P is more rapid than the transformation. Then many nuclei distributed over the whole system may form and give the transformation a gradual appearance. In the slow isothermal case we may assume that the transformation starts at the surface or very close to it. The transformation will then be microstructurally sharp if Gibbs' phase rule predicts that the dimensionality of the $\alpha + L$ phase field should be zero. For a unary system we get

$d = c + 2 - p - n_s + n_m = 1 + 2 - 2 - 1 + 0 = 0$.

11.4   *Phase transformations in alloys*

Diffusion is usually much slower than heat conduction and may thus give a very severe kinetic restriction on the phase transformations in alloys. This is true even if we would decide to control the experimental conditions by keeping all the potentials constant except for T, which is varied gradually. The complications are not immediately evident from the T,μ_B phase diagram but are clearly demonstrated by the T,x_B phase diagram at constant P in Fig. 11.4.

If we could keep μ_B constant during an increase of T, we would move through the T,x_B phase diagram according to the broken arrow in Fig. 11.4(b). This corresponds to the straight arrow in the T,μ_B phase diagram and represents a sharp transformation $\alpha \rightarrow \gamma$. However, this would require an exchange of atoms with the surroundings and, due to the low rate of diffusion compared to heat conduction, the system would rather move along the straight arrow in the T,x_B phase diagram, when T is increased, and the composition rather than the chemical potential would stay constant if the time of the experiment is not very long. One would not manage to keep μ_B constant except in very special cases. The composition would not have time to change much and the system would move into the $\alpha + \gamma$ two-phase field. A gradual phase transformation would result. The transformation would be microstructurally gradual and the system would show a mixture of α and γ and the fraction of γ would gradually increase on increasing T.

The effect of slow diffusion discussed here is the reason why most experimental

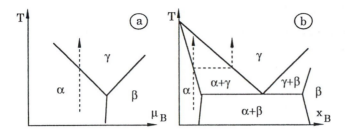

Figure 11.4 Illustration of difficulty for a phase transformation in a binary system to occur under constant μ_B when T is increased under constant P.

conditions can be approximated by assuming that the composition is constant during a change of T (or P).

Exercise 11.4.1

Suppose the temperature is increased gradually under constant pressure. Consider a phase transformation which would be sharp if a particular chemical potential were kept constant. However, due to slow diffusion it is difficult to study a phase transformation under a constant chemical potential. Instead, one keeps the molar content of the same component constant. Is it thus possible to minimise the role played by diffusion?

Hint

If $d = 0$ when the particular chemical potential is kept constant, then $d = 1$ if the corresponding molar content is kept constant because n_m increases by one unit. The transformation may thus be classified as gradual.

Solution

Since this transformation is now gradual, there will be a gradual change of the fractions of various phases. One will thus get a mixture of phases. If they have different compositions, diffusion is required over distances related to the coarseness of the microstructure. However, such diffusion distances will normally be much shorter than those which are necessary under conditions of constant potential.

11.5   *Classification of sharp phase transformations*

Sharp phase transformations in alloys at constant P involving few phases have been classified into various groups. For unary systems there is only one type, $\alpha \to \beta$, and it is called an **allotropic** transformation; melting may be regarded as a special case. In binary systems there are two main types,

$$\gamma \to \alpha + \beta \qquad \text{eutectoid transformation}$$
$$\gamma + \alpha \to \beta \qquad \text{peritectoid transformation}$$

The $\alpha + \beta$ mixture resulting from the first transformation is often called a eutectoid structure or simply a eutectoid. In order to identify a particular eutectoid it is sometimes denoted by the name of the parent phase. In the present case it would thus be called γ-eutectoid. In addition there are special names depending upon the role played by the liquid phase. The following names refer to transformations occurring on cooling but the same names are often applied to the corresponding features in the phase diagram.

$$L \to \alpha + \beta \qquad \text{eutectic}$$
$$L_1 \to \alpha + L_2 \quad \text{monotectic}$$
$$\alpha \to L + \beta \qquad \text{metatectic}$$
$$L + \alpha \to \beta \qquad \text{peritectic}$$
$$L_1 + \alpha \to L_2$$
$$L_1 + L_2 \to \alpha \quad \text{syntectic}$$
$$\alpha + \beta \to L$$

The first transformation to be given a name was the eutectic transformation $L \to \alpha + \beta$. The word 'eutectic' is taken from Aristotle who used it as meaning 'easily melted' and that was the definition when first used by Guthrie (1884). He was not yet aware of the regular microstructure usually formed in such alloys on solidification, a lamellar example of which is sketched in Fig. 11.5. Today, when we speak of a 'eutectic' we tend to imply this type of microstructure. The eutectoid transformation has come to mean $\gamma \to \alpha + \beta$ independent of whether it occurs on heating or cooling or under isothermal conditions and independent of how the phase diagram looks. It is interesting to note that the eutectic type of microstructure can form on partial melting of an intermetallic phase from a peritectic phase diagram.

    The growth conditions for a eutectic transformation can be illustrated by Fig. 11.6, where two two-phase regions have been extrapolated to a transformation temperature below the equilibrium temperature for $L + \alpha + \beta$. Figure 11.6(b) and (c) shows the variation of composition within the parent phase, L, in front of β and α, respectively. Diffusion of B may thus occur inside the L phase from the α interface to the β interface and growth is thus made possible. As illustrated by the arrows in Fig. 11.5, diffusion may

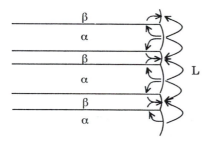

Figure 11.5 Cooperative growth of two phases in a eutectic transformation. The arrows indicate possible diffusion paths.

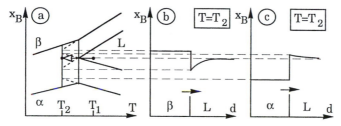

Figure 11.6 Conditions for a eutectic transformation, $L \rightarrow \alpha + \beta$, in a binary system at constant P and a constant overall composition. The temperature was decreased from T_1 to T_2.

also occur inside the two growing phases, α and β, when they grow side by side, but that diffusion is generally much slower than diffusion in the liquid.

It is interesting to note that eutectoid transformations often result in a rather regular arrangement of the two new phases. The reason is that such arrangements give short diffusion paths. It is called cooperative growth.

The peritectoid transformations derive their name from the peritectic transformation, $L + \alpha \rightarrow \beta$, which occurs on solidification. The name 'peritectic' means that a phase formed by such a reaction grows along the interface, i.e. along the periphery of the primary solid phase as illustrated in the sketched microstructure of Fig. 11.7.

Normally, all peritectoid transformations give the same type of geometric arrangement. The growth conditions during a transformation can be illustrated by Fig. 11.8.

The diffusion distance is shortest close to the β tip advancing along the L/α interface in Fig. 11.7. That growth process is thus rapid. The subsequent thickening of β, can occur only by diffusion through β itself. It grows slower the thicker it gets and a peritectoid reaction seldom goes to completion. On continued cooling, β can also grow into the matrix phase as an ordinary primary precipitation but it is common that some of the primary solid phase, α, remains.

In ternary systems there are three kinds of sharp phase transformations.

$\alpha \rightarrow \beta + \gamma + \delta$ Four-phase eutectoid transformation or class I four-phase transformation

$\alpha + \beta \rightarrow \gamma + \delta$ Four-phase peritectoid transformation or class II four-phase transformation

$\alpha + \beta + \gamma \rightarrow \delta$ Class III four-phase transformation.

The four-phase transformations are illustrated in Fig. 11.9.

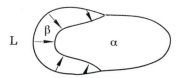

Figure 11.7 Geometric arrangement of the growing β phase during a peritectic transformation, obtained by growth along the previous phase interface, L/α, and by subsequent thickening.

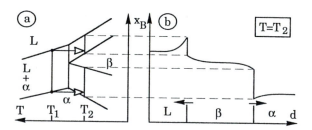

Figure 11.8 Conditions for an L $+ \alpha \rightarrow \beta$ transformation in a binary system at constant P and a constant over-all composition. The temperature was changed from T_1 to T_2. The arrows above diagram (b) indicate the migration of the new phase interfaces during thickening.

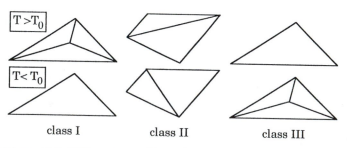

Figure 11.9 Different types of four-phase reactions in a ternary system, represented in a compositional coordinate system.

Exercise 11.5.1

Vertical sections through two different ternary T, x_B, x_C diagrams at constant P are reproduced. Discuss what type of sharp four-phase transformations the four alloys indicated go through on cooling. Show projections of the four-phase planes and draw lines representing the two sections.

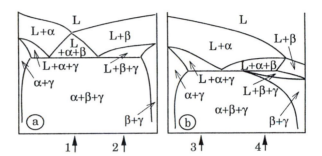

Exercise 11.5.1

Hint

In both diagrams all four three-phase fields connected to the invariant four-phase field are shown in the section. It is thus possible to know the type of transformation. In diagram (a), three of the four fall above the invariant one and these three all contain liquid. It is evident that this is a four-phase eutectic transformation. Both alloys, 1 and 2, give $L \rightarrow \alpha + \beta + \gamma$. In diagram (b) there are two three-phase fields on each side of the four-phase horizontal. This must be a class II transformation and both alloys, 3 and 4, give $L + \alpha \rightarrow \beta + \gamma$.

Solution

See diagram.

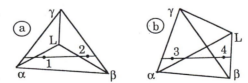

Solution 11.5.1

Exercise 11.5.2

When a composite material, composed of many thin alternating layers of pure Zr and pure Co, was heat treated at 573 K for 10 hours, an amorphous Co–Zr phase was formed. Study the Co–Zr phase diagram and explain how this transformation can be thermodynamically possible. How should one classify this reaction?

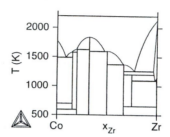

Exercise 11.5.2

Hint

We are not asked to discuss how it can be kinetically possible to avoid the formation of a series of stable intermetallic phases. We can accept the experimental result and simply consider the three phases under discussion, solid Zr, solid Co and the liquid (which is described as an amorphous phase because of its high viscosity due to the low temperature).

Solution

The reaction observed is the reverse of a eutectic one, Zr + Co → L and may be called eutectic melting but is really of a peritectoid type. A sketch of the diagram with all the other phases excluded would show a deep eutectic. Even though it is uncertain how to extrapolate the two liquidus curves below the glass transition (where the viscosity rises to very high values), it is evident that the eutectic point may very well fall below 573 K. From the experimental result we may conclude that this is actually the case.

11.6   *Applications of Schreinemakers' projection*

Schreinemakers' T,P diagram can be very useful in a discussion of phase transformations, in particular for higher-order systems where all other methods of reducing the number of axes to two would yield much more complicated pictures. As an introduction, consider the diagram in Fig. 9.1(b). It shows that there are three two-phase surfaces covering the angle between the (α) and (β) lines. By keeping P and T constant at values within that angle and varying μ_B we could expect the transformations $\alpha \leftrightarrow \delta$, $\gamma \leftrightarrow \delta$ and

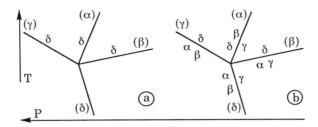

Figure 11.10 Method to decide the type of transformation. For instance, the δ phase only exists on the other side of the invariant point, counted from the δ-absent line, here denoted (δ).

$\beta \leftrightarrow \gamma$. Of course, they should occur one after the other and only two reaction sequences are possible, $\beta \to \gamma \to \delta \to \alpha$ or the reverse. By this consideration we can immediately conclude that the phases are arranged in this order along the composition axis, z_B. This is confirmed by Fig. 9.10. Similar considerations based on the other angles will give less complete answers. Unfortunately, this very simple way of deciding the relative compositions of the phases taking part in an invariant equilibrium gets much more complicated in higher-order systems. The method based on coincidence, described in Section 9.4, may be more powerful.

Next, let us consider a transformation occurring by changing T or P and keeping the other constant. If the composition is also constant, then the phase field rule would yield

$$d = c + 2 - p - n_s + n_m = c + 2 - p - c + c - 1 = c + 1 - p$$

because $n_s = 1 + c - 1 = c$ and $n_m = c - 1$. A sharp transformation should be obtained for $d = 0$, i.e. $p = c + 1$, and should thus occur if the system would cross a univariant line (for which $p = c + 1$, see Section 9.1). This can be accomplished by a suitable choice of composition. For illustration, see the dashed arrow in Fig. 9.9. Then the question is, what type of sharp transformation will it be. From Fig. 9.1 we would only know that the (δ) line should give a transformation between α, β and γ. However, the following method can be used to give more detailed information.

Since δ does not exist along the (δ) line, it can only exist on the other side of the invariant point. It will thus exist on the upper sides of the other univariant lines but not on their lower sides (see Fig. 11.10(a)). Using the same kind of information from the other lines we get the results shown in Fig. 11.10(b). By crossing the (δ) line from left to right, i.e. by decreasing P, under a suitable constant value of z_B, we thus get the transformation $\alpha + \beta \to \gamma$. It should be emphasized that the transformations described by the positions of the greek letters in Fig. 11.10 only occur when a line is crossed in the plane of the projected diagram. It gives no information on the transformations in any other direction.

As we have already seen, the $\alpha + \gamma$ surface is situated in the angle between the (β) and (δ) lines. The fact that α and γ are on opposite sides of the (δ) line but on the same side of the (β) line, is of no consequence in this connection.

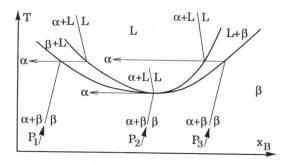

Figure 11.11 Detail of a binary T,x phase diagram with a point of minimum for the $\beta + L$ equilibrium. There the transformation $\beta \to L$ will be congruent. Three sets of lines representing equilibrium with α at different P are given.

In Section 9.8 it was concluded that one can include in Schreinemakers' projection a curve showing where an equilibrium with $p = c$ phases degenerates by the phases falling on the same point for a binary system, on the same line for a ternary system, etc. Such a singular curve may originate from a univariant line, as demonstrated in Figs. 11.11 and 11.12, using a binary system for illustration.

In order to simplify the construction, it was here assumed that the $\beta + L$ equilibrium is not affected by P but an increased P will increase the stability of α. Lines for equilibrium with α are presented for three P values. With the lowest P value, P_1, the $\alpha + \beta + L$ equilibrium is of the peritectic type. With the highest P value, P_3, it is of the eutectic type and the intermediate P value, P_2, shows the transition where α does not take part in the transformation of L to β. That will give a singular point on the univariant line for $\alpha + \beta + L$ in Fig. 11.12 and that is where the singular curve for $\beta + L$ starts.

At low P (to the left of the transition point in Fig. 11.12) an alloy of suitable composition would transform by $L + \alpha \to \beta$ on the univariant line if its composition is such that it reaches the three-phase horizontal in Fig. 11.11 on cooling. Otherwise, it would transform by $L \to \beta$ at lower T. That would happen if the composition is to the right of the L point for P_1 in Fig. 11.11. In any case, the transformation $L \to \beta$ would be completed at or before the point of minimum in Fig. 11.11, i.e. the singular curve in Fig. 11.12. At high P (to the right of the transition point) an alloy of suitable composition would transform by $L \to \alpha + \beta$ on the univariant line. That would happen for compositions on both sides of the L point for P_3 in Fig. 11.11, but usually after a proeutectic precipitation of α or β. If the liquid alloy can be undercooled by α not nucleating, it may solidify by $L \to \beta$ according to the part of the $L + \beta$ phase field below the eutectic temperature. The lowest temperature of solidification by $L \to \beta$ according to the phase diagram is again the minimum. However, this part of the $L + \beta$ phase field is only metastable at P_3. That is why the singular curve in Fig. 11.12 has been drawn with a dashed line to the right of the transition point.

Figure 9.25 illustrated a congruent point in a ternary system and it was

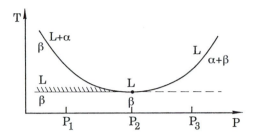

Figure 11.12 Schreinemakers' P,T diagram corresponding to Fig. 11.11. The univariant line changes character at the filled circle. That is where the singular curve originates and there is a compositional degeneracy in that point. The α phase does not take part in the reaction there.

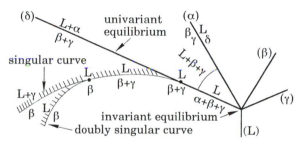

Figure 11.13 Schematic Schreinemakers' projection of a ternary system, illustrating a possible arrangement of a univariant line, a singular curve and a doubly singular curve.

concluded that the position of such points could also be illustrated by a line in Schreinemakers' projection. The name 'doubly singular' was proposed. Such a curve can originate from a transition point on a singular curve, much in the same way as the singular curve originates from a univariant line. This is illustrated in Fig. 11.13 where an invariant equilibrium is also included. The L + β + γ surface covers the area between the (α) and (δ) lines. However, to the left of the singular point L/β + γ that surface, when coming from the (α) line, will overshoot the (δ) line, reach the singular curve and the bend back and end up on the (δ) line. If the composition is suitable, then the alloy will not transform by L + $\alpha \to \beta$ + γ on the univariant line because there will be no α present. Such an alloy will solidify by L $\to \beta$ + γ below the univariant line but in any case not later than on the singular curve. However, if there is a transition point on the singular curve, to the left of which the solidification reaction is L + $\gamma \to \beta$, then the solidification can only occur by L $\to \beta$ if there is no γ present. The alloy may then pass the singular curve on cooling but in any case it should have solidified before passing the doubly singular curve.

Exercise 11.6.1

What transformation would occur on crossing the (α) line in Fig. 9.14 by increasing the value of Y^* at constant values of Y^l, X_m^m and X_m^n?

Hint

Use the method illustrated by Fig. 11.10.

Solution

$\beta + \gamma \rightarrow \delta + \varepsilon$.

Exercise 11.6.2

The following diagram gives a detail of Schreinemakers' projection of a quinary system. It shows a univariant line for the $\alpha + \beta + \gamma + \delta + \varepsilon + \phi$ equilibrium and a singular curve for the $\alpha + \beta + \gamma + \delta + \varepsilon$ singular equilibrium. What transformation can be expected when the univariant line is crossed?

Exercise 11.6.2

Hint

Use the fact that the singular curve is stable only on the indicated side of the singular point.

Solution

To the left of the transition point one can avoid the univariant reaction and reach the singular curve if ϕ is not present before the univariant line is reached. Thus, ϕ does not form by the univariant reaction but would be consumed if it were present. The reaction must be $\alpha + \beta + \gamma + \phi \rightarrow \delta + \varepsilon$. To the right of the transition point, the univariant reaction cannot be suppressed, not even if ϕ is absent, and the reaction must be $\alpha + \beta + \gamma \rightarrow \delta + \varepsilon + \phi$.

11.7 *Scheil's reaction diagram*

In many types of systems, P has a negligible effect and without any loss of information one can section at $P = 1$ bar. For a binary system one can thus construct the usual T,x diagram. For a ternary system there is one dimension more but one could project in the T direction and use x_B and x_C as axes. Such diagrams are useful but tend to be overloaded with phase boundaries if many phases are solutions because there will be lines showing the compositional changes of all those phases. A simpler diagram would be obtained by using μ_B and μ_C (or a_B and a_C) as axes. However, much information would be missing. Using the method illustrated in Fig. 11.10 one could easily find what transformation would occur on crossing a univariant line but that would be of little use. In order to hit the line one must now work with a constant heat content because the projected axis is T. Furthermore, one would have to vary μ_B or μ_C which is rarely very practical.

A rather useful method was proposed by Scheil (1949) for ternary systems. His reaction diagram shows how the lines representing three-phase equilibria are connected to form four-phase equilibria as a function of T but with no regard for composition. His diagram also shows what three-phase equilibria originate from the binary sides. In addition, the reactions occurring on cooling through the four-phase equilibria are given explicitly in boxes. The diagram for a simple eutectic system is presented in Fig. 11.14. Of course, similar diagrams can be constructed for quaternary systems, showing four- and five-phase equilibria.

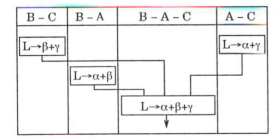

Figure 11.14 Scheil's reaction scheme for a simple ternary system with eutectic reactions.

Exercise 11.7.1

Part of Scheil's diagram for the Al–Fe–Ni system is shown, reproduced from a publication. A mistake was made by joining the binary $(L + \lambda \to \kappa)$ with $(L \to \kappa + \tau_1 + (Al))$. Try to correct it.

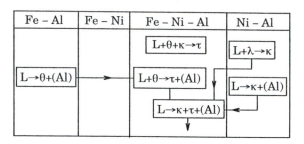

Exercise 11.7.1

Hint

What phases are common for the four-phase equilibria? What one-dimensional equilibria should connect them?

Solution

See diagram.

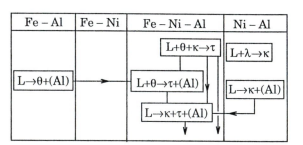

Solution 11.7.1

11.8   *Gradual phase transformations at fixed composition*

If a new component is added to a system where a sharp transformation with $p = c + 1$ has been found at constant P (see Fig. 11.15(a)), then the value of c increases by one unit and for the same transformation one will now have $p = c$. The dimensionality of the corresponding phase field will thus increase by one unit. This case may be illustrated by an x_C section at a low value of x_C (see Fig. 11.15(b)).

It is evident that the phase transformation between γ and $\alpha + \beta$, occurring when T is changed, can no longer be sharp but is somewhat gradual. However, if the addition of the new component is small, its effect on the actual phase transformation should also be small and one may still recognize its characteristic features, for instance in the resulting microstructure, in particular if the temperature has been changed enough

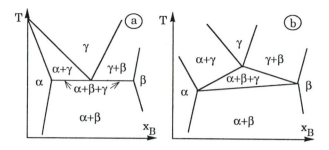

Figure 11.15 A section at constant P and x_C through (b) a ternary phase diagram, compared with (a) a binary phase diagram.

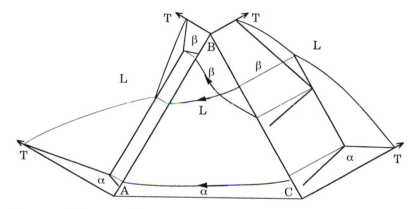

Figure 11.16 A temperature projection of the isobaric A–B–C phase diagram.

to move the system from the γ phase field to the $\alpha + \beta$ phase field before the transformation has started. The transformation may thus appear as sharp even though it is classified as a gradual transformation on thermodynamic grounds. As an example, we shall now examine a case involving three phases and three components. Figure 11.16 shows the T projection of such a phase diagram under constant P.

In this particular case the same three-phase equilibrium occurs in two of the binary systems but it has different character, being eutectic on one side and peritectic on the other. Evidently, there must be a transition between the two types somewhere inside the ternary system. In order to decide where the transition is situated we must first examine how we can recognize the two types when the compositions of the phases change during the transformation. This is fairly easy if we consider a system which consists of an L phase only and if it has the correct composition for equilibrium with the α and β phases (see the cross in Fig. 11.17(a)). As the temperature is lowered slightly, the three-phase triangle moves and covers the composition of the system. Evidently, we should expect the reaction $L \rightarrow L + \alpha + \beta$. Here we have included the L phase on both sides because it has different compositions and it would be impossible to satisfy the mass balance condition if that is not taken into account.

The dashed line in Fig. 11.17(a) is the extrapolation of the direction in which

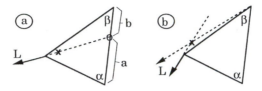

Figure 11.17 Conditions for (a) a eutectic transformation and (b) a peritectic transformation in a ternary system under constant pressure and a gradual decrease of the temperature.

the L phase is moving. It goes through the L corner of the triangle and the average composition of the system, and it intersects the opposite side, a and b being the intercepts. From Fig. 3.5 it is easy to see that the α and β phases must form in the proportion $b:a$ and this will be the ratio between them in the microstructure. As far as α and β are concerned they have formed from material corresponding to the circle. It may not be very important that this material has been drawn from a phase with a different composition. From this point of view, the reaction is clearly of the eutectic type. The result will be quite different if the extrapolation does not intersect the opposite side. An example is given in Fig. 11.17(b). The composition of the system will then fall outside the new three-phase triangle and inside a two-phase field. The reaction will simply be $L \rightarrow L + \beta$ and L will not move in the direction of the solid arrow but straight away from the β phase (see dashed arrow).

The limiting case is found when the extrapolation coincides with the side of the three-phase triangle. Using that criterion one may find the point of transition in the phase diagram in Fig. 11.16. Even though the criterion was derived by considering an alloy composed of an L phase only, it is more general because, in practice, it may often be justified to neglect the diffusion inside the solidified material in comparison with the rapid diffusion in the liquid phase. The progress of the reaction at each stage is thus mainly determined by the momentary composition of the L phase and in which direction it is moving. For a reaction, where three solid phases are involved, it may be necessary to make a detailed analysis of the diffusion of all the elements in all the phases. In the next section we shall consider a special case where one component diffuses much faster than the others.

Exercise 11.8.1

Alkemade's rule states that L moves away from the α–β line for a reaction $L \rightarrow \alpha + \beta$ in a ternary system. Examine if the same rule, or maybe its opposite, applies to L in a peritectic reaction, $L + \alpha \rightarrow \beta$.

Hint

If L precipitates a β phase, its composition must move away from β. If an α phase is dissolved, that reaction will move the composition of L towards α.

Solution

The net change in composition of L must fall within the angle between the two changes caused by the interactions with α and β, respectively. A sketch shows that the net change may be in a direction of (1) away from or (2) towards the extension of the α–β line.

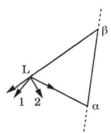

Solution 11.8.1

Exercise 11.8.2

The sketched detail of an isobarothermal section of a ternary phase diagram shows how all the phases in a three-phase equilibrium change on cooling.

(a) Test how a melted specimen with the composition of the L point will react on cooling. Give a reaction formula.

(b) Test how a β phase specimen with the composition of the β point will react on cooling. Give a reaction formula.

(c) Compare the two results. Discuss anything that may seem surprising.

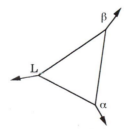

Exercise 11.8.2

Solution

(a) L → α + β (+ L);   (b) β → L + α (+ β);   (c) The first reaction looks like a
eutectic reaction and the second one like a metatectic one. The reason is that this is
not a sharp transformation and the concepts developed for sharp transformations
cannot be strictly applied.

11.9   ## *Phase transformations controlled by a chemical potential*

It is sometimes possible to contain a system inside a wall which allows some components to penetrate but not others. In alloy systems it sometimes happens that one
component diffuses much faster than the others. In other cases, one or a few components
are volatile and can easily be exchanged with the surroundings. In these cases it is
possible to produce a phase transformation by gradually changing the chemical potential of the mobile component but keeping constant the content of all the other components and also T and P. The conditions may be illustrated by the pair of x_C, x_B and μ_C, x_B
phase diagrams in Fig. 11.18(a) and (b), respectively, for a case where $p = c$ and the
mobile component is denoted by C. The arrow in Fig. 11.18(a) represents a discontinuous change of the C content and is pointing towards the C corner.

It is evident that the binary A–B alloy represented by a cross will eventually
undergo a phase transformation

$$\gamma \rightarrow \alpha + \beta$$

if the C content is gradually increased. This may be indicated in the following way using
a reaction formula

$$\gamma + C(source) \rightarrow \alpha + \beta$$

The μ_C, x_B diagram demonstrates that the transformation will be sharp if the μ_C potential
can be controlled experimentally and there are no kinetic restrictions. In fact, the result
of such a transformation would be very similar to the result of the well-known pearlite
transformation taking place on a gradual change of temperature in an iron–carbon
alloy. As a consequence, one should expect γ to transform to an intimate mixture of the
two new phases, α and β, a so-called eutectoid microstructure. This has actually been
observed in many carbon-containing alloyed steels when carburized.

The same transformation is predicted to be gradual if the C content is increased. However, when the supply of C comes from the surroundings, there must be a
chemical potential difference driving the diffusion of C. A growing surface layer of $\alpha + \beta$
will thus form and the transformation will behave as a microscopically sharp one. A
region has either transformed completely to $\alpha + \beta$ or is still pure γ.

When one of the components is much more mobile than the others, it may be

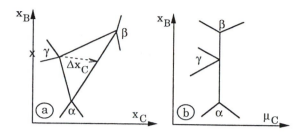

Figure 11.18 Conditions of a transformation under changing C content in a ternary system. T and P are kept constant. The arrow in (a) points towards the C corner.

useful to express compositions with the composition variable u_j defined as $u_j = x_j/(1 - x_C) = x_j/\Sigma'x_i = N_j/\Sigma'N_i$ where Σ' does not include the mobile component C. What happens when the content of the mobile component is changed is then shown on a horizontal line (see Figs. 3.1 and 3.2).

Exercise 11.9.1

The following diagram shows a very rough sketch of the Ni–O–Be phase diagram at 1623 K and 1 bar. The hyperbolic solubility curve for BeO in the Ni-rich phase comes very close to the Ni corner. It is known that pure Ni oxidizes to NiO in air at 1623 K. Construct a reasonable concentration profile for O from the surface and into the interior of the Ni–Be alloy denoted by the filled circle, after some time in air at 1623 K.

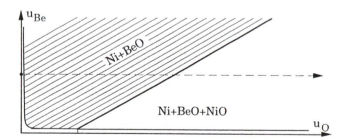

Exercise 11.9.1

Hint

The composition of all layers must lie on the horizontal line through the initial alloy composition because the u_{Be} axis has been used and the diffusion of Be is slow compared to that of O. Remember that the inward diffusion of O requires a continuous decrease of the O potential or, more conveniently in the present case, a

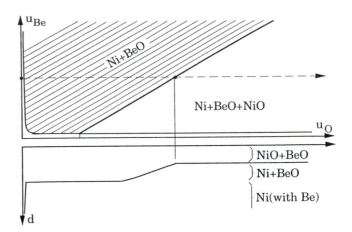

Solution 11.9.1

continuous decrease of the O content of the Ni phase in the corner of the diagram (because its Be content is too low to affect the O potential).

Solution

Suppose an oxide scale of NiO + BeO will form on the surface and also an inner layer of Ni + BeO. Between them there will be a sharp interface because a three-phase layer of Ni + NiO + BeO could not exist in a potential gradient. It can exist at a particular O potential, only. Furthermore, the average O content in the layer of Ni + BeO varies very quickly close to its inner side where the O solubility in the Ni phase is low (the solubility line is almost vertical in the phase diagram). Otherwise, practically no O could diffuse through it. After passing the point closest to the Ni corner, the solubility of O will vary very rapidly. It will there give a slow variation of the average O content and the composition gradient will be much smaller.

Exercise 11.9.2

The micrograph (a) (*opposite page*) shows the structure of an Fe–20 mass% Mo–1 mass% C alloy which has been carburized further at 1273 K and then quenched. The lower part shows the original structure of M_6C particles (black) in a matrix of γ (now martensite after quenching and the upper part the new structure. The surface is above this picture. Explain the microstructure using the phase diagram (b) for 1273 K.

Hint

From the composition given we calculate $u_{Mo} = 0.13$ and $u_C = 0.05$. The value of

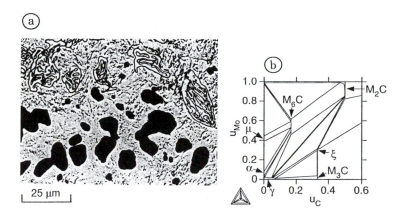

Exercise 11.9.2 Micrograph (a) and phase diagram (b).

u_{Mo} does not change when we add more carbon. The alloy will thus move along a horizontal line to the right.

Solution

The alloy is initially in the $\gamma + M_6C$ phase field. Moving to the right in the phase diagram the alloy may enter the three-phase triangle $\gamma + M_6C + M_2C$ and approach the $\gamma + M_2C$ phase field. We can thus understand that M_6C must transform. A horizontal line from the M_6C corner to the $\gamma + M_2C$ side of the triangle would illustrate the reaction $M_6C + carbon \rightarrow M_2C + \gamma$. This may be regarded as a eutectoid transformation where carbon plays the role usually played by heat. The conclusion is confirmed by the upper part of the picture showing regions of a eutectic-like two-phase mixture, evidently $M_2C + \gamma$ formed from M_6C particles by the above reaction formula.

12 Transformations at constant composition

12.1 The phase field rule at constant composition

Most of the discussion in the preceding chapter concerned transformations in systems of constant composition, so-called closed systems. We shall now examine that case in more detail.

According to our way of reasoning we keep the composition constant and use T and P as the independent variables. This means that we section at constant values for all independent molar contents but we do not section the T and P axes. Thus, $n_s = c - 1 = n_m$ to be inserted in the phase field rule.

In Section 9.6 we have already examined how to calculate the dimensionality of a phase field in mixed diagrams and found different expressions for high and low numbers of phases. In the present case of constant composition the expressions hold one on either side of the following p value

$$p = 1 + n_{pr} + n_m = 1 + 0 + (c - 1) = c$$

For all $p \le c$ the expression is

$$d = r = c + 1 - n_{pr} - n_s = c + 1 - 0 - (c - 1) = 2$$

and these two variables are T and P. We can vary both T and P within some T,P region without changing the phase assemblage. For all $p \ge c$ the expression is

$$d = c + 2 - p - n_s + n_m = c + 2 - p - (c - 1) + (c - 1) = c + 2 - p \le 2$$

This expression resembles Gibbs' phase rule but it should be emphasized that for systems with constant composition it only holds for $p \ge c$, and in all such cases we obtain $d \le 2$. For $p = c + 1$ we find $d = 1$ and we can only vary one of T and P independently without losing one of the phases.

In Sections 12.9 and 12.10 it will be shown that one must take special account of the presence of congruent transformations. They were neglected in Section 8.6.

Exercise 12.1.1

In Section 9.6 we derived Duhem's theorem which says that the state of a closed system is uniquely defined by choosing values for T and P. That seems to be in agreement with the present result for $p \leq c$ which is $d = 2$, i.e. T and P. However, how can it be reconciled with the result for $p > c$ yielding $d = c + 2 - p < 2$, i.e. only one of T and P or none of them?

Solution

Duhem's theorem concerns the situation where one has not fixed the number of phases in advance. From the present result we can conclude that by fixing composition, T and P one should not expect to get more than c phases.

Exercise 12.1.2

Consider the equilibrium $CH_4 \leftrightarrow C + 2H_2$ at a constant pressure of 1 bar. Can it exist at one temperature only or in a range of temperatures?

Hint

C is solid(graphite), CH_4 and H_2 are both gaseous but there can be only one gas phase which is thus a mixture of them.

Solution

We have two components, C and H, $c = 2$. We have two phases, graphite and gas, $p = 2$, and thus $p = c$ and $d = c + 2 - p = 2 + 2 - 2 = 2$. We may vary P and T, i.e. under any value chosen for P we can still vary T.

Exercise 12.1.3

Consider the equilibrium $CaCO_3 \leftrightarrow CaO + CO_2$ in an atmosphere, initially composed of pure N_2. Can the equilibrium exist in a range of temperature if the pressure is kept constant at 1 bar?

Hint

$CaCO_3$ and CaO are two different solid phases. If CO_2 forms, it will go into the gas and may form several species, CO_2, CO and O_2, mixed with N_2, but there will still be only one gas phase.

Solution

We have four components, Ca, C, O and N, $c = 4$. We have three phases, $CaCO_3$, CaO and gas, $p = 3$. Thus $p < c$ and the phase field rule gives $d = 2$. For any chosen value of P we can still vary T. It should be emphasized that without N_2 we could not vary T at a chosen P.

12.2   *Reaction coefficients in sharp transformations for* $p = c + 1$

Keeping composition and P constant we have $n_s = c$ and $n_m = c - 1$, obtaining

$$d = c + 2 - p - n_s + n_m = c + 2 - p - c + (c + 1) = c + 1 - p$$

With $p = c + 1$ phases, we will thus get a sharp transformation ($d = 0$) by changing T. The result would be the same by keeping T constant and varying P. This is why we shall now discuss the case $p = c + 1$ in more detail.

Figure 12.1 shows conditions for a sharp transformation in (a) a binary and (b) a ternary system. For the binary case ($c = 2$, $p = 3$) we can write the reaction formula for the sharp transformation as follows if we omit any part of an initial phase that remains when the reaction is completed.

$$\alpha + \beta \rightarrow \gamma$$

This is independent of whether one passes from $\alpha + \beta$ to $\alpha + \gamma$ or from $\alpha + \beta$ to $\gamma + \beta$, i.e. independent of whether some α or β will remain.

It is common to write chemical reaction formulas with **reaction coefficients**, v. Accepting this procedure we can modify the reaction formula for the phase transformation and make it more quantitative,

$$v^\alpha \alpha + v^\beta \beta = v^\gamma \gamma$$

It expresses the fact that v^α moles of the α phase react with v^β moles of the β phase to form

Figure 12.1 Conditions for (a) a three-phase transformation in a binary system and (b) a four-phase transformation in a ternary system at a low constant value of x_C. P is constant in both cases.

Hillert: Phase Equilibria, Phase Diagrams and Phase Transformations

ISBN 0 521 56270 8 hardback
ISBN 0 521 56584 7 paperback

Erratum

Fig 12.1, reproduced below, is missing from page 314

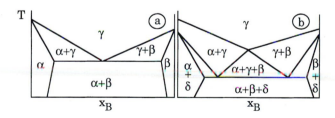

v^γ moles of the γ phase. As an example we may consider the oxidation of solid Ag by gaseous O_2

$$4Ag + 1O_2 = 2Ag_2O$$

In this simple case the reaction coefficients can be given as small integers. In the general case this is not possible since the phases are not always stoichiometric.

By making the reaction coefficients negative for all the reactants and positive for all the products we can simply write the formula as $\Sigma v^j J = 0$. The v^j values will represent the relative amounts of the phases taking part in the reaction, for instance expressed as formula units. Naturally, the v^j values must be such that mass balance is fulfilled for each component i,

$$\sum_j v^j a_i^j = 0 \text{ for each } i$$

where a_i^j is the number of i atoms per formula unit of phase j. We are considering a sharp phase transformation with $p = c + 1$ and we thus have a system of c equations with $p = c + 1$ coefficients each and in the form of a $(c + 1) \times c$ matrix. By excluding the jth column of coefficients one obtains a $c \times c$ determinant and the value of each v^j is given by such a determinant.

$$v^j = (-1)^{j-1} |a_1^\alpha \ldots a_{j-1}^{j-1} a_j^{j+1} \ldots a_c^\varepsilon|$$

It is easy to see that the above condition is fulfilled by this expression because we find

$$\sum_j a_i^j v^j = a_i^\alpha |a_1^\beta a_2^\gamma \ldots a_c^\varepsilon| - a_i^\beta |a_1^\alpha a_2^\gamma \ldots a_c^\varepsilon| + \ldots = |a_i^\alpha a_1^\alpha a_2^\gamma \ldots a_c^\varepsilon| = 0$$

due to the fact that two columns have identical elements because i is one of the numbers 1 to c. It should be noticed that in the calculation of v^j one makes no distinction between reactants and products. Some of the v values will turn out positive and others negative and one may thus identify the members of each group. If the value for a selected phase turns out with the wrong sign, according to the direction chosen for the reaction, then one should simply change all the signs.

When non-stoichiometric phases are involved it may be convenient to identify the a_i coefficients with the molar contents x_i. We get, for instance,

$$v^\alpha = (-1)^{1-1} |x_1^\beta x_2^\gamma \ldots x_c^\varepsilon| = +|x_1^\beta x_2^\gamma \ldots x_c^\varepsilon|$$

The reaction coefficients of a sharp phase transformation can be used for evaluating the change of any molar quantity, X_m, during the transformation. We obtain

$$\Delta X_m = \Sigma X_m^j v^j = |X_m^\alpha a_1^\beta a_2^\gamma \ldots a_c^\varepsilon|$$

This value refers to 1 mole of the reaction formula, as defined by the reaction coefficients.

It must be emphasized that the present discussion only applies to phase equilibria with $p = c + 1$, i.e. phase equilibria which are univariant in the complete phase diagram. That is exactly the case considered in Section 7.8. There, a relation was derived which can be written as

$$| H_m^\alpha\, x_1^\beta\, x_2^\gamma \ldots x_c^\varepsilon |\, dT/T = | V_m^\alpha\, x_1^\beta\, x_2^\gamma \ldots x_c^\varepsilon |\, dP$$

It can now be transformed into the simpler form

$$\frac{dP}{dT} = \frac{\Delta H_m/T}{\Delta V_m}$$

Consequently, this simple expression holds for any univariant equilibrium and not only for the two-phase equilibrium considered initially in Section 7.3.

Exercise 12.2.1

Prove Kirchhoff's law for a reaction between well-defined substances, $(\partial \Delta H/\partial T)_P = \Delta C_p$.

Hint

Express ΔH in terms of H_m for the various substances and the reaction coefficients.

Solution

$\Delta H = \Sigma v^j H_m^j$; $(\partial \Delta H/\partial T)_P = (\partial \Sigma v^j H_m^j/\partial T)_P = \Sigma v^j (\partial H_m^j/\partial T)_P = \Sigma v^j C_P^j = \Delta C_P$ but only because $(\partial v^j/dT)_P = 0$ for well-defined substances.

Exercise 12.2.2

From dilatometric measurements on the pearlite transformation in the Fe–C system at 1 bar we know $\Delta V_m = 0.047\ \text{cm}^3/\text{mol}$ and from calorimetric measurements we know that $\Delta H_m = -4540\ \text{J/mol}$. Calculate the pressure dependence of the transformation temperature.

Hint

The pearlite transformation is $\gamma \to \alpha + Fe_3C$. First check how many degrees of freedom this equilibrium has in a binary system. Then use an equation derived for that particular case.

Solution

We have $p = 3$ and $c = 2$ and, thus, $p = c + 1$, i.e. a sharp transformation at constant P. For that case we get the slope $dT/dP = T\Delta V_m/\Delta H_m = 1000\cdot0.047\cdot10^{-6}/(-4540) = -1.04\cdot10^{-8}$ K/Pa $= -10^{-3}$ K/bar.

Exercise 12.2.3

Prove that $\Sigma v^j = 0$ for the case $p = c + 1$, when the molar contents x_i are inserted as a_i in the expression for v^j.

Solution

$$\Sigma v^j = |x_1^\beta x_2^\gamma \ldots x_c^\varepsilon| - |x_1^\alpha x_2^\gamma \ldots x_c^\varepsilon| + \cdots = |1\, x_1^\beta x_2^\gamma \ldots x_c^\varepsilon| = 0 \text{ since } \sum_i x_i^j = 1. \text{ The}$$

result simply means that the number of moles of atoms does not change by the reaction.

12.3

Graphical evaluation of reaction coefficients

The reaction coefficients for a sharp transformation can also be evaluated graphically using the lever rule. For $c = 2$, $p = 3$ one of the phases transforms into a mixture of the other two. The composition of the first phase is thus equal to the average composition of the other two and the lever rule can be applied directly. For $c = 3$, $p = 4$ there are three different cases as illustrated by Fig. 11.9. In class I and class III reactions one of the phases may transform into a mixture of the other three and, again, the composition of the first phase is equal to the average composition of the others. If the reaction coefficient of the first phase is taken as -1, the coefficients of the other phases are obtained as the fractions of the subsystems using one of the methods described in Fig. 3.5.

Class II can be handled by considering that a mixture of two phases will transform into a mixture of the remaining two phases. Evidently, the compositions of the two mixtures must be equal and should thus fall on the point of intersection between the two diagonals, point 'i' in Fig. 12.2.

If the average composition of the system does not coincide with the first phase discussed for class I and III, then it falls inside one of the three smaller triangles (see the diagram for $\delta \to \alpha + \beta + \gamma$ in Fig. 12.3(a)). The composition of δ will be adjusted by precipitation of first one and later two of the other phases in the small triangle as the four-phase plane is approached. There the rest of it will fall on the δ point in the diagram and will transform to a mixture of the other three phases. The microstructure will show a matrix with a characteristic pattern of the three-phase mixture in which one can see

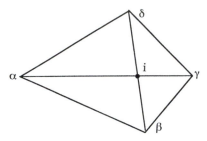

Figure 12.2 Compositions of phases in a class II reaction in a ternary system. The average of the two reacting phases, say α and γ, must fall on the intersection between the diagonals and so must the average of the two product phases, say β and δ.

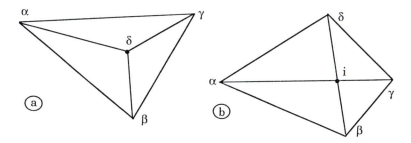

Figure 12.3 Relation between phase compositions in (a) a class I or III transformation; and in (b) a class II transformation.

imbedded one-phase regions of the first phase to precipitate and eutectoid regions of the two co-precipitated phases. If the average composition falls outside the triangular four-phase plane, then the δ phase will never reach the four-phase plane.

For class II there are four alternatives and it is interesting to note that the range of existence of each one of the four phases extends to both sides of the four-phase plane. In Fig. 12.3(b) the three-phase fields α + β + δ and β + δ + γ extend to one side and α + β + γ and α + γ + δ extend to the other side. If a specimen with an average composition falling inside the α–β–i triangle is approaching the four-phase plane from the first side, then it will contain a mixture of α + β + δ when reaching the four-phase plane. From a mass balance point of view it may be regarded as a mixture of β + δ falling on point i, some extra amount of β and also some α. The mixture of β + δ will transform to α + γ when the system crosses the four-phase plane. However, since the extra amount of β is present in the β + δ mixture over the whole specimen, there are no particular β regions predestined not to take part in the β + δ → α + γ transformation. The progress of the transformation will determine which parts of β will not transform and, afterwards, they will be found scattered all over the specimen. The α, present before the four-phase reaction, may indirectly take part in the reaction by providing favourable sites for the precipitation of α.

Exercise 12.3.1

Suppose the δ phase in Fig. 12.3(b) is a liquid and that the average composition of the system is such that the very last part of the liquid just reaches the four-phase plane on cooling. What phases will the system then contain?

Hint

Remember that the amount of a phase in a three-phase assemblage is given by the position in the three-phase triangle.

Solution

At an earlier stage the composition would fall inside the $\alpha + \beta + \delta(L)$ or $\beta + \gamma + \delta(L)$ triangle. Just below the four-phase plane, it will fall on the $\alpha + \gamma + \delta(L)$ triangle but very close to the $\alpha + \gamma$ side. The average composition of the system was thus chosen somewhere on the line between α and γ but the $\alpha + \gamma$ state is not established until the system crosses the four-phase plane and one of the phases α and γ does not form until then.

12.4   *Reaction coefficients in gradual transformations for $p = c$*

Let us now consider a gradual transformation in a system with $p = c$ by keeping composition and pressure constant and changing the temperature. In order to write a reaction formula with the mass balance conserved it is now necessary also to include the change in composition of regions not taking part directly in the phase transformation. As a simple example of $p = c = 2$, consider the precipitation of Al_2Cu from α phase, a solution of Cu in fcc-Al. The solubility decreases with decreasing temperature at constant pressure and there will thus be a gradual precipitation of Al_2Cu. One way of writing this reaction would be

$$\alpha + Cu(\text{from remaining } \alpha) \rightarrow Al_2Cu$$

The reaction coefficients can then be evaluated with the same method used for sharp transformations with $p = c + 1$ but with the extra supply of Cu introduced instead of the missing phase $c + 1$. However, it should be emphasized that this way of writing the transformation is not unique. Another possibility would be

$$\alpha \rightarrow Al_2Cu + Al(\text{to the remaining } \alpha)$$

In order to define a unique way one would have to specify some special criterion. If one

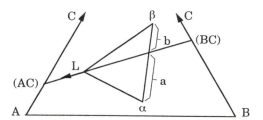

Figure 12.4 Gradual three-phase transformation in a ternary system. The arrow shows the direction of change in composition of the L phase.

is interested in the mechanism of the transformation, then one should consider exchange of both Cu and Al but in proportions balanced according to their rates of diffusion in the α phase.

A three-phase transformation in a ternary system is another case of gradual transformation, now with $p = c = 3$. An example was given in Fig. 11.17 but in Fig. 12.4 it is reproduced with the composition triangle included. Two of the ways of writing this transformation are

$$L + (BC) \rightarrow \alpha + \beta$$
$$L \rightarrow \alpha + \beta + (AC)$$

where (BC) and (AC) represent the compositions one can read on the two sides of the composition triangle. In each case one can calculate the reaction coefficients by including (BC) or (AC) instead of the missing phase $c + 1$. The ratio between α and β will indeed be independent of whether one includes (BC) or (AC). It will be b/a according to the lever rule.

Exercise 12.4.1

In an Al–Cu–Si specimen at 1 bar and 803 K one finds the equilibrium: $\alpha(0.025Cu;0.006Si) + L(0.16Cu;0.05Si) + Al_2Cu$. When the temperature is decreased, L changes in the direction away from the point 0.83Al;0.17Cu. Calculate the relative amounts of α and Al_2Cu in the eutectic structure formed by $L \rightarrow \alpha + Al_2Cu$ on further cooling. (The numbers given above are molar contents.)

Hint

We have $c = 3$ and $p = 3$ but we should write the transformation in a way resembling a sharp transformation for $p = c + 1$. The relative amounts of the two phases are then obtained as the ratio of their reaction coefficients.

Solution

Write the reaction as $L + (0.83Al;0.17Cu) \rightarrow \alpha + Al_2Cu$.

$$v^\alpha = (-1)^{3-1} \begin{vmatrix} -0.79 & -0.16 & -0.05 \\ -0.83 & -0.17 & 0 \\ 2 & 1 & 0 \end{vmatrix} = +0.0245$$

$$v^{Al_2Cu} = (-1)^{4-1} \begin{vmatrix} -0.79 & -0.16 & -0.05 \\ -0.83 & -0.17 & 0 \\ 0.969 & 0.025 & 0.006 \end{vmatrix} = +0.00719$$

$$v^{Al_2Cu}/v^\alpha = 0.00719/0.0245 = 0.2935 = 0.23 : 0.77$$

12.5   *Driving force for sharp phase transformations*

The driving force for the precipitation of a new phase in a gradual transformation was discussed in Section 6.7. As an introduction to a discussion of the driving force for a sharp transformation we shall now consider a eutectoid transformation in a binary alloy. We have seen that it usually gives rise to an intimate mixture of the two new phases, illustrated by Fig. 11.5. It may give the material advantageous properties. The most famous example is pearlite, the eutectoid formed from the austenite phase in steel.

Because the rate of transformation is controlled by slow diffusion and evolution of the heat of transformation will thus be slow, it is often possible to control the temperature and it makes sense to discuss the transformation under isothermal conditions, for instance at T_2 in Fig. 11.6. The character of the transformation as sharp is evident from its progress. A region has either been completely transformed or is not at all affected. The transformation occurs by the growth of colonies composed of an intimate mixture of the two new phases and, under isothermal conditions, the growth continues until the whole system has transformed.

It is well known that the mixture will be the finer, the lower the temperature of formation is. The reason is that the interfaces in the mixture have surface energy and cannot form without the supply of a corresponding amount of driving force. According to an approximate treatment, one-half of the available driving force goes into surface energy and the other half is used for driving the diffusion.

The conditions for cooperative growth of the two new phases can be illustrated by extrapolating the phase boundaries in the T,x_B phase diagram as shown in Fig. 11.6. This kind of construction is given again in Fig. 12.5(a) but a solid phase, γ, has been substituted for the liquid phase and the diagram has been rotated. The transformation temperature is now denoted by T_1. The diagram shows how one can evaluate the

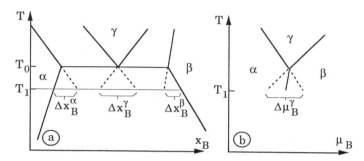

Figure 12.5 Evaluation of the differences in composition (a) and chemical potential (b) driving diffusion in a three-phase transformation in a binary system.

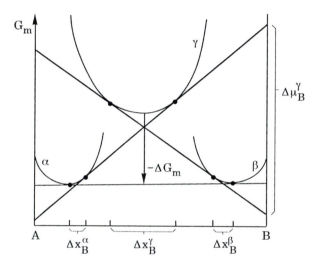

Figure 12.6 Gibbs energy diagram illustrating the eutectoid transformation $\gamma \to \alpha + \beta$ at T_1 in Fig. 12.5.

composition difference driving the diffusion in each one of the phases (see the arrows in Fig. 11.5).

A similar construction in the T,μ_B phase diagram in Fig. 12.5(b) yields a difference in chemical potential of B, $\Delta\mu_B^\gamma$, which may be used in a treatment of the rate of diffusion of B in γ, although the composition difference is usually used for that purpose. The conditions for the eutectoid reaction $\gamma \to \alpha + \beta$ may also be illustrated with a molar Gibbs energy diagram at T_1 (see Fig. 12.6).

As explained in Section 6.7, the driving force for the precipitation of a new phase in a gradual transformation decreases during its growth but for the whole reaction one can define and evaluate an integrated driving force along the reaction path. This problem is absent if the phase transformation is a sharp one because the parent phase which has not yet transformed has not changed at all. Under constant T and P the driving force is thus constant and the integrated driving force is equal to the momentary driving force if they are both expressed per mole of the transformed structure. Under constant T and P we obtain, by identifying the extent of the transformation, ξ, with the number of moles of the products which is also equal to the number of moles of transformed reactants,

$$D = -\left(\frac{\partial G}{\partial \xi}\right)_{T,P,N_i} = -G_m(\text{products}) + G_m(\text{reactants}) = -\Delta G_m$$

This quantity is illustrated in Fig. 12.6 for a eutectoid transformation $\gamma \to \alpha + \beta$ in a binary system.

The value of the driving force may be calculated using the reaction coefficients for a sharp phase transformation at fixed composition. Per mole of the reactant, γ, we obtain

$$D = -\Delta G_m = -\Sigma G_m^j v^j/(-v^\gamma) = -|G_m^\alpha\, x_1^\beta\, x_2^\gamma|/|x_1^\beta\, x_2^\gamma|$$

Even though this expression looks quite simple, it may sometimes be difficult to evaluate all the molar contents to be inserted. A useful approximation would be to assume that all the phases have the same compositions they have at the equilibrium temperature. If it is further assumed that the resulting value of ΔG_m varies linearly with temperature, we could use the method introduced in Section 4.4 for a transformation that we now recognize as a sharp one. Since D stays constant for a sharp transformation, we get

$$\Delta S(T_1 - T_0) = \Delta H(T_1 - T_0)/T_0 = -\Delta G = \int D\mathrm{d}\xi = D\int \mathrm{d}\xi = D\Delta\xi$$

Let $\Delta\xi$ be the number of moles transformed. The driving force for the transformation of one mole is thus

$$D = \Delta S_m(T_1 - T_0) = \Delta H_m(T_1 - T_0)/T_0$$

Here, T_0 is the equilibrium temperature and T_1 is the actual temperature of the transformation. For small ΔT the heat of transformation may be taken as the value at T_0. It may be available from direct measurements. For larger undercoolings one may have to consider variations of ΔH_m with temperature, for instance due to changes on the compositions of various phases.

Let us now examine the situation below the equilibrium temperature in more detail. Fig. 12.6 demonstrates the complexity found in a eutectoid transformation in a binary system. Each one of the phases is in contact with the other two phases and two different compositions are thus defined for each phase. Figure 12.5 shows how they are obtained from the phase diagram by extrapolating the phase boundaries to the temperature of transformation T_1.

The situation gets even more complicated if one tries to analyze how the driving force is consumed during the transformation. This is illustrated in Fig. 12.7. This diagram demonstrates several complications. Firstly, α and β grow under an increased pressure because the interfaces to the parent γ are curved (as illustrated in Fig. 11.5). The corresponding $-\Delta G_1$ is consumed by the creation of all the α/β interfaces in the

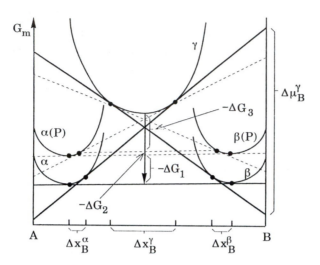

Figure 12.7 Molar Gibbs energy diagram for a binary system with a eutectoid transformation, showing three ways of consuming the driving force, ΔG_1 ΔG_2 and ΔG_3.

eutectoid structure. Secondly, the two new phases are not formed with their final compositions, which are governed by the α/β equilibrium, because they grow from the γ phase. The corresponding $-\Delta G_2$ is consumed by diffusion from the interior of α to the interior of β behind the reaction front. The remaining part of the driving force, $-\Delta G_3$, is consumed by diffusion at the reaction front.

As already mentioned, according to an approximate treatment of the rate of transformation, the highest growth rate is obtained when one-half of the total driving force goes into the surface energy of all the α/β interfaces.

Exercise 12.5.1

The difference in chemical potential of B driving the diffusion in the γ phase during the eutectoid transformation, shown in Fig. 11.5, was identified in Fig. 12.5(b). Find the corresponding differences for α and β. How are the three related?

Solution

$\Delta\mu_B^{\alpha}$ is found between the lines for γ/α and β/α and $\Delta\mu_B^{\beta}$ is found between the lines for β/α and β/γ. Evidently, $\Delta\mu_B^{\gamma} = \Delta\mu_B^{\alpha} + \Delta\mu_B^{\beta}$.

Exercise 12.5.2

The heat of formation of pearlite from austenite is $\Delta H_m = -4.5\,\text{kJ/mol}$ and the equilibrium temperature is $T_0 = 1000\,\text{K}$. Estimate the coarseness of pearlite formed at $T_1 = 950\,\text{K}$, assuming that all the driving force goes into interfacial energy between the two phases of pearlite. Suppose that the interfacial energy is approximately $\sigma = 1\,\text{J/m}^2$ and the molar volumes of all the phases are approximately $V_\text{m} = 7 \cdot 10^{-6}\,\text{m}^3/\text{mol}$. Compare with an experimental value of the coarseness, $w = 0.14\,\mu\text{m}$.

Hint

w is the total thickness of one lamella of each phase in pearlite. One mole of pearlite then contains an area of $2V_\text{m}/w$ of interfaces.

Solution

The total interfacial energy is $2\sigma V_\text{m}/w$ J/mol. The driving force is $(\Delta H_\text{m}/T_0)(T_1 - T_0)$. Thus, $2\sigma V_\text{m}/w = (\Delta H_\text{m}/T_0)(T_1 - T_0)$; $w = 2\sigma V_\text{m} T_0/(-\Delta H_\text{m})(T_0 - T) = 2 \cdot 1 \cdot 7 \cdot 10^{-6} \cdot 1000/4500 \cdot 50 = 6 \cdot 10^{-8}\,\text{m} = 0.06\,\mu\text{m}$. The observed value is about twice as large, which is expected if only one-half of the driving force should go into interfacial energy.

12.6

Driving force under constant chemical potential

In the preceding section it was shown how the driving force for a sharp transformation can be estimated from the undercooling ΔT at which the transformation occurs. In the same way, the driving force for a γ-eutectoid transformation in a ternary A–B–C system, controlled by the chemical potential of a mobile component C under constant T and P, should depend upon the difference in chemical potential of C during the transformation and at equilibrium, $\Delta\mu_\text{C}$. We can illustrate the conditions by extrapolations in the μ_C, x_B phase diagram in Fig. 11.18 or in the corresponding $\mu_\text{C}, \mu_\text{B}$ phase diagram (see Fig. 12.8 where the diagram has been rotated in order to emphasize the similarity with the binary case in Fig. 12.5).

Since μ_C is assumed to be constant instead of N_C, we must evaluate the driving force from a new alternative of the combined first and second law,

$$D\text{d}\xi = V\text{d}P - S\text{d}T + \sum_{j \neq \text{C}} \mu_j \text{d}N_j + \mu_\text{C}\text{d}N_\text{C} - \text{d}G + N_\text{C}\text{d}\mu_\text{C} - N_\text{C}\text{d}\mu_\text{C}$$

$$= V\text{d}P - S\text{d}T + \sum_{j \neq \text{C}} \mu_j \text{d}N_j - N_\text{C}\text{d}\mu_\text{C} - \text{d}(G - N_\text{C}\mu_\text{C})$$

where $G - N_\text{C}\mu_\text{C}$ is a new characteristic state function. At constant T, P, N_j and μ_C we get

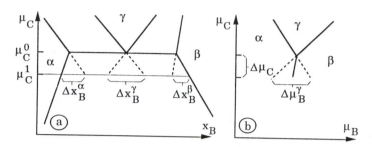

Figure 12.8 Evaluation of the driving force, $\Delta\mu_C$, for an isobarothermal transformation under changing C content in a ternary system.

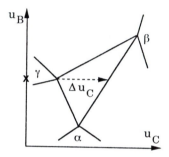

Figure 12.9 Sharp transformation $\gamma \to \alpha + \beta$ in a ternary system with a mobile component, C. The increase of C content, Δu_C, can be read directly.

$$D d\xi = -d(G - N_C\mu_C) = \mu_C dN_C - dG$$

By identifying the extent of the transformation, ξ, with the increased content of C, N_C, we get

$$D = \mu_C - (\partial G/\partial N_C)_{T,P,N_j}$$

By definition, the second term is the chemical potential of C at equilibrium. Therefore,

$$D = \mu_C - \mu_C^o = \Delta\mu_C$$

This is the driving force per mole of C added to the system. It should be multiplied by the amount of C required by the transformation. That quantity is conveniently expressed in terms of the u_C fraction, the amount of C per mole of A + B. Figure 12.9 demonstrates how the increase Δu_C can be evaluated graphically and the driving force for the transformation $\gamma \to \alpha + \beta$ expressed per mole of A + B is given by

$$D = \Delta u_C \cdot \Delta\mu_C$$

Figure 12.9 resembles Fig. 11.18 but the choice of u axis makes the arrow horizontal. The analytical evaluation of Δu_C is described in the next section.

Exercise 12.6.1

In the Fe–C system the γ phase exists above 1000 K and at lower temperatures it transforms to a lamellar aggregate of α and cementite (Fe_3C) which is called pearlite. The following phase diagram for 923 K shows that it is possible to stabilize γ by the addition of Mn. Consider an alloy composed of γ with a composition falling exactly on the γ corner of the $\alpha + \gamma +$ cementite triangle at 923 K. By carburizing such an alloy one can form a surface layer of pearlite. Estimate its coarseness if one carburizes with an atmosphere having a carbon activity of 0.9 and the carbon activity is 0.7 in the initial alloy. Use the values $V_m = 7 \cdot 10^{-6} \, m^3/mol$ and surface energy $\sigma = 1 \, J/m^2$.

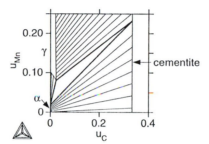

Exercise 12.6.1

Hint

It should first be realized that we do not know how much of the difference in carbon activity is used to drive the diffusion of carbon through the carburized layer. In order to get a numerical result, let us assume that all the driving force acts at the reaction front.

The increase of the C content is obtained from a horizontal construction. The increase in C potential is obtained from the activity through $\mu_C = {}^\circ\mu_C + RT\ln a_C$. Assume that half of the driving force goes into surface energy. Express the coarseness with w, the total width of one lamella of α and one of cementite. The area of α/cementite interfaces is then $(2V_m/w)$ per mole of the material and we thus get the relation $2V_m\sigma/w = 0.5 \cdot \Delta u_C \Delta \mu_C$.

Solution

By measuring the horizontal distance of the γ point from the $\alpha +$ cementite side of the triangle we get $\Delta u_C = 0.08$. By comparing the C activities we get $\Delta \mu_C = RT\ln(0.9/0.7) = 1930 \, J/mol$ and using half of this we find $w = 2V_m\sigma/\Delta u_C \Delta \mu_C = 2 \cdot 7 \cdot 10^{-6} \cdot 1/0.5 \cdot 0.08 \cdot 1930 = 2 \cdot 10^{-7} \, m = 0.2 \, \mu m$.

12.7   *Reaction coefficients at a constant chemical potential*

Previously we have been able to calculate the fractions of the various phases taking part in a sharp transformation in a system of constant composition because p was equal to $c + 1$. Now we shall consider a sharp transformation in the case $p = c$ where one of the components is very mobile and is controlled through its potential. The total contents of all the other components in the system will be kept constant. The mobile component will be denoted C and will be given the number c. By not considering that component we get the same condition as before but must now express the molar contents without regard for the mobile component. Thus we must use the u variable instead of the ordinary molar content x, and we obtain for instance

$$v^\gamma = |u_1^\alpha u_2^\beta u_3^\delta \ldots u_{c-1}^\varepsilon|$$

The mobile component is not included in the determinant. The amount of the γ phase taking part in the reaction, v^γ, is here expressed without regard for the mobile component. The change in content of the mobile component can be evaluated just like the change of any other molar quantity using the method outlined in Section 12.2.

$$\Delta u_C = |u_C^\alpha u_1^\beta u_2^\gamma \ldots u_{c-1}^\varepsilon|$$

This is the increase of C per mole of units of the reaction formula as given by the v^j values. If γ is the only reactant and all the other phases are products, it is interesting to evaluate Δu_C per mole of all the other components in γ. It is obtained by dividing with $-v^\gamma$.

$$\Delta u_C = |u_C^\alpha u_1^\beta u_2^\gamma u_3^\delta \ldots u_{c-1}^\varepsilon| / |u_1^\alpha u_2^\beta u_3^\delta \ldots u_{c-1}^\varepsilon|$$

If a mixture of α and β in the ternary A–B–C system is treated under conditions of a low chemical potential for C, the system will move from right to left in Fig. 12.9 and one should expect the reverse transformation

$$\alpha + \beta \rightarrow \gamma + C(\text{sink})$$

This is a peritectoid transformation and one should primarily expect the new γ to form at the α/β interfaces. If the transformation is not inhibited due to slow diffusion of the two sluggish elements, it will look almost as a sharp phase transformation.

Exercise 12.7.1

When an Fe–Mo alloy is carburized at 1273 K, it may transform by a eutectoid transformation $\alpha \rightarrow \gamma + M_6C$. The ideal composition of the initial Fe–Mo alloy for this transformation is defined by the α corner of the $\alpha + \gamma + M_6C$ triangle. Consider such an alloy but try to find a construction showing that it could transform to only γ if the carbon activity is high enough. Use the enlarged detail of the phase diagram at 1273 K. Estimate that critical value of the carbon activity.

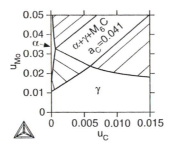

Exercise 12.7.1

Hint

First try to find how one can evaluate the fractions of γ and M_6C if the transformation is caused by a very small driving force. Then investigate what happens to the fraction of M_6C if the driving force is higher.

Solution

The fraction of M_6C is proportional to $|u_{Fe}^{\alpha} u_{Mo}^{\gamma}| = |1 \; u_{Mo}^{\gamma}| = u_{Mo}^{\gamma} - u_{Mo}^{\alpha}$. It will go to zero if u_{Mo}^{γ} approaches u_{Mo}^{α}. The construction in the diagram shows that this will happen at a value of approximately $a_C \cong 0.041 \cdot 0.0095/0.006 = 0.065$ if the carbon activity is proportional to the carbon content of γ and independent of the Mo content. The dashed line is an extrapolation of γ/α.

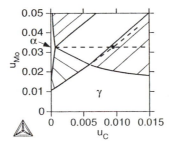

Solution 12.7.1

Exercise 12.7.2

The equation for Δu_C, the addition of a mobile component C consumed by a sharp transformation, can be applied to the reaction $\gamma + C(\text{source}) \rightarrow \alpha + \beta$ in a ternary system at constant T and P and the result will be $\Delta u_C/v^{\gamma} = |u_C^{\alpha} u_1^{\beta} u_2^{\gamma}| / |u_1^{\alpha} u_2^{\beta}|$. Show that the same result is obtained if C(source) is regarded as a phase taking part in the transformation.

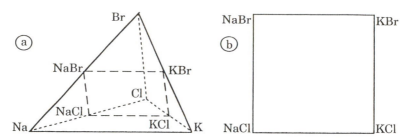

Figure 12.11 The composition space for a quaternary, reciprocal system reduced to a plane due to a stoichiometric constraint, common to all the phases.

The mass balance is satisfied because for each component i from 2 to c we find,

$$\Sigma x_i^j v^j = x_i^\alpha |x_2^\beta x_3^\gamma \ldots x_c^\varepsilon| - x_i^\beta |x_2^\alpha x_3^\gamma \ldots x_c^\varepsilon| + \ldots = |x_i^\alpha x_2^\beta \ldots x_c^\varepsilon| = 0$$

because there are two identical columns, since i has a value from 2 to c.

In this example, the first column of elements in the composition matrix dropped out because the additional phase was taken as pure component 1. For a different choice, another column would have dropped out. We may summarize the result of this section as follows: If the composition determinant for a phase transformation with $p = c$ is equal to zero, then it is a sharp transformation and there is a compositional degeneracy. The reaction coefficients can be calculated from the determinants obtained by first omitting any column from the composition matrix and then, in turn, the row corresponding to each phase. In addition, a minus sign must be added for the second, fourth, etc., phases.

Figure 12.10 illustrates a case where a compositional degeneracy occurs in a particular place in the phase diagram where the phases happen to fall on a line. There is another very important case where the compositions of all the phases are subject to a stoichiometric constraint that results in a compositional degeneracy in the whole phase diagram. An example is an ionic system where each element has a fixed valency (see Fig. 12.11 which gives the composition space for the Na–K–Cl–Br system). It is evident that all possible compositions will fall on a plane inside the three-dimensional space. The phase relations can thus be plotted in a diagram with one dimension less. Such a diagram is called a **quasi-ternary** diagram. In the same way, a ternary system can sometimes be represented with a **quasi-binary** diagram. In practice, one often uses the word quasi-binary to describe an isopleth section of a ternary diagram when many or the most important tie-lines fall in or close to the section. In the present case a composition square can be used and, except for the different outer shape, the diagram would have the same properties as a diagram for a ternary system. This is often called a **reciprocal system** because the amounts of the four components are not independent but are related by a reciprocal reaction

$$NaCl + KBr \rightarrow NaBr + KCl$$

All ionic phases in a reciprocal system will fall in the composition square and the composition of each phase can only move inside the square. As an example, the liquid phase covers the whole square and each solid covers a small area close to its corner. However, it should be realized that the chemical system under consideration may contain other phases which are not subject to the same constraint. In the present case there may be metallic phases of Na and K and a gas phase composed mainly of Cl_2 and Br_2. They fall outside the plane and they can be shown only by the use of the three-dimensional quaternary diagram.

Exercise 12.8.1

Illustrate how a compositional degeneracy may be used to locate a change from a eutectic to a peritectic reaction in a ternary system. Use an isobarothermal section and show the tie-triangle.

Hint

Consult Fig. 11.17.

Solution

If at any temperature $v^\alpha = 0$, then α does not take part in the reaction at that particular temperature. That is where $L \rightarrow \alpha + \beta$ can change to $L + \alpha \rightarrow \beta$.

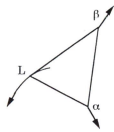

Solution 12.8.1

Exercise 12.8.2

Check if $Al_2O_3 + Si_3N_4 = SiO_2 + AlN$ can be a sharp transformation (neglecting the possible formation of intermediary phases). If so, evaluate the reaction coefficients.

Hint

We can see that $p = c$. It is thus necessary to test the complete determinant.

Solution

$$
\begin{array}{c}
 & \text{Al} \quad \text{O} \quad \text{Si} \quad \text{N} \\
\begin{array}{c}
\text{Al}_2\text{O}_3 \\
\text{Si}_3\text{N}_4 \\
\text{SiO}_2 \\
\text{AlN}
\end{array}
\begin{vmatrix}
2 & 3 & 0 & 0 \\
0 & 0 & 3 & 4 \\
0 & 2 & 1 & 0 \\
1 & 0 & 0 & 1
\end{vmatrix} = 0
\end{array}
$$

This is thus a sharp transformation and we now add an additional phase, e.g. pure Al. The whole composition matrix will then be

$$
\begin{array}{c}
 & \text{Al} \quad \text{O} \quad \text{Si} \quad \text{N} \\
\begin{array}{c}
\text{Al}_2\text{O}_3 \\
\text{Si}_3\text{N}_4 \\
\text{SiO}_2 \\
\text{AlN} \\
c + 1
\end{array}
\begin{bmatrix}
2 & 3 & 0 & 0 \\
0 & 0 & 3 & 4 \\
0 & 2 & 1 & 0 \\
1 & 0 & 0 & 1 \\
1 & 0 & 0 & 0
\end{bmatrix}
\end{array}
$$

and we obtain the reaction coefficients as

$$
v^{\text{Al}_2\text{O}_3} = +
\begin{vmatrix}
0 & 0 & 3 & 4 \\
0 & 2 & 1 & 0 \\
1 & 0 & 0 & 1 \\
1 & 0 & 0 & 0
\end{vmatrix}
= -
\begin{vmatrix}
0 & 3 & 4 \\
2 & 1 & 0 \\
0 & 0 & 1
\end{vmatrix} = +6
$$

$$
v^{\text{Si}_3\text{N}_4} = -
\begin{vmatrix}
2 & 3 & 0 & 0 \\
0 & 2 & 1 & 0 \\
1 & 0 & 0 & 1 \\
1 & 0 & 0 & 0
\end{vmatrix}
= +
\begin{vmatrix}
3 & 0 & 0 \\
2 & 1 & 0 \\
0 & 0 & 1
\end{vmatrix} = +3
$$

$$
v^{\text{SiO}_2} = +
\begin{vmatrix}
2 & 3 & 0 & 0 \\
0 & 2 & 1 & 0 \\
1 & 0 & 0 & 1 \\
1 & 0 & 0 & 0
\end{vmatrix}
= -
\begin{vmatrix}
3 & 0 & 0 \\
0 & 3 & 4 \\
0 & 0 & 1
\end{vmatrix} = -9
$$

$$
v^{\text{AlN}} = -
\begin{vmatrix}
2 & 3 & 0 & 0 \\
0 & 0 & 3 & 4 \\
0 & 2 & 1 & 0 \\
1 & 0 & 0 & 0
\end{vmatrix}
= +
\begin{vmatrix}
3 & 0 & 0 \\
0 & 3 & 4 \\
2 & 1 & 0
\end{vmatrix} = -12
$$

We may give the arbitrary constant a value of 1/3 and obtain the coefficients

$+ 2, + 1, - 3$ and $- 4$ and the reaction formula will thus be $2Al_2O_3 + Si_3N_4 = 3SiO_2 + 4AlN$.

Exercise 12.8.3

Suppose we have a computer program for the calculation of phase equilibria. When trying to calculate the equilibrium temperature for $SiO_2 + Al_2SiO_5 + Al_2O_3$ at a pressure of 1 bar we get the message, 'cannot calculate because $v \neq 0$'. What action could we take?

Hint

Evidently, the program is constructed to calculate sharp phase transformations. It may thus require that $p = c + 1$ at constant temperature. We should start by checking if our transformation is sharp, although $p = c$. Otherwise, we have a gradual transformation and cannot expect to calculate a unique value of T.

Solution

The composition determinant is

$$\begin{array}{ccc} Si & O & Al \\ \end{array}$$
$$\begin{vmatrix} 1 & 2 & 0 \\ 1 & 5 & 2 \\ 0 & 3 & 2 \end{vmatrix} = 10 - 4 - 6 = 0$$

The transformation thus has a compositional degeneracy and is sharp. The reason is that all the c phases fall on the straight line going from SiO_2 to Al_2O_3. It is thus possible to calculate a unique transformation temperature. Our program seems to need $p = c + 1$ phases. We can solve the problem by introducing any phase in the Al–O–Si system which is outside the straight line, e.g. pure Al.

Exercise 12.8.4

We have seen the following. For $p = c + 1$ we get a sharp transformation by gradually changing T, keeping the composition and P constant. For $p = c$ we get a sharp transformation by gradually changing μ for a mobile component, keeping the composition constant except for the mobile component, and keeping P and T constant.

Then we saw that for $p = c$ it may happen that one gets a sharp transformation by gradually changing T and keeping the composition and P constant.
Discuss whether it is possible also to get a sharp transformation in a system where
$p = c - 1$ by gradually changing μ for a mobile component, keeping the composition constant, except for the mobile component, and keeping P and T constant. If
so, what should be the expression for the reaction coefficients, v^α, etc.

Hint

Accept that equations for the case of a mobile component are obtained by using u_i
instead of x_i.

Solution

For $p = c$ we can get a sharp transformation in the ordinary case by gradually
changing T if $|x_1^\alpha \, x_2^\beta \ldots x_c^\varepsilon| = 0$ for constant P. For $p = c - 1$ we would also get a
sharp transformation by gradually changing μ_c if $|u_1^\alpha \, u_2^\beta \ldots u_{c-1}^\varepsilon| = 0$ for constant P
and T. For the relative fraction of phase we get $v^\alpha = |u_2^\beta \, u_3^\gamma \ldots u_{c-1}^\varepsilon|$.

Exercise 12.8.5

Consider the reaction $NH_3(gas) + HCl(gas) \rightarrow NH_4Cl$. It is trivial to find the
reaction coefficients but, nevertheless, show how they can be calculated with the
method presented in this section.

Hint

Treat NH_3 and HCl as different phases although in reality there is only one gas
phase. Then we have $p = c = 3$.

Solution

The reaction would correspond to a sharp transformation if there is one
compositional degeneracy. This is confirmed because the composition determinant
gives

$$
\begin{array}{c|ccc|}
 & N & H & Cl \\
\hline
NH_3 & 1 & 3 & 0 \\
HCl & 0 & 1 & 1 \\
NH_4Cl & 1 & 4 & 1 \\
\end{array} = 0.
$$

In calculating the reaction coefficients let us omit the second column in order
to simplify the calculations as much as possible. Then,

$$v^{NH_3} = \begin{vmatrix} 0 & 1 \\ 1 & 1 \end{vmatrix} = -1; \quad v^{HCl} = -\begin{vmatrix} 1 & 0 \\ 1 & 1 \end{vmatrix} = -1; \quad v^{NH_4Cl} = \begin{vmatrix} 1 & 0 \\ 0 & 1 \end{vmatrix} = 1$$

12.9   *Effect of two compositional degeneracies for $p = c - 1$*

Let us now consider the case $p = c - 1$. It may then be suggested that we need two compositional degeneracies in order to get a sharp transformation at constant composition and pressure. If this is correct, we could treat this case by introducing two additional phases, c and $c + 1$, and then require that their reaction coefficients are both zero. We can try this suggestion by first letting phase c be pure component 1. For phase $c + 1$ we then obtain

$$v^{c+1} = (-1)^c \begin{vmatrix} x_1^\alpha & x_2^\alpha & \cdots & x_c^\alpha \\ \vdots & \vdots & \vdots & \vdots \\ x_1^\varepsilon & x_2^\varepsilon & \cdots & x_c^\varepsilon \\ 1 & 0 & 0 & 0 \end{vmatrix} = (-1)^c |x_2^\alpha x_3^\beta \ldots x_c^\varepsilon| = 0$$

By letting phase $c + 1$ be pure component 2 we obtain, for phase c,

$$v^c = (-1)^{c-1} |x_1^\alpha x_3^\beta \ldots x_c^\varepsilon| = 0$$

We have thus found that two compositional degeneracies can be defined for a system which has a sharp transformation between $c - 1$ phases at constant pressure and composition. In Section 9.9 we called the corresponding equilibrium doubly singular. The compositional degeneracies may be obtained by forming two determinants from the composition matrix by omitting first one and then another column and putting to zero the two determinants thus obtained. The same set of two equations was obtained in Section 8.9 when extrema in T and P were discussed and it was concluded that they imply that the compositions of the phases fall on the same point for $p = 2$, same line for $p = 3$, etc. The same is true here, of course.

We may also look at the situation from the other side and conclude that there is a sharp transformation at constant composition and pressure in the case $p = c - 1$ if there are two compositional degeneracies. Then we may evaluate the reaction coefficients from the determinants obtained by omitting two columns from the composition matrix and then the row corresponding to each phase, one at a time

$$v^\alpha = \begin{vmatrix} x_1^\beta & x_2^\beta & x_3^\beta & \cdots & x_c^\beta \\ \vdots & \vdots & \vdots & \vdots & \vdots \\ x_1^\varepsilon & x_2^\varepsilon & x_3^\varepsilon & \cdots & x_c^\varepsilon \\ 1 & 0 & 0 & 0 & 0 \\ 0 & 1 & 0 & 0 & 0 \end{vmatrix} = |x_3^\beta x_4^\gamma \ldots x_c^\varepsilon|$$

We may summarize the result of this section as follows: If the composition determinants, obtained in a case of $p = c - 1$ by excluding one column at a time, are equal to zero, it is a sharp transformation. The reaction coefficients can be calculated from the determinants obtained by excluding any two columns from the composition matrix and, in turn, the row corresponding to each phase. In addition, a minus sign must be added for the second, fourth, etc. phase.

We may generalize the above result. A transformation involving p phases in a system with c components will be sharp if the composition determinants, obtained by omitting $c - p$ columns from the composition matrix, are all zero. The reaction coefficients can then be evaluated from the determinants obtained by omitting $c + 1 - p$ columns and then omitting in turn the row corresponding to each phase.

When a chemical reaction involving compounds and species is written in the form $\Sigma v^j J = 0$, it is implied that all the compounds and species with negative v^j disappear completely by the reaction and all with positive v^j appear suddenly. This is equivalent to assuming that there is a sharp transformation and the rule for calculating the reaction coefficients, derived here, applies if each compound and species is regarded as a phase.

Exercise 12.9.1

On heating solid NH_4Cl it will evaporate to gaseous NH_3 and HCl. Examine if this will happen at a particular T if P is kept at 1 bar or if it will happen over a range of T.

Hint

There can be only one gas phase and even though it has three components, N, H and Cl, it is evident that its composition cannot vary at all. It must have the same composition as the initial solid phase from which it is formed. We may thus conclude that the whole system behaves as if there is only one component, NH_4Cl. Justify this conclusion by using the method derived in this section.

Solution

We have the case $p = 2$ and $c = 3$ and thus $p = c - 1$ but both phases have the composition NH_4Cl and the composition matrix will thus be

$$
\begin{array}{c}
 \quad \text{Si} \quad \text{O} \quad \text{Al} \\
\begin{array}{c} \text{solid} \\ \text{gas} \end{array}
\begin{bmatrix} 1 & 4 & 1 \\ 1 & 4 & 1 \end{bmatrix}
\end{array}
$$

By omitting one column at a time we find $\begin{vmatrix} 1 & 4 \\ 1 & 4 \end{vmatrix} = 0$ and $\begin{vmatrix} 1 & 1 \\ 1 & 1 \end{vmatrix} = 0$

and there are thus two compositional degeneracies. The transformation solid → gas will thus be sharp at any chosen P value. It will then occur at a particular T.

Exercise 12.9.2

Evaluate the reaction coefficients for the transformation involving Ca_2SiO_4, $Ca_3Mg(SiO_4)_2$ and $Ca_5Mg(SiO_4)_3$ at constant P.

Hint

We should first check if the transformation may be sharp although $p = 3$ and $c = 4$ and thus $p = c - 1$ and not $c + 1$.

Solution

The composition matrix is

$$
\begin{array}{cccc}
\text{Ca} & \text{Mg} & \text{Si} & \text{O} \\
\end{array}
$$
$$
\begin{bmatrix}
2 & 0 & 1 & 4 \\
3 & 1 & 2 & 8 \\
5 & 1 & 3 & 12
\end{bmatrix}
$$

By omitting one column at a time we obtain

$$
\begin{vmatrix}
0 & 1 & 4 \\
1 & 2 & 8 \\
1 & 3 & 12
\end{vmatrix} = 0
\qquad
\begin{vmatrix}
2 & 1 & 4 \\
3 & 2 & 8 \\
5 & 3 & 12
\end{vmatrix} = 0
\qquad
\begin{vmatrix}
2 & 0 & 4 \\
3 & 1 & 8 \\
5 & 1 & 12
\end{vmatrix} = 0
\qquad
\begin{vmatrix}
2 & 0 & 1 \\
3 & 1 & 2 \\
5 & 1 & 3
\end{vmatrix} = 0
$$

We may conclude that there are two compositional degeneracies and the transformation is sharp at any chosen value of P. The reaction coefficients are obtained as

$$
\nu^{Ca_2SiO_4} = \begin{vmatrix} 1 & 2 \\ 1 & 3 \end{vmatrix} = 1
$$

$$
\nu^{Ca_3Mg(SiO_4)_2} = - \begin{vmatrix} 0 & 1 \\ 1 & 3 \end{vmatrix} = 1
$$

$$
\nu^{Ca_5Mg(SiO_4)_3} = \begin{vmatrix} 0 & 1 \\ 1 & 2 \end{vmatrix} = - 1
$$

Exercise 12.9.3

Suppose we have a computer program for the calculation of phase equilibria. When trying to calculate the equilibrium temperature at 1 bar for the equilibrium discussed in the preceding exercise, we get the message, 'cannot calculate because $v \neq 0$'. What action should we take?

Hint

We have three phases and four components, $p = c - 1$, but we have just shown that there are two compositional degeneracies. The transformation should thus be sharp and it should be possible to calculate a unique equilibrium temperature at 1 bar.

Solution

The program seems to need $c + 1$ phases. We should thus introduce two new phases, e.g. pure O and pure Mg.

12.10   *Compositional degeneracies and the phase field rule*

In Section 12.1 we found that the phase field rule, modified for the condition of constant composition, yields the same expression as Gibbs' phase rule as long as $p \geq c$. However, from the preceding section it is evident that the system will lose one degree of freedom for each compositional degeneracy. We should thus modify the phase field rule for constant composition as follows:

$$d = c + 2 - p - n_c \text{ for } d \leq 2$$

where n_c is the number of compositional degeneracies. For all $p < c + 1$ one will thus get $d = 2$ unless the number of compositional degeneracies $n_c \geq c + 1 - p$. A smaller number of compositional degeneracies will have no effect on d.

An $\alpha \rightarrow \beta$ transformation is called congruent if α and β have the same composition. The elements in the two rows of the composition matrix will thus be equal, $x_j^\alpha = x_j^\beta$, and all the determinants obtained by omitting $c - p$ (i.e. $c - 2$) columns are zero. According to the preceding discussion, this means that there are $c - p + 1$ (i.e. $c - 1$) compositional degeneracies and the phase field rule modified for constant composition yields

$$d = c + 2 - p - n_c = c + 2 - 2 - (c - 1) = 1$$

Again we see that congruent transformations have one degree of freedom and will be sharp if P is kept constant and T is varied and vice versa.

Exercise 12.10.1

In Exercise 12.1.3 we considered the variance of the equilibrium $CaCO_3 \leftrightarrow CaO + CO_2$ in an N_2 atmosphere. Now, discuss the same transformation without N_2 present.

Hint

We have $p = 3$ and $c = 3$ and may thus expect that $d = c + 2 - p = 3 + 2 - 3 = 2$. However, we should check if there is a compositional degeneracy.

Solution

The composition determinant yields

$$
\begin{array}{c|ccc}
 & \text{Ca} & \text{C} & \text{O} \\
\hline
CaCO_3 & 1 & 1 & 3 \\
CaO & 1 & 0 & 1 \\
CO_2 & 0 & 1 & 2 \\
\end{array} = 3 - 1 - 2 = 0.
$$

There is thus one compositional degeneracy and $d = c + 2 - p - n_c = 3 + 2 - 3 - 1 = 1$. Under constant P we get $d = 0$ and we thus have a sharp phase transformation at any chosen value of P.

Exercise 12.10.2

Discuss the variance of the equilibrium between pure water, H_2O, and a gas composed of H_2O, H_2 and O_2 molecules in a system of constant composition.

Hint

We have $p = 2$ and $c = 2$ (H and O) and may thus have expected that $d = c + 2 - p = 2 + 2 - 2 = 2$. However, we should check if there is a compositional degeneracy.

Solution

The above expectation is justified if the composition is defined by starting with arbitrary amounts of H_2O, H_2 and O_2. On the other hand, if we start with pure water, H_2O, then the gas will have the same composition even if it dissociates into a mixture of H_2O, H_2 and O_2 molecules. In that case we have a composition determinant with two identical rows. There is thus a compositional degeneracy and we get $d = c + 2 - p - n_c = 2 + 2 - 2 - 1 = 1$. At constant P the equilibrium can be established at a single temperature, only, and by changing the temperature gradually through this value we will get a sharp phase transformation. (Here we have neglected the solubilities of H_2 and O_2 in water which are very low but, in principle, give the two phases slightly different compositions if we start from only water. Thus, $d = 2$.)

Exercise 12.10.3

Pure Zn is produced from ZnO by reduction with C under the formation of CO and CO_2. Examine the variance of the equilibrium between these phases.

Hint

The temperature is high and Zn evaporates. There are three phases, gas, solid ZnO and solid C, and three components, Zn, O and C. We may thus expect that $d = c + 2 - p = 3 + 2 - 3 = 2$ but showld check if there is a compositional degeneracy.

Solution

If we start from ZnO and C and no gas, then gas must form from ZnO and C and its composition must fall on the straight line between ZnO and C in the composition triangle. Thus, there is a compositional degeneracy. Mathematically, we may show this by considering x formula units of the reaction $2ZnO + C \rightarrow CO_2 + Zn$ and y formula units of $ZnO + C \rightarrow CO + Zn$. The gas will thus contain $x + y$ moles of C, $2x + y$ moles of O and $2x + y$ moles of Zn. The composition determinant is obtained as

$$
\begin{array}{ccc}
C & O & Zn
\end{array}
$$
$$
\begin{vmatrix}
1 & 0 & 0 \\
0 & 1 & 1 \\
x + y & 2x + y & 2x + y
\end{vmatrix} = 0
$$

Thus, there is one compositional degeneracy and we get $d = c + 2 - p - n_c = 3 + 2 - 3 - 1 = 1$. At constant P the three phases can be present at a single T, only. At any given T, the transformation will continue until a certain P value has been reached. On the other hand, if one starts to remove Zn from the gas by condensation, the degeneracy is relaxed and the variance will be $d = 2$.

Exercise 12.10.4

Discuss the variance of the equilibrium between the three solid phases NH_4Cl, K_2CO_3, $(NH_4)_2CO_3$, and a liquid phase formed from them by melting.

Hint

We have $p = 4$ and $c = 6$ (N, H, Cl, K, C, and O) and may thus expect the value for $p < c$, i.e. $d = 2$. However, we should check on compositional degeneracies. In order to get the result $d = 1$ we need $1 = c + 2 - p - n_c = 6 + 2 - 4 - n_c = 4 - n_c$; $n_c = 3$, i.e. we need three compositional degeneracies. This is tested directly by studying the composition matrix. Notice that the liquid phase is not stoichiometric but due to the condition of electroneutrality we may represent its composition with $(NH_4)_xK_yCl_z(CO_3)_{(x+y-z)/2}$.

Solution

The composition matrix is

$$
\begin{array}{cccccc}
N & H & Cl & K & C & \quad\quad O \\
\begin{bmatrix}
1 & 4 & 1 & 0 & 0 & 0 \\
0 & 0 & 0 & 2 & 1 & 3 \\
2 & 8 & 0 & 0 & 1 & 3 \\
x & 4x & z & y & (x+y-z)/2 & 3(x+y-z)/2
\end{bmatrix}
\end{array}
$$

It can be shown that all the determinants obtained by omitting any two columns are zero. Thus there are three compositional degeneracies and $d = c + 2 - p - n_c = 6 + 2 - 4 - 3 = 1$. This is a sharp transformation at any chosen value of P.

12.11   *Overlapping transformations*

In Section 11.2 we found that constant composition and pressure yields $d = -1$ for $p = c + 2$ for an invariant equilibrium and such an equilibrium is possible only at a particular value of P, which is unlikely to be established in such an experiment. Furthermore, we found that this transformation has the character of two or more overlapping transformations. After the introduction of compositional degeneracies in the phase field rule, it is evident that overlapping transformations may occur even if $p < c + 2$.

For the case $p = c + 1$ one can test for a compositional degeneracy by omitting a row in the composition matrix and requiring that the resulting determinant is zero. If

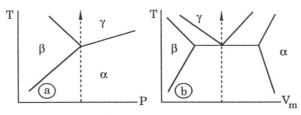

Overlapping transformations   Sharp transformation

$$\alpha \longrightarrow \gamma \quad \text{or} \quad \alpha \longrightarrow \beta \longrightarrow \gamma \qquad \alpha + \beta \longrightarrow \gamma$$

Figure 12.12 Illustration of the fact that a case of overlapping sharp transformations, obtained when T is varied under a constant P, can change to an ordinary sharp transformation if T is varied under constant V_m.

that is the case, then the phases remaining in the matrix are $p = c$ and they can take part in a sharp transformation. If the same result is obtained when another row is omitted, then the phases now remaining in the matrix can take part in another sharp phase transformation. Thus, this is a case of overlapping transformations.

For the case $p = c$ one would need two compositional degeneracies in order to get overlapping transformations. In order to test if this is the case, one can first omit a row and check if the remaining $c - 1$ phases can take part in a sharp transformation by omitting any column. Then one should instead omit another row and in the same way check if the new set of $c - 1$ phases can take part in a sharp transformation. If this is also the case, then each one of the two different sets of phases can take part in a sharp phase transformation and it is a case of overlapping phase transformations. Another way of making this test is to start by checking that the complete composition determinant is zero. Then it would be sufficient to check one of the minor determinants.

The experimental conditions we have considered here are constant composition and pressure and a gradual change of temperature. Let us now consider a system with $p = c + 2$, which gives overlapping transformations ($d = -1$) under such conditions, and examine what happens if we keep the volume constant instead of the pressure. Since we have thus introduced one more molar quantity, we have increased the variance by one unit to $d = 0$ and can expect an ordinary sharp transformation but not overlapping transformations. Its reaction coefficients can be calculated from

$$v^j = (-1)^{j-1} | V_m^\alpha x_1^\beta \ldots x_{j-1}^{j-1} x_j^{j+1} \ldots x_c^\varepsilon |$$

This is illustrated in Fig. 12.12 for the simple case of $c = 1$ and $p = 3$. In agreement with the lever rule we find

$$v^\alpha = | V_m^\beta 1 | = V_m^\beta - V_m^\gamma$$

$$v^\beta = - | V_m^\alpha 1 | = - (V_m^\alpha - V_m^\gamma)$$

$$v^\gamma = | V_m^\alpha 1 | = V_m^\alpha - V_m^\beta$$

Exercise 12.11.1

Consider the equilibrium between four phases in the A–B–C–D system. The molar contents of B, C and D are for each phase (0.20;0.20;0.40), (0.15;0.10;.0.25), (0.14;0.08;0.22) and (0.13;0.06;0.19).

Hint

The phase field rule for constant composition yields $d = c + 2 - p = 4 + 2 - 4 = 2$. However, we should test for compositional degeneracies.

Solution

The composition matrix is

$$
\begin{bmatrix}
0.20 & 0.20 & 0.20 & 0.40 \\
0.50 & 0.15 & 0.10 & 0.25 \\
0.56 & 0.14 & 0.08 & 0.22 \\
0.62 & 0.13 & 0.06 & 0.19
\end{bmatrix}
$$

The complete determinant is zero, but we do not need to take the trouble to show this if we suspect that there is another compositional degeneracy. Then the whole test can be done on minor determinants, obtained by omitting one row and then one column at a time. We get, for instance,

$$
\begin{vmatrix}
0.20 & 0.20 & 0.40 \\
0.15 & 0.10 & 0.25 \\
0.14 & 0.08 & 0.22
\end{vmatrix} = 0, \text{ and }
\begin{vmatrix}
0.20 & 0.20 & 0.20 \\
0.50 & 0.15 & 0.10 \\
0.62 & 0.13 & 0.06
\end{vmatrix} = 0
$$

There are thus two compositional degeneracies and $d = c + 2 - p - n_c = 4 + 2 - 4 - 2 = 0$ and at a constant P we have $d = -1$. This is a case of overlapping transformations.

Exercise 12.11.2

Consider the equilibrium NaCl + ice + water + vapour.

Hint

There are four phases and four components and we could thus expect $d = c + 2 - p = 4 + 2 - 4 = 2$. However, we should check for compositional degeneracies. Water dissolves much NaCl, ice dissolves some but very little and there is practically no NaCl in the vapour. We may give the composition of the water as $x_{Na} = x_{Cl} = y$, $x_O = x_H/2 = (1 - 2y)/3$ and of ice in a similar way.

Solution

The composition determinant is zero because two columns are identical,

$$
\begin{array}{cc}
 & \begin{array}{cccc} \text{Na} & \text{Cl} & \text{H} & \text{O} \end{array} \\
\begin{array}{c} \text{NaCl} \\ \text{ice} \\ \text{water} \\ \text{vapour} \end{array} &
\begin{vmatrix}
1 & 1 & 0 & 0 \\
z & z & 2(1-2z)/3 & (1-2z)/3 \\
y & y & 2(1-2y)/3 & (1-2y)/3 \\
0 & 0 & 2 & 1
\end{vmatrix} = 0
\end{array}
$$

This proves that there is at least one compositional degeneracy. We can test if there is another one by omitting one row and one column. We get, for instance,

$$
\begin{vmatrix}
1 & 0 & 0 \\
z & 2(1-2z)/3 & (1-2z)/3 \\
0 & 2 & 1
\end{vmatrix} =
\begin{vmatrix}
2(1-2z)/3 & (1-2z)/3 \\
2 & 1
\end{vmatrix} = 0
$$

Thus there are two compositional degeneracies and the phase field rule yields $d = c + 2 - p - n_c = 4 + 2 - 4 - 2 = 0$ and for constant P we get $d = -1$. This is a case of overlapping transformations.

Exercise 12.11.3

Suppose we have a computer program for the calculation of phase equilibria. When trying to calculate the equilibrium temperature at a pressure of 1 bar for four phases in a ternary system, we get the message, 'cannot calculate because $d \neq 0$'. What action could we take?

Hint

Evidently, the program is constructed to calculate sharp phase transformations. We have $p = 4$ and $c = 3$ and thus $p = c + 1$ and we thus expected that the calculation could be carried out. However, the message indicates that there is some kind of compositional degeneracy which decreases d by one unit.

Solution

If $n_c = 1$, we have $d = c + 2 - p - n_c = 3 + 2 - 4 - 1 = 0$ and at constant P we have $d = -1$. This may thus be a case of overlapping transformations. We should simply relax the condition $P = 1$ bar and can expect to calculate T as well as P. Then we could introduce a constant P equal to that value and calculate the reaction coefficients for any transformation involving three of the phases.

13

Partitionless transformations

13.1   *Deviation from local equilibrium*

As discussed in Section 6.8 it is common to assume that the rate of a phase transformation in an alloy is controlled by the rate of diffusion. The local compositions at the phase interfaces are then used as boundary conditions for the diffusion problem and they are evaluated by assuming local equilibrium at the interfaces. That is a very useful approximation but there are important exceptions. It is necessary to realize that the exceptions are of two different types and they have opposite effects. The first type of exception is caused by a limited mobility of the interface. In order to keep pace with the diffusion, the interface requires a driving force which is subtracted from the total driving force and decreases the driving force for the diffusion process. Due to this effect, a partitionless transformation, which would otherwise be diffusion-controlled, requires an increased supersaturation of the parent phase, as shown in Section 6.8. Formally, this case was treated by assuming a pressure difference between the two phases as if the interface were curved more than it actually is, and the local equilibrium assumption was modified to this case.

The other type of exception will instead decrease the driving force needed by decreasing the need for diffusion and will thus result in a higher rate of transformation and make it possible for an alloy with a lower supersaturation to transform. It is primarily caused by a low atomic mobility in the migrating interface. The present chapter will discuss such cases but also related cases of full local equilibrium. Naturally, such phenomena cannot be described by assuming local equilibrium under a pressure difference which would increase the driving force needed. Instead, the local equilibrium seems to be constrained in some way. Sometimes one talks about partial equilibrium or deviation from local equilibrium.

In general, the rate of migration of an interface during a phase transformation is limited by the mobility of the interface itself and by the transport of various extensive quantities, i.e., contents of various components by diffusion, heat content (i.e. enthalpy) by heat conduction and volume by elastic and plastic flow. We shall not consider the latter problem but presume that there is some efficient mechanism for the accommodation of changes in volume. However, there are many interesting thermodynamic features that could have been discussed. In general, we shall also neglect the need for heat conduction and assume that isothermal conditions can be maintained in spite of the heat of transformation. However, we shall start with that problem because it has much in common with diffusion and may be used to demonstrate important principles.

13.2   ## *Adiabatic phase transformation*

For a process taking place under adiabatic and isobaric conditions, $dQ = 0$ and $dP = 0$, we have from the first law

$$dH = d(U + PV) = dU + PdV + VdP = dQ + VdP = 0$$

For a system which is closed to exchange of matter as well as heat we also have $dN_i = 0$ and it is convenient to use the combined law in the following form,

$$T \cdot d_{ip}S = TdS + VdP + \Sigma\mu_i dN_i - dH = TdS$$

It should be emphasized that here we have not represented $T \cdot d_{ip}S$ with the driving force $Dd\xi$ because the reaction is not isothermal (see Section 4.3). The condition for a reversible reaction is $d_{ip}S = 0$ and thus $dS = 0$. In order for the reaction to proceed with a measurable rate it is necessary that $d_{ip}S > 0$ and thus $dS > 0$.

   A homogeneous reaction (e.g. a reaction between molecules in a gas) occurs gradually in the whole system and one can usually presume that it has proceeded to the same extent ξ in all parts of the system. It is evident that such a reaction can occur under adiabatic and isobaric conditions. The situation is different for a heterogeneous reaction which takes place by nucleation and growth. Let us examine the simple case of a sharp phase transformation which goes to completion instantaneously at any point as an interface passes by. The extent of reaction can be measured as the fraction of the system which has undergone the transformation. Thus, ξ would go from 0 to 1. Alternatively, ξ can be given as the number of atoms in the transformed part. Let us suppose that this reaction can occur under adiabatic conditions. Due to the heat of transformation, this should mean that the transformed part of the system is at a different temperature than the rest and heat would thus flow between the different parts unless the transformation is extremely rapid and leaves no time for the flow of heat. Such high transformation rates are not very common. An explosion may come close. We may thus conclude that the transformation in a material cannot normally be truly adiabatic even if it occurs inside a thermally insulated system. Before considering the effect of the heat transfer we shall nevertheless examine the conditions for a hypothetical transformation which is truly adiabatic. Since we realize that the transformation will cause a change of temperature we must start by defining the thermal properties.

   Let us consider a unary system with two phases, α and β, and let us suppose that the difference in heat capacity, ΔC_P, is independent of temperature. Then the difference in molar enthalpy ΔH_m and entropy ΔS_m at any temperature is also indepen-

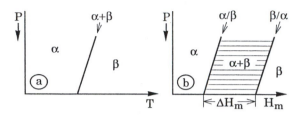

Figure 13.1 Ordinary phase diagrams for a unary system.

dent of temperature but may vary with pressure. The equilibrium temperature at a given pressure will be $T^e = \Delta H_m / \Delta S_m$. Schematic T,P and H_m,P phase diagrams are given in Fig. 13.1. The boundary between α and $\alpha + \beta$ is denoted α/β because it represents α in equilibrium with β. We have here taken β as the high-temperature phase and it is evident from the H_m,P diagram that $\Delta H_m = H_m^\beta - H_m^\alpha$ is positive and then $\Delta S_m = S_m^\beta - S_m^\alpha$ must also be positive because T^e is positive.

On the other hand, if the phases are at different temperatures we get

$$\Delta H = \Delta H_m + C_P(T^\beta - T^\alpha)$$
$$\Delta S = \Delta S_m + C_P \ln(T^\beta / T^\alpha)$$

Suppose it were possible to transform β of T^β to α of T^α under adiabatic and reversible conditions, i.e. under isentropic conditions, $\Delta S = 0$. Suppose the pressure is also kept constant. Then $\Delta H = 0$ and we have two equations from which we can evaluate T^α and T^β,

$$T^\beta = T^e \cdot \frac{\Delta S_m / C_P}{\exp(\Delta S_m / C_P) - 1} \cong T^e - \Delta H_m / 2C_P$$

$$T^\alpha = T^e \cdot \frac{\Delta S_m / C_P}{1 - \exp(-\Delta S_m / C_P)} \cong T^e + \Delta H_m / 2C_P$$

where the approximation is justified for $\Delta S_m \ll C_P$, only. These results are plotted in two new diagrams, see Fig. 13.2. In this case there is only one line in the P,H_m diagram and it shows where α and β have equal values of H_m and also equal values of S_m. This results in the α and β one-phase fields overlapping in the T,P diagram. The interpretation is that, on cooling under these conditions, a β phase would not transform to α of the same temperature when cooled to the equilibrium temperature T^e, but it would transform at $T^\beta = T^e - \Delta H_m / 2C_P$ and the α phase would be at a higher temperature $T^\alpha = T^e + \Delta H_m / 2C_P$ when it forms. The two-phase boundaries in the T,P diagram have thus separated by $T^\alpha - T^\beta = \Delta H_m / C_P$. This diagram would predict that β, if super-cooled to reach the $(\alpha + \beta)_H$ line, could transform instantaneously and completely to α if there were no kinetic obstacles. It would be a sharp transformation at the constant H_m value. It has been speculated that this kind of reaction could occur in solidification of very rapidly cooled liquid droplets.

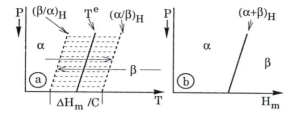

Figure 13.2 Phase diagrams for isobaric and adiabatic conditions in a unary system. Notice that the α and β phase fields in the P,T diagram overlap in a region around the equilibrium temperature.

However, there are two major objections. Firstly, the reaction must be extremely fast in order to prevent heat flowing from the warmer, growing α into the colder parent β, $T^\alpha - T^\beta$ being positive (equal to $\Delta H_m/C_P$). Secondly, even if α of the temperature T^α could form, it would not be stable because it is inside the stable one-phase field for β according to the T,P diagram in Fig. 13.1. Part of α could thus transform back to β. In the next section we shall examine a more realistic model.

Exercise 13.2.1

Estimate the internal entropy production when 1 mole of a pure substance transforms adiabatically from α to β when the temperature of α is ΔT above the value where there is no entropy production.

Hint

Of the two conditions used in the text, only one holds here, $\Delta H = 0$, but $\Delta_{ip}S$ is still equal to ΔS and can be calculated from $\Delta S_m + C_P\ln(T^\beta/T^\alpha)$, if C_P is constant and equal in the two phases.

Solution

$T^\alpha = T^e + \Delta H_m/2C_P + \Delta T = T^e + T^e\Delta S_m/2C_P + \Delta T$. From $0 = \Delta H = \Delta H_m + C_P(T^\beta - T^\alpha) = T^e\Delta S_m + C_P(T^\beta - T^\alpha)$ we get $T^\beta = T^\alpha - T^e\Delta S_m/C_P = T^e - T^e\Delta S_m/2C_P + \Delta T$; $\Delta_{ip}S = \Delta S_m - C_P\ln(T^\beta/T^\alpha) = \Delta S_m - C_P\ln\{[1 - T^e(\Delta S_m/2C_P)/(T^e + \Delta T)]/[1 - T^e(\Delta S_m/2C_P)/(T^e + \Delta T)]\} \cong \Delta S_m - C_P T^e(\Delta S_m/C_P)/(T^e + \Delta T) \cong \Delta S_m\Delta T/T^e$.

13.3   *Quasi-adiabatic phase transformation*

Let us now examine if there are conditions under which the transformation can occur by a steady-state process, i.e. without a gradual change of the conditions at the migrating

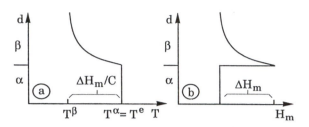

Figure 13.3 Steady-state conditions for a quasi-adiabatic transformation $\beta \to \alpha$ under constant P. The growing α has the same enthalpy as the initial β.

interface. The growing α should then have a uniform temperature, T^{α}, but the temperature may vary inside the parent β. The temperature profile can be illustrated by Fig. 13.3(a) which has been drawn under the assumption of local equilibrium at the interface. All of the α phase must be at the equilibrium temperature, T^{e}. In order for this to be a steady-state process it is necessary that α has the same heat content as the bulk of the β phase. This is illustrated in Fig. 13.3(b) and the following equations are obtained

$$\Delta H = \Delta H_m + C_P(T^{\beta} - T^{\alpha}) = 0$$
$$T^{\beta} = T^{\alpha} - \Delta H_m/C_P = T^e - \Delta H_m/C_P$$

The reaction can thus be essentially adiabatic if it is possible to change the temperature of the whole β system to $T^e - \Delta H_m/C_P$ before the nucleation of α occurs. This model thus requires twice as large a ΔT as the truly adiabatic model. This is a demonstration of the fact that a deviation from local equilibrium results in less need of driving force. After a transient period, during which an enthalpy spike of height ΔH_m will develop in the β phase at the migrating interface, steady-state conditions will be maintained towards the end of the reaction. The duration of the transient period and the width of the temperature spike in the parent phase will depend upon the rate of transformation and the rate of heat flow. We can use the phase diagrams, given above, for a summary of our conclusions (see Fig. 13.4).

Point 1 in Fig. 13.4 is the isothermal transformation temperature for β. A β phase cooled just below that point could start to transform to α but the progress of the transformation would be directly controlled by the further extraction of heat from the system. The growing α phase will be at point 2. If a β phase could be cooled to point 3 before the transformation starts, then the transformation could, in principle, occur very quickly and adiabatically and the α phase would be at point 4. However, if the phase interface does not move with an extremely high velocity, there will be heat conduction away from the remaining parent phase and it will no longer be able to transform adiabatically but would depend upon further heat extraction. Finally, if a β phase could be cooled to point 5 before the transformation starts, then the transformation could occur without any further heat extraction even if there is time for heat conduction. All of the α formed would be at point 2 and β at the interface would be at point 1. The transformation could occur in a steady-state fashion where the growing α phase forms at

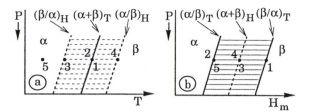

Figure 13.4 Phase diagrams illustrating the conditions for a quasi-adiabatic transformation in a unary system. Subscript T indicates isothermal conditions. Subscript H indicates adiabatic conditions because P is kept constant.

a higher temperature than the initial β phase (compare point 2 in the P,T phase diagram with point 5) but with the same enthalpy (compare point 2 in the P,H_m phase diagram with point 5). This type of reaction could be called a quasi-adiabatic transformation. It is interesting to note that it occurs when the parent phase, by cooling, is entering into the field for the new phase in the P,H_m diagram.

It is usually assumed that a transformation starts from a stationary nucleus and picks up speed during an initial transient period. It would then be natural to expect a situation somewhat similar to Fig. 13.3 to be established after some short time. It is an interesting question whether it could later develop into a truly adiabatic mechanism. The requirement would be that the speed becomes so fast that the temperature spike in the parent phase becomes so thin that it disappears between the atoms. A simple calculation would show that the thickness should be less than about $\mathscr{D}/v$ where $\mathscr{D}$ is the diffusion coefficient and v is the growth rate. This transition turns out to be very unlikely.

Exercise 13.3.1

To what temperature must a liquid metal be cooled before solidification starts, in order for the solidification to take place by a steady-state reaction controlled by heat conduction over short distances, i.e. quasi-adiabatic conditions? Use Richard's rule and $C_P \cong 3R$.

Hint

Find inspiration from Fig. 13.3. Richard's rule says $H_m^L(T^e) - H_m^\alpha(T^e) = RT^e$, i.e. $S_m^L - S_m^\alpha = R$.

Solution

Under quasi-adiabatic conditions $H_m^L(T^L) = H_m^\alpha(T^e)$. We can relate $H_m^L(T^L)$ to $H_m^L(T^e)$ by integration from T^L to T^e: $H_m^L(T^e) = H_m^L(T^L) + \int C_P dT = H_m^L(T^L) +$

$3R(T^e - T^L)$. Insert $H_m^\alpha(T^e)$ and $H_m^L(T^L)$ in Richard's rule: $H_m^L(T^L) + 3R(T^e - T^L) - H_m^L(T^L) = RT^e$; $2T^e = 3T^L$; $T^L = (2/3)T^e$.

Exercise 13.3.2

Evaluate the entropy production for quasi-adiabatic solidification of 1 mole of a pure, liquid metal.

Hint

Compare with Exercise 13.2.1.

Solution

According to Exercise 13.3.1 we now have $T^\alpha = T^e$ and $T^L = (2/3)T^e$. Thus $\Delta_{ip}S = S_m^\alpha - S_m^L + 3R\ln(T^\alpha/T^L) = R[-1 + 3\ln(3/2)] = 0.216R$.

Exercise 13.3.3

Estimate how fast a transformation should be in order to reach truly adiabatic conditions.

Hint

The diffusion coefficient for heat conduction is about $10^{-5}\,\mathrm{m^2/s}$. The atomic distances are about $10^{-8}\,\mathrm{m}$.

Solution

$\mathscr{D}/v < 10^{-8}$ yields $v > 10^{-5}/10^{-8} = 10^3\,\mathrm{m/s}$. This is comparable to the speed of sound.

13.4 *Partitionless transformations in binary system*

We shall now examine a two-phase reaction in a binary system under constant composition, i.e. a partitionless transformation, and we shall find striking similarities with the

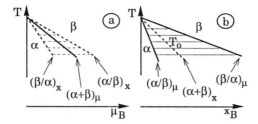

Figure 13.5 Phase diagrams for a binary system illustrating the conditions for diffusionless and quasi-diffusionless transformations. The full lines show the phase boundaries under equilibrium (constant T,P and μ_i). The dashed lines hold if there is no diffusion.

adiabatic case. We shall use the combined law in the following form because temperature and pressure will be kept constant in addition to composition.

$$D d\xi = -SdT + VdP + \Sigma\mu_i dN_i - dG = -dG$$

In this case G plays the same role as S under adiabatic conditions. The driving force for the reaction is $D = -dG/d\xi$ and for $D = 0$ we would have a reversible reaction which would occur without a change of the Gibbs energy but infinitely slow. The reaction could proceed with a measurable rate if $D > 0$, i.e. $dG < 0$. Partitionless transformations were discussed in Section 6.8. We shall now examine such a reaction in more detail and discuss two limiting cases that we may describe as **diffusionless** and **quasi-diffusionless** transformations, respectively. The similarity with the adiabatic cases becomes obvious if we examine corresponding phase diagrams. In the present case we shall keep P constant and introduce axes for μ_B and x_B (see Fig. 13.5 which can be compared with Fig. 13.4). The dashed line in the T,x_B diagram, denoted by T_o, corresponds to the dashed line in the P,H_m diagram, which showed where α and β have equal values of H_m and equal values of S_m. On the T_o line α and β have equal values of x_B and equal values of G_m. We can thus follow the same reasoning and conclude that there are two limiting cases, the true diffusionless case described by the dashed lines and the quasi-diffusionless case described by the full lines. Both can be classified as partitionless because α would form with the composition of the initial β.

The true diffusionless transformation is easy to understand. If a β alloy is cooled below the T_o line where α of the same composition has the same G_m value, then G_m may decrease by the $\beta \rightarrow \alpha$ transformation even without any change of composition. The molar Gibbs energy diagram in Fig. 13.6 demonstrates that μ_B will increase by that transformation, which corresponds to the increase of T during the true adiabatic transformation.

The quasi-diffusionless transformation was actually discussed in Section 6.8 and illustrated by Fig. 6.14. For slow transformations the quasi-diffusionless transformation is the most likely alternative. In order to get a transition to the true diffusionless case, it is necessary to avoid local equilibrium at the interface. How this could happen

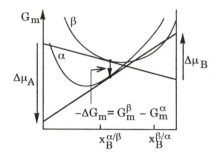

Figure 13.6 Conditions for a true diffusionless transformation.

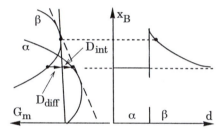

Figure 13.7 Model for describing a deviation from local equilibrium.

may be illustrated with a very simple picture, shown in Fig. 13.7 which resembles Fig. 6.14 but now it is supposed that the growth rate is so high that the composition spike will be so thin that its top falls between neighbouring atoms. The black dot on the composition profile is an attempt to identify the highest composition in β that actually exists. A tangent to the corresponding point on the G_m curve for β has been drawn and is supposed to represent the thermodynamic properties of β at the interface.

Only part of the total driving force will now be dissipated by diffusion in the spike. The remaining part can be used on processes at and in the interface, as illustrated by Fig. 6.15. Now, however, the β alloy does not need to be on the other side of the equilibrium composition of α. It can even fall inside the ordinary $\alpha + \beta$ two-phase field but not on the other side of the intersection between the tangent and the α curve because then the total driving force would be too small to drive the diffusion in the spike. Exactly how and where one can find the growth conditions in the diagram for an actual case depends upon the properties of the interface. If the interface requires only a small driving force, then the rate will be high, most of the spike will fall between the neighbouring atoms and only a small driving force is dissipated by diffusion. The partitionless transformation can then occur close to T_o, i.e. the composition of the initial β and the growing α can fall just to the left of the intersection between the two G_m curves and the β composition at the interface can fall just to the right of the intersection. That would represent a case very close to the true diffusionless case.

In metallic materials there are two well-known partitionless transformations called 'martensitic' and 'massive', respectively. The martensitic transformation comes close to the true diffusionless case but its interface migrates with an atomic mechanism that creates strong stresses which require a high driving force. This type of transformation can very well occur far inside the $\alpha + \beta$ two-phase field but only at a considerable distance below the T_0 line. The massive transformation probably progresses by individual atoms crossing the interface and the growth rate cannot be extremely high. At relatively high temperatures it seems that the deviation from local equilibrium is small and this transformation comes close to the quasi-diffusionless case. However, at lower temperatures the atomic mobility in the crystal lattice decreases faster than in the interface. The rate of diffusion thus decreases faster than the rate of migration of the interface. It is thus possible that more and more of the spike falls between neighbouring atoms and the massive transformation can occur further inside the two-phase field.

An interesting problem should be mentioned in this connection. Figure 13.6 demonstrates that the chemical potential of a B atom increases as it crosses the interface and gets incorporated in the growing α phase. One should ask what forces the B atoms to cross when the driving force on them seems to go in the other direction. In the quasi-diffusionless case this problem is solved by the composition spike which lifts up μ_B at the expense of μ_A. Just at the interface the B atoms have reached the necessary μ_B to become incorporated in the α phase. On the other hand, if there is a deviation from local equilibrium at the interface, then some other mechanism inside the interface must do what the spike did at local equilibrium. This phenomenon is sometimes called solute trapping.

Exercise 13.4.1

Given the phase diagram in Fig. 13.5, mark the regions where one could expect the massive or the martensitic transformations $\beta \rightarrow \alpha$. Suppose that the martensitic transformation requires an undercooling below T_0 which is independent of the composition and that the massive transformation occurs with some small deviation from equilibrium.

Hint

Martensite will normally grow much faster because it requires no diffusion. Martensite would thus predominate in a region where both types of transformation could occur, in principle.

Solution

See diagram.

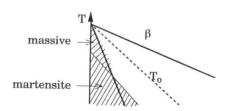

massive

β

martensite

T_o

Solution 13.4.1

13.5   *Partial chemical equilibrium*

In a ternary alloy it could very well happen that one of the elements diffuses very much faster than the other two, for example if it is an interstitial solute. It is then possible that a new phase forms with a different content of the mobile element but without a change of the relative contents of the other two. Such a transformation would be partly partition-less. Hultgren (1947) proposed that it could even occur without any diffusion of the latter two elements and used the term **paraequilibrium** to describe the local equilibrium at the phase interface under such a transformation. We shall now examine that kind of local equilibrium. Hultgren studied systems with Fe, C and a metallic element which we shall denote by M. We shall keep those symbols but Fe could represent any element, C any mobile element and M any element as sluggish as Fe.

Under full local equilibrium at a phase interface, there is no driving force on the interface as shown by the following form of the combined law (see last form derived in Section 3.3),

$$Dd\xi = -SdT + VdP - \Sigma N_i d\mu_i = 0$$

because T, P and all μ_i have the same values on both sides of the interface. When a transformation occurs under paraequilibrium, μ_C has the same value on both sides because C is very mobile, but μ_{Fe} and μ_M have different values. Instead, u_{Fe} and u_M have the same values if u_i is defined as $N_i/(N_{Fe} + N_M)$. It is thus useful to consider the combined law in a new form which can be derived as follows,

$$Dd\xi = -SdT + VdP - \Sigma N_i d\mu_i \pm (\mu_{Fe} dN_{Fe} + \mu_M dN_M)$$
$$Dd\xi/(N_{Fe} + N_M) = -S_{m12} dT + V_{m12} dP - u_C d\mu_C +$$
$$\mu_{Fe} du_{Fe} + \mu_M du_M - d(u_{Fe}\mu_{Fe} + u_M\mu_M)$$

Under paraequilibrium $dT = dP = d\mu_C = du_{Fe} = du_M = 0$ and we find

$$Dd\xi/(N_{Fe} + N_M) = -d(u_{Fe}\mu_{Fe} + u_M\mu_M)$$

The driving force should be zero for a transformation occurring under paraequilibrium conditions because paraequilibrium is supposed to be a kind of local equilibrium. It is thus necessary that $u_{Fe}\mu_{Fe} + u_M\mu_M$ has the same value on both sides of the interface. Of course, T, P and μ_C must also have the same values on both sides. The new quantity that must have the same value in both phases is simply an average value for Fe and M, as if

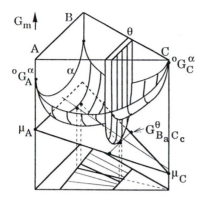

Figure 13.8 Molar Gibbs energy diagram for a ternary system illustrating the paraequilibrium conditions. The common tangent line from the C axis is situated above the common tangent plane.

they together have formed a new element. The quantity can be written in various ways because $G_{m12} = \Sigma u_i\mu_i$.

$$u_{Fe}\mu_{Fe} + u_M\mu_M = G_{m12} - u_C\mu_C = \frac{G_m - x_C\mu_C}{1 - x_C}$$

Suppose the three elements can form a line compound θ of the formula $(Fe,M)_aC_c$. For paraequilibrium between θ and a solution phase, γ, we find

$$\left(\frac{G_m - c\mu_C}{a}\right)^\theta = \left(\frac{G_m - x_C\mu_C}{1 - x_C}\right)^\gamma$$

It should be noted that the same form of the combined law was actually derived in Section 12.6 but in a slightly different way and written in a slightly different but equivalent form.

In a molar Gibbs energy diagram the tie-line between the two phases in paraequilibrium is directed towards the C corner. It falls on a common tangent line to the two Gibbs energy surfaces but not on the common tangent plane. Figure 13.8 demonstrates that the common tangent line for paraequilibrium, which must go through the C axis, is situated above the common tangent plane that holds for full equilibrium.

Figure 13.9(a)–(d) gives various versions of the phase diagram showing the equilibrium between two solution phases, α and γ, at some convenient T and P values. Instead of using the chemical potentials, μ_C and μ_M, as axes the chemical activities, a_C and a_M have been used in order to make the diagram show low contents of C and M where the chemical potentials would approach negative infinity. It is interesting to note that Figs. 13.9(a) and (b) are very similar to Fig. 13.5 but T has been replaced by a_C. In fact, the two reactions are very similar because the additional component in the present case is compensated by the temperature being kept constant.

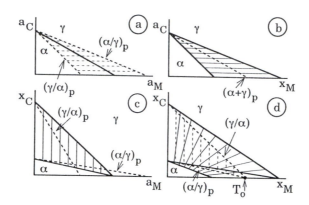

Figure 13.9 The phase diagram for a ternary system at constant T and P, drawn with different sets of axes in order to illustrate the paraequilibrium conditions (dashed lines), assuming that C is the only mobile component.

The point of equal Gibbs energy in α and γ on the binary Fe–M side has been marked as T_o in Fig. 13.9(d) because it belongs to the T_o line in the binary T,x_M diagram. It is important to note that the two paraequilibrium phase boundaries fall inside the full equilibrium two-phase field. This is a general rule.

As in the binary case, discussed in the preceding section, we should also examine the possibility of obtaining a partitionless (here of Fe and M) transformation under full local equilibrium. That should be possible if there is a composition spike in front of the migrating interface. The critical limit for a $\gamma \to \alpha$ transformation to take place under such **quasi-paraequilibrium** conditions is that the initial γ phase falls on the α phase boundary in the a_C,x_M phase diagram (see Fig. 13.10(b)). Again the conclusions are very similar to the previous case. However, in the present case it is more common to use an x_C,x_M phase diagram (see Fig. 13.10(d)). It should be noticed that the critical limit for a quasi-paratransformation will not fall on the α phase boundary in such a diagram because γ and α must have the same μ_C (i.e. a_C) and that requires different x_C. The position of the critical limit in all four kinds of diagrams are shown in Fig. 13.10.

Finally, we may compare the critical limit for the two partitionless kinds of growth by means of Fig. 13.11. It is interesting to note that paraequilibrium with its deviation from full local equilibrium requires less supersaturation of the parent γ. This is in agreement with a more general principle mentioned in Section 13.1. In practice, one should expect something between quasi-paraequilibrium and paraequilibrium depending on the mobilities of Fe and M, especially inside the interface, relative to the rate of migration of the interface. In the next section we shall give a more detailed account of some phase transformations assuming that they take place under quasi-paraequilibrium.

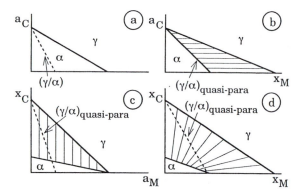

Figure 13.10 Dashed lines show the condition for quasi-paratransformation $\gamma \rightarrow \alpha$ in a ternary system with a very mobile component C.

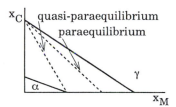

Figure 13.11 Comparison of two partitionless growth conditions for the $\gamma \rightarrow \alpha$ transformation in a ternary system with a very mobile component C.

Exercise 13.5.1

The paraequilibrium condition $[(G_m - x_C\mu_C)/(1 - x_C)]^\alpha = [(G_m - x_C\mu_C)/(1 - x_C)]^\gamma$ is not valid if one has a measurable atomic mobility within the phase interface. Discuss the reason.

Hint

Examine under what conditions the equation was derived.

Solution

The equation was derived for $dN_{Fe} = dN_M = 0$ which is not true during the integration across the interface if there is an atomic mobility.

Exercise 13.5.2

In Fig. 13.9 there are two diagrams with dashed tie-lines. They hold for paraequilibrium. In the other two diagrams the corresponding tie-lines have not been drawn. Indicate where they should fall.

Hint

Find tie-lines in Fig. 13.9(c) by projection from Fig. 13.9(a) and (d).

Solution

The $\alpha + \gamma$ two-phase field under paraequilibrium is a line in Fig. 13.9(b) and there are no tie-lines. The modified Fig. 13.9(c) is shown below.

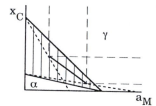

Solution 13.5.2

13.6 ***Transformations in steel under quasi-paraequilibrium***

In a steel with carbon and some substitutional alloying elements it often happens that a new phase forms with the same alloy content as the parent phase but with a different carbon content. Such a phase transformation may occur under local paraequilibrium at the migrating interface, or under quasi-paraequilibrium, or in between. In this section we shall examine the quasi-paraequilibrium case, using results from the preceding section. As a simple example we shall first discuss the $\gamma \rightarrow \alpha$ transformation. In a u_M, u_C phase diagram all products of a paratransformation or a quasi-paratransformation will fall on the same horizontal line as the parent phase. In the phase diagram we can easily find a point representing the composition of a growing phase, because it must fall on the correct level of alloy content but also on the appropriate phase boundary, the α/γ phase boundary in the present case. Having located that intersection we have found a tie-line representing the equilibrium conditions at the migrating α/γ interface. We can thus construct a concentration profile for the alloying element. In order to do this we shall plot u_M on the ordinate axis (see Fig. 13.12).

From the concentration profile we may conclude that there is a thin spike of the alloying element in front of the migrating interface. This is similar to the quasi-diffusionless case. According to the mathematics of diffusion, we can estimate the width of the spike to $\mathcal{D}_M/v$ where v is the rate of migration of the interface and $\mathcal{D}_M$ is the diffusion coefficient of the alloying element M relative to Fe. The width is usually

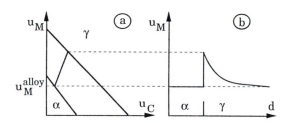

Figure 13.12 Construction giving the conditions at the phase interface for a quasi-paratransformation $\gamma \to \alpha$ in a ternary system with a mobile component C.

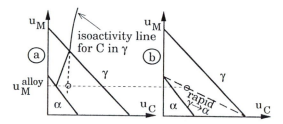

Figure 13.13 Construction giving the critical limit for a quasi-paratransformation $\gamma \to \alpha$ in a ternary system with a mobile component C.

extremely small and the alloy content in the spike originates from a transient stage of growth at a very early time.

One may normally expect that the rate is governed by the rate of carbon diffusion. In order for the transformation to proceed, carbon must diffuse from the α/γ interface and into the interior of the γ phase because the growing α phase has less carbon than γ. It is thus necessary that the carbon potential is higher at the interface than in the interior of the γ phase. The critical limit for the position of the initial γ phase is thus found on the intersection of the level of alloy content, u_M^{alloy}, and the isoactivity line for carbon in γ which goes through the γ end-point of the tie-line. This construction is shown in Fig. 13.13. Naturally, this isoactivity line must be extrapolated below the $\gamma/(\alpha + \gamma)$ boundary. The critical limit is represented by a circle in the diagrams and falls on the line for quasi-paraequilibrium in Fig. 13.11. The rapid, quasi-paratransformation can only occur on the left-hand side of that critical line, i.e. in this case below the line.

Suppose the carbon activity for a γ phase is initially lower than for the isoactivity line in Fig. 13.13. The $\gamma \to \alpha$ transformation can then start in the way described above. However, during the growth of α, the γ phase will accumulate more and more carbon. Its carbon activity will increase and eventually reach the value for the isoactivity line in Fig. 13.13. The rapid growth will then stop and the transformation can only continue at a much slower rate which permits the sluggish alloying element M to be redistributed. During this stage of slow growth there will be sufficient time for the mobile carbon to equilibrate inside the system and all the γ phase present may fall on practically the same isoactivity line for carbon in γ. If we know the composition of the interior of the

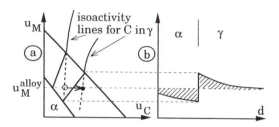

Figure 13.14 Growth conditions of α from γ at a stage where diffusion of the sluggish alloying element M is required.

γ phase we can easily find the γ end-point of a tie-line representing the equilibrium conditions on the migrating α/γ interface. We can thus construct the composition profile for the alloying element M during this stage of very slow growth (see Fig. 13.14).

The filled circle in Fig. 13.14 represents the present composition of the interior of the γ phase and the arrow indicates that its carbon content is gradually increasing during this growth. In the composition profile the spike has now widened to a considerable thickness, which may be evaluated by using the fact that the two shaded areas represent the same amount of M. The rate of growth is now governed by diffusion of the alloying element down the spike.

Let us now examine the more complicated case where the γ phase transforms to the eutectoid mixture of α and cementite, $(Fe,M)_3C$, which is called pearlite. In order for that reaction to be governed by the rate of carbon diffusion it is necessary for both α and cementite to inherit the alloy content of the parent γ. Each one of the two growing phases must fall on the correct side of its critical line. We shall first illustrate this by two separate phase diagrams in Fig. 13.15, (a) showing the α/γ equilibrium and (b) showing the γ/cementite equilibrium. The position of the region for rapid precipitation of cementite, relative to the full equilibrium phase boundary, has been drawn in agreement with the general rule that the paraequilibrium phase boundaries lie inside the full equilibrium two-phase field.

If the alloy is at a temperature where the γ → α + cementite transformation is possible, the two two-phase regions α + γ and γ + cementite must overlap to some degree and there must be a three-phase α + γ + cementite region, as illustrated in Fig. 13.16.

This phase diagram resembles the Fe–Ni–C phase diagram because Ni prefers to dissolve in γ rather than in α or cementite. The metastable parts of the phase boundaries have been drawn with dotted lines. The two critical lines form a small triangular region which we may regard as a critical triangle. Rapidly growing quasi-parapearlite can be expected to form from a γ phase situated inside the critical triangle. A γ phase situated to the left of the triangle should first precipitate so-called proeutectoid α and thus move into the triangle where pearlite can start forming. A γ phase situated to the right of the triangle should first precipitate proeutectoid cementite and then pearlite. However, a requirement is that the level of alloy content falls below the top of the triangle. Otherwise, the rapid, proeutectoid precipitation will stop at its critical

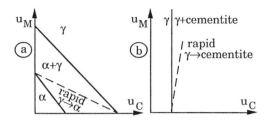

Figure 13.15 The critical limits for the quasi-paratransformation of γ to (a) α; and (b) cementite in an Fe–M–C alloy.

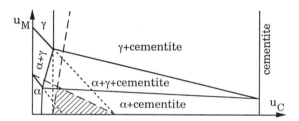

Figure 13.16 Phase diagram for an Fe–M–C system, illustrating the conditions for a rapid formation of α + cementite (so-called pearlite) under quasi-paraequilibrium (see shaded triangle).

line. The transformation can continue only by the slow rate of diffusion of the alloying element into the interior of the austenite because both α and cementite must dispose of nickel into the parent γ. That should normally be a very slow reaction and pearlite formed under such conditions has not been reported. In solidification, the growth conditions for a eutectic may be similar but the rate of diffusion in the liquid phase is so rapid that the eutectic reaction is not inhibited above the triangle for the rapid eutectic transformation.

Exercise 13.6.1

An Fe–Ni alloy with $u_{Ni} = 0.01$ is first in the state of γ (fcc) at 1273 K but then it is carburized and cementite (Fe_3C) forms isothermally. Suppose one would like to try to produce homogeneous cementite with $u_{Ni} = 0.01$. Estimate what carbon activity is required during the carburization. It is known that the γ + cementite equilibrium in the binary Fe–C system is 1.01 at 1273 K. It is also known that the distribution coefficient for Ni between cementite and γ is $K_{Ni}^{cementite/\gamma} = 0.26$.

Hint

Inspiration can be obtained from Fig. 13.15(b) because formation of cementite with the initial Ni content can only occur under a rapid reaction. However, in order to calculate the necessary carbon activity we can go to Section 10.3. The carbon content of cementite is $u_C = 1/3$ and for γ it is about 0.073.

Solution

If the growing cementite has $u_{Ni} = 0.01$ and there is local equilibrium (quasi-paraconditions) then the adjoining γ has $u_{Ni} = 0.01/0.26 = 0.038$. An equation from Section 10.3 gives $\ln(a_C/a_C^{binary}) = (1 - 0.26)/[(1/3) - 0.073] \cdot 0.038 = 0.108$; $a_C = 1.01 \cdot 1.114 = 1.125$. The carbon activity must be higher than this value.

13.7 *Transformations in steel under partitioning of alloying elements*

In the preceding section we concluded for Fe–Ni–C that, for Ni contents falling above the top of the critical triangle, pearlite could grow only by nickel diffusing into the remaining γ because both growing phases, α and cementite, require that the nickel content is lower than in the adjoining γ. The situation will be quite different if one of the phases, α or cementite, can grow with a higher alloy content than the adjoining γ. Then it would be sufficient that the alloying element diffuses side-wise and distributes itself between the two growing phases. This process can take place with an observable speed by diffusion inside the pearlite/γ interface. Interfacial diffusion can be orders of magnitude faster than volume diffusion.

Examples of alloying elements allowing pearlite to grow under partitioning between α and cementite are manganese and silicon. Cementite will attract manganese and α will attract silicon.

Exercise 13.7.1

Suppose a steel is first transformed to homogeneous γ at a high temperature and then cooled to 1023 K where it is represented by the cross in the Fe–Si–C phase diagram. The thin line is an isoactivity line for γ going through the cross. Examine if there would be a positive difference in Si content inside γ, Δu_{Si}^γ, to drive the redistribution of Si between α and cementite and allow γ to transform to pearlite. If that is the case, evaluate the fractions of α and cementite in that pearlite.

Hint

First find the intersections with the isoactivity line and the extrapolated phase boundaries for γ. Then try to find the other end-points on the corresponding tie-lines.

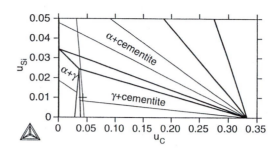

Exercise 13.7.1

Solution

The construction is shown in the following phase diagram (a) and again, with an expanded u_C scale, in diagram (b). The intersection of the isoactivity line with the two phase boundaries will give the compositions of γ at the interfaces to α and cementite if there is full local equilibrium. The difference $\Delta u_{Si}^{\gamma} = u_{Si}^{\gamma/cementite} - u_{Si}^{\gamma/\alpha}$ is positive and will thus make silicon diffuse away from cementite and to α. Pearlite can thus grow but it must have the same average Si content as the initial γ, $u_{Si} = 0.01$. Since cementite has practically no Si and α has about 0.035, we find that the fraction of α is $0.01/0.035 = 0.3$. The fraction of cementite is thus 0.7 and the carbon content of this pearlite is extremely high. It requires a large supply of carbon to be drawn from the interior of the γ phase.

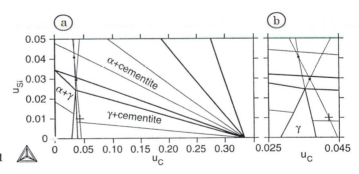

Solution 13.7.1

Limit of stability, critical phenomena and interfaces

Transformations and transitions

In Chapter 11 we were mainly concerned with the question whether a transformation is sharp or gradual. The difference between them is very practical and straight-forward. It is based on a one-dimensional phase diagram where the only axis represents the quantity that is being varied. If that diagram shows a two-phase field of some extension between the two one-phase fields, then the transformation between the two phases will be gradual. If the two-phase field has no extension, then the transformation will be sharp. For a unary system this will happen if one varies a potential, e.g. T. The Gibbs energy is a continuous function of T across the sharp transformation but its derivatives, yielding S and V, show discontinuous jumps. This is why the phase boundaries separate when a molar axis is introduced (see Figs. 8.1 and 8.2). In other cases there is no such separation because the first-order derivatives are zero. A typical example is found in a ferromagnetic substance which gradually loses its magnetization as the temperature is increased. At the Curie temperature it reaches zero and the substance has thus become paramagnetic. There is no temperature where ferromagnetic and paramagnetic regions coexist in a pure substance, not even if one varies a molar quantity. As a consequence, there is not really a two-phase field between the two one-phase fields and this fact is indicated by the use of a dashed line to separate the one-phase fields (see Fig. 14.1 where the two phases are denoted by β and β' in order to emphasize their close relationship). If there is no discontinuous jump in the first-order derivatives but there is one in the second-order derivatives, then one calls this a second-order transition as distinguished from a first-order transition when there is a jump in the first-order derivatives.

When considering a T,P diagram and calculating the locus of a sharp transformation between two phases in Section 7.3, we applied the Gibbs–Duhem relation to each of the phases, obtaining two equations for $d\mu_A$. By requiring that μ_A must change in the same way for the two phases along their line of coexistence, it was possible to calculate the slope of that line, (see Fig. 7.6). In this way dP/dT was obtained as a function of discontinuous jumps in the two first-order derivatives, ΔS and ΔV. This method does not work in the present case because the two equations become identical on the line we want to calculate, and ΔS and ΔV both go to zero there. This problem was

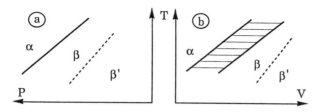

Figure 14.1 Characteristics of a second-order transition $\beta \leftrightarrow \beta'$. The $\alpha \leftrightarrow \beta$ transition is first order.

solved by Ehrenfest (1911) who instead used the fact that on the line there is no difference between V of the two states. The expression for dV,

$$dV = \left(\frac{\partial V}{\partial T}\right)_P dT + \left(\frac{\partial V}{\partial P}\right)_T dP = V\alpha dT - V\kappa_T dP$$

must give the same value in both states. Because V itself also has the same value in both states but α and κ_T do not, we get, by taking the difference between the two states,

$$\frac{dP}{dT} = \frac{\Delta\alpha}{\Delta\kappa_T}$$

It is also possible that the discontinuous jumps first occur in the third-order derivatives of G and the corresponding transition would be of third order, etc. In practice, it is often difficult to decide by experimental measurements whether $\Delta\alpha$ and $\Delta\kappa_T$ differ from zero, as they should for a second-order transition. Sometimes the individual values of the second-order derivatives appear to go to infinity at the transition point and it is not meaningful to try to evaluate their difference. Ehrenfest's expression for dP/dT is then of little practical use. Of course, it can be used when one investigates a particular model which gives definite values for second-order derivatives. In view of these complications, it is common to call all transitions with continuous first-order derivatives second order.

It should be emphasized that nothing really happens in a system when it passes a second-order transition point. It does not really transform. The only difference is that it starts behaving in a new way. The real changes in the system occur gradually as the system moves away from the point of transition. In that sense, the second-order transition is just the start of a gradual transformation. Figure 14.2 illustrates different possibilities for a pure substance at constant pressure. The internal variable ξ is some measure of the arrangements of the atoms or electrons. Curve (a) with its discontinuous jump represents a first-order transition. Curve (b) shows a discontinuous jump but also a gradual change. In curve (c) the jump has disappeared but the curve is horizontal at the break point. In curve (d) the curve never turns horizontal.

The words transformation and transition are often used as synonyms. One word is favoured by the experts in some fields and the other word in other fields. There is

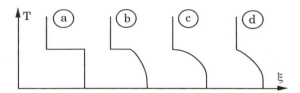

Figure 14.2 Different cases of transition and transformation. (a) A first-order transition and a sharp transformation. (b) A first-order transition where the transformation is partly sharp, partly gradual. (c) and (d) Second-order transitions and completely gradual transformations.

a need for two words with different meanings. For our purposes, it would seem most natural to use transition in much the same way as above but use transformation to describe the progress of the real changes occurring in a system. The break points in all the curves in Fig. 14.2 would thus be regarded as the transition point and the transition is the change occurring at the transition point, whether it is a real change or a change in behaviour that will reveal itself as the system moves away from the transition point. In curve (a) the first-order transition yields a sharp transformation. In curve (b) part of the transformation would be sharp and occur at the transition point, and another part would be gradual and occur below the transition point. In curves (c) and (d) the transformation would be completely gradual and start as the transition point is crossed. Thus, a transition point yielding a transformation that is at least partly sharp, would be of first order. Otherwise, it would be of second order. It should be emphasized, however, that there are many cases where the concept of first- and second-order transition does not appear to be very useful. This will be demonstrated in Section 14.4.

The strict difference between first-order and second-order transitions is of considerable theoretical interest but from a practical point of view it may sometimes be of less importance whether there is a small discontinuous jump or no jump at all. At what temperature a substance would precipitate from a particular solution would depend on how its Gibbs energy varies with temperature which, in turn, is affected by how the transformation proceeds, independent of whether the first part of it occurred with a jump or not.

In the previous chapters, only the word transformation has been used. From now on, an attempt will be made to apply both terms and with the definitions given here. In view of the conclusion drawn in Section 11.2, a sharp transformation, i.e. a first-order transition, will turn gradual when the variable is changed from a potential to a molar quantity. From the theoretical point of view this is a trivial effect and should not affect the classification of the transition. The theoretical study of phase transitions is thus carried out without involving any molar quantity.

A phase transition is often caused by a tendency of an ordered arrangement to disorder. Such transitions are called **order–disorder transitions** and the driving force comes primarily from the increasing configurational entropy. In other cases, the cause may be the lowering of the energy by deformation of the structure, e.g. by decreasing the tetragonality, without changing the configurational entropy. Such transitions are called **displacive transitions**. Of course, the characteristics of order–disorder and displacive can be applied to the corresponding transformation, as well. In both cases, the progress of the change can be expressed by some internal variable, e.g. the degree of long-range order or the tetragonality. For simplicity, all such variables are sometimes called 'order parameters' and, in principle, all internal variables could play this role.

Another method of classifying phase transitions is based on what happens to the atoms during the transition. A **reconstructive** transition involves a reorganization of the atomic arrangement with the breaking of atomic bonds and the formation of new bonds. The opposite case would be a displacive transition which involves only small adjustments of the atomic positions without the atoms ever losing contact with their neighbours. This classification thus depends on the nature of the interface migrating through the parent crystal. If the two crystal structures are closely related, one could imagine an interface so highly coherent that the atoms find their positions in the new crystal (phase) with only small adjustments of their positions relative to each other. However, it is possible in the same material that another interface is incoherent and one could not predict exactly where an atom from the parent crystal will end up in the growing crystal. The transition would then be regarded as reconstructive even if the structures of the two crystals (phases) are identical. That is the case in ordinary grain growth where large crystals consume small ones of the same phase and composition.

Furthermore, we may define a **diffusional transition** as a transition where the new phase has a different composition and can grow only under long-range diffusion. Evidently, a diffusional transition can at the same time be reconstructive or displacive. It can even be an order–disorder transition.

Exercise 14.1.1

Derive an expression for dP/dT for a second-order transition by considering the variation of S along the line of coexistence.

Hint

A Maxwell relation can be used to transform the result into well-known parameters.

Solution

$dS = (\partial S/\partial T)_P dT + (\partial S/\partial P)_T dP = (C_P/T)dT - V\alpha dP$ since $(\partial S/\partial P)_T = -(\partial V/\partial T)_P$. On the transition line, where $\Delta(dS) = 0$, we get $dP/dT = \Delta C_P/VT\Delta\alpha$.

14.2   *Order–disorder transitions*

Let us consider an ordering phenomenon in a phase with a crystal symmetry such that the properties can be expressed as even functions of the order parameter ξ. As demonstrated by Landau and Lifshitz (1958) the simplest form of the Gibbs energy expansion in the neighbourhood of the transition from disordered to ordered state would be

$$G_m = g_o + \tfrac{1}{2}g_{\xi\xi}\xi^2 + \tfrac{1}{24}g_{\xi\xi\xi\xi}\xi^4$$

where $g_{\xi\xi} = \partial^2 G_m/\partial\xi^2$, etc., and g_o, $g_{\xi\xi}$ and $g_{\xi\xi\xi\xi}$ may vary with temperature and composition although variations in composition will not be considered yet. In order to place the minimum of G_m in the region close to $\xi = 0$, where the G_m expression is supposed to hold, it is necessary to make $g_{\xi\xi\xi\xi} > 0$. In order to predict an ordered state at low temperatures but not at high, it is necessary to assume that $g_{\xi\xi}$ is negative at low temperatures and positive at high. The equilibrium value ξ_e can be found from

$$dG_m/d\xi = g_{\xi\xi}\xi + \tfrac{1}{6}g_{\xi\xi\xi\xi}\xi^3 = 0$$

The disordered, high-temperature state is described by $\xi_e = 0$. At low temperature there are two other solutions

$$\xi_e = \pm(-6g_{\xi\xi}/g_{\xi\xi\xi\xi})^{1/2}$$

By symmetry these two solutions are physically equivalent. They only exist as long as $g_{\xi\xi} < 0$ and they approach $\xi_e = 0$ as $g_{\xi\xi}$ approaches 0. The transition point would thus be given by $g_{\xi\xi} = 0$. Below the temperature where this occurs, the solution $\xi_e = 0$, representing a disordered state, would give a G_m maximum and the disordered state would thus be unstable here. Figure 14.3(a) demonstrates the shape of G_m at temperatures above and below the transition point, T_{tr}. Figure 14.3(b) shows how ξ_e, obtained from the minima, varies with temperature.

Let us now examine how we can calculate the limit of stability for the disordered state. The condition would be simply

$$\left(\frac{\partial^2 G_m}{\partial\xi^2}\right)_{T,P,\text{comp.},N_1} = 0$$

since we have decided not to consider variations in composition yet. We get directly

$$g_{\xi\xi} + \tfrac{1}{2}g_{\xi\xi\xi\xi}\xi^2 = 0$$

but the stability condition can only be applied to states of equilibrium. Thus, we must insert the equilibrium value, which is $\xi_e = 0$ for the disordered state, yielding the limit of stability at $g_{\xi\xi} = 0$ for the disordered state when cooled from a high temperature. This limit of stability thus falls on the transition point. When inserting the expression for ξ_e in the ordered state we find

$$g_{\xi\xi} = -\tfrac{1}{2}g_{\xi\xi\xi\xi}\xi_e^2 = 3g_{\xi\xi}; \quad \text{Thus, } g_{\xi\xi} = 0$$

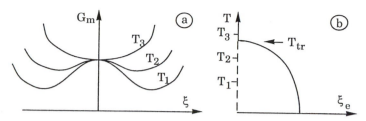

Figure 14.3 Molar Gibbs energy diagram illustrating the properties of a substance showing a second-order transition. The dashed line in (b) represents unstable states of equilibrium, shown as points of maximum in (a).

The limit of stability for the ordered state when heated from a low temperature also falls on the transition point. That is typical of second-order transitions.

It is interesting to insert the equilibrium value ξ_e in the expression of G_m and thus to obtain Gibbs energy expressions at equilibrium for ξ.

$$G_m^{dis} = g_o \text{ in disordered state}$$
$$G_m^{ord} = g_o - \tfrac{3}{2}g_{\xi\xi}^2/g_{\xi\xi\xi\xi} \text{ in ordered state}$$

At the transition point $G_m^{dis} = G_m^{ord}$ and $dG_m^{dis}/dT = dG_m^{ord}/dT$ because $g_{\xi\xi} = 0$ there, but the second-order derivatives are different, confirming that this transition is of second-order.

In order to model a first-order transition one can either remove the symmetry by introducing a ξ^3 term or one can keep the symmetry but introduce a ξ^6 term. With the latter alternative we obtain

$$G_m = g_o + \tfrac{1}{2}g_{\xi\xi}\xi^2 + \tfrac{1}{24}g_{\xi\xi\xi\xi}\xi^4 + \tfrac{1}{720}g_{\xi\xi\xi\xi\xi\xi}\xi^6$$

In this case we must take $g_{\xi\xi\xi\xi\xi\xi} > 0$ and $g_{\xi\xi\xi\xi} < 0$. Equilibrium requires that

$$dG_m/d\xi = g_{\xi\xi}\xi + \tfrac{1}{6}g_{\xi\xi\xi\xi}\xi^3 + \tfrac{1}{120}g_{\xi\xi\xi\xi\xi\xi}\xi^5 = 0$$

One solution is the disordered, high-temperature state, $\xi_e = 0$, but one also finds low-temperature states

$$\xi_e^2 = -10g_{\xi\xi\xi\xi}/g_{\xi\xi\xi\xi\xi\xi} \pm [100(g_{\xi\xi\xi\xi}/g_{\xi\xi\xi\xi\xi\xi})^2 - 120g_{\xi\xi}/g_{\xi\xi\xi\xi\xi\xi}]^{1/2}$$

The '+' sign gives a new minimum and the '−' sign gives a maximum in between. Figure 14.4 illustrates how G_m varies with ξ above and below a temperature of equilibrium between the two states, the transition temperature, T_{tr}, where the minima fall on the same level.

It is evident that the low-temperature minimum exists only as long as

$$100(g_{\xi\xi\xi\xi}/g_{\xi\xi\xi\xi\xi\xi})^2 - 120g_{\xi\xi}/g_{\xi\xi\xi\xi\xi\xi} \geq 0$$
$$g_{\xi\xi} \leq g_{\xi\xi\xi\xi}^2/1.2g_{\xi\xi\xi\xi\xi\xi}$$

and ξ_e does not approach zero at any temperature. Wherever the transition occurs, it must occur with a discontinuous jump in ξ and will thus be of first order. The ordered

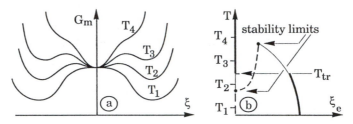

Figure 14.4 Molar Gibbs energy diagram illustrating the properties of a substance showing a first-order transition. The filled circles in (b) represent the two limits of stability. Metastable states (exhibiting a minimum with a higher G_m value than another minimum at the same temperature) are represented by thin lines, and unstable states (shown as points of maximum) by dashed line.

state can exist as a metastable state above the point of transition. The limit of stability for the ordered state is obtained from

$$d^2G_m/d\xi^2 = g_{\xi\xi} + \tfrac{1}{2}g_{\xi\xi\xi\xi}\xi^2 + \tfrac{1}{24}g_{\xi\xi\xi\xi\xi\xi}\xi^4 = 0$$

By inserting the equilibrium value ξ_e for the ordered state and solving for $g_{\xi\xi}$ we find the limit of stability where

$$g_{\xi\xi} = g_{\xi\xi\xi\xi}^2/1.2g_{\xi\xi\xi\xi\xi\xi}$$

As expected, the limit of stability occurs when the low-temperature minimum disappears by merging with the maximum and forming a point of inflexion.

By inserting the equilibrium value of ξ for the disordered state, $\xi_e = 0$, we find another limit

$$d^2G_m/d\xi^2 = g_{\xi\xi} = 0$$

The disordered state thus becomes unstable at the point where $g_{\xi\xi}$ turns negative. Between the two limits, $g_{\xi\xi} = 0$ and $g_{\xi\xi\xi\xi}^2/1.2g_{\xi\xi\xi\xi\xi\xi}$, one of the states is stable and the other is metastable. The first-order transition between the states occurs where they change roles. The exact position can be evaluated from the condition that G_m has the same value for the two states, somewhere between T_2 and T_3 in Fig. 14.4. On the other hand, in a system showing a second-order transition, a state is never metastable on the wrong side of the transition point because that is also the limit of stability and there is only one such limit.

It is worth emphasizing that for a second-order transition Landau's approach is not a special model because it only applies at small ξ values and it says nothing about the temperature dependencies of the coefficients. Any analytical model can be represented by a Taylor series expansion near the transition point and will thus predict the temperature dependencies. If $g_\xi = g_{\xi\xi\xi} = 0$ and $g_{\xi\xi\xi\xi} > 0$ at all T, and if $g_{\xi\xi}$ goes through zero at some value of T, then the model predicts a second-order transition and all the results for transition obtained from Landau's approach apply. On the other hand, if $g_{\xi\xi\xi\xi} < 0$ then the model does not predict a second-order transition but maybe a first-order transition. However, in that case the characteristics of the transition are not given completely by the properties at low values of ξ. In order to examine an analytical model of this kind, it is not enough to retain just one more term, ξ^6, in the series expansion and the result obtained above does not apply in all its details. For a first-order transition, Landau's approach with the choice of only three terms, ξ^2, ξ^4 and ξ^6, represents a special model and should be regarded just as a means of demonstrating schematically the characteristics of such a transition.

Exercise 14.2.1

Use the mathematical description of the first-order transition and calculate exactly where the transition point falls.

Hint

The two minima must have the same G_m value.

Solution

$g_o = g_o + (1/2)g_{\xi\xi}\xi^2 + (1/24)g_{\xi\xi\xi\xi}\xi^4 + (1/720)g_{\xi\xi\xi\xi\xi\xi}\xi^6$. We also have $dG_m/d\xi = g_{\xi\xi}\xi + (1/6)g_{\xi\xi\xi\xi}\xi^3 + (1/120)g_{\xi\xi\xi\xi\xi\xi}\xi^5 = 0$. Divide by ξ and subtract the first expression $(1/12)g_{\xi\xi\xi\xi}\xi^2 + (2/360)g_{\xi\xi\xi\xi\xi\xi}\xi^4 = 0$. Divide by ξ^2 and insert ξ_e^2: $15g_{\xi\xi\xi\xi}/g_{\xi\xi\xi\xi\xi\xi} - 10g_{\xi\xi\xi\xi}/g_{\xi\xi\xi\xi\xi\xi} + [100(g_{\xi\xi\xi\xi}/g_{\xi\xi\xi\xi\xi\xi})^2 - 120g_{\xi\xi}/g_{\xi\xi\xi\xi\xi\xi}]^{1/2} = 0$; $g_{\xi\xi} = (g_{\xi\xi\xi\xi})^2/1.6g_{\xi\xi\xi\xi\xi\xi}$. This is between the two limits of stability.

Exercise 14.2.2

Try to describe a second-order transition with the asymmetric expression $G_m = g_o + (1/2)g_{\xi\xi}\xi^2 + (1/6)g_{\xi\xi\xi}\xi^3 + (1/24)gg_{\xi\xi\xi\xi}\xi^4$.

Hint

Calculate the equilibrium value of ξ for the ordered state and examine if it can approach zero gradually.

Solution

The equilibrium value is obtained from $dG_m/d\xi = g_{\xi\xi}\xi + (1/2)g_{\xi\xi\xi}\xi^2 + (1/6)g_{\xi\xi\xi\xi}\xi^3 = 0$. For the ordered state we get $\xi_e = -(3/2)g_{\xi\xi\xi}/g_{\xi\xi\xi\xi} \pm [(9/4)(g_{\xi\xi\xi}/g_{\xi\xi\xi\xi})^2 - 6g_{\xi\xi}/g_{\xi\xi\xi\xi}]^{1/2}$. It is evident that ξ_e cannot approach zero gradually unless $g_{\xi\xi\xi} = 0$ which would make G_m symmetric. The asymmetric G_m expression can only describe a first-order transition. To describe a second-order transition we need a symmetric G_m function. On the other hand, a symmetric G_m function can describe a second-order or a first-order transition, as demonstrated above.

14.3   *Miscibility gaps*

It sometimes happens that a two-phase coexistence line in the T,P phase diagram ends at a critical point and this always happens for the liquid–vapour line. Above the critical point one can move continuously from a high density, characteristic of a liquid, to a low density, characteristic of a vapour (see Fig. 14.5). A similar phenomenon can occur in binary systems under constant pressure (see Fig. 14.6). Such phase fields are often called miscibility gaps and the top is called a **consolute point**.

In the binary case the condition for the stability limit would be $(\partial^2 G_m / \partial x_B^2)_{T,P} \equiv g_{BB} = 0$. However, for most compositions this would give a point falling inside the miscibility gap where the homogeneous state is not the most stable one. As explained in Section 6.2, the stability condition here defines inflexion points and Fig. 14.7 gives G_m curves for a series of temperatures demonstrating that the two inflexion points move together to a point at the top of the miscibility gap. The consolute point can thus be found by applying the stability condition already given and a new condition

$$\left(\frac{\partial^3 G_m}{\partial x_B^3}\right)_{T,P} = 0$$

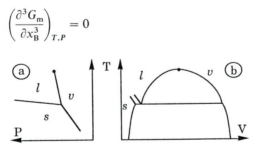

Figure 14.5 Phase diagrams for a unary system showing the liquid + vapour miscibility gap.

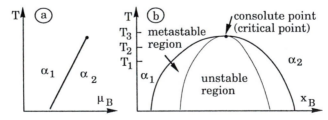

Figure 14.6 Phase diagram of a binary system at constant P, showing a solid miscibility gap.

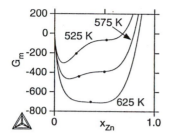

Figure 14.7 Gibbs energy curves at a series of temperatures through the fcc miscibility gap in the Al–Zn system at constant P. The inflexion points are marked with filled circles. They represent the spinodal. Just below 625 K they coincide and form a critical point.

We thus have two equations and can evaluate two unknowns, the temperature and the composition of the consolute point. For the miscibility gap in a unary system the top can be found in two ways because the limit of stability can be expressed in two ways,

$$F_{VV} = -\left(\frac{\partial P}{\partial V}\right)_T = \frac{1}{V\kappa_T} = 0 \text{ and } F_{VVV} = 0$$

$$H_{SS} = -\left(\frac{\partial T}{\partial S}\right)_P = \frac{T}{C_P} = 0 \text{ and } H_{SSS} = 0$$

The thin line in Fig. 14.6(b) is the locus of points representing the stability limit and the diagram confirms that it touches the top of the miscibility gap. The consolute point is thus the only point where the stability limit can be reached by a stable system. It is regarded as a **critical point** because two coexisting states become identical there. For all other compositions the homogeneous system turns metastable on cooling before reaching the stability limit. It should be noted that the transition point for a second-order transition is not a critical point in this sense because the ordered and the disordered states never coexist as two different phases if the transition is second-order.

The line representing the stability limit in a miscibility gap is called a **spinodal curve** or simply a spinodal or a spinode because it falls on a sharp point (*spine* meaning *thorn*) in property diagrams with potential axes. An example is shown in Fig. 14.8. In this connection it is common to call the phase boundary of the miscibility gap a **binodal**.

The critical point on a miscibility gap extends into a line when a third component is added, into a surface when a fourth component is added, etc. According to Section 5.6 the limit of stability of a multicomponent system, i.e. the spinodal, is defined by

$$\left(\frac{\partial g_c}{\partial x_c}\right)_{T,P,g_2,\ldots,g_{c-1},N} = 0$$

and the critical point is found by also applying the condition

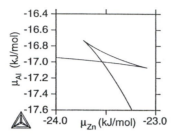

Figure 14.8 Property diagram at constant P and T for a binary system with a miscibility gap, the fcc phase in Al–Zn at 525 K.

$$\left(\frac{\partial^2 g_c}{\partial x_c^2}\right)_{T,P,g_2,\ldots,g_{c-1},N} = 0$$

It should be remembered that g_c is the notation for $(\partial G_m/\partial x_c)_{x_2,x_3,\ldots,x_{c-1}}$. In Section 5.6 it was shown that the stability condition can be transformed into such quantities, using the Jacobian method. For a ternary system the result can be written as

$$\begin{vmatrix} g_{22} & g_{23} \\ g_{32} & g_{33} \end{vmatrix} / g_{22} = g_{33} - (g_{23})^2/g_{22} = 0$$

Using the same technique the condition for a critical point can be transformed, but the result will be more complicated (see, for instance, Münster, 1970). In the ternary case it can be written as

$$g_{33} - 3g_{233}(g_{23}/g_{22}) + 3g_{223}(g_{23}/g_{22})^2 - g_{222}(g_{23}/g_{22})^3 = 0$$

For the binary case the result is simply $g_{222} = 0$ which is just a notation for $\partial^3 G_m/\partial x_B^3 = 0$.

The reason why one cannot find the critical point in a ternary case by applying $g_{22} = 0$ and $g_{33} = 0$ is that the most dangerous fluctuation may not be parallel to any of the two composition axes. It should be noted that this is the reason why the condition for the stability limit is primarily given under constant potentials, not contents.

For the binary miscibility gap it may be instructive to introduce an internal variable in order to describe the progress of the reaction as a function of temperature in a system with fixed composition. We can define an internal variable having the following equilibrium value

$$\xi_e(T) = [x^\beta(T) - x^\circ][x^\circ - x^\gamma(T)]$$

where $x^\beta(T)$ and $x^\gamma(T)$ are the equilibrium compositions on the two sides of the miscibility gap and x° is the average composition. In Fig. 14.9 this variable is plotted against temperature for three values of the average composition, x°, equal to x^1, x^2 and x^3, respectively.

It is evident that this will be a gradual transformation for all compositions and

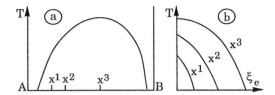

Figure 14.9 Progress of reaction in a miscibility gap for a series of compositions, demonstrating that this is a gradual transformation for all compositions, but it is not a second-order transition.

an examination of Gibbs energy would confirm that it is also a second-order transition. However, this fact does not say much about the physical character of the reaction because the same conclusion holds for any gradual transformation.

With only potential variables we would find a sharp transformation, $\alpha_1 \rightarrow \alpha_2$, on crossing the two-phase line in Fig. 14.6, for instance. However, under such conditions it would be meaningless to try to classify the transition occurring on crossing the critical point because, firstly, the chance of hitting that point would be negligible, and secondly, if we were to hit the point we would probably find that we just touch the point but stay inside the one-phase field all the time.

Exercise 14.3.1

The following part of a phase diagram was obtained when one used a thermodynamic database to calculate the hcp + liquid(L) equilibrium in a binary system. The result looks strange. Try to find the explanation.

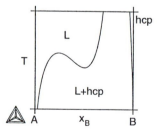

Exercise 14.3.1

Hint

The whole liquid boundary (the so-called liquidus) represents equilibrium with almost pure B. Use that fact in order to examine how μ_B varies at the points of maximum and minimum.

Solution

At the points of maximum and minimum we find $(d\mu_B/dx_B)_{T,P,N} = 0$. This looks like a stability condition. We can find μ_i in the first group of conjugate variables in Table 8.2, Section 8.2. From its first row we can formulate the following stability condition: $(d\mu_B/dz_B)_{T,P,N_A} = 0$. It is evident that the maximum and minimum would fall at the same compositions even if we had plotted the results with the z_B axis instead of the x_B axis. We may conclude that there is a liquid miscibility gap and its spinodal goes through the points of maximum and minimum.

Exercise 14.3.2

Consider a unary system with a liquid(l) + vapour(v) miscibility gap in the T,V_m phase diagram. Within the gap there is a spinodal curve, representing the limit of stability. Examine what happens to the spinodal in the diagram if P is introduced instead of V_m. Furthermore, sketch a μ_A,P property diagram at constant T.

Hint

The spinodal, has two branches. Denote the stability limit of liquid by s_l and of vapour s_v. Each one represents the end of a metastable range and should thus be situated on the 'wrong' side of the line of coexistence in the P,T diagram.

In the property diagram each phase is represented by a line and they intersect in such a way that the stable phase always has the lowest μ_A value. They only extend to their limits of stability.

Solution

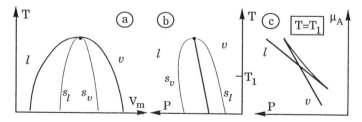

Solution 14.3.2

Exercise 14.3.3

Transform the condition for a critical point in a ternary system, $(\partial^2 g_3/\partial x_3^2)_{g_2} = 0$, to the variables x_2 and x_3 using Jacobians in order to confirm the expression given above.

Hint

The derivative should first be expressed as $\left(\dfrac{\partial}{\partial x_3}\left(\dfrac{\partial g_3}{\partial x_3}\right)_{g_2}\right)_{g_2}$ where $\left(\dfrac{\partial g_3}{\partial x_3}\right)_{g_2} =$

$\begin{vmatrix} g_{33} & g_{32} \\ g_{23} & g_{22} \end{vmatrix} / g_{22} = g_{33} - (g_{23})^2/g_{22}.$

Solution

$$\left(\frac{\partial}{\partial x_3}\left(\frac{\partial g_3}{\partial x_3}\right)_{g_2}\right)_{g_2} = \begin{vmatrix} \dfrac{\partial(g_{33} - (g_{23})^2/g_{22})}{\partial x_3} & \dfrac{\partial(g_{33} - (g_{23})^2/g_{22})}{\partial x_2} \\ \dfrac{\partial g_2}{\partial x_3} & \dfrac{\partial g_2}{\partial x_2} \end{vmatrix} \Big/ \begin{vmatrix} \dfrac{\partial x_3}{\partial x_3} & \dfrac{\partial x_3}{\partial x_2} \\ \dfrac{\partial g_2}{\partial x_3} & \dfrac{\partial g_2}{\partial x_2} \end{vmatrix}$$

$$= [g_{22}g_{333} - g_{22}\cdot 2g_{23}g_{233}/g_{22} + g_{22}(g_{23})^2 g_{223}/(g_{22})^2 - g_{23}g_{233}$$
$$+ g_{23}\cdot 2g_{23}g_{223}/g_{22} - g_{23}(g_{23})^2 g_{222}/(g_{22})^2]/g_{22}$$
$$= g_{333} - 3g_{233}(g_{23}/g_{22}) + 3g_{223}(g_{23}/g_{22})^2 - g_{222}(g_{23}/g_{22})^3 = 0$$

14.4 *Spinodal decomposition*

Thermodynamically, a system inside the spinodal is unstable with respect to compositional fluctuations and one could expect the system to decompose to a mixture of regions with the two stable compositions, one on each side of the miscibility gap. This process is called **spinodal decomposition** (Cahn, 1961). However, a fluctuation will be surrounded by a matrix of a different composition and the interfacial region where the composition varies will add some extra energy to the system. It is sometimes described as a *gradient energy*. As a result, the driving force for the process will be diminished by some amount and there may not even be a positive driving force if the fluctuation is too localized. The interfacial area-to-volume ratio must not be too large. In order to simplify the mathematics one may consider sinusoidal fluctuations in composition and then one will find a critical wavelength above which the driving force is positive. The rate of reaction will have its maximum somewhere above the critical wavelength but not very much above because the longer the wavelength, the longer the diffusion distances will be.

A mathematical treatment of this phenomenon is based on the condition for the stability limit $d^2 G_m/dx^2 = 0$. As demonstrated in Section 6.7, $d^2 G_m/dx^2$ appears in the expression for the diffusion coefficient. Thus, inside the spinodal, where d^2G_m/dx^2 is negative, the diffusion should go in the wrong direction and small fluctuations should grow. This is called up-hill diffusion. However, we should also include the contribution from the gradient energy. For small fluctuations in composition one may use the following simple approach,

$$G_m = G_m(x) + K\cdot(dx/dy)^2$$

where y is the length coordinate and dx/dy is the composition gradient. As shown by Cahn (1961), this yields the following expression to be inserted in Fick's first law and a related expression for the second law.

$$\frac{d^2G_m}{dx^2} = \frac{d^2G_m(x)}{dx^2} - 2K\frac{d^3x/dy^3}{dx/dy}$$

If the composition x varies proportional to $\sin(ky)$, the limit of stability would be found where

$$\frac{d^2G_m}{dx^2} = \frac{d^2G_m(x)}{dx^2} - 2Kk^2 = 0$$

A sinusoidal fluctuation could thus grow in amplitude if its wavelength $\lambda(=2\pi/k)$ is longer than a critical value

$$\lambda_{\text{crit.}} = 8\pi^2 K / \left(-\frac{d^2G_m(x)}{dx^2}\right)$$

Shorter wavelengths could not grow in amplitude but would shrink.

In reality, one should expect some random fluctuation in composition throughout the whole system. It is possible to describe it with a spectrum of wavelengths. In view of the above result, one could expect those that are longer than the critical wavelength to grow in amplitude and one could guess that the fastest growth may occur at about twice the critical wavelength. This value is what one could expect to find in early observations of spinodal decomposition. At longer times, one could expect a continuous coarsening.

Any sinusoidal fluctuation with a wavelength longer than the critical value would spontaneously grow in amplitude and approach a stable state which would still be periodic but no longer sinusoidal. The maxima and minima would be much flatter and could be expected to fall close to the two equilibrium compositions of the miscibility gap. In principle, such states of equilibrium can be calculated by applying the equilibrium condition to each point in the system (Hillert, 1961). If the gradient energy is included and only one direction is considered, then one will find solutions represented by a periodic variation of composition, characterized by a wavelength and an amplitude. As an example, Fig. 14.10 illustrates schematically all the solutions for a system inside the spinodal.

The critical wavelength will approach infinity as the average composition is

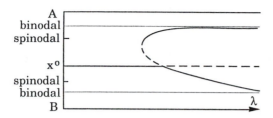

Figure 14.10 Stable (full lines) and unstable (dashed lines) equilibrium states with a periodic variation in composition. T and P are constant and the average composition is inside the spinodal.

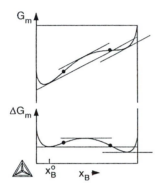

Figure 14.11 Illustration of the fact that small fluctuations in composition are not stable and would disappear if the average composition is outside the spinodal (marked with filled circles). x_B^o is the initial composition. The diagram is easier to read when the tangent to x_B^o is turned horizontally, as in the (*lower*) ΔG_m versus x_B diagram.

chosen closer and closer to the spinodal. Outside the spinodal the homogeneous system will be metastable. All small fluctuations in composition will increase the Gibbs energy as illustrated by the molar Gibbs energy diagram in Fig. 14.11. It is constructed without considering the gradient energy that would increase the Gibbs energy of the fluctuations even more.

Figure 14.12 illustrates schematically the effect of the gradient energy for the same case. The inverse of the wavelength, $1/\lambda$, is used here in order to include an infinite wavelength ($1/\lambda = 0$) in the diagram. The end-points at infinite wavelength are particularly interesting. The upper one is close to the binodal and represents the stable state where all the surplus of the minor component is concentrated in a single, local enrichment surrounded by a diffuse interface. It represents a system with a precipitated second-phase particle. The other point falls on the line for unstable equilibria and represents a system with a local critical fluctuation, termed a 'critical nucleus'.

An additional factor of importance to spinodal decomposition in crystalline phases should be mentioned. The lattice parameter of a crystalline structure usually varies with composition. Fluctuations in composition will thus give rise to internal

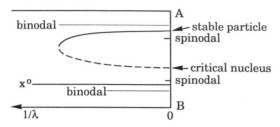

Figure 14.12 Same as Fig. 14.10 but for an average composition outside the spinodal.

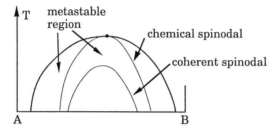

Figure 14.13 A solid-phase miscibility gap showing two spinodals.

stresses (coherency stresses) and these will increase the energy of a fluctuation. As long as the crystal is fully coherent, this effect is independent of the wavelength. The effect will be denoted by a constant C and should be added to d^2G_m/dx^2 yielding

$$\frac{d^2G_m}{dx^2} = \frac{d^2G_m(x)}{dx^2} - 2Kk^2 + C = 0$$

for the limit of stability. C is always positive and will thus stabilize the homogeneous state. It will displace the spinodal to lower temperatures. In this connection one talks about two spinodals, the coherent one, and the incoherent (or chemical) spinodal (see Fig. 14.13). In practice, the region of metastability will thus be extended.

Exercise 14.4.1

Au and Ni both have the simple fcc structure. The Au–Ni system has a miscibility gap in the fcc phase with a consolute temperature of 1083 K. A homogeneous alloy of the consolute composition, cooled from 1150 to 900 K, will decompose by a microscopically sharp eutectoid-like transformation $fcc_0 \rightarrow fcc_1 + fcc_2$ and not by spinodal decomposition. Suggest an explanation.

Solution

The Au atoms are much larger than the Ni atoms. The molar volumes of pure Au and Ni are 10.2 and 6.59 cm^3/mol, respectively. Fluctuations in composition will thus give rise to high coherency stresses. The coherent spinodal will be depressed to lower temperatures by several hundred kelvins.

Tri-critical points

In the discussion of order–disorder transitions we only considered a single composition but it is self-evident that one can represent the points for a second-order transition at all compositions in a binary system by a line. For a second-order transition, which is known to occur in one of the components, we would expect a diagram like the one in Fig. 14.14(a). However, now we should also consider the possibility of obtaining a miscibility gap by separation into regions of different compositions. In order to explore this possibility we can use Landau's simple mathematical model, discussed in Section 14.2, where all the parameters may be functions of T and x. In the disordered state $G_m = g_0$ and we shall assume that g_0 does not contain any factor favouring the formation of a miscibility gap. Then there would be no spinodal inside the region for the disordered state, i.e. above the transition line. In the ordered state we have from before

$$G_m^{ord} = g_0 - \tfrac{3}{2}(g_{\xi\xi})^2/g_{\xi\xi\xi\xi}$$

and we should consider the possibility of a spinodal reaching the transition line from below. Remembering that $g_{\xi\xi} = 0$ on the transition line, we find there

$$\partial^2 G_m^{ord}/\partial x^2 = \partial^2 g_0/\partial x^2 - 3(g_{\xi\xi x})^2/g_{\xi\xi\xi\xi} = 0$$

There would thus be a miscibility gap with its consolute point on the transition line if this expression is zero which could very well happen. The temperature and composition of the consolute point would be found by combination with $g_{\xi\xi} = 0$. The resulting phase diagram would look like the diagram in Fig. 14.14(b). This consolute point is regarded as a **tri-critical point** which is an unfortunate name. It may remind us of a triple point between three phases in a T,μ_B diagram but it has that shape only in a T,x diagram. In the T,μ_B diagram it would just be a point on a line.

It should be emphasized that we have two internal variables or 'order parameters', one describing the ordering and the other describing the separation into

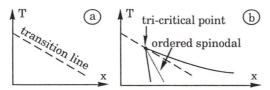

Figure 14.14 The formation of a miscibility gap around a transition line for a second-order transition.

different compositions. We could thus test the stability with respect to variations in one or the other. When testing for variations in ordering we found the transition line. However, the test of the real stability limit must take into account simultaneous variations in both internal variables. That would be the most severe test. We should look for the possibility that a system encounters the real stability limit and transforms *before* reaching the transition line. We could then find one spinodal on each side of the transition line but it would be quite a different case from that illustrated in Fig. 14.4. That case concerned a first-order transition where a system may cross the transition line and reach a limit of stability *on the other side*.

Figure 14.14(b) shows a spinodal below the transition line and it applies to homogeneous, ordered states coming from lower temperatures. We should also look for a spinodal above the transition line, applicable to the disordered state, cooled from a high temperature. It should be given by

$$\partial^2 G_m^{dis}/\partial x^2 = \partial^2 g_o/\partial x^2 = 0$$

because $\xi = 0$. As a consequence, our model which yields this relation cannot predict such a spinodal unless g_o contains a factor promoting a miscibility gap. Otherwise, $\partial^2 g_o/\partial x^2 > 0$. That is why Fig. 14.14 was constructed without a disordered spinodal. The transition line itself acts as the limit of stability line for disordered systems coming from higher temperatures. However, it is not the same type of spinodal as the lower one because it does not represent the limit of stability against compositional fluctuations. On the other hand, as soon as the system starts to order, it will find that it is above the spinodal for ordered states and is no longer stable against compositional fluctuations. It may thus be regarded as a **conditional spinodal**.

When looking for spinodals and trying to find the tri-critical point, we have here applied the condition of stability limit to the function $G_m(T,x)$ obtained with the equilibrium value of ξ inserted in $G_m(T,x,\xi)$. We could instead have used $G_m(T,x,\xi)$ directly by applying the stability condition from Section 5.8, reduced to two variables

$$\begin{vmatrix} g_{11} & g_{12} \\ g_{21} & g_{22} \end{vmatrix} = 0; \quad g_{11}g_{22} - (g_{12})^2 = 0$$

Identify Δx with variable 1 and ξ with variable 2. Then G_m from Section 14.2 gives

$$(\partial^2 g_o/\partial x^2 + \tfrac{1}{2}g_{\xi\xi xx}\xi^2 + \tfrac{1}{24}g_{\xi\xi\xi\xi xx}\xi^4)(g_{\xi\xi} + \tfrac{1}{2}g_{\xi\xi\xi\xi}\xi^2) - (g_{\xi\xi x}\xi + \tfrac{1}{6}g_{\xi\xi\xi\xi x}\xi^3)^2 = 0$$

This relation should be applied to the equilibrium value of ξ which is equal to $(-6\delta g_{\xi\xi}/g_{\xi\xi\xi\xi})^{1/2}$ in the ordered region. Neglecting the ξ^2 and ξ^4 terms in comparison to $\partial^2 g_o/\partial x^2$, and $g_{\xi\xi\xi\xi x}\xi^2$ in comparison to $6g_{\xi\xi x}$ close to the transition line, we get

$$\partial^2 g_o/\partial x^2 = (g_{\xi\xi x})^2/(g_{\xi\xi}/\xi^2 + \tfrac{1}{2}g_{\xi\xi\xi\xi}) = 3(g_{\xi\xi x})^2/g_{\xi\xi\xi\xi}$$

in full agreement with the previous result.

In the disordered region the equilibrium value is $\xi_e = 0$ and we find

$$\partial^2 g_o/\partial x^2 \cdot g_{\xi\xi} = 0$$

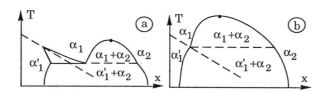

Figure 14.15 Different types of interaction between an ordering transition and a usual miscibility gap.

Above the transition line $g_{\xi\xi} > 0$ and we would find a spinodal only if $\partial^2 g_o/\partial x^2$ turns negative before the transition line is approached on cooling. That would yield a usual miscibility gap, which would interact with the one formed due to the tendency of ordering. This is illustrated in Fig. 14.15(a).

If the usual miscibility gap is larger, it may cover the other one, as illustrated in Fig. 14.15(b). At the intersection with the transition line the phase boundary shows an angle. The reason is that g_{22} appears in the denominator of the expression for $(dx_2^3/dT)_{coex}$, given in Section 10.4, and the phase boundary will thus be less steep (smaller dT/dx) below the transition line because g_{22} (i.e. $\partial^2 G_m/\partial x_2^2$) is smaller in the ordered region.

In our first calculation of the tri-critical point we started with a function $G_m(T,x,\xi)$ where ξ is an internal variable. An expression for its equilibrium value was then inserted and a function $G_m(T,x)$ was obtained which had different expressions above and below the transition line. They were then used in the calculation. Of course, one could just as well have used the results of experimental measurements in such a calculation. As a demonstration, let us suppose the effect G_m^p of an ordering reaction on the Gibbs energy has been measured across the transition line, including the effects of short- as well as long-range order. Suppose further that this effect is found to be approximately the same function of $T - T_{tr}$ for all compositions and the transition temperature, T_{tr}, varies linearly with temperature. Then

$$\frac{\partial G_m^p}{\partial x} = \frac{\partial G_m^p}{\partial T_{tr}}\cdot\frac{dT_{tr}}{dx} = -\frac{\partial G_m^p}{\partial T}\cdot\frac{dT_{tr}}{dx}$$

$$\frac{\partial^2 G_m^p}{\partial x^2} = \frac{\partial^2 G_m^p}{\partial T^2}\cdot\left(\frac{dT_{tr}}{dx}\right)^2 = -\frac{\partial C_P^p}{T}\cdot\left(\frac{dT_{tr}}{dx}\right)^2$$

where C_P^p is the effect of the ordering reaction on the heat capacity. Suppose the solution is otherwise ideal, $\partial^2 G_m^{ideal}/\partial x^2 = RT_{tr}/x(1 - x)$. The intersection of a spinodal with the transition line is found where

$$\frac{\partial^2 G_m}{\partial x^2} = \frac{RT_{tr}}{x(1 - x)} - \frac{C_P^p}{T_{tr}}\cdot\left(\frac{dT_{tr}}{dx}\right)^2 = 0$$

$$x(1 - x) = (RT_{tr}^2/C_P^p)\cdot\left(\frac{dx}{dT_{tr}}\right)^2$$

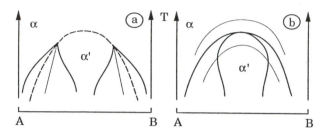

Figure 14.16 Two types of ordering miscibility gap. (a) The ordering transition is of second order and the dashed line is the transition line. The spinodals are drawn with thin lines. (b) The transition is of first order and the thin lines represent limits of stability calculated for homogeneous systems and without considering compositional fluctuations.

This would be a tri-critical point. It is thus demonstrated that the tri-critical point will be closer to the T axis of the system, the larger the heat effect is. This is an expected result but the direct role played by C_P is very interesting in view of the many measurements indicating that C_P goes to very high values close to T_{tr} and theoretical models of ordering predicting that C_P actually should approach infinity at T_{tr}. The tri-critical point would thus approach the T axis of the binary system but the miscibility gap would there be extremely thin. It is also worth noting that C_P may approach different values on the two sides of T_{tr}. The spinodals on the two sides may thus intersect the transition line at different points. The point of intersection for the upper spinodal could fall much below the other one but would move up along the transition line if there is short-range order in the disordered state above the transition line. However, it could not reach the point of intersection for the lower spinodal because the heat effect of long-range order is larger.

It is also interesting to note the role of the slope of the transition line, dT_{tr}/dx. Its effect is demonstrated in Fig. 14.16(a) which shows an ordering reaction that does not occur in the pure components but has its ideal composition in the middle of the system. Miscibility gaps with tri-critical points may appear on both sides where the transition line is steep enough. It should be emphasized that this case is not related to the case of a first-order transition which forms a complete two-phase field in a binary diagram (see Fig. 14.16(b)). This two-phase field can be regarded as two connected miscibility gaps but there is no consolute point where the two phases become identical. Instead, the point of maximum is here a congruent point where the ordered phase can transform into the disordered phase by a first-order transition as it would do if the two phases were not related structurally.

The two points representing the limit of stability for an ordering reaction of first-order, indicated in Fig. 14.4(b), also extend into lines and they also demonstrate that the congruent point is not a critical point. The disordered state is metastable well below and the ordered state is metastable well above the temperature where the ordered and disordered states of the same composition have the same Gibbs energy. However, in order to find the real limits of stability, the spinodals, one should also consider the simultaneous variation in composition. That would make no difference when cooling the disordered state because g_{12}, which is defined as $\partial^2 G_m/\partial x \partial \xi$, is zero for $\xi = 0$. On the other hand, when heating the ordered state, one may encounter a spinodal before the limit of stability calculated without considering fluctuations in composition. This possibility is not indicated in Fig. 14.14(b).

Exercise 14.5.1

The following diagram looks like a violation of the 180° rule. Try to find an explanation.

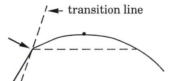

transition line

Exercise 14.5.1

Hint

Compare with Fig. 14.15(b) and apply the same type of argument to the present case.

Solution

In the present case, the lower part of the binodal is situated *above* the transition line. In both diagrams, the binodal is steeper above the transition lines than below and the explanation is the same.

Exercise 14.5.2

Examine the possibility of having a tri-critical point in the T,P phase diagram of a unary system by trying to model a change from first-order to second-order transition.

Hint

From Exercise 14.2.2 we know that we must use a symmetric G_m function in order to describe a second-order transition. It would thus be interesting to examine
$G_m = g_0 + (1/2)g_{\xi\xi}\xi^2 + (1/24)g_{\xi\xi\xi\xi}\xi^4 + (1/720)g_{\xi\xi\xi\xi\xi\xi}\xi^6$. In Section 14.2 it was used

to describe a first-order transition. Examine if the parameters can be adjusted to make ξ_e for the ordered state approach zero which could change the transition to second-order.

Solution

The expression for ξ_e^2 in Section 14.2 can approach zero if $g_{\xi\xi\xi\xi} = 0$ but the expression for $dG_m/d\xi = 0$ shows that it will happen only if $g_{\xi\xi} = 0$ at the same time. Suppose $g_{\xi\xi}$ and $g_{\xi\xi\xi\xi}$ are both functions of T and P. It should then be possible that they both go through zero in a point in the T,P diagram. On the side where $g_{\xi\xi\xi\xi} < 0$ we have a first-order transition (see thick line in the diagram) as already described in Section 14.2. Where $g_{\xi\xi\xi\xi} > 0$ the result will be much like the first case where we did not use the ξ^6 term and were able to describe a second-order transition (dashed line).

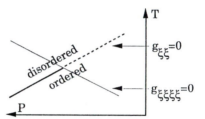

Solution 14.5.2

14.6   *Interfaces*

As an introduction to the treatment of nucleation, we shall examine the effect of an interface between two phases, α and β, in equilibrium. It will be assumed that the interface has a constant specific interfacial energy, σ, independent of composition and curvature. At equilibrium the interface should be spherical, with a radius r, and the surface energy would be $4\pi r^2\sigma = 4\pi\sigma(3V^\beta/4\pi)^{2/3}$ since $V^\beta = (4\pi/3)r^3$.

This is a convenient opportunity to introduce the use of Lagrange's multipliers for finding maxima and minima. It will be demonstrated that they can be a powerful tool in the calculation of equilibria. For a completely closed (isolated) system with $dN_i = dV = dQ = 0$, and thus $dU = 0$, the combined law yields

$$dS = (1/T)dU + (P/T)dV - \Sigma(\mu_i/T)dN_i + (D/T)d\xi = (D/T)d\xi$$

The condition of equilibrium is thus that the total system is at a maximum of S but we must find that maximum under the constant values of N_i, V and U.

$$N_i^\alpha + N_i^\beta = N_i \,(\text{constant})$$
$$V^\alpha + V^\beta = V \,(\text{constant})$$
$$U^\alpha + U^\beta + 4\pi\sigma(3V^\beta/4\pi)^{2/3} = U \,(\text{constant})$$

According to Lagrange's method we should form a new function which must have its maximum at the same time because the additional terms are always zero.

$$L = S^\alpha + S^\beta + \lambda[U - U^\alpha - U^\beta - 4\pi\sigma(3V^\beta/4\pi)^{2/3}] + v(V - V^\alpha - V^\beta)$$
$$+ \Sigma\eta_i(N_i - N_i^\alpha - N_i^\beta)$$

Here, λ, v and η_i are Lagrange multipliers and their values will be determined by maximizing L. We get six equations,

$$\partial L/\partial U^\alpha = \partial S^\alpha/\partial U^\alpha - \lambda = 0$$
$$\partial L/\partial U^\beta = \partial S^\beta/\partial U^\beta - \lambda = 0$$
$$\partial L/\partial V^\beta = \partial S^\beta/\partial V^\beta - \lambda\sigma(32\pi/3V^\beta)^{1/3} - v = 0$$
$$\partial L/\partial V^\alpha = \partial S^\alpha/\partial V^\alpha - v = 0$$
$$\partial L/\partial N_i^\alpha = \partial S^\alpha/\partial N_i^\alpha - \eta_i = 0$$
$$\partial L/\partial N_i^\beta = \partial S^\beta/\partial N_i^\beta - \eta_i = 0$$

All the derivatives of S are well known from the combined law applied to one phase at a time. We obtain

$$1/T^\alpha = \partial S^\alpha/\partial U^\alpha = \lambda = \partial S^\beta/\partial U^\beta = 1/T^\beta$$
$$\mu_i^\alpha/T^\alpha = -\partial S^\alpha/\partial N_i^\alpha = -\eta_i = -\partial S^\beta/\partial N_i^\beta = \mu_i^\beta/T^\beta$$

We have thus derived the well-known conditions of equilibrium, $T^\alpha = T^\beta = T$, and $\mu_i^\alpha = \mu_i^\beta = \mu_i$. The remaining two equations give

$$P^\alpha/T^\alpha = \partial S^\alpha/\partial V^\alpha = v = \partial S^\beta/\partial V^\beta - \lambda\sigma(32\pi/3V^\beta)^{1/3} = P^\beta/T^\beta - \lambda\sigma(32\pi/3V^\beta)^{1/3}$$
$$P^\beta - P^\alpha = \sigma(32\pi/3V^\beta)^{1/3} = 2\sigma/r$$

This is a well-known equation from physics which can be derived from a purely mechanical consideration and applies whether or not there is chemical equilibrium across the interface. For a non-spherical interface it can be written $P^\beta - P^\alpha = \sigma(1/\rho_1 + 1/\rho_2)$ where ρ_1 and ρ_2 are the principle radii of curvature.

In view of the fact that we found $T^\alpha = T^\beta$ but $P^\alpha \neq P^\beta$ it would often be wise to write the condition $\mu_i^\alpha = \mu_i^\beta$ as $\mu_i^\alpha(P^\alpha) = \mu_i^\beta(P^\beta)$. This equilibrium condition can be applied to the local equilibrium at any piece of interface. The pressure difference at equilibrium can be calculated from information on the properties of the two phases. If they are incompressible, we would find

$$\mu_i^\alpha(P^\alpha) = \mu_i^\beta(P^\beta) = \mu_i^\beta(P^\alpha) + \int V_i^\beta dP = \mu_i^\beta(P^\alpha) + (P^\beta - P^\alpha)V_i^\beta$$

and for a unary system

$$G_m^\alpha(P^\alpha) = G_m^\beta(P^\alpha) + (P^\beta - P^\alpha)V_m^\beta$$

The effect of a pressure difference was illustrated with molar Gibbs energy diagrams in Section 6.6.

Another important aspect is segregation of one or some components to the interface. In order to describe this phenomenon we shall use a very crude model.

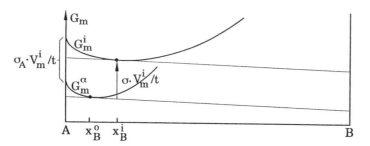

Figure 14.17 The parallel tangent construction to find the interface composition.

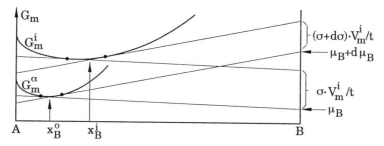

Figure 14.18 Derivation of Gibbs' adsorption equation.

Suppose the interface can be approximated as a thin layer of a homogeneous phase of constant thickness and with its own Gibbs energy function. We shall also assume that the partial molar volumes of all the phases, including the interfacial phase, are independent of composition. It is then easy to see that the composition of the interface can be found by a parallel tangent construction when the volume of the interface is constant. We cannot consider the addition of N_A but the exchange of N_A for N_B. Thus, it is the slopes of the tangents that must be equal, not their intersections with the component axes. Figure 14.17 shows a reasonable molar Gibbs energy diagram for a one-phase material with an interface between two crystals, a so-called grain boundary. The distance between the two curves on the left-hand side is equal to $\sigma_A \cdot V_m^i/t$ where σ_A is the specific interfacial energy of pure A, V_m^i is the molar volume of the interfacial phase and t is its thickness. Now, suppose the material has the composition x_B^o. The composition of the interface, x_B^i, is obtained from the parallel tangent construction and the length of the vertical arrow is equal to $\sigma \cdot V_m^i/t$ because it is the energy required for the formation of an interface of composition x_B^i from an α reservoir of composition x_B^o.

Let us now vary the composition of the material and, thus, its chemical potential for B from μ_B to $\mu_B + d\mu_B$. By comparing triangles in Fig. 14.18 one can derive an equation, called Gibbs' adsorption equation,

$$-\,d\sigma = \frac{x_B^i - x_B^o}{1 - x_B^o}\cdot\frac{t}{V_m^i}\cdot d\mu_B$$

Exercise 14.6.1

Consider two solid (α) spherical particles of a pure element, floating in a melt (L) of the same element. Derive an equation for the difference in temperature between them if they have different radii, r_1 and r_2.

Hint

The difference in radii gives them a difference in pressure. We can assume that the pressure in the melt is uniform. P_o and thus $P^\alpha = P_o + 2\sigma/r$. Start with the general equation which simplifies greatly for a pure element.

Solution

For one particle $0 = 1 \cdot (V_m^L dP^L - V_m^\alpha dP^\alpha) - 1 \cdot (H_m^L - H_m^\alpha) dT/T$ where $dP^L = 0$. We can easily integrate if all the parameters are essentially constant in the range of T and P to be considered and for small ΔT we get $\Delta T = V_m^\alpha (P^\alpha - P_o) T/(H_m^\alpha - H_m^L)$. Taking the difference between the two particles we get $T_1 - T_2 = - V_m^\alpha (P_1^\alpha - P_2^\alpha) T/(H_m^L - H_m^\alpha) = - 2\sigma V_m^\alpha (1/r_1 - 1/r_2) T/(H_m^L - H_m^\alpha)$.

Exercise 14.6.2

A liquid substance A under a low pressure (say $P = 0$) has an equilibrium vapour pressure of P_A^0. The pressure in a droplet of the same substance is higher by $\Delta P = 2\sigma/r$. How much would that increase its vapour pressure?

Hint

For a vapour it is often a good approximation to use $\mu_A^v \equiv G_m = K + RT \ln P_A$. Remember that the total pressure on the substance inside the droplet, P, is $\Delta P + P_A$. Also approximate the liquid substance as incompressible.

Solution

$\mu_A^{liq}(P) = \mu_A^v$ yields $\mu_A^{liq}(P) = L + (\Delta P + P_A) \cdot V_A^{liq} = K + RT \ln P_A$ and for $\Delta P = 0$ it gives $L + P_A^0 \cdot V_A^{liq} = K + RT \ln P_A^0$ and $P_A \cong P_A^0 \cdot \exp(\Delta P \cdot V_A^{liq}/RT)$.

Exercise 14.6.3

Consider the same problem again but suppose that the gas phase also contains an inert gas B of pressure P_B.

Hint

For a very accurate calculation one should realize that the total gas pressure is $P_A + P_B$.

Solution

Now the total pressure in the liquid A will be $\Delta P + P_A + P_B$, yielding

$$P_A = P_A^0 \cdot \exp[(\Delta P + P_B)V_A^{liq}/RT].$$

Exercise 14.6.4

Apply the regular solution model and use the parallel tangent construction to calculate the composition of a grain boundary. Examine what factor can give strong segregation.

Hint

The regular solution model gives $G_m^\alpha = x_A^\alpha \cdot {}^\circ G_A^\alpha + x_B^\alpha \, {}^\circ G_B^\alpha + RT(x_A^\alpha \ln x_A^\alpha + x_B^\alpha \ln x_B^\alpha) + L^\alpha x_A^\alpha x_B^\alpha$. We should apply the same type of model to the interface. Remember that ${}^\circ G_A^i - {}^\circ G_A^\alpha = \sigma_A V_m^i/t$ and ${}^\circ G_B^i - {}^\circ G_B^\alpha = \sigma_B V_m^i/t$. The tangent construction gives $G_A^\alpha - G_B^\alpha = G_A^i - G_B^i$.

Solution

$dG_m/dx_B = {}^\circ G_B^\alpha - {}^\circ G_A^\alpha + RT\ln(x_B^\alpha/x_A^\alpha) + L^\alpha(x_A^\alpha - x_B^\alpha) = {}^\circ G_B^i - {}^\circ G_A^i + RT\ln(x_B^i/x_A^i) + L^i(x_A^i - x_B^i); \quad RT\ln(x_A^\alpha x_B^i/x_B^\alpha x_A^i) = (\sigma_A - \sigma_B)V_m^i/t + L^\alpha(x_B^\alpha - x_A^\alpha) - L^i(x_B^i - x_A^i)$. For ordinary metals $\sigma \cong 1 \, \text{J/m}^2$, $V_m \cong 7 \, \text{cm}^3/\text{mol}$, $t \cong 10^{-7} \, \text{cm}$. At $T = 1000 \, \text{K}$ the first term on the right-hand side will be less than RT even if $\sigma_B = 0$. Strong segregation must be due to the L terms and, in particular, to a large negative value of L^i, i.e. to the tendency of A and B to mix in the interface. However, the L^i term will go to zero at $x_B^i = 0.5$. Thus, the regular solution model predicts that $x_B^i < 0.5$ even for the strongest segregation. The strongest segregation is thus predicted to give not much more than what corresponds to a monolayer.

Exercise 14.6.5

Gibbs' adsorption equation is usually written as $-\,d\sigma = \Gamma_{B(A)} \cdot d\mu_B$ where $\Gamma_{B(A)}$ is a notation for $\Gamma_B - \Gamma_A x_B^0/(1 - x_B^0)$ and Γ_B and Γ_A are the excess amounts of B and A per unit area, due to segregation. Show that it gives the same result as the equation derived graphically.

Hint

Remember that we assumed that the volume of the interfacial phase is constant. Thus, $-\Gamma_A = \Gamma_B$.

Solution

$-\Gamma_A = \Gamma_B = (x_B^i - x_B^o)\cdot t/V_m^i$; $\Gamma_{B(A)} = (x_B^i - x_B^o)/(1 - x_B^o)\cdot(t/V_m^i)$ in full agreement with our result.

14.7

Nucleation

When a system is no longer stable, it can be expected to transform to another state. In Section 14.4 we discussed how this could happen if a binary system with a miscibility gap passes the limit of stability and is no longer even metastable. Then a spontaneous reaction can be expected, started by fluctuations, but it was stressed that it may take very extensive fluctuations if the system is still close to the limit of stability. There is another mechanism possible within the metastable range and it uses fluctuations that are small in size but very intensive, i.e. represent a large change in structure or composition. Such a localized fluctuation is called a **nucleus** and the first question to ask is how large it must be in order to start growing spontaneously. In analyzing this problem we shall assume that the nucleus is a spherical volume of homogeneous composition and of a new structure, β, and also that it is separated from a homogeneous α matrix by an interface with an interfacial energy proportional to its area.

It is evident that the interfacial energy will resist the growth of a nucleus and the driving force for growth will initially be negative. If it, by some mechanism, could reach a critical size, the driving force would go through zero and then become positive and cause spontaneous growth. The system nucleus + matrix would have its maximum Gibbs energy when the nucleus has the critical size. The system would thus be in a state of unstable equilibrium and the critical radius r^* is immediately obtained from $P_e^\beta - P^\alpha = 2\sigma/r^*$. For a pure substance A, P_e^β and P^α must be such that $G_m^\beta(P_e^\beta) \equiv \mu_A^\beta(P_e^\beta) = \mu_A^\alpha(P^\alpha) \equiv G_m^\alpha(P^\alpha)$. For an alloy where the composition of the β nucleus may differ from the composition of the parent phase, one must also evaluate the composition of the nucleus $(x_A, x_B, \ldots)$ and can use the tangent construction illustrated in Fig. 6.11. Then one can evaluate P_e^β from the condition

$$G_m^\beta(P_e^\beta) = \Sigma x_i \mu_i^\alpha(P^\alpha)$$

and, finally, calculate the critical radius r^*. In the following we shall need a quantity defined as

$$\Delta G_{\mathrm{m}} = G_{\mathrm{m}}^{\beta}(P^{\alpha}) - \Sigma x_i \mu_i^{\alpha}(P^{\alpha}) = G_{\mathrm{m}}^{\beta}(P^{\alpha}) - G_{\mathrm{m}}^{\beta}(P_{\mathrm{e}}^{\beta})$$

The negative of this quantity, $-\Delta G_{\mathrm{m}}$, is sometimes called the chemical driving force for the reaction.

The second question to ask concerns the driving force for the process of nucleation. We would like to evaluate the integrated driving force from zero size up to the critical size. In principle, this will depend upon how the system is controlled. However, for the formation of a small nucleus in a large system, the differences will be negligible. We shall consider the condition of constant temperature, pressure and composition which is the most common one in experiments. Then the integrated driving force is equal to the difference in Gibbs energy between the initial and final states.

$$\int D \mathrm{d}\xi = G^{\mathrm{initial}} - G^{\mathrm{final}} = N\Sigma x_i \mu_i^{\alpha} - NG_{\mathrm{m}}^{\beta} - A\sigma$$

where $A\sigma$ is the contribution from the interfacial energy. It is important to realize that Gibbs energy is defined as $U - TS + PV$ where T and P are external variables but U, S and V must be evaluated at the actual state of compression, i.e. at P^{β} for the β phase. With the external pressure $P = P^{\alpha}$ we could thus write

$$G_{\mathrm{m}}^{\beta} = U_{\mathrm{m}}^{\beta}(P^{\beta}) - TS_{\mathrm{m}}^{\beta}(P^{\beta}) + PV_{\mathrm{m}}^{\beta}(P^{\beta}) \pm P^{\beta}V_{\mathrm{m}}^{\beta}(P^{\beta}) = G_{\mathrm{m}}^{\beta}(P^{\beta}) - (P^{\beta} - P^{\alpha})V_{\mathrm{m}}^{\beta}(P^{\beta})$$

$$\int D \mathrm{d}\xi = N\Sigma x_i \mu_i^{\alpha}(P^{\alpha}) - NG_{\mathrm{m}}^{\beta}(P^{\beta}) + (P^{\beta} - P^{\alpha}) \cdot NV_{\mathrm{m}}^{\beta}(P^{\beta}) - A\sigma$$

Inserting $\sigma = r/2(P^{\beta} - P^{\alpha})$, $A = 3V^{\beta}/r$ and $V^{\beta} = NV_{\mathrm{m}}^{\beta}$ we find

$$\int D \mathrm{d}\xi = N[\Sigma x_i \mu_i^{\alpha}(P^{\alpha}) - G_{\mathrm{m}}^{\beta}(P^{\beta})] + (P^{\beta} - P^{\alpha}) \cdot NV_{\mathrm{m}}^{\beta} - \tfrac{3}{2}(P^{\beta} - P^{\alpha}) \cdot NV_{\mathrm{m}}^{\beta}$$

$$= N[\Sigma x_i \mu_i^{\alpha}(P^{\alpha}) - G_{\mathrm{m}}^{\beta}(P^{\beta})] - \tfrac{1}{2}(P^{\beta} - P^{\alpha}) \cdot NV_{\mathrm{m}}^{\beta}$$

For the critical size, P^{β} will have the value P_{e}^{β}, and NV_{m}^{β}, which is always equal to $(4\pi/3)r^3$, is obtained by inserting the r value yielding equilibrium at P_{e}^{β}

$$N^* V_{\mathrm{m}}^{\beta} = \frac{(32\pi/3)\sigma^3}{(P_{\mathrm{e}}^{\beta} - P^{\alpha})^3}$$

The first term in $\int D \mathrm{d}\xi$ will vanish and we obtain for the **activation energy** for nucleation

$$W^* = -\int D \mathrm{d}\xi = \tfrac{1}{2}(P_{\mathrm{e}}^{\beta} - P^{\alpha}) \cdot V^* = \tfrac{1}{2}(P_{\mathrm{e}}^{\beta} - P^{\alpha}) \cdot N^* V_{\mathrm{m}}^{\beta} = -\frac{(16\pi/3)\sigma^3}{(P_{\mathrm{e}}^{\beta} - P^{\alpha})^2}$$

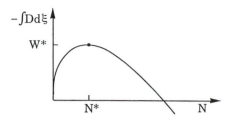

Figure 14.19 Illustration of the nucleation barrier.

As expected, the integrated driving force to form a critical nucleus is negative. There is thus a barrier to the formation of a nucleus instead of a driving force. The height of that barrier, i.e. $W^* = -\int D d\xi$, is regarded as the activation energy, W^*, because it must be provided by thermal fluctuations in order for a critical nucleus to form. The probability that this will occur is proportional to $\exp(-W^*/kT)$ and in most cases the probability is negligible unless there are preferred sites in the material where W^* has a much lower value. This phenomenon is called heterogeneous nucleation. It may even happen that the activation barrier disappears completely at a very favourable site and the reaction can then start spontaneously.

It is common to illustrate the nucleation process with a diagram showing $-\int D d\xi$ as a function of the size (see Fig. 14.19 where the size is expressed through the number of atoms, N moles). Such a diagram is easily calculated for an incompressible β phase where

$$\Delta G_m = G_m^\beta(P^\alpha) - G_m^\beta(P_e^\beta) = -(P_e^\beta - P^\alpha)V_m^\beta$$
$$\Sigma x_i \mu_i^\alpha(P^\alpha) - G_m^\beta(P^\beta) = G_m^\beta(P_e^\beta) - G_m^\beta(P^\beta) = G_m^\beta(P^\alpha) - \Delta G_m - G_m^\beta(P^\beta)$$
$$= -\Delta G_m - (P^\beta - P^\alpha)V_m^\beta$$

Using $P^\beta - P^\alpha = 2\sigma/r$ and $N V_m^\beta = (4\pi/3)r^3 = (32\pi/3)\sigma^3/(P^\beta - P^\alpha)^3$ we get

$$\int D d\xi = -N\Delta G_m - N(P^\beta - P^\alpha)V_m^\beta - (1/2)N(P^\beta - P^\alpha)V_m^\beta$$

$$= -N\Delta G_m - (3/2)N(P^\beta - P^\alpha)V_m^\beta$$
$$= -N\Delta G_m - (3/2)N V_m^\beta(32\pi/3)^{1/3}\sigma/(N V_m^\beta)^{1/3}$$
$$= -N\Delta G_m - (N V_m^\beta)^{2/3}(36\pi)^{1/3}\sigma$$

The first term represents the driving force due to the phase transformation itself (the chemical driving force) and it is a positive driving force because ΔG_m is negative. The second term is negative and comes from the resistance to the reaction caused by the interfacial energy. The point of extremum falls at

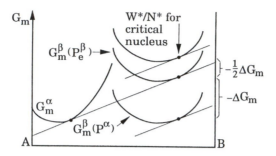

Figure 14.20 Molar Gibbs energy diagram for a critical nucleus.

$$N^* V_m^\beta = \frac{(32\pi/3)\sigma^3}{(-\Delta G_m/V_m^\beta)^3}$$

We have already seen that the activation energy for nucleation can always be written as

$$W^* = -\int D\mathrm{d}\xi = \tfrac{1}{2}(P_e^\beta - P^\alpha)\cdot N^* V_m^\beta$$

For an incompressible β phase we now obtain

$$W^* = \tfrac{1}{2}(-\Delta G_m)N^* = \tfrac{1}{2}[G_m^\beta(P_e^\beta) - G_m^\beta(P^\alpha)]N^*$$

Let us now illustrate our results with a molar Gibbs energy diagram, Fig. 14.20. First, we should draw the tangent to the α curve for the composition of the matrix. The curve for the β phase falls below that tangent by a distance $-\Delta G_m$, evaluated with the common tangent construction. The material in a critical nucleus is under the pressure P_e^β which lifts the β curve by $-\Delta G_m$ and places it right on the initial tangent. This is because the critical nucleus is in equilibrium with the matrix, even though it is an unstable equilibrium.

One may then ask how the activation energy can be illustrated in the same diagram. The answer is that it cannot be done because the activation energy is not a molar quantity. The best we can do is to illustrate W^*/N^* which happens to be $\tfrac{1}{2}(-\Delta G_m)$. We thus draw a curve by lifting the β curve a distance $\tfrac{3}{2}(-\Delta G_m)$. Of course, the parallel tangent to that curve seems to indicate that the nucleus has higher values of μ_A and μ_B than the α matrix. However, in this case the intercepts on the axes do not represent the chemical potentials. When evaluating $\mu_B = \partial G_m/\partial N_B$ for the nucleus, we must remember that N_A is kept constant, by definition. The addition of B thus makes the nucleus larger. Its internal pressure will fall and the nucleus moves down to a lower G_m curve. This is illustrated in Fig. 14.21.

When discussing spinodal decomposition in Section 14.4, we examined spontaneous initiation of the decomposition and found that an activation barrier appears as the average composition is changed across the spinodal and moves from the unstable region into the metastable region. In Fig. 14.12 the most favourable nucleus is represen-

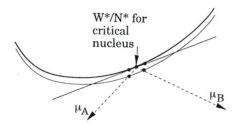

Figure 14.21 Enlarged detail of Fig. 14.20, illustrating that the critical nucleus has the correct values of the chemical potentials.

ted by a point on the line for unstable states of equilibrium at an infinite wavelength because it represents a single local fluctuation in composition. It resembles the kind of nucleus discussed in the present section but its composition cannot be obtained by the parallel tangent construction. In fact, there are two different parallel tangent constructions and they are both illustrated in Fig. 14.11. The minimum in the lower part of that diagram would give the composition of a critical nucleus if the interfacial energy is independent of composition. That point would give the largest chemical driving force which could be used to overcome the effect of interfacial energy. The maximum would give the composition of a critical nucleus if there were no interfacial energy. That point would then represent the whole activation barrier per mole of atoms in a critical nucleus. In reality, one must consider the interfacial energy and one could guess that it should vary approximately proportional to $(x^{nucleus} - x^o)^2$ and the composition of the nucleus should fall somewhere between the points of maximum and minimum. (It should be noted that the interfacial energy could hardly be proportiona to $(x^{nucleus} - x^o)$ because that would give negative values for $x^{nucleus} < x^o$.

Exercise 14.7.1

In text-books one often finds the following simple treatment of nucleation. The increase of the Gibbs energy of a system due to a nucleus is given as $\Delta G = (4\pi/3)r^3(\Delta G_m/V_m) + 4\pi r^2\sigma$ which yields: $\partial\Delta G/\partial r = 4\pi r^2(\Delta G_m/V_m) + 8\pi r\sigma = 0$; $r^* = 2\sigma/(-\Delta G_m/V_m)$; $W^* \equiv \Delta G^* = (4\pi/3)r^3(\Delta G_m/V_m) + 2\pi r^3(-\Delta G_m/V_m) = (2\pi/3)r^3(-\Delta G_m/V_m) = (16\pi/3)\sigma^3/(-\Delta G_m/V_m)^2$.

   Try to apply this result to the nucleation of a gas bubble at a temperature above the boiling point of a liquid.

Hint

Approximate the gas as ideal. The vapour pressure of the liquid, corresponding to P_e^β in our treatment, must be larger than the pressure of the liquid P^α.

Solution

$\Delta G_{\mathrm{m}} = G_{\mathrm{m}}^{\mathrm{gas}}(P^{\alpha}) - G_{\mathrm{m}}^{\mathrm{gas}}(P_{\mathrm{e}}^{\beta}) = -RT\ln(P_{\mathrm{e}}^{\beta}/P^{\alpha})$ and $V_{\mathrm{m}} = RT/P_{\mathrm{e}}^{\beta}$; $W^{*} = (16\pi/3)\sigma^{3}/[P_{\mathrm{e}}^{\beta}\ln(P_{\mathrm{e}}^{\beta}/P^{\alpha})]^{2}$. For $P_{\mathrm{e}}^{\beta} - P^{\alpha} \ll P_{\mathrm{e}}^{\beta}$ this reduces to the correct expression $W^{*} = (16\pi/3)\sigma^{3}/(P_{\mathrm{e}}^{\beta} - P^{\alpha})^{2}$ but not for larger $P_{\mathrm{e}}^{\beta} - P^{\alpha}$. The simple treatment holds for incompressible phases, only.

Methods of modelling

General principles

By 'modelling' we shall understand the selection of some assumptions from which it is possible to calculate the properties of a system. Sometimes it is possible to obtain a close mathematical expression giving a property as a function of interesting variables. In this chapter and several of the following ones we shall mainly concern ourselves with such models. However, in many cases the model cannot be expressed in a closed mathematical form but results can be obtained by numerical calculations using some iterative method. When the iteration in some way resembles the behaviour of a real physical system one talks about 'simulation'. Such methods are becoming increasingly more powerful thanks to access to more and more powerful computers.

The purpose of modelling is two-fold. From a scientific point of view one likes to learn how nature functions. One way of gaining knowledge is to define some hypothesis resulting in a model and test it by comparing the predictions from the model with experimental information. Then, it does not matter much if the predictions are made by an analytical calculation or by some numerical method. From a more techno-logical point of view one likes to predict the properties of a particular system in order to put it to efficient use in some practical construction or operation. Then it is often most convenient to have a model which yields an analytical expression.

In the simplest case, modelling is just the selection of a mathematical form which has proved useful, whether it is based on some physical model or not. However, experience shows that a model is usually more powerful if it is based upon physically sound principles. With such a model one can hope to make predictions outside the tested range with some confidence.

The first question to discuss is what thermodynamic function to model. That question will be addressed in the next section.

Exercise 15.1.1

A P,V projection of the L + gas region in the property diagram in Exercise 7.2.2 is given. There is a critical point, c. An isotherm is given and through the miscibility gap it is a straight, horizontal line representing mixtures of liquid (L₁) and gas₁ in various proportions. By the application of some model one may connect the curves for the single liquid and single gaseous phases with a curve for metastable and unstable single phase states representing a gradual change from liquid to gas (see the dashed line). Show that the model must be constructed in such a way that the two areas A_1 and A_2 are equal.

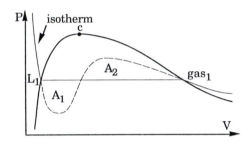

Exercise 15.1.1

Hint

Evaluate the change in Gibbs energy along the dashed line. It is an isothermal change and thus $dG = VdP - SdT = VdP$. Then remember that the Gibbs energy of L_1 and g_1 must be equal because they represent two states of the same substance in equilibrium with each other at given values of P and T.

Solution

$G(g_1) - G(L_1) = \int VdP = A_1 - A_2$. But $G(g_1) = G(L_1)$. Thus, $A_1 = A_2$. This result demonstrates that satisfactory modelling is more than just choosing a mathematical expression.

15.2   *Choice of characteristic state function*

From a practical point of view we are most interested in models giving the Gibbs energy which has temperature and pressure as its natural variables. They are usually the most convenient experimental variables. On the other hand, the physical model itself may make a different choice more natural. An example is the effect of thermal vibrations of the atoms. The frequencies depend upon the forces between atoms which in turn, depend upon the atomic distances. In that case it is most straight-forward to consider the effect under a constant volume in spite of the fact that the vibrations themselves tend to expand the system. It is thus natural to consider the Helmholtz energy. The formation of thermal vacancies is quite different. There the physical picture is that an atom is removed from the interior of a crystal and placed on the surface. The volume is thus increased by one lattice site. If the atomic distances are to be kept constant, it is now necessary to let the volume increase. It will thus be more straight-forward to consider this process under a constant pressure and to work with the Gibbs energy. It may be emphasized that one should choose to model the internal energy only for a case where it is natural to consider entropy and volume as constant and that may be very rare. On the other hand, there may be cases where the internal energy and the volume are kept constant and one could then model the entropy.

A different question is how can one model the simultaneous effects of two different phenomena. It is true that the law of additivity applies to all the extensive state functions as far as contributions from different parts of the system are concerned. However, it must be remembered that the internal energy, volume or entropy of the whole system is then the sum of the values for the parts, whereas the variables T and P are often the same in all the parts and in the whole system. If there are curved interfaces between different parts then the parts may be under different pressures and it must be remembered that the variable P in the Gibbs energy for the whole system refers to the externally applied pressure. It may then be more straight-forward to consider the Helmholtz energy but, even so, one must take into account the effect of the actual pressure inside each part because it affects the molar volume.

The situation is quite different if each one of two phenomena apply to the whole system. Each phenomenon may contribute to the total pressure which may be regarded as the sum of two partial pressures but only if the two phenomena do not interact with each other. On the other hand, the two phenomena in this case refer to the same volume.

In the present chapter we shall mainly discuss modelling by the use of Gibbs energy.

Exercise 15.2.1

What would be the most natural quantity to use in a model of evaporation of a solid?

Hint

Consider what state variables should be kept constant in order to leave the bulk of the solid unaffected by the process of evaporation.

Solution

It is not convenient to require that the volume should stay constant because then one must compress the solid to make room for the vapour. The bulk will be unaffected if the pressure is kept constant. One should thus model the Gibbs energy.

15.3   *Reference states*

As an introduction to our discussion of the modelling of the Gibbs energy it is instructive to examine some real cases. The Gibbs energy is always given relative to some reference. As an example, if one inquires about the Gibbs energy for the various forms of pure iron one may find information on all the other forms relative to the fcc form. Figure 15.1 shows how these Gibbs energy differences vary with temperature. It is a striking feature that such curves are often rather straight but in some regions the curvature is strong.

Since the negative slope of these curves is identical to the entropy difference between the two phases, we may conclude that the entropy difference is often rather constant. In regions of strong curvature there is a strong variation of the entropy difference and this should in turn be due to some special physical effect. The strong curvature in $^\circ G_{Fe}^{bcc} - {}^\circ G_{Fe}^{fcc}$ above 1000 K is due to the magnetic transition in bcc-Fe with a Curie temperature at 1043 K. It is evident that in order to model this curve one must include a description of the magnetic contribution to the Gibbs energy. Furthermore, all the curves start out parallel to the T axis at absolute zero in accordance with the third law which states that all ordered forms of a substance should approach the same entropy at absolute zero. On heating from absolute zero, a difference in entropy between the various forms develops at fairly low temperatures resulting in strong curvatures. This is due to differences in the vibration frequencies, a factor which evidently must be taken into account in modelling these curves at low temperatures. These two physical factors (magnetic and vibrational) will be discussed in some detail in the next chapter. Except for such specific physical phenomena the Gibbs energy difference is often described with fairly simple mathematical expressions. The use of a power series in T will be described in the next section.

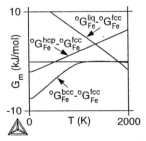

Figure 15.1 The Gibbs energy of various forms of iron given relative to the fcc form at the same temperature. The pressure is 1 bar.

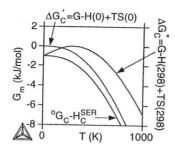

Figure 15.2 The Gibbs energy of graphite given relative to various references. The pressure is 1 bar.

Before discussing the use of a power series it is useful to examine a different way of choosing the reference for the Gibbs energy. When representing the Gibbs energy for a substance as a function of T, one can use as reference the same substance at some constant T and P. As an example, Fig. 15.2 shows the Gibbs energy of graphite as a function of T at 1 bar, and three different curves are presented because three different references have been used.

It is evident that changing the choice of reference does not always displace the curve vertically; the slope may also be affected. This is due to the fact that Gibbs energy may be regarded as composed of an enthalpy part and an entropy part and neither of these quantities has a natural zero point. In reality, one must define references for both. This can be done by choosing graphite of some temperature, e.g. 0 or 298.15 K. The quantities plotted with these choices are

$$\Delta^{\circ}G_{C}' = {}^{\circ}G_{C}(T) - {}^{\circ}H_{C}(0) + T{}^{\circ}S_{C}(0)$$
$$\Delta^{\circ}G_{C}'' = {}^{\circ}G_{C}(T) - {}^{\circ}H_{C}(298) + T{}^{\circ}S_{C}(298)$$

where 298.15 has been abbreviated as 298. The slope of the curves is obtained as

$$d\Delta^{\circ}G_{C}'/dT = - {}^{\circ}S_{C}(T) + {}^{\circ}S_{C}(0)$$
$$d\Delta^{\circ}G_{C}''/dT = - {}^{\circ}S_{C}(T) + S_{C}(298)$$

These are both satisfactory results.

It is also possible to define the references for H and S using different states. A rather popular choice is the following

$$\Delta^{\circ}G_{C} = {}^{\circ}G_{C}(T) - {}^{\circ}H_{C}(298) + T{}^{\circ}S_{C}(0)$$

It is usually combined with the convention to set ${}^{\circ}S_{C}(0)$ to zero. The last term is thus omitted. Furthermore, ${}^{\circ}H_{C}(298)$ is often chosen as the value for the element in its most stable form at 298 K and 1 bar, a quantity which is sometimes denoted by H^{SER} for Stable Element Reference. This is why the third curve in Fig. 15.2 is identified as ${}^{\circ}G_{C} - H_{C}^{SER}$. When no reference entropy is given, it means that one has in fact chosen ${}^{\circ}S_{i}(0)$ and set this quantity to zero. In any case, it is not necessary to decide on the choice of reference until one wants to put in numerical values. In the present text we shall use the notation H_i^{REF} and it mainly serves the purpose of reminding us that for each element one should normally decide on a particular choice and then stick to it. However, it is only after one has started to put in numbers that one must not change the reference.

Before leaving the discussion of reference states we should mention one more alternative. One sometimes uses $G(298)$ as reference for $G(T)$. However, $G(298)$ is just a number and could be interpreted as a reference for enthalpy chosen in such a way that $\Delta^{\circ}G_C = 0$ at 298 K when $^{\circ}S_C(0) = 0$ is the reference for entropy. This alternative may lead to misunderstandings.

For compounds we shall use the weighted average of H^{REF} for the components

$$H^{REF} = \Sigma v_i H_i^{REF}$$

and thus plot a quantity ΔG_m which is defined as $G_m - H^{REF}$. The v_i parameters are the stoichiometric coefficients for the substance, i.e. the number of moles of each component i in one mole of the substance.

Exercise 15.3.1

Find an argument why it would be less convenient to give the Gibbs energies for various forms of pure Fe relative to the bcc form instead of the fcc form.

Hint

These values can be read as the differences between the curves in Fig. 15.1.

Solution

All the curves would show the very strong curvature at about 1300 K, which is due to the properties of bcc-Fe.

Exercise 15.3.2

All the curves in Fig. 15.1 and two of the three curves in Fig. 15.2 are parallel to the T axis at the left-hand side of the diagram. What determines the slope of the third curve in Fig. 15.2?

Hint

$\Delta S = - d\Delta G/dT$.

Solution

For the third curve we get, for the slope at $T = 0$, $d\Delta G/dT = d[G - H(298) + TS(298)]/dT = dG/dT + S(298) = S(298) - S(0)$.

Exercise 15.3.3

Calculate the change of Gibbs energy on heating a substance from 273 to 373 K under a constant pressure of 1 bar. Assume C_P has a constant value.

Hint

Start by evaluating the changes in H and S separately.

Solution

$H'' - H' = \int C_P dT = 100 C_P$; $S'' - S' = \int (C_P/T) dT = C_P \ln(373.15/273.15) = 0.312 C_P$; $G'' - G' = H'' - 373.15 S'' - H' + 273.15 S' = H'' - H' - 373.15(S'' - S') - 100 S' = 100 C_P - 373.15 \cdot 0.312 C_P - 100 S' = -16.4 C_P - 100 S'$. We thus find that it is necessary to have a numerical value of S at some temperature. It is not sufficient to be able to calculate the changes in H and S. The change in G also depends upon the choice of reference for S.

15.4

Representation of Gibbs energy of formation

The molar Gibbs energy of formation of a stoichiometric compound Θ from the pure elements at any temperature and pressure is obtained by combining expressions for the compound and the component elements, respectively, if they are available,

$$\Delta_f^o G_m^\Theta = {}^o G_m^\Theta - \Sigma v_i \, {}^o G_i^?$$

This quantity is usually regarded as the standard Gibbs energy of formation of Θ and was illustrated in Fig. 6.3. The pressure is chosen as 1 bar but any one temperature can be chosen. Often one evaluates this quantity at the actual temperature of interest. ${}^o G_m^\Theta$ for the compound can be given for any type of formula unit. v_i denotes the stoichiometric coefficients of the compound according to the formula unit chosen and ${}^o G_i^?$ usually represents G_m for pure element i in the stable state at 1 bar and the temperature chosen. The quantity H^{REF} which refers to a particular temperature and pressure does not enter into the equation and one could thus describe experimental information on $\Delta_f^o G_m^\Theta$

without involving H^{REF}. However, in order to evaluate $^{\circ}G_m^{\Theta}$ from a tabulated value of $\Delta_f^{\circ}G_m^{\Theta}$ we must introduce H^{REF} through the following modification of the equation,

$$^{\circ}G_m^{\Theta} - H^{REF} = \Delta_f^{\circ}G_m^{\Theta} + \Sigma v_i(^{\circ}G_i^{\alpha} - H_i^{REF})$$

The temperature dependence of $^{\circ}G_i^{\alpha} - H_i^{REF}$ may be known for all the components through mathematical expressions, e.g. power series in T, and stored in that way in a data bank. For the Θ compound one may thus store $\Delta_f^{\circ}G_m^{\Theta}$ and the set of v_i values, or one may store an expression for the temperature dependence of the whole right-hand side of the equation, i.e. of $^{\circ}G_m^{\Theta} - H^{REF}$. A drawback with the first method is that one must also store information on the particular state α for each element that $\Delta_f^{\circ}G_m^{\Theta}$ refers to. Furthermore, $\Delta_f^{\circ}G_m^{\Theta}$ will contain all the peculiarities of the states of the component elements and may thus require a complicated mathematical representation. The second method may thus be more convenient. It may be argued that for many compounds the properties are only known in a narrow range of temperatures and could thus be adequately represented by $\Delta_f^{\circ}G_m^{\Theta}$ using a few parameters. On the other hand, the representation of $^{\circ}G_m^{\Theta} - H^{REF}$ provides a better means of extrapolation because it only involves the temperature dependence of the compound itself. It thus seems that this second method should be recommended for general use although the first method may occasionally be used, especially if the experimental information is meagre. One may then use the Neumann–Kopp rule stating that the heat capacity of a substance can be estimated as an average of the values of the components. This leads to the simple expression

$$\Delta_f^{\circ}G_m^{\Theta} = A + BT$$

In the field of oxides it is common to talk about the Gibbs energy of formation of a complex oxide from its component oxides and apply the same type of expression

$$\Delta_f^{\circ}G_m^{\text{complex oxide}} = {}^{\circ}G_m^{\text{complex oxide}} - \Sigma v_i \, {}^{\circ}G_i^{\text{component oxide}}$$

What has here been said about compounds also applies to various states of a pure element. Using a different notation for this case we can write the equation as,

$$^{\circ}G_i^{\beta} - H_i^{REF} = \Delta^{\circ}G_i^{\beta/\alpha} + {}^{\circ}G_i^{\alpha} - H_i^{REF}$$

The quantity $\Delta^{\circ}G_i^{\beta/\alpha}$ is the Gibbs energy of formation of β from α and is often called **lattice stability**. In view of the discussion above it is recommended that $^{\circ}G_i^{\beta} - H_i^{REF}$ be stored rather than $\Delta^{\circ}G_i^{\beta/\alpha}$.

Solution phases differ from stoichiometric phases by having variable composition instead of the constant stoichiometric coefficients v_i. As a consequence, it is not practical to store the properties of solution phases as $G_m - H^{REF}$. It is more convenient first to compare with the standard states chosen for the components at the same T and P. Usually one chooses the pure components in the same structure as the solution, the so-called end-members, and the quantity thus defined for the solution is the Gibbs energy of mixing

$$^{M}G_m^{\alpha}(x_i) = G_m^{\alpha}(x_i) - \Sigma x_i \, {}^{\circ}G_i^{\alpha}$$

Then one can add information on the T and P dependence of the standard states, using for instance

$$G_m^\alpha(x_i) - H^{REF} = {}^M G_m^\alpha(x_i) - \Sigma x_i({}^\circ G_i^\alpha - H_i^{REF})$$

Exercise 15.4.1

Give the lattice stability of hcp-Fe relative to fcc-Fe as a linear function of T for 800–1800 K.

Hint

The information can be obtained from Fig. 15.1 by fitting a straight line.

Solution

$$\Delta^\circ G_{Fe}^{hcp/fcc} = {}^\circ G_{Fe}^{hcp} - H^{REF} - {}^\circ G_{Fe}^{fcc} + H^{REF} = {}^\circ G_{Fe}^{hcp} - {}^\circ G_{Fe}^{fcc} = -2400 + 5T$$

Exercise 15.4.2

From a database using H^{SER} as reference, we get the following Gibbs energy values in J/mole of atoms at 1000 K: For bcc-Cr $- 36\,694$, for C as graphite $- 12\,659$, for Cr_7C_3 $- 47\,633$. Calculate the standard Gibbs energy of formation of Cr_7C_3 at 1000 K.

Hint

We must trust that a database is self-consistent and always uses the same references, whether H^{SER} or another kind. We can thus forget what references this particular database uses.

Solution

$$\Delta_f^\circ G_m^{Cr_7C_3} = {}^\circ G_m^{Cr_7C_3} - 0.7 H_{Cr}^{SER} - 0.3 H_C^{SER} - 0.7 {}^\circ G_{Cr}^{bcc} + 0.7 H_{Cr}^{SER} - 0.3 {}^\circ G_C^{graphite} +$$
$$0.3 H_C^{SER} = {}^\circ G_m^{Cr_7C_3} - 0.7 {}^\circ G_{Cr}^{bcc} - 0.3 {}^\circ G_C^{graphite} = -18\,150 \text{ J/mol.}$$

15.5   Use of power series in T

Before discussing more sophisticated models for various types of substances, it may be useful to consider the use of a power series in T and P. Let us start with terms in T. Using an ordinary power series we get

$$G_m - H^{REF} = a + bT + dT^2 + \ldots$$
$$C_P = -T\partial^2 G_m/\partial T^2 = -2dT$$

Comparison with experimental data shows that this expression for C_P is not very satisfactory and the addition of higher-power terms like T^3 to G_m does not improve the situation much. There are strong experimental indications that one should first of all add a constant term in C_P in order to describe information from above room temperature. That can be done by adding a term in $T\ln T$ to G_m:

$$G_m = a + bT + cT\ln T + dT^2$$
$$S_m = b - c - c\ln T - 2dT$$
$$H_m - H^{REF} = a - cT - dT^2$$
$$C_P = -c - 2dT$$

It could be suggested that we should have written the new term as $cT\ln(T/T_0)$ in order to make the argument dimensionless. However, it is generally agreed to express T and T_0 in kelvin and to include $-cT\ln T_0$ in the bT term.

We may conclude that in the expression for a quantity, representing a contribution to the Gibbs energy, one should normally use $cT\ln T$ as the first term after a and bT. This may simply be regarded as a mathematical model for Gibbs energy which has proved itself useful. It may also be possible to justify the $T\ln T$ term by a physical model predicting that the leading term in the heat capacity should be a constant.

When higher-power terms are needed it may seem natural to continue with a T^2 term, possibly followed by even higher powers. However, the coefficients are usually fitted to information from room temperature and up and sometimes one likes to extrapolate to temperatures above the experimental range. Terms in T^2 and T^3 may then give difficulties because they increase rapidly with temperature. For this reason, it is often preferred to use T^{-2} instead of T^2. Of course, T^{-2} will give the same kind of difficulty in extrapolations below room temperature. However, the power series is already quite inadequate at low temperatures because of the $\ln T$ term. It is thus necessary to use at least two different mathematical descriptions, one for low temperatures and one for high. The description for low temperatures will be discussed in the next chapter. For practical reasons, it would often be advantageous to choose 298 K as the break point. It must be noticed that special care must be taken to make H_m, S_m and C_P continuous at the break point.

Exercise 15.5.1

Suppose $C_P = A + BT$ where A and B are independent of P. Derive an expression for G_m.

Hint

Remember that $C_P = T(\partial S/\partial T)_P;\ \ S = -(\partial G/\partial T)_P.$

Solution

$dS = (A/T + B)dT;\ \ S(T) - S(298) = A\ln T + BT - A\ln 298 - 298B;\ \ dG =$
$- SdT = - [S(298) + A\ln T + BT - A\ln 298 - 298B]dT;\ \ G(T) -$
$G(298) = T[- S(298) + A\ln 298 + 298B] - A(- T + T\ln T) - 0.5BT^2 +$
$298S(298) - 298A - 0.5(298)^2 B.$ But $G(298) = H(298) - 298S(298);$ we thus
obtain $G(T) = H298 - TS(298) - 298A - 0.5(298)^2 B + T(A + A\ln 298 +$
$298B) - 0.5BT^2 - AT\ln T.$

Exercise 15.5.2

The modelling of heat capacity at low temperatures is based on C_V but we have
discussed the use of C_P. In order to get a feeling for the difference between C_P and
C_V it may be instructive to estimate roughly the difference between C_P and C_V at
1500 K for an element with a thermal expansion of $\alpha = 3 \cdot 10^{-5}\,\mathrm{K}^{-1}$.

Hint

Suppose this temperature is so high that $C_V = 3R$. The Grüneisen constant γ may
be estimated as 2.

Solution

$C_P - C_V = C_V T\gamma\alpha = 3R \cdot 1500 \cdot 2 \cdot 3 \cdot 10^{-5} = 0.27R.$

15.6 # *Representation of pressure dependence*

Let us now define a mathematical model for the pressure dependence by adding terms in
P to the power series representation of the Gibbs energy of a substance

$$G_m - H^{\mathrm{REF}} = a + bT + cT\ln T + dT^2 + \ldots + eP + fTP + gP^2 + \ldots$$

It yields the following expressions for other quantities, if the power series is truncated.

$C_P = - c - 2dT$
$V_m = e + fT + 2gP$
$S_m = - b - c - c\ln T - 2dT - fP$

$$H_m - H^{REF} = a - cT - dT^2 + eP + gP^2$$
$$F_m - H^{REF} = a + bT + cT\ln T + dT^2 - gP^2$$
$$U_m - H^{REF} = a - cT - dT^2 - fTP - gP^2$$

$$\alpha = f/V_m = \frac{f}{e + fT + 2gP}$$

$$\kappa_T = - 2g/V_m = \frac{- 2g}{e + fT + 2gP}$$

If we only use G_m terms up to P^2, we can invert $V_m(P)$ to $P(V_m)$ and thus replace the variable P and express the Helmholtz energy, F_m, as a function of its natural variables:

$$F_m - H^{REF} = a + bT + cT\ln T + dT^2 - (e + fT - V_m)^2/4g$$

On the other hand, we cannot replace T by an expression in terms of S_m if we use terms higher than bT. In general, it is thus impossible to get a closed mathematical expression for H_m or U_m as functions of their natural variables, which are (S_m, P) and (S_m, V_m), respectively.

From F_m we get

$$C_V = - T(\partial^2 F/\partial T^2)_V = - c - 2dT + f^2T/2g$$

The term gP^2 in G_m causes severe difficulties at high P. From the expression for V_m it is evident that g must be negative and V_m will go through zero at some high P, a result which is non-physical. Like the power series in T, a power series in P can thus be used only in a limited range.

As an example of the many alternative models suggested for the representation of the P dependence up to very high P, the following expression may be mentioned for a special reason

$$G_m - H^{REF} = a + bT + cT\ln T + dT^2$$
$$+ A[(1 + nPK)^{1 - 1/n} - 1]\exp(\alpha_0 T + 0.5\alpha_1 T^2)/K(n - 1)$$

It yields the following expression for the molar volume, which is a form of an equation named after Murnaghan (1944).

$$V_m(T,P) = A(1 + nPK)^{- 1/n}\exp(\alpha_0 T + 0.5\alpha_1 T^2)$$

It is evident that the parameter A is formally equal to the molar volume at zero T and P. Furthermore, it can be shown that K is equal to the isothermal compressibility at zero pressure and $\alpha_0 + \alpha_1 T$ can be used to represent the thermal expansivity. This model correctly predicts that the volume should decrease monotonously with increasing pressure but the volume is predicted to approach zero at infinitely high P, which is not realistic. To compensate for this one may add a constant V_0 to the expression for $V_m(T,P)$, which means that one should add a term $V_0 P$ to the G_m expression:

$$G_m - H^{REF} = a + bT + cT\ln T + dT^2 + V_0 P$$
$$+ [(1 + nPK)^{1-1/n} - 1]\exp(\alpha_0 T + 0.5\alpha_1 T^2)/K(n-1)$$

The interesting property of Murnaghan's expression for $V_m(T,P)$ is that it can be analytically solved for $P(T,V_m)$,

$$P(T,V_m) = \{A^n(V_m - V_0)^{-n}\exp[n(\alpha_0 T + 0.5\alpha_1 T^2)] - 1\}/nK$$

This is thus a rare case where $F_m(T,V_m)$ can be derived analytically from $G_m(T,P)$,

$$F_m(T,V_m) = G_m - PV_m$$

$$= H^{SER} + a + bT + cT\ln T + dT^2 + \frac{A^n(V_m - V_0)^{1-n}}{Kn(n-1)}\exp[n(\alpha_0 T + 0.5\alpha_1 T^2)]$$

$$- \frac{A}{K(n-1)}\exp(\alpha_0 T + 0.5\alpha_1 T^2) + \frac{(V_m - V_0)}{Kn}$$

Many models for the effect of P are formulated as an equation for P as a function of V_m. Integration yields an expression for the Helmholtz energy. If the equation can give the same P value for two different values of V_m, then it is, in principle, impossible to invert the equation to get $V_m(T,P)$ and by integration to get the Gibbs energy. There are substances which can thus be modelled by the use of the Helmholtz energy but not the Gibbs energy which will always give a unique V_m for each P. The critical phenomenon at the gas–liquid transition is a case which cannot be modelled by the use of a Gibbs energy expression.

Exercise 15.6.1

Examine where the power series representation of the Gibbs energy of a pure substance, given as the first equation in this section, should be truncated in order for the sum of the P terms to be PV_m.

Hint

Evaluate PV_m from the expression given for V_m.

Solution

$PV_m = eP + fTP + 2gP^2$. However, the G_m expression contains $eP + fTP + gP^2$. It is thus necessary to omit gP^2 and higher terms.

Exercise 15.6.2

Show that K in Murnaghan's equation is equal to the isothermal compressibility at zero P.

Hint

Section 2.6 gives $\kappa_T = -(\partial V_m/\partial P)_T/V_m = -(\partial \ln V_m/\partial P)_T$

Solution

$\ln V_m = \ln A - (1/n)\ln(1 + nPK) + \alpha_0 T + 0.5\alpha_1 T^2$; $\kappa_T = (1/n)\cdot 1/(1 + nPK)\cdot nK = K/(1 + nPK) \to K$ when $P \to 0$. However, this is no longer true if we add a constant V_0 to V_m.

15.7 *Application of physical models*

When a particular physical effect can be identified, it could be better described with some special mathematical expression than with a power series. The power series could still be retained for the purpose of describing other effects occurring simultaneously. We may thus divide the molar Gibbs energy into two parts, one due to the special physical effect, G_m^p, and one describing a hypothetical state without that effect, G_m^h,

$$G_m = G_m^p + G_m^h$$

or

$$G_m - H^{REF} = G_m^h - H^{REF} + G_m^p$$

We may apply some special expression for G_m^p and a power series for $G_m^h - H^{REF}$ and adjust the values of a, b, c, etc., to give a satisfactory representation of the experimental information on $G_m - H^{REF}$.

In the following chapters we shall examine a large number of different substances or phases and discuss mathematical models based upon some information on their physical or structural properties. The superscript 'p' in G_m^p will sometimes be replaced by letters referring to the particular effect under consideration.

Exercise 15.7.1

Let us examine a very simple physical model. Measurements on alloys indicate that metallic melts undergo a glass transition if undercooled to a temperature, T_{glass}, which may be about one-third of the melting point, $T_{m.p.}$. On cooling down to that temperature the melt has lost most of its excess entropy relative to the crystalline state. Below the glass temperature C_P of the amorphous phase is close to C_P of the crystalline phase. Just above the glass temperature it is much larger. At higher temperatures it decreases gradually from this large value and we shall assume that it approaches C_P of the (superheated) crystalline phase at high temperatures, a behaviour which can be modelled by the following expression: $G_m^L = G_m^{cryst} + a + bT + c\exp(-T/T_{glass})$.

Use this crude model to evaluate the difference in enthalpy between the glassy and crystalline phases at absolute zero. Evaluate the model parameters from the information given above.

Hint

Suppose the last two terms apply only above T_{glass}. Thus, $H_m^L - H_m^{cryst}$ is constant up to T_{glass} and there is practically no entropy difference. $H_m^L - H_m^{cryst}$ at absolute zero can thus be estimated as $G_m^L - G_m^{cryst}$ at the glass temperature. We must first determine the model parameters. The information on S at the glass temperature, $T_{glass} = T_{m.p.}/3$, gives one equation. The information on phase equilibrium at $T_{m.p.}$ gives a second equation and Richard's rule that the entropy of melting of ordinary metals is equal to about R gives a third one.

Solution

$S_m^L - S_m^{cryst} = -b + (c/T_{glass})\exp(-T/T_{glass}) = -b + (3c/T_{m.p.})\cdot\exp(-3T/T_{m.p.})$.
At $T = T_{m.p.}/3$: $-b + (3c/T_{m.p.})\exp(-1) = 0$.
At $T = T_{m.p.}$: $-b + (3c/T_{m.p.})\exp(-3) = R$. Therefore, $c = -(RT_{m.p.}/3)/[\exp(-1) - \exp(-3)] = -1.048RT_{m.p.}$; $b = (3c/T_{m.p.})\exp(-1) = -1.157R$.
$G_m^L = G_m^{cryst}$ at $T_{m.p.}$ gives $a + bT_{m.p.} + c\exp(-3) = 0$; $a = 1.209RT_{m.p.}$.
At $T = T_{m.p.}/3$: $G_m^L - G_m^{cryst} = 0.438RT_{m.p.} \cong H_m^L - H_m^{cryst}$ at 0 K.

15.8   Ideal gas

In order to demonstrate the principles of modelling, we shall now consider gases. The simplest model of a gaseous element A is defined by the following expression,

$$G_m = {}^\circ G_A(T,P_0) + RT\ln(P/P_0)$$

This expression is defined for one mole of gas molecules. ${}^\circ G_A(T,P_0)$ is the value of G_m at any temperature but at a reference pressure usually chosen as $P_0 = 1$ bar $= 100\,000$ Pa. ${}^\circ G_A(T,1 \text{ bar})$ is usually expressed as a power series $K(T)$, including the term $-RT\ln P_0$ which is equal to $-RT\ln 10^5$ if one uses the SI unit, pascal (Pa).

$$G_m - H^{REF} = K(T) + RT\ln P$$

It should be remembered that one must express P in thé term $RT\ln P$ in the same unit that was used in evaluating $K(T)$.

Using the standard procedures we find that this mathematical model yields

$$V_m = RT/P$$
$$S_m = -K'(T) - R\ln P$$
$$F_m - H^{REF} = K(T) + RT\ln P - PV_m = K(T) - RT + RT\ln P$$
$$U_m - H^{REF} = K(T) - RT + RT\ln P + TS_m$$
$$= K(T) - RT + RT\ln P - TK'(T) - TR\ln P$$
$$= K(T) - RT - TK'(T)$$

where $K' = dK/dTV$. This is a very useful model because it has been found that many gases have an internal energy which is a function of T but varies very little with P or V under constant T, and they also satisfy the expression for V_m very well. In fact it seems that all gases approach this model at low enough pressures.

This model is regarded as the model for an **ideal gas** and $PV_m = RT$ is called the ideal gas law. It is usually written for N moles of gas molecules (not N moles of atoms),

$$PV = NRT$$

We can easily express the Helmholtz energy as a function of its natural variables

$$F_m - H^{SER} = K(T) - RT + RT\ln(RT/V_m)$$

It is sometimes convenient to express $K(T)$ as a power series in T and from Section 15.5 we remember that it should include a $T\ln T$ term. We may thus write the model for an ideal gas as

$$G_m - H^{REF} = a + bT + cT\ln T + dT^2 + eT^3 + \dots + RT\ln P$$

and by standard procedures we obtain

$$S_m = -b - c - c\ln T - 2dT - \dots - R\ln P$$
$$V_m = RT/P$$
$$F_m - H^{REF} = a + (b - R)T + cT\ln T + dT^2 + eT^3 + \dots + RT\ln P$$
$$H_m - H^{REF} = a - cT - dT^2 - 2eT^3 - \dots$$
$$C_P = -c - 2dT - 6eT^2 - \dots$$
$$U_m - H^{REF} = a - (c + R)T - dT^2 - 2eT^3 - \dots$$
$$C_V = -c - R - 2dT - 6eT^2 - \dots = C_P - R$$

Sometimes one defines an **ideal classical gas** by further requiring that C_V should be independent of T, which means that d, e and all higher coefficients must be zero.

For monatomic gases the 'ideal classical value' of C_V is $1.5R$ and for diatomic gases it is $2.5R$. Values found experimentally for diatomic gases confirm that they can often be approximated as ideal but not as ideal classical.

Exercise 15.8.1

Suppose a numerical expression for $K(T)$ has been evaluated for a gas which obeys the ideal gas model very well. The evaluation was made with P measured in pascal. How should one change the numerical expression if one would like to give P in bar (1 bar = 100 000 pascal)?

Hint

P(in pascal) = 100 000·P(in bar)

Solution

$G_m - H^{SER} = K(T) + RT\ln[P(\text{in pascal})] = K(T) + RT\ln[100\,000 \cdot P(\text{in bar})] = K(T) + 11.51RT + RT\ln[P(\text{in bar})]$

Exercise 15.8.2

A thermally insulated container has two compartments of volumes V_1 and V_2. A gas is contained in V_1 but V_2 is empty. Suddenly, the wall between the two compartments is removed. Calculate the change in T of the gas when it has come to rest in the whole volume. Assume that the gas is ideal.

Hint

There is no exchange of heat or work with the surroundings.

Solution

The internal energy has not changed, and since the internal energy is only a function of T and not of P, we realize that T cannot change.

15.9   *Real gases*

The properties of a real gas can sometimes be approximated by a model obtained by adding a power series in P to the ideal gas model,

$$G_m - H^{REF} = K(T) + RT\ln P + LP + MP^2 + NP^3 + \ldots$$

where L, M, N etc., may depend on T. By standard procedures we obtain

$$V_{\mathrm{m}} = RT/P + L + 2MP + 3NP^2 \ldots$$
$$F_{\mathrm{m}} - H^{\mathrm{REF}} = G_{\mathrm{m}} - H^{\mathrm{REF}} - PV_{\mathrm{m}} = K(T) + RT\ln P - RT - MP^2 - 2NP^3 - \ldots$$

It is evident that the V_{m} expression cannot be inverted to $P(V_{\mathrm{m}})$ and it is thus impossible to derive an analytical expression for $F_{\mathrm{m}}(T,V_{\mathrm{m}})$. However, by introducing the LP term as the only pressure term in G_{m} one obtains an expression for V_{m} which can be inverted

$$P = RT/(V_{\mathrm{m}} - L)$$

and we can express F_{m} in its natural variables,

$$F_{\mathrm{m}} - H^{\mathrm{REF}} = K(T) - RT + RT\ln\frac{RT}{(V_{\mathrm{m}} - L)}$$

A gas obeying this model is sometimes called a **slightly imperfect gas**.

It is sometimes convenient to treat real gases by introducing a new quantity f, called **fugacity**, through the expression

$$G_{\mathrm{m}} - H^{\mathrm{SER}} = K^f(T) + RT\ln f$$

where $K^f(T)$ has been chosen in such a way that f approaches P for small P. Using this concept one can temporarily treat any gas as if it were ideal and postpone the introduction of its real properties until later. Sometimes one introduces the **fugacity coefficient**, f/P, and it is evident that it approaches the value 1 at low P. However, if this approach is applied to a gas with molecules, e.g. O_2, it should be realized that there may be some dissociation into atoms $O_2 \rightarrow 2O$ which would increase the pressure. This effect would be more pronounced at low pressures. The concept of fugacity should thus be applied to each gas species separately.

So far we have based the modelling of gases on the Gibbs energy, using T and P as the variables. However, it is sometimes more convenient to use T and V_{m} as the variables and thus to define the model using the Helmholtz energy. One may for instance define a model with the following expression where $K(T)$ is not the same as before,

$$F_{\mathrm{m}} - H^{\mathrm{REF}} = K(T) + RT[-\ln V_{\mathrm{m}} + B_2/V_{\mathrm{m}} + B_3/2V_{\mathrm{m}}^2 + \ldots]$$

which gives

$$P = RT[1/V_{\mathrm{m}} + B_2/V_{\mathrm{m}}^2 + B_3/V_{\mathrm{m}}^3 + \ldots]$$

B_2, B_3, etc., are called virial coefficients. There is no exact relation between these coefficients and those introduced through the G_{m} expression, but in order to compare them we can write the expressions in the following forms

$$PV_{\mathrm{m}} = RT + LP + 2MP^2 + 3NP^3 + \ldots$$
$$PV_{\mathrm{m}} = RT + RTB_2/V_{\mathrm{m}} + RTB_3/V_{\mathrm{m}}^2 + \ldots$$

For small values of the coefficients, we may first approximate $1/V_{\mathrm{m}}$ by P/RT. By introducing this in the second term and dropping the last term we get

$$PV_m = RT + RTB_2P/RT = RT + B_2P$$

and we can use

$$1/V_m = 1/(RT/P + B_2) = (P/RT)/(1 + B_2P/RT) \cong P/RT - B_2(P/RT)^2$$

as a better approximation. By introducing this we get

$$PV_m = RT + B_2P + (B_3 - B_2^2)P^2/RT$$

when omitting higher-order terms. We may thus approximate L as B_2 and M as $(B_3 - B_2^2)/2RT$. Near the critical point for the transition between a gas and its liquid, the coefficients are too large for such approximations. There the modelling with $G_m(T,P)$ becomes increasingly more difficult and below the critical point it breaks down because a function $V_m(T,P)$ cannot give the two different V_m values required for given T and P, one for the gas and one for the liquid. This is a case where one must define the model through $F_m(T,V_m)$.

Another model based on $F_m(T,V_m)$ is the following

$$F_m - H^{REF} = K(T) - a/V_m - RT\ln(V_m - b)$$

It gives

$$P = -a/V_m^2 + RT/(V_m - b)$$

which can be rearranged into

$$(P + a/V_m^2)(V_m - b) = RT$$

This expression was proposed by van der Waals. The terms a/V_m^2 and $-b$ may be regarded as corrections to the ideal gas law. This expression correctly predicts a critical point below which a system decomposes into a liquid and a gas. However, it does not describe real systems with any accuracy. Many improvements of the van der Waals equation have been suggested but they will not be discussed here.

Exercise 15.9.1

In a so-called throttling or Joule–Thomson experiment a gas of a constant pressure P_1 is allowed to flow through a porous plug into a cylinder kept at a constant pressure P_2. It is found that the temperatures may then be different if the two subsystems are thermally insulated from the surroundings and from each other. Calculate the temperature difference in terms of the parameters used to describe the properties of a slightly imperfect gas. Suppose the change $P_2 - P_1$ is small.

Hint

Start by applying the first law to derive a general expression for the change in the

state of the gas. The only exchange of energy with the surroundings comes from work done on the gas at pressure P_1 and by the gas at pressure P_2. In order to simplify the expressions use $(\partial H/\partial T)_P = C_P$ and
$(\partial H/\partial P)_T = \partial^2(G_m/T)/\partial(1/T)\partial P = L - TdL/dT$ for $G_m = K(T) + RT\ln P + LP$.

Solution

Consider an amount of gas moving from state 1 to state 2 since P_1 and P_2 are kept constant during the experiment. We get $U_2 - U_1 = P_1V_1 - P_2V_2$ and thus $H(P_1,T_1) = H(P_2,T_2)$. The enthalpy of the gas is thus constant and for a small change in P we get
$dH = (\partial H/\partial T)_P dT + (\partial H/\partial P)_T dP = C_P dT + (L - TdL/dT)dP = 0$, so that

$$\frac{dT}{dP} = \frac{-L + TdL/dT}{C_P} = \frac{-1}{C_P}\frac{d(L/T)}{d(1/T)}.$$

Exercise 15.9.2

Derive an expression for the fugacity of a gas obeying the van der Waals equation of state. Then calculate the fugacity coefficient, f/P.

Hint

$G_m - H^{REF} = F_m - H^{REF} + PV_m = K(T) - a/V_m - RT\ln(V_m - b) - a/V_m + RTV_m/(V_m - b)$ should be compared with $G_m - H^{REF} = K^f(T) + RT\ln f$. At small P, and thus large V_m, the two expressions should be equal if f is replaced by P. For large V_m we find that $G_m - H^{REF}$ goes towards $K(T) - RT\ln RT + RT\ln P + RT$. We should thus identify $K^f(T)$ with $K(T) - RT\ln RT + RT$.

Solution

For any P we can write $K(T) - RT\ln RT + RT + RT\ln f = K(T)2a/V_m - RT\ln(V_m - b) + RTV_m/(V_m - b)$ and we get $RT\ln f = -RT\ln[(V_m - b)/RT] + RTb/(V_m - b) - 2a/V_m$. Thus we obtain $f = RT/(V_m - b)\cdot\exp[b/(V_m - b) - 2a/RTV_m]$. By dividing with $P = -a/V_m^2 + RT/(V_m - b)$ we finally find $f/P = [1 - a(V_m - b)/RTV_m^2]^{-1}\cdot\exp[b/(V_m - b) - 2a/RTV_m]$.

15.10   *Mixtures of gas species*

Gases dissolve in each other so readily and with so little interaction that the resulting phase is regarded as a mixture of the component species (molecules). In principle, it may also be regarded as a solution. We shall first discuss a mixture of several species, each one of which can form an ideal gas. Let y_k be the fraction of species k and thus $\Sigma y_k = 1$. Let us consider a mixture of one mole of species. Each gas will fill the complete volume V, but in line with the ideal gas concept we shall assume that there is no interaction between them. We shall thus assume that the ideal gas law applies to the mixture, $PV_m = RT$, and that the chemical potential of a component k, μ_k, has the same value it would have, had it been alone in the same volume V. The pressure of that component would then be $P_k = y_k RT/V = y_k P$ where P is the pressure of the mixture

$$\mu_k = G_m(k) = {}^\circ G_k(T,P_0) + RT\ln P_k$$
$$= H_k^{\mathrm{REF}} + K_k(T) + RT\ln P_k = H^{\mathrm{REF}} + K_k(T) + RT\ln y_k + RT\ln P$$

and we get for the mixture

$$G_m = \Sigma y_i\mu_i = \Sigma y_i{}^\circ G_i(T,P_0) + RT\Sigma y_i\ln P_i$$
$$= H^{\mathrm{REF}} + \Sigma y_i K_i(T) + RT\Sigma y_i\ln y_i + RT\ln P$$
$$G_m - H^{\mathrm{REF}} = \Sigma y_i K_i(T) + RT\ln P + RT\Sigma y_i\ln y_i$$

The last term may be regarded as the contribution from the mixing of different molecules. It is proportional to T and is thus of pure entropy character. $-R\Sigma y_i\ln y_i$ may be regarded as the ideal entropy of mixing.

For a mixture of ideal gases it can be imagined that gas k actually has the pressure P_k in the gas mixture and that the total pressure of the mixture is the sum of the pressures of the individual gases. We have thus used

$$\Sigma P_i = \Sigma y_i P = P$$

P_k is regarded as the **partial pressure** of gas k in the mixture. For real gases it is no longer possible to define the partial pressure in a strict sense but it is common to use this concept and the relation $P_i = y_i P$ even in such cases. As a first attempt to model G_m for a mixture of real gases, one could add the terms for slightly imperfect gases in the form $\Sigma y_i L_i P_i$ which can be written as $\Sigma y_i^2 L_i P$. The second power of y_i indicates that these terms describe the interaction of molecules with other molecules of the same kind. It would be natural also to add terms representing interactions between different kinds of molecules, $\Sigma\Sigma y_i y_j L_{ij} P$. We thus get

$$G_m - H^{\mathrm{REF}} = \Sigma y_i K_i(T) + RT\ln P + RT\Sigma y_i\ln y_i + P\Sigma\Sigma y_i y_j L_{ij}$$

If all interactions are equal, the last sum of terms becomes $PL\Sigma\Sigma y_i y_j = PL$ which is an attractive feature of this model. It gives the following expression for the molar volume

$$V_m = RT/P + \Sigma\Sigma y_i y_j L_{ij}$$

It is interesting to calculate what pressure, P', the same amount of gas k would have if it were alone in the same volume. For pure k we get

$$°G_k - H_k^{REF} = K_k(T) + RT\ln P' + P'L_{kk}$$

The molar volume would be $RT/P' + L_{kk}$ and the volume of the actual amount y_k would be

$$V = y_k(RT/P' + L_{kk})$$

However, this is supposed to be equal to the molar volume of the mixture,

$$y_k(RT/P' + L_{ki}) = RT/P + \Sigma\Sigma y_i y_j L_{ij}$$

We thus find

$$1/P' = 1/y_k P + (\Sigma\Sigma y_i y_j L_{ij} - y_k L_{kk})/RTy_k$$
$$= 1/P_k + (\Sigma\Sigma y_i y_j L_{ij} - y_k L_{kk})/RTy_k$$

by introducing $P_k = y_k P$. It is evident that the partial pressure defined as $y_k P$ is no longer equal to the pressure of the same amount of pure k, unless all the interaction are zero. The concept of partial pressure is thus less useful for real gas mixtures than for ideal ones. The concept of fugacity is more useful. For a component k in a mixture it is defined from

$$\mu_k - H^{REF} = K_k(T) + RT\ln f_k$$

where $K_k(T)$ is formulated in such a way that f_k approaches $y_k P$ at low P. In order to evaluate the fugacity from a particular model one must first derive an expression for μ_k from the model. For the present case we would get

$$\mu_k - H^{REF} = K_k(T) + RT\ln(y_k P) + P[2y_k\Sigma L_{kj} - \Sigma\Sigma y_i y_j L_{ij}]$$

and for the fugacity coefficient we obtain

$$f_k/P_k = f_k/y_k P = \exp(P[2y_k\Sigma L_{kj} - \Sigma\Sigma y_i y_j L_{ij}]/RT)$$

For an ideal gas mixture we get

$$\mu_k - H^{REF} = K_k(T) + RT\ln P + RT\ln y_k$$
$$f_k/P_k = f_k/y_k P = 1$$

A similar model based upon the Helmholtz energy is generally written as

$$P = RT[1/V_m + (\Sigma y_i^2 B_{ii} + \Sigma\Sigma y_i y_j B_{ij})/V_m^2 + \ldots]$$
$$= RT[1/V_m + (\Sigma y_i B_{ii} + \Sigma\Sigma y_i y_j(B_{ij} - (B_{ii} + B_{jj})/2))/V_m^2]$$

and it is often found that $B_{ij} - (B_{ii} + B_{jj})/2$ is small.

Exercise 15.10.1

Show how one can calculate the heat of reaction for C(graphite) +

O_2(gas) → CO_2(gas) where O_2 and CO_2 are taken from a gas mixture which is ideal.

Hint

We want $\Delta H_m = H_{CO_2} - {}^{\circ}H_C^{gr} - H_{O_2}$ where H_{CO_2} and H_{O_2} are partial quantities for the gas.

Solution

$G_m^{gas} = y_{CO_2}{}^{\circ}G_{CO_2} + y_{O_2}{}^{\circ}G_{O_2} + RT[y_{CO_2}\ln y_{CO_2} + y_{O_2}\ln y_{O_2}) + RT\ln P$ yields
$H_m^{gas} = \partial(G_m^{gas}/T)/\partial(1/T) = y_{CO_2}{}^{\circ}H_{CO_2} + y_{O_2}{}^{\circ}H_{O_2}$ and $y_{O_2} = 0$ and 1 yields
$H_{CO_2} = {}^{\circ}H_{CO_2}$ and $H_{O_2} = {}^{\circ}H_{O_2}$, respectively. The result is
$\Delta H_m = H_{CO_2} - {}^{\circ}H_C^{gr} - H_{O_2} = {}^{\circ}H_{CO_2} - {}^{\circ}H_C^{gr} - {}^{\circ}H_{O_2}$.

Exercise 15.10.2

Examine a slightly imperfect gas with two kinds of molecules, A and B. Is there any relation between the L coefficients which would make the fugacity coefficients independent of the composition?

Hint

Express the composition dependence in y_B only.

Solution

$\ln(f_A/P_A) = (P/RT)\cdot(2y_AL_{AA} + 2y_BL_{AB} - y_Ay_AL_{AA} - 2y_Ay_BL_{AB} - y_By_BL_{BB}) = (P/RT)\cdot[(2 - 2y_B - 1 + 2y_B - y_{BB}^2)L_{AA} + (2y_B - 2y_B + 2y_B^2)L_{AB} + (-y_B^2)L_{BB}] = (P/RT)\cdot[L_{AA} - (L_{AA} + L_{BB} - 2L_{AB})y_B^2]$. We thus find that $L_{AA} + L_{BB} - 2L_{AB} = 0$ makes f_A/P_A independent of composition. By symmetry reasons, it also makes f_B/P_B independent of composition.

15.11   *Black-body radiation*

It is illustrative to compare the ideal gas behaviour with the properties of black-body radiation and with an electron gas. For the former we may start from a result from quantum mechanics saying that the energy of black-body radiation is proportional to the volume and T^4. The constant of proportionality, a, is called the Stefan constant and it has a value of $8\pi^5 k^4/15c^3 h^3 = 7.57 \cdot 10^{-16}$ J/m^3 K^4, where k is Boltzmann's constant, c is the speed of light and h is Planck's constant.

$$U = aT^4 V$$

However, in order to derive other quantities we need $U(S,V)$. By simple manipulations we obtain

$$C_V = \left(\frac{\partial U}{\partial T}\right)_V = 4aT^3 V$$

$$S = \int_0^T \frac{C_V}{T} dT = \frac{4}{3} aT^3 V; \quad T^3 = 3S/4aV$$

$$U(S,V) = a(3S/4aV)^{4/3} V = (3S/4)^{4/3}/(aV)^{1/3}$$

We have thus derived a characteristic state function for black-body radiation and can now calculate any thermodynamic quantity.

$$F = U - ST = -\frac{1}{3} aT^4 V$$

$$P = -\left(\frac{\partial U}{\partial V}\right)_S = -(3S/4)^{4/3} \cdot \frac{-1}{3} (aV)^{-4/3} = \frac{1}{3}(3S/4aV)^{4/3} = \frac{1}{3} aT^4$$

$$PV = \frac{1}{3} aT^4 V$$

$$G = U + PV - TS = aT^4 V + \frac{1}{3} aT^4 V - \frac{4}{3} aT^4 V = 0$$

The last result may seem surprising but is natural because P and T are not independent variables in the particular case of black-body radiation. If the container of the radiation is kept at a certain temperature T and an external pressure P is applied, then the container will collapse if $P > aT^4/3$ and all the radiation will vanish. If $P < aT^4/3$ then the container will expand indefinitely and new radiation will be created with no limitation.

Exercise 15.11.1

Derive an expression for the isothermal compressibility of black-body radiation.

Hint

It may be easier to derive $(\partial P/\partial V)_T$ than $(\partial V/\partial P)_T$.

Solution

We get $\kappa_T = -(1/V_m)\cdot(\partial V_m/\partial P)_T = -(1/V)\cdot(\partial V/\partial P)_T$ from the definition. From $P = aT^4/3$ we get $(\partial P/\partial V)_T = 0$ and $(\partial V/\partial P)_T = \infty$ and $\kappa_T = \infty$. This is in complete agreement with the last two sentences in the text.

Exercise 15.11.2

Consider a container filled with N moles of an ideal gas and black-body radiation. Derive an expression for G_m as a function of its natural variables.

Hint

Suppose that the expressions for Helmholtz's energy F are additive. Then use $G = F + PV$. Finally, eliminate V by introducing P.

Solution

$F_1 = aT^4V/3; \quad F_2 = NK(T) + NRT\ln(NRT/V) - NRT; \quad F = F_1 + F_2 = -aT^4V/3 + NK(T) + NRT\ln(NRT/V) - NRT; \quad P = -\partial F/\partial V = aT^4/3 + NRT/V; \quad G = F + PV = -aT^4V/3 + NK(T) + NRT\ln(NRT/V)NRT + aT^4V/3 + NRT = NK(T) + NRT\ln(NRT/V) = NK(T) + NRT\ln(P - aT^4/3)$

15.12   *Electron gas*

The heat capacity for the electron gas in a metal is often given as

$$C_V = \gamma_e T$$

In real metals, the coefficient γ_e has different values at low and high T and a single expression cannot be used for the whole range of T. The low-T value is often about 1.4 times the high-T value. However, in a free electron gas γ_e is independent of T. We shall now examine the properties of such a gas. Its γ_e value depends upon the density which can be expressed through V_m. According to the electron theory, γ_e is actually proportional to $V_m^{2/3}$. For one mole of electrons we obtain

$$U_m = \int_0^T C_V dT = U_m(0\ \text{K}) + \gamma_e T^2/2$$

$$S_m = \int_0^T \frac{C_V}{T} dT = \gamma_e T$$

$$F_m = U_m - TS_m = U_m(0\,\mathrm{K}) - \gamma_e T^2/2$$

In order to derive an expression for G_m we must calculate P from F_m as a function of T, V_m. Then we must remember that the electron theory predicts that γ_e is proportional to $V_m^{2/3}$ and we should treat $\gamma_e/V^{2/3}$ as a constant.

$$F_m = U_m(0\,\mathrm{K}) - (\gamma_e/V_m^{2/3}) \cdot V_m^{2/3} T^2/2$$

$$P = -\left(\frac{\partial F_m}{\partial V_m}\right)_T = \frac{1}{2}\frac{\gamma_e T^2}{V_m^{2/3}} \frac{2}{3} V_m^{-1/3} = \frac{1}{3}\gamma_e T^2/V_m$$

$$G_m = F_m + PV_m = U_m(0\,\mathrm{K}) - \gamma_e T^2/2 + \gamma_e T^2/3 = U_m(0\,\mathrm{K}) - \gamma_e T^2/6$$

We have neglected to discuss the possible dependence of $U_m(0\,\mathrm{K})$ on V_m but that could only contribute a T-independent term in G_m.

If we want to apply the treatment of an electron gas to a metal we should realize that there is a strong interaction between the electrons and the ionized atoms which more or less eliminates the pressure of the electron gas. It is thus common to give their contribution to G_m as $-\gamma_e T^2/2$.

It should be realized that published values of γ_e for metals refer to one mole of atoms, and not electrons. The values of γ_e are relatively small and its contribution to C_V only grows proportional to T. Even though it predominates at very low T it will soon be negligible in comparison with the contribution due to thermal vibrations of the atoms which initially increases proportional to T^3. However, at very high T the latter contribution levels off at the value of $3R$ and the electronic contribution again becomes important.

Exercise 15.12.1

Estimate the temperature at which the electronic contribution to C_V of pure Ni will be one-half the contribution from thermal vibrations, if $\gamma_e = 54.4 \cdot 10^{-4}\,\mathrm{J/mol\,K^2}$.

Hint

The temperature will be so high that the vibrational contribution can be estimated as $3R$ and the γ_e value may be taken as the tabulated value divided by 1.4.

Solution

$(54.5 \cdot 10^{-4}/1.4)T = 3R/2;\;\; T = 3200\,\mathrm{K}.$

Modelling of disorder

Introduction

In this chapter we shall model the thermodynamic effect of some physical phenomena. In each case we shall start by defining an internal variable representing the extent of the physical phenomenon to be discussed. We shall proceed by deriving an expression for one of the characteristic state functions in terms of the internal variable together with a set of external variables. The choice of characteristic state function depends upon what set of external variables is most convenient. Then we shall calculate the equilibrium value of the internal variable by putting the driving force for its change equal to zero. Finally, we shall try to eliminate the internal variable by inserting the expression for its equilibrium value in the characteristic state function.

Our derivation of an expression for the characteristic state function will usually be based upon two separate evaluations, one concerned with the entropy due to the disorder created by the physical phenomenon and the other concerned with what may be called the non-configurational contribution. The entropy will be evaluated from Boltzmann's relation which is here preferred because it is felt that it gives a better physical insight than the more general and elegant method of statistical thermodynamics based upon the use of partition functions. The purpose of statistical thermodynamics is to model the thermodynamic properties of various types of systems from statistical considerations on the atomic level. The relation proposed by Boltzmann can be derived from such considerations.

Exercise 16.1.1

Boltzmann's relation relates the configurational entropy, ΔS, to the number of ways, W, the system can be arranged under the conditions given, $\Delta S = k \ln W$, where k is Boltzmann's constant. Use this relation to prove that the configurational entropy of a system, consisting of two parts, is equal to the sum of the configuration entropy for each part.

Hint

If each part can be arranged in W_1 and W_2 ways, respectively, then the whole system can be arranged in $W = W_1 \cdot W_2$ ways.

Solution

$$\Delta S = k\ln W = k\ln(W_1 W_2) = k\ln W_1 + k\ln W_2 = \Delta S_1 + \Delta S_2.$$

16.2   *Thermal vacancies in a crystal*

We shall start by considering vacancies in a crystal of a pure element. A convenient internal variable would be the number of vacancies, N_v, in a crystal with N atoms. This internal variable is sufficient for defining the contribution of the vacancies to the internal energy if it is assumed that the interaction between the vacancies can be neglected. This may be a reasonable approximation for low concentrations of vacancies and we shall construct a model from this approximation.

The state with N_v vacancies can be realized in a number of ways by placing the vacancies in different arrangements on the lattice sites. The number of different arrangements is

$$W = \frac{(N + N_v)!}{N! N_v!}$$

since there must be $N + N_v$ lattice sites and we cannot distinguish between two atoms, nor between two vacancies. From Boltzmann's relation we can now calculate the contribution to the entropy from the vacancies and we can use the following approximation if we consider a system where N and N_v are both large numbers,

$$\Delta S/k = \ln W = \ln(N + N_v)! - \ln N! - \ln N_v!$$
$$\cong (N + N_v)\ln(N + N_v) - N\ln N - N_v\ln N_v$$

$$= -N\ln\frac{N}{N + N_v} - N_v\ln\frac{N_v}{N + N_v} = -N\left[\ln(1 - y_v) + \frac{y_v}{1 - y_v}\ln y_v\right]$$

where $y_v = N_v/(N + N_v)$, the fraction of sites that are vacant.

When choosing a characteristic state function we should realize that it is very awkward to treat the energy of a vacancy, u, as a function of S_m and V_m. To treat u as a function of T and V_m is also inconvenient because V_m is usually defined for one mole of atoms, not lattice sites. Thus V_m will naturally increase, or the lattice must be compressed, when a vacancy is introduced by the creation of a new lattice site. It is most convenient to treat u as a function of T and P and we should thus choose G as our characteristic state function. In order to be rigorous in the derivation to follow, we should actually introduce a Gibbs energy of formation of a vacancy, g, instead of the energy, u, and obtain for the contribution due to N_v vacancies, when added to N atoms,

$$\Delta G = N_v g + kNT\left(\ln\frac{N}{N + N_v} + \frac{N_v}{N}\ln\frac{N_v}{N + N_v}\right)$$

The quantity g may be regarded as the non-configurational Gibbs energy per vacancy. We have thus achieved our first goal. It is worth emphasizing that we managed to derive an expression for ΔG without introducing any further internal variable. Before proceeding it is important to realize that N_v is in fact an internal variable because we do not need an external reservoir from which to take vacancies. Alternatively, we may imagine that we have an external reservoir of vacancies with a chemical potential equal to zero. We may thus regard g as the non-configurational Gibbs energy of formation of a vacancy or of dissolution of a vacancy from an external reservoir.

We have found an expression for the contribution to the Gibbs energy from a given number of vacancies, N_v. It can be used to evaluate the equilibrium number of vacancies under constant T and P. If there is a mechanism, or 'reaction', by which the number of vacancies can vary, then there should be a spontaneous change as long as the driving force is positive. Assuming that g is independent of N_v under constant T, P and N, we obtain

$$-D = (\partial/G\partial\xi)_{T,P,N} \equiv (\partial G/\partial N_v)_{T,P,N}$$

$$= g + kNT\left[\frac{-1}{N + N_v} + \frac{1}{N}\ln\frac{N_v}{N + N_v} + \frac{N_v}{N}\cdot\frac{1}{N_v} - \frac{N_v}{N}\cdot\frac{1}{N + N_v}\right]$$

$$= g + kT\ln\frac{N_v}{N + N_v}$$

By putting this equal to zero we obtain, for the equilibrium number of vacancies,

$$y_v^{eq} = \exp(-g/kT)$$

This is an expression characteristic of so-called Boltzmann statistics. It reveals how the fraction of vacant sites varies with temperature. The higher the temperature, the larger the fraction is. This is why such vacancies are often called **thermal vacancies**.

If we instead focus our interest upon the number of vacancies per mole of atoms we obtain, for its equilibrium value,

$$\frac{y_v^{eq}}{1 - y_v^{eq}} = \frac{N_v^{eq}}{N} = \frac{1}{\exp(g/kT) - 1}$$

This kind of expression is usually connected with the so-called Bose–Einstein statistics.

In the present case it will be possible to eliminate the internal variable at equilibrium,

$$\Delta G = N_v g + kTN\ln(1 - y_v^{eq}) + kTN_v\ln y_v^{eq}$$
$$= N_v g + kTN\ln(1 - y_v^{eq}) + kTN(-g/kT)$$
$$= kTN\ln(1 - y_v^{eq})$$

For a system containing one mole of atoms, i.e. with N equal to Avogadro's number, we obtain

$$\Delta G_m = RT \ln[1 - \exp(-g/kT)]$$

The final expression is surprisingly simple. It gives the complete information on ΔG at equilibrium as function of T and, if one is only interested in thermodynamic properties at equilibrium, this expression will be sufficient. On the other hand, the model can also be applied for evaluating the number of vacancies at equilibrium and that is of importance because of the effect of vacancies on other properties like volume or diffusivity. The model can even be applied for evaluating ΔG when the number of vacancies differs from the equilibrium number. In such cases one must use the basic equation of ΔG as a function of T and N_v. One may, for instance, be interested in the process by which the number of vacancies could approach the equilibrium value. There are many possibilities. If the actual number of vacancies is much higher than the equilibrium value, then the driving force could be large enough for the nucleation of a pore. If it is lower, mechanisms involving dislocations could still operate. For such considerations it is essential to know the driving force. It can be obtained from the expression for ΔG as a function of N_v. Since we are now considering a decrease and would like to identify $d\xi$ with $-dN_v$, we write

$$D = -(\partial G/\partial N_v)_{T,P,N} \equiv (\partial G/\partial N_v)_{T,P,N} = g + kT \ln \frac{N_v}{N + N_v}$$

Expressing g through the equilibrium number of vacancies yields

$$D = -kT \ln y_v^{eq} + kT \ln y_v = kT \ln(y_v/y_v^{eq})$$

This is the driving force for the disappearance of one vacancy. The driving force per mole of disappeared vacancies is obtained as

$$D = RT \ln(y_v/y_v^{eq})$$

For a typical metal we can estimate the value of g from the information that the fraction of vacant sites is between 10^{-3} and 10^{-4} close to the melting temperature, $T_{m.p.}$. We thus obtain $g/kT_{m.p.} = 2.3 \cdot (3 \text{ to } 4) \cong 8$. For such low values of y_v^{eq} we obtain

$$-\Delta G_m/RT = -\ln(1 - y_v^{eq}) \cong +y_v^{eq} = 10^{-4} \text{ to } 10^{-3}$$

This contribution is usually too small to have a decisive influence on the thermodynamic stability of a crystalline phase.

Exercise 16.2.1

Suppose a solid metal in equilibrium has 10^{-3} vacancies per atom at the melting point. How much lower would the melting point be if there were no vacancies?

Hint

Apply Richard's rule, $\Delta S \cong R$ on melting. Close to the melting point it yields $\Delta G \cong (T_{\text{m.p.}} - T)R$.

Solution

With vacancies we have $G^{\text{liquid}} - G^{\text{solid}} = \Delta G = (T_{\text{m.p.}} - T)R$ and without we get
$G^{\text{liquid}} - G^{\text{solid}} = (T_{\text{m.p.}} - T)R + RT\ln(1 - y_v^{\text{eq}}) \cong (T_{\text{m.p.}} - T)R - RTy_v^{\text{eq}} = 0$;
$T = T_{\text{m.p.}}/(1 + y_v^{\text{eq}}) = T_{\text{m.p.}}/1.001 = 0.999T_{\text{m.p.}}$.

Exercise 16.2.2

Consider a pure metal A with an equilibrium amount of vacancies. Derive expressions for the chemical potentials of vacancies and A, respectively.

Hint

We have $\Delta G = N_v g_v + kNT\{\ln[N/(N + N_v)] + (N_v/N)\ln[N_v/(N + N_v)]\}$ but that is only the contribution from the vacancies. We want the total G and should then start with pure A without any vacancies. That G is proportional to N, say Ng_A.

Solution

$G = Ng_A + N_v g_v + kNT\{\ln[N/(N + N_v)] + (N_v/N)\ln[N_v/(N + N_v)]\}$; $\mu_v = \partial G/\partial N_v = g_v + kNT\{-1/(N + N_v) + (1/N)\ln[N_v/(N + N_v)] + (N_v/N)(1/N_v) - (N_v/N)[1/(N + N_v)]\} = g_v + kT\ln[N_v/(N + N_v)]$.
At equilibrium we get $\mu_v = g_v - g_v = 0$; $\mu_A = \partial G/\partial N = g_A + kT\{\ln[N/(N + N_v)] + N/N - N/(N + N_v) - N_v[1/(N + N_v)]\} = g_A + kT\ln[N/(N + N_v)] = g_A + kT\ln(1 - y_v) \cong g_A - kTy_v$. Notice that this is the chemical potential per atom of A.

Exercise 16.2.3

The temperature dependence of the entropy due to thermal vacancies in a metal is sometimes given by the expression $\Delta S/R = -(T_{\text{m.p.}}/T)\cdot(y_0)^{T_{\text{m.p.}}/T}\cdot\ln y_0$ where y_0 is the value of y_v^{eq} at $T = T_{\text{m.p.}}$. Check if this expression is correct.

Hint

We are asked to compare an expression in y_0 with our known expression in y_v^{eq}. For equilibrium at all T we know $-g = kT\ln y_v^{eq}$ and thus $y_v^{eq} = (y_0)^{T_{m.p.}/T}$ if g is independent of T, because $y_v = y_0$ at $T = T_{m.p.}$.

Solution

The new expression gives $\Delta S/R = y_v^{eq}\ln y_v^{eq}$ which is the approximate value of $\Delta S/R = -\ln(1 - y_v^{eq}) - [y_v^{eq}/(1 - y_v^{eq})]\cdot\ln y_v^{eq}$ for small y_v^{eq}.

16.3 *Topological disorder*

Even though the structure of a melt may intuitively be imagined as completely disordered, in reality the short-range arrangement of the atoms in a liquid metal is rather similar to that in the crystalline state. It may thus be useful to describe the liquid as a crystalline phase with so much disorder that all the long-range order has been destroyed. In this connection one talks about **topological disorder**. A very simple and crude model of the topological disorder in a liquid metal is based upon the assumption that the amount of thermal vacancies increases discontinuously during melting. This model can be supported by many semi-quantitative considerations involving such properties as density, compressibility, thermal expansion and diffusivity. X-ray measurements have indicated that the coordination number in liquid metals is about 11, as compared to 12 for the close-packed fcc and hcp structures. This result immediately suggests that the vacancy concentration would be about 1/12, i.e. 8%.

This value compares favourably with a value we obtain by evaluating how many vacancies can form when the heat of melting is added to the solid. According to Richards' rule the entropy of melting is approximately R and the heat of melting is $RT_{m.p.}$. With the approximate value of 8 for $g/kT_{m.p.}$ given above, we get

$$N_v = RT_{m.p.}/g = RT_{m.p.}/8kT_{m.p.} = N/8$$

$$y_v = \frac{N_v}{N + N_v} = \frac{1}{9} = 11\%$$

It does not compare so nicely with a value obtained from the entropy of melting

$$\Delta S/k = -N\ln\frac{N}{N + N_v} - N_v\ln\frac{N_v}{N + N_v} = R/k = N$$

$$\ln(1 - y_v) + \frac{y_v}{1 - y_v}\ln y_v = -1$$

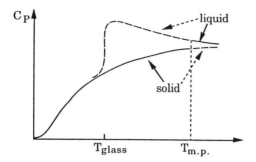

Figure 16.1 Schematic diagram comparing the heat capacities of a crystallized metal and its amorphous state.

By solving this equation numerically one finds approximately $y_v = 1/3$ or 35%. It is evident that this simple model is too crude to give a quantitative description of melting.

At the melting point of a metal, C_P is often larger in the melt than in the solid. This difference grows for the undercooled melt and the difference in entropy between the two phases will thus decrease with decreasing temperature. Extrapolations have indicated that one may approach a temperature below which the undercooled melt would have a lower entropy than the crystalline solid. This is called Kauzmann's paradox (Kauzmann, 1948) and it is not regarded as possible. It has instead been suggested that the point where the difference in entropy disappears is the glass transition where the melt turns into an amorphous solid with almost the same entropy as the crystalline solid. According to this picture, the topological disorder in the amorphous solid is rather limited. It does not make the phase liquid and it has a rather low entropy. On heating above the glass transition temperature a large amount of defects are created. They make the amorphous phase more liquid and contribute to the entropy and heat capacity. An increasing amount of defects are created as the temperature is raised towards the melting point. From the thermodynamic point of view, the melting of the amorphous solid in thus a gradual process which continues to or even above the melting point. Figure 16.1 illustrates this behaviour schematically.

Exercise 16.3.1

Estimate the difference in H at 0 K between the amorphous and crystalline states of a metal from the following simplifying assumptions. The two states have the same entropy at 0 K. Up to T_{glass} the two states have the same C_P. T_{glass} falls at $T_{m.p.}/3$ and above $T_{m.p.}/3$ the difference in C_P is linear in T. Let $\Delta C_P = 0$ at $T_{m.p.}$.

Hint

Estimate the difference in enthalpy and entropy at $T_{m.p.}$ from Richards' rule, $\Delta H_m/T_{m.p.} = \Delta S_m \cong R$. The assumptions give $\Delta S_m = 0$ at $T_{m.p.}/3$.

Solution

$\Delta C_P = a + bT$ where $a + bT_{m.p.} = 0$. A second relation between a and b is obtained by integrating for the entropy between $T_{m.p.}/3$ and $T_{m.p.}$: $R = \int(\Delta C_P/T)dT = a\ln 3 + (2/3)bT_{m.p.}$; $a = R/(\ln 3 - 2/3) = 2.3R$; $b = -2.3R/T_{m.p.}$. ΔH_m changes between $T_{m.p.}/3$ and $T_{m.p.}$ by $\int \Delta C_P dT = a(T_{m.p.} - T_{m.p.}/3) + 0.5b(T_{m.p.}^2 - T_{m.p.}^2/9) = 2.3RT_{m.p.}(2/3 - 4/9) = 0.51RT_{m.p.}$. Thus we get, at $T_{m.p.}/3$ as well as at 0 K, $\Delta H_m = RT_{m.p.} - 0.51RT_{m.p.} = 0.49RT_{m.p.}$.

16.4   *Heat capacity due to thermal vibrations*

According to quantum mechanics, the energy of heating can only be added in quanta. For an harmonic oscillator of frequency v, the magnitude of the quanta is hv where h is Planck's constant. Einstein constructed a simple model of a crystal by assuming that each atom vibrates independent of all the others and has three directions of movement. A crystal with N atoms could thus be regarded as consisting of $3N$ linear oscillators, all of them with a frequency v. The question, how many quanta such a crystal should have at equilibrium, is closely related to our previous question how many vacancies a crystal should have at equilibrium. If there are n quanta, the energy increase is $\Delta U = nhv$ compared to the conditions at absolute zero. In order to evaluate the entropy contribution we must find in how many ways n quanta can distribute themselves on $3N$ oscillators. By numbering the oscillators $a_1 a_2 a_3 \dots a_{3N}$ and the quanta $k_1 k_2 k_3 \dots k_n$ we can describe a particular distribution by first giving the number of a certain oscillator and then the quanta which are placed there, e.g. $a_4 k_3 k_5 a_6 k_2 a_1 k_7 k_9 a_7 a_2 \dots$ If all the permutations of these elements represent possible distributions there should be $(3N + n)!$ different distributions. However, we must start with an oscillator and the first factor should thus be $3N$ instead of $3N + n$. Furthermore, it would be impossible to distinguish between many of the distributions since all the oscillators are identical and so are all the quanta. The number of distinguishable distributions is thus

$$W = \frac{3N}{3N + n} \frac{(3N + n)!}{(3N)!\, n!}$$

This is almost identical to the result for thermal vacancies except for the first factor, which is of no importance for large N and n when we take the logarithm, and except for the fact that N has been replaced by $3N$. We can thus use the final result from the previous derivation but this time we shall apply it to the Helmholtz energy because the frequency, and thus the energy hv, will vary with the distance between the atoms, i.e. the volume. Thus we should like to keep the volume constant. We obtain

$$\Delta F = nhv + 3kNT\left(\ln\frac{3N}{3N + n} + \frac{n}{3N}\ln\frac{n}{3N + n}\right)$$

The equilibrium number of quanta under constant T and V is obtained from $\partial\Delta F/\partial n = 0$,

$$\frac{n}{3N + n} = \exp(-h\nu/kT)$$

Again it will be possible to eliminate the internal variable at equilibrium and we find

$$\Delta F = 3RT\ln[1 - \exp(-h\nu/kT)]$$

By standard methods we can calculate the heat capacity due to thermal vibrations,

$$C_V = -T(\partial^2\Delta F/\partial T^2)_V = 3R\left(\frac{h\nu}{kT}\right)^2 \cdot \frac{\exp(h\nu/kT)}{[\exp(h\nu/kT) - 1]^2}$$

The only parameter characteristic of the particular material under consideration is the frequency, ν. It always appears in the dimensionless combination $h\nu/kT$. The combination of constants $h\nu/k$ is of the dimension temperature and we can thus introduce a new material constant instead of the frequency, the Einstein temperature $\Theta = h\nu/k$.

$$C_V = 3R(\Theta/T)^2 \cdot \exp(\Theta/T)[\exp(\Theta/T) - 1]^2$$

There have been many attempts to improve Einstein's model by removing the assumption that the atoms vibrate independent of each other. An elegant method was proposed by Debye. If the atoms cooperate when they vibrate, it means that the vibrating units are larger and the frequency should be lower. For each possible frequency, one should be able to apply Einstein's model and by a summation over all the frequencies one may obtain the proper result. Debye considered a whole spectrum of oscillators down to the mechanical vibrations of the crystal. His spectrum thus extends all the way from the high frequency of individual atoms and down to the acoustic range. He further assumed that the number of oscillators is still $3N$ and that they distribute themselves over the range of frequencies proportional to ν^2. From Debye's model one does not get C_V as an analytical expression of T but C_V is available in tables using the parameter $x = \Theta/T$ where Θ is equal to $h\nu_{max}/k$ and ν_{max} is the maximum frequency, i.e. the vibrational frequency of an atom. One often tabulates not only C_V but also S and U. Debye's model agrees fairly well with measurements for many materials. Θ is evaluated as a material constant to give the best agreement between model and experiment.

We can calculate C_P from C_V by means of the Grüneisen constant. With the approximate value $\gamma = 2$ we would get

$$C_P = C_V(1 + 2\alpha T)$$

S as a function of T at constant volume is obtained by integrating C_V/T. On the other hand, if one should like to have S as a function of T at constant pressure it could be obtained by integrating C_P/T which yields

$$S = S(x) + 2\alpha U(x)$$

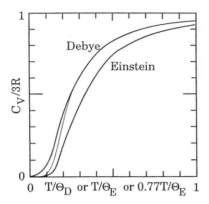

Figure 16.2 Comparison between the Einstein and Debye theories of C_V, made at temperatures given relative to the characteristic temperature Θ as defined in each theory. In addition, the thin line is for Einstein's theory plotted versus $0.77T/\Theta_E$.

if α is a constant. Furthermore, after some manipulations one obtains approximately

$$H = U(x) + (5/2)R\alpha T^2 + (9/8)R\Theta + H_o$$

Notice that H_o is obtained as an unknown constant of integration, and it represents the cohesive energy. The term $(9/8)R\Theta$ comes from quantum mechanics which says that a linear oscillator of frequency v has a zero point energy of $hv/2$.

A comparison between Einstein's and Debye's models is given in Fig. 16.2. Both models predict a C_V value of $3R$ at very high temperatures but the diagram seems to indicate that the theories give quite different results up to fairly high temperatures. However, if Θ is regarded as an empirical constant, to be evaluated from experimental information, then the two models would work with different Θ values. These empirical constants are called Einstein's and Debye's temperatures. Reasonably good agreement can thus be obtained from high down to fairly low temperatures and for many applications it is reasonable to use the equations derived by Einstein. The value of the Einstein temperature can be estimated from tabulated values of the Debye temperature by multiplication with a factor. The factor is sightly different depending on which quantity one is most interested in. For heat capacity, internal energy, entropy and Helmholtz energy one should use the values 0.77, 0.73, 0.71 and 0.72, respectively. The thin line in Fig. 16.2 shows how well Einstein's theory agrees with Debye's if the same temperature scale is used but different Θ values. The difference is negligible above $0.3\Theta_D$.

The fact that C_V approaches the same value of $3R$ at high temperatures, independent of the material constant Θ, does not mean that the relative stability of all crystalline substances would be independent of temperature at high temperatures. For a given element with two possible structures, the structure with the highest S value will increase its relative stability towards high temperatures more than the other one. The S value is the result of an integration over C_V/T from absolute zero. S will thus be larger the faster C_V approaches the asymptotic value of $3R$, i.e. the lower Θ is. A difference in C_V between two structures at low temperatures will be very important at high temperatures because T appears in the denominator of the integrand. A structure with a low Θ will thus have high stability at high temperature. The fact that the S term dominates over the U term in F at high temperatures can be demonstrated by a calculation for some temperature T_1.

$$F = U - T_1 S = U_0 + \int_0^{T_1} C\mathrm{d}T - T_1 \int_0^{T_1} (C/T)\mathrm{d}T = U_0 - \int_0^{T_1} [C(T_1 - T)/T]\cdot\mathrm{d}T$$

The integrand is always positive since C and $T_1 - T$ are both positive. We can make a similar calculation for G and obtain a similar result.

At low temperatures where $T \ll \Theta$ one would obtain, according to Einstein

$$C_V = 3R\left(\frac{\Theta}{T}\right)^2 \cdot \exp\left(-\frac{\Theta}{T}\right)$$

This would imply that C_V decreases rapidly as one approaches the absolute zero. Experiments showed a slower decrease and that was the reason why one wanted to improve Einstein's theory. Debye's theory instead yields the following expression at low temperature, and it agrees fairly well with experiments if the electronic contribution is subtracted for metallic conductors.

$$C_V = 234R(T/\Theta)^3$$

Exercise 16.4.1

Derive the expression for S as a function of T at constant P by integrating $C_P = C_V(1 + 2\alpha T)$ assuming that α is a constant. Examine if there are any other approximations.

Hint

V changes during this integration since P is kept constant.

Solution

Debye gave $C_V(x)$ where $x = \Theta/T$ and Θ was treated as independent of T. Now we must assume that Θ is also independent of P which cannot be quite true. Then we get, under constant P: $S = \int(C_P/T)\mathrm{d}T = \int(C_V/T)\mathrm{d}T + 2\alpha\int C_V\mathrm{d}T = S(x) + 2\alpha U(x)$.

Exercise 16.4.2

Derive the term in H originating from the zero-point energy, $(9/8)R\Theta$.

Hint

First neglect the difference between U and H. According to Einstein the zero-point energy for all oscillators is the same, yielding $\Delta U = 3N(h\nu/2) = 3Nk\cdot\Theta/2 = (3/2)R\Theta$. However, according to Debye there is a distribution of frequencies $z(\nu) = K\nu^2$ with $3N = \int z(\nu)d\nu = \int K\nu^2 d\nu = K\nu_{max}^3/3$ and thus $K = 9N/\nu_{max}^3$.

Solution

$\Delta H \cong \Delta U = \int (h\nu/2)\cdot z(\nu)d\nu = \int (h\nu/2)\cdot(9N/\nu_{max}^3)\cdot\nu^2 d\nu = (9N/\nu_{max}^3)\cdot(h\nu_{max}^4/8) = (9/8)\cdot Nk\cdot(h\nu_{max}/k) = (9/8)\cdot R\Theta_{Debye}$

Exercise 16.4.3

Einstein's treatment of lattice vibrations gave the following contribution to the Helmholtz energy $\Delta F = 3RT\ln[1 - \exp(-\Theta/T)]$. Debye's treatment is much more complicated but, in principle, the result can also be expressed as $\Delta F = T\cdot\Phi(\Theta/T)$. In a simple approach we may assume that Θ is independent of T but it certainly depends upon V. Show that the following expression for the Grüneisen parameter can then be derived, $\gamma = -\,\mathrm{d}\ln\Theta/\mathrm{d}\ln V$.

Hint

Neglect all other contributions to F. Then we have F as a function of its natural variables T and V and should be able to calculate any thermodynamic quantity. Calculate α/κ_T from $-F_{VT}$. Denote $\mathrm{d}\Phi/\mathrm{d}(\Theta/T)$ with Φ', etc.

Solution

$F = T\Phi(\Theta/T)$; $F_T = \Phi + T\cdot\Phi'\cdot(-\Theta/T^2) = \Phi - (\Theta/T)\cdot\Phi'$;
$C_V = -TF_{TT} = -T\cdot[\Phi'\cdot(-\Theta/T^2) + (\Theta/T^2)\cdot\Phi - (\Theta/T)\cdot\Phi''\cdot(-\Theta/T^2)] = -(\Theta^2/T^2)\cdot\Phi''$; $\alpha/\kappa_T = (\partial V/\partial T)_P/(-\partial V/\partial P)_T = (\partial P/\partial T)_V = (-\partial F_V/\partial T) = -F_{VT} = -F_{TV} = -[\Phi'\cdot(1/T)\cdot(1/T)\cdot\mathrm{d}\Theta/\mathrm{d}V - \mathrm{d}\Theta/\mathrm{d}V\cdot(1/T)\cdot\Phi' - (\Theta/T)\cdot\Phi''\cdot\mathrm{d}\Theta/\mathrm{d}V\cdot(1/T)] = (\Theta/T^2)\cdot\mathrm{d}\Theta/\mathrm{d}V\cdot\Phi''$; $\gamma = V\alpha/C_V\kappa_T = V\cdot(\Theta/T^2)\cdot\mathrm{d}\Theta/\mathrm{d}V\cdot\Phi''/(-\Phi''\Theta^2/T^2) = V\cdot(\mathrm{d}\Theta/\mathrm{d}V)/(-\Theta) = -\,\mathrm{d}\ln\Theta/\mathrm{d}\ln V$.

16.5 *Relation between vibrational energy and elastic properties*

It is self-evident that the vibrational frequency of the atoms, and thus the values of Θ and C_V, depend upon the strength of the atomic bonds. This strength will also determine the value of the elastic modulus. By applying the following simple model we can derive an approximate relation. Suppose that the crystal has a simple cubic atomic arrangement, held together by springs between the atoms. The spring constant K can be estimated from the macroscopic modulus of elasticity E and one finds $K = Ed$ if d is the atomic distance. If each atom oscillates under the influence of such a spring, the frequency will be

$$v = \frac{\sqrt{Ed/M}}{2\pi}$$

if M is the mass of an atom. From $hv = k\Theta$ we obtain

$$\Theta = \frac{h\sqrt{Ed/M}}{2\pi k}$$

We can see that a high E value results in high values of v and Θ. Substances with covalent bonds like diamonds have high E values and thus high Θ values. Through the relation between Θ and S (via C) such substances would thus have unusually low entropy. Their stability relative to other possible structures should thus decrease at high temperatures.

Zener has suggested that one can apply the same reasoning for a comparison between the typical metallic phases. He emphasized that the atomic arrangement in the simple bcc structure is such that the elastic modulus is low in a certain direction, namely for shear in the $1\bar{1}0$ direction on the 110 plane. This would result in an abnormally low value of Θ and thus a high entropy. The bcc structure should thus become more stable at high temperatures relative to other typical metallic structures like fcc and hcp. According to Zener, this explains why bcc is a typical high-temperature phase. Typical examples are β-Ti, β-Zr, δ-Fe and δ-Mn and also β-brass.

Exercise 16.5.1

Tin has two stable, solid phases. One is a metallic phase and is called white tin. The other is non-metallic and has the same cubic structure as diamond. It is called grey tin. Which one should we guess is the high-temperature phase?

Hint

We may suspect that grey tin has thermodynamic properties similar to those of diamond.

Solution

Supposing that the bonds in grey tin are similar to those in diamond, we guess that grey tin has a higher Debye temperature than white tin. It will thus have a lower entropy and be less stable the higher the temperature is. We should guess that white tin is the high-temperature phase, which is actually the case.

16.6   *Magnetic contribution to thermodynamic properties*

The atoms in a ferromagnetic substance contain unpaired electron spins which give the atom a certain magnetic moment. It can have different directions. According to the localized spin model of magnetism, a disordered arrangement gives a contribution to the entropy which can be evaluated from Boltzmann's relation. Due to its simplicity we shall only use this model. W is the number of different arrangements for the whole system. Let w be the number of possible directions for each atom. If the direction of each atom in the disordered state is independent of all the others, we can write $W = w^N$ for one mole of the substance. Thus

$$S = k\ln W = k\ln w^N = kN\ln w = R\ln w$$

According to quantum mechanics,

$$w = \beta + 1 = 2s + 1$$

where β is the number of unpaired electron spins and s is the resulting spin. For a free electron $s = 1/2$.

There is a critical temperature, the Curie temperature T_C, below which the spins will position themselves parallel to each other in a ferromagnetic substance. For a perfectly ordered state, the above contribution to the entropy should disappear completely but, in practice, there will be some disorder left. However, it will disappear as one approaches absolute zero and the saturation magnetization will thus increase (see Fig. 16.3). The degree of magnetic order can be measured by measuring the saturation magnetization. Note that an externally applied magnetic field usually does not change the ordering appreciably. It simply makes the magnetization of the various magnetic domains align along one direction. From the saturation magnetization at low temperatures, one can get direct information on the magnetic moment of the atoms, i.e. the values of s and β. The magnetic moment per atom is usually given in units of Bohr magnetons which is identical to $2s$, i.e. to β according to our simple model. Such measurements indicate that β is seldom an integer. This casts some doubt on the localized spin model and one may wonder if the entropy of magnetic disorder can be evaluated as shown above. Despite this, the method is often used and may be regarded as a convenient approximation.

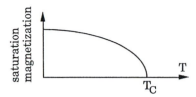

Figure 16.3 Schematic diagram showing the variation of the saturation magnetization as a function of temperature.

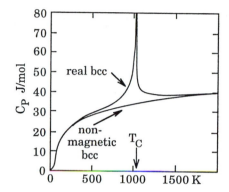

Figure 16.4 Heat capacity (C_P) of bcc-Fe. Notice that a considerable part of the magnetic effect occurs above T_C.

The disordered state above the Curie temperature is called paramagnetic. It is evident that it has a high entropy and should thus grow even more stable at higher temperatures. The reason why the ferromagnetic state becomes stable below a critical temperature must be a lower energy for that state. When the disorder increases as the temperature is raised, energy must thus be added and the magnetic transformation is revealed in the curve for the heat capacity (see Fig. 16.4).

The curve shows that the transformation occurs gradually up to T_C but a small part of the order still exists at T_C. It does not disappear until well above T_C and it thus yields a contribution to the heat capacity above T_C. We can evaluate the magnetic enthalpy by integrating the abnormal contribution to the curve. As always, the transformation temperature is determined by the balance between enthalpy and entropy. This is best demonstrated by approximating the magnetic transformation (which is actually a second-order transition (see Section 14.1)) with a sharp transformation. For such a transformation we have

$$\Delta G = \Delta H - T\Delta S$$

With $\Delta G = 0$ we find a transformation temperature $T^1 = \Delta H/\Delta S$. The larger the energy gain is at the magnetic ordering, the higher the transformation temperature will be.

Figure 16.5 illustrates the difference in the Gibbs energy between the most stable state at each temperature, G^{eq}, and the completely disordered state.

The slope of the curve at low temperature gives the entropy decrease due to

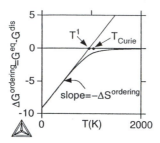

Figure 16.5 The effect of magnetic ordering on the Gibbs energy of bcc-Fe. The reference state is the completely disordered state at each temperature.

complete order of the spins. The thin line gives the difference between the completely ordered and the completely disordered states at each temperature. The intersection with the abscissa gives the temperature, T^1, for the hypothetical, sharp transformation. In reality, the ferromagnetic state is stable to a higher temperature, T_C, because it lowers its Gibbs energy by disordering gradually.

Exercise 16.6.1

Evaluate β for bcc-Fe from an estimate of the magnetic entropy of bcc-Fe using Fig. 15.1.

Hint

The sharp bend in the curve for $^\circ G_{Fe}^{bcc} - {}^\circ G_{Fe}^{fcc}$ is due to the magnetic entropy of bcc-Fe and is not influenced by the magnetic entropy of fcc-Fe which changes at much lower temperatures.

Solution

The slopes for the paramagnetic and ferromagnetic states can be estimated where the curve is reasonably flat. We obtain $\Delta S = - \mathrm{d}(^\circ G_{Fe}^{paramagn.bcc} - {}^\circ G_{Fe}^{fcc})/\mathrm{d}T + \mathrm{d}(^\circ G_{Fe}^{ferromagn.bcc} - {}^\circ G_{Fe}^{fcc})/\mathrm{d}T = 2.4 + 9.1 = 11.5 = 1.4R; \quad R\ln(\beta + 1) = 1.4R; \quad \beta = 3.$

Exercise 16.6.2

Use information from Fig. 15.1 in order to estimate how many unpaired electrons hcp-Fe has if fcc-Fe has 0.5. Assume that fcc- and hcp-Fe both have very low critical temperatures and that they have approximately the same Debye temperature.

Hint

The slope of the curve for $°G_{Fe}^{hcp} - °G_{Fe}^{fcc}$ at 500 K should represent the difference in magnetic entropy when fcc and hcp are almost completely disordered.

Solution

The slope gives an entropy difference of 5 J/mol K. If fcc-Fe has 0.5 unpaired electrons this would give a magnetic entropy of $R\ln 1.5 = 3.4$. The magnetic entropy of hcp-Fe would be $3.4 - 5 = -1.6$ but should not be negative. It seems reasonable to accept that hcp-Fe has no unpaired electrons and that our negative value is due to a difference in Debye temperature between hcp- and fcc-Fe.

16.7

A simple physical model for the magnetic contribution

The magnetic disorder remaining below the Curie temperature can be treated in about the same way as we treated vacancies. We have already discussed the contribution to the entropy. In order to develop a complete model for the magnetic transformation we must also formulate the contribution to the enthalpy. The calculations will be especially simple if the magnetic moment of the atoms is 1/2 since this value gives two possible orientations with a spin difference of 1, according to the requirements of quantum mechanics. Let us suppose that n atoms in a system of totally N atoms have their spins directed opposite to the majority which consists of $N - n$ atoms. We may guess that the extra energy connected with the misorientation of the n atoms may be represented by the following simple expression,

$$\Delta H = K_1 n(N - n) = K_1 x(1 - x)N^2$$

where x represents the fraction of atoms with the wrong orientation, $x = n/N$. This expression gives ΔH a maximum at $x = 1/2$, i.e. for the disordered, demagnetized state. Since $x = 0$ represents the fully magnetized state, we can write

$$\Delta H = x(1 - x)\cdot 4\Delta H^{dis}$$

where ΔH^{dis} is the enthalpy increase on complete disordering.

Let us now turn to the configurational entropy. The two types of atoms can mix with each other in a number of ways,

$$W = \frac{N!}{n!(N-n)!}$$

Boltzmann's relation will thus yield

$$\Delta S/k = \ln N! - \ln n! - \ln(N-n)! = N\ln N - n\ln n - (N-n)\ln(N-n)$$

$$= - n\ln\frac{n}{N} - (N-n)\ln\frac{N-n}{N}$$

$$\Delta S/R = - \frac{n}{N}\ln\frac{n}{N} - \frac{N-n}{N}\ln\frac{N-n}{N} = - x\ln x - (1-x)\ln(1-x)$$

By combination of ΔH and ΔS we obtain for the Gibbs energy of disordering

$$\Delta G = x(1-x)\cdot 4\Delta H^{\text{dis}} + RT[x\ln x + (1-x)\ln(1-x)]$$

We want to determine the equilibrium value of x under constant T and P and shall thus consider a process by which x is increased. By identifying x with an internal variable ξ we obtain, at equilibrium,

$$- D = \left(\frac{dG}{d\xi}\right)_{T,P} = \frac{d\Delta G}{dx} = (1-2x)\cdot 4\Delta H^{\text{dis}} + RT\left[\ln x + \frac{x}{x} - \ln(1-x) - \frac{1-x}{1-x}\right]$$

$$\ln\frac{x}{1-x} = - (1-2x)\cdot 4\Delta H^{\text{dis}}/RT$$

The solutions of this equation are plotted against temperature in Fig. 16.6. Below a critical temperature there are two identical states, $x_2 = 1 - x_1$. They represent ordered states but the degree of order varies from perfect at absolute zero ($x = 0$ or 1) and decreases gradually as the temperature is raised. Finally it disappears completely at the critical temperature where $x = 0.5$. The states above the critical temperature are disordered because $x = 0.5$. The value of the critical temperature is obtained by inserting the value of $x = 0.5$,

$$T_C = \frac{4\Delta H^{\text{dis}}}{R}\cdot\frac{2x-1}{\ln[x/(1-x)]} = \frac{4\Delta H^{\text{dis}}}{R}\cdot\frac{2}{1/x + 1/(1-x)}$$

$$= \frac{4\Delta H^{\text{dis}}}{R}\cdot 2x(1-x) = 2\Delta H^{\text{dis}}/R$$

Above the critical temperature there is only one solution showing that the disordered state ($x = 0.5$) is the stable state. The equation also has the solution $x = 0.5$ below the critical temperature but there it represents an unstable state. That is why it has been drawn with a dashed line. The fact that the disordered state is here unstable, can be tested with a stability condition from Section 5.8. We obtain, for $x = 0.5$ and $T < T_C = 2\Delta H^{\text{dis}}/R$,

$$B = \left(\frac{\partial^2 G_m}{\partial x^2}\right)_{T,P} = - 2\cdot 4\Delta H^{\text{dis}} + RT(1/x + 1/(1-x)) = - 8\Delta H^{\text{dis}} + 4RT < 0$$

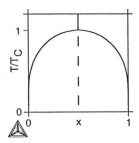

Figure 16.6 Magnetic ordering as a function of temperature according to a simple model. $x = 0.5$ represents the disordered state. The dashed line represents unstable equilibria.

Let us now assume incorrectly that the magnetic transformation is a sharp one and that a completely disordered state becomes completely ordered at T^1. The entropy of the completely disordered state is $\Delta S^{dis} = R\ln 2$ and for a sharp transformation we should thus have expected the following temperature,

$$T^1 = \Delta H^{dis}/\Delta S^{dis} = \Delta H^{dis}/R\ln 2 = 1.44\Delta H^{dis}/R = 0.72T_C$$

The fact that T_C is larger than T^1 was illustrated in Fig. 16.5. This result demonstrates that the possibility for some of the atoms to disorder within the ordered state increases the stability of the ordered state and raises the transformation temperature from the value predicted for a sharp transformation. On the other hand, it should also be pointed out that the stability of the disordered state is increased by the existence of some short-range order which has not been considered in our model. This is the factor that produces the tail above the C_P maximum which was shown in Fig. 16.4. It makes the difference between T_C and T^1 smaller than indicated by the above model.

The model we have used here for the magnetic order–disorder transformation is based upon the assumption that each individual atom has its own magnetic moment. This is called the localized spin model and is regarded as a rather crude approximation. Magnetism can be treated in a more satisfactory way by the electron band theory. However, it is much more complicated and thus difficult to apply, in particular at high temperature and in alloys.

Exercise 16.7.1

Use the C_P curve in Fig. 16.4 to evaluate the enthalpy of magnetic disordering. Then estimate the number of unpaired electrons in bcc-Fe.

Hint

The number of unpaired electrons could be estimated from the magnetic entropy which, in turn, could be evaluated from the enthalpy and the estimated transformation temperature for a sharp transformation.

Solution

A rough graphical integration yields $H^{dis} = 7000\,\text{J/mol}$. We known $T_C = 1043\,\text{K}$ and accepting the result of the simple model we can estimate $T^1 = 0.72T_C$, and thus we get $\Delta S^{dis} = \Delta H^{dis}/T^1 = 7000/(0.72\cdot1043) = 9.3$; $\ln(\beta + 1) = \Delta S/R = 1.12$; $\beta = 2.1$.

16.8   *The physical background of magnetism*

So far we have limited the discussion to ferromagnetic substances. Antiferromagnetism can be treated in a very similar way. From a thermodynamic point of view it makes a very small difference if there is a tendency for the magnetic moment of an atom to arrange itself parallel to its neighbours or anti-parallel. We do not need to discuss the reason why materials with unpaired spins behave in these two different ways. From a thermodynamic point of view it may be regarded as self-evident that one of the two types of ordering should give the lowest possible energy and that state should form at a sufficiently low temperature.

There are no unpaired electron spins in the electron gas of a metal and all the electrons in the outer shell go into the electron gas. However, unpaired electron spins can be found in an inner shell if it is not completely filled. This is the case for the 3d shell in the first series of transition metals and for the 4f shell in the rare earth metals. Table 16.1 shows the number of unpaired spins in the first series of transition metals according to magnetic measurements (Weiss 1963). It is evident that the number of unpaired electrons per atom is rarely close to an integer.

Let us now examine these values using the very naive model that the electrons in the 3d shell should have a tendency to arrange their spins in parallel as far as possible. If we further assume that all these metals should have one valence electron, then we predict the solid line in Fig. 16.7.

Experimental values from the table have also been plotted in the diagram and it is now interesting to discuss the disagreement. Up to V it is simply due to the fact that the number of valence electrons increases up to 4. For Ni, Co and α-Fe we have a rather satisfactory agreement. On the other hand, the existence of two points for Fe, one referring to α(bcc) and the other to γ(fcc), indicates that the full explanation must be based upon a much deeper insight. It may be pointed out that it has even been suggested that the atoms in γ-Fe are divided on two different energy levels with different numbers of unpaired spins.

It is also possible to evaluate the number of unpaired electrons from the thermodynamic effect of the magnetic transition, using the relation $\Delta S^{dis} = R\ln(\beta + 1)$.

Table 16.1 *Magnetic properties of some element*

Element	Number of unpaired electrons per atom	Magnetic structure	$T_C(K)$
bcc-Cr	~0.4	anti-ferro	313
α-Mn		anti-ferro	~100
fcc-Mn	2.6	anti-ferro	580
bcc-Fe	2.22	ferro	1043
fcc-Fe	~0.5	anti-ferro	80
hcp-Co	1.73	ferro	1388
fcc-Co	1.8	ferro	
fcc-Ni	0.6	ferro	627

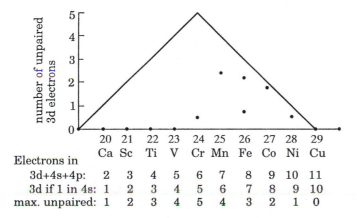

Figure 16.7 Diagram illustrating a simple method of estimating the number of unpaired electrons.

This method can only be applied after separation of ΔS^{dis} from other contributions to the entropy and is thus subject to considerable uncertainties. The results obtained are presented in Table 16.2. In the modelling of thermodynamic properties it is necessary to use such 'thermodynamic' values.

The situation is more complicated in phases with different kinds of sites. As an example one may take magnetite, Fe_3O_4, where one Fe atom occupies sites in a tetrahedral sublattice and two occupy sites in an octahedral sublattice. In this case there is a tendency for anti-parallel arrangement of neighbouring atoms but they are situated in different sublattices. The net result will thus be a parallel arrangement within each sublattice but the magnetic moment of the two sublattices will act against each other. The saturation magnetization will thus be given by the difference between them. On the other hand, all the atoms will contribute to the thermodynamic effect. If this model is correct, then the 'thermodynamic' value will be considerably larger than the value obtained from magnetic measurements.

Table 16.2 *Estimated number of unpaired electrons*

Element	Physical information	Thermodynamic information
bcc-Cr	0.4	0.008
bcc-Fe	2.22	2.22
fcc-Co	1.8	1.35
fcc-Ni	0.6	0.52

Exercise 16.8.1

Estimate the change in saturation magnetization of magnetite as Zn is dissolved. The structure of magnetite can be roughly described by the formula $(Fe^{+3})_1(Fe^{+2},Fe^{+3})_2(O^{-2})_4$. Suppose Zn goes into the first sublattice.

Hint

For convenience, abbreviate Fe^{+2} as 2 and Fe^{+3} as 3 and Zn^{+2} as Zn. By applying the condition of electroneutrality to pure magnetite we find $y_2'' = y_3'' = 0.5$, if y_i'' is used to denote site fractions on the second sublattice. This relation must be modified when there is Zn on the first sublattice, $y_{Zn}' > 0$. The scheme in Fig. 16.7 shows that Fe^{+2} should have 4 unpaired electrons, Fe^{+3} should have 5.

Solution

Electroneutrality yields
$$0 = 1 \cdot [3(1 - y_{Zn}') + 2y_{Zn}'] + 2 \cdot [3y_3'' + 2(1 - y_3'')] + 4 \cdot (-2)$$
$$= 3 - y_{Zn}' + 2y_3'' + 4 - 8 = 2y_3'' - y_{Zn}' - 1. \text{ Thus, } y_3'' = 0.5 + 0.5y_{Zn}'$$
and $y_2'' = 0.5 - 0.5y_{Zn}'$ and the magnetic moment is $-5y_3' + 2 \cdot [5y_3'' + 4y_2''] =$
$-[5(1 - y_{Zn}')] + 2 \cdot [5(0.5 + 0.5y_{Zn}') + 4(0.5 - 0.5y_{Zn}')] = 4 + 6y_{Zn}'.$ For each
Zn atom the magnetic moment will increase by 6 Bohr magnetons. The highest value seems to be 10 and would be obtained for $(Zn^{+2})_1(Fe^{+3})_2(O^{-2})_4$. However, with no moments on the first sublattice there will be no anti-ferromagnetic coupling and the magnetic order may be completely different from what we have here assumed. This model is thus of doubtful value, especially at high Zn contents.

16.9   ***Random mixture of atoms***

Before leaving the discussion of models of disordering phenomena we should mention disorder in solution phases. Actually, most substances can have a variable composition and we shall call such substances solution phases. On the atomic scale they consist of a mixture of different species, in the simplest case atoms. In a crystal the atomic sites are arranged in a regular pattern, a lattice, but the distribution of different kinds of atoms on the sites is generally determined by chance to some extent. This situation is often described as chemical disorder or configurational disorder. For a complete description of such a case one needs a model with an internal variable representing the degree of order. Such a model will be further discussed in the Chapters 18 and 19. In order to prepare for that discussion it is convenient now to discuss the entropy of solution phases with the maximum chemical disorder, so called **random mixtures**. The degree of order can be used as an internal variable but will not be introduced until later.

Let us consider a substance where all the sites are equivalent. Since all the various atoms dissolved in the substance can substitute for each other, such a substance is called a substitutional solution. The number of different ways in which N_A atoms of A, N_B atoms of B, N_C atoms of C, etc., can be arranged is

$$W = \frac{N!}{N_A! N_B! N_C! \dots}$$

This randomness will give the following contribution to the entropy of the system according to Boltzmann's relation,

$$\Delta S/k = \ln W = \ln N! - \Sigma \ln N_i!$$
$$\cong N \ln N - \Sigma N_i \ln N_i = - N\Sigma (N_i/N) \ln(N_i/N) = - N\Sigma x_i \ln x_i$$

since N is equal to ΣN_i and $x_i = N_i/N$.

We may construct a model for an ideal substitutional solution by requiring that there is no energy change on mixing the atoms. The only effect on the Gibbs energy comes from the configurational disorder,

$$\Delta G = - T\Delta S = NkT\Sigma x_i \ln x_i = NRT\Sigma x_i \ln x_i$$

where N is the number of moles of atoms. For one mole of atoms $\Delta G_m = \Delta G/N$ and we obtain, by using the relation between partial and integral quantities derived in Section 3.5,

$$\Delta G_m = RT\Sigma x_i \ln x_i$$

$$\Delta \mu_j = \Delta G_m + \frac{\partial \Delta G_m}{\partial x_j} - \Sigma x_i \frac{\partial \Delta G_m}{\partial x_i} = RT \ln x_j$$

Figure 16.8 gives ΔG_m and $\Delta \mu_B$ for a binary A–B system. It is worth emphasizing that $\Delta \mu_B$ goes to $-\infty$ as x_B goes to 0. As a consequence, the ΔG_m curve should be

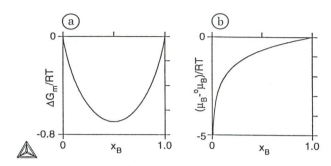

Figure 16.8 (a) The Gibbs energy of mixing for an ideal, binary solution. (b) The variation of a chemical potential in the same solution. In both cases the values are normalized by dividing with RT.

vertical at its end-points. However, this tendency starts to develop so close to the end-points that it is hardly discernible.

In the model for an ideal solution there is no internal variable because the situation has been completely fixed by the assumption of random mixing. The situation will be different if all the arrangements do not have the same energy. The disorder will then be incomplete and one can introduce internal variables describing short- and long-range order. This phenomenon will be described in Section 19.6. However, there is a class of solution phases where all the sites are not equivalent and they may be, at the same time, close to random in one sense and well ordered in another. They will be discussed in the next sections.

Exercise 16.9.1

Stirling's approximation to three terms is: $\ln N! \cong N\ln N - N + \ln(2\pi N)^{1/2}$. Derive S for a mixture of A and B using one, two and three terms.

Hint

Only the first term was used in the text.

Solution

$\ln N! - \ln N_A! - \ln N_B! = N\ln N - N_A\ln N_A - N_B\ln N_B - N + N_A + N_B + 0.5\ln[2\pi N(2\pi N_A \cdot 2\pi N_B)] = - N(x_A\ln x_A + x_B\ln x_B) - (1/2)\ln(2\pi N_A N_B/N) \cong - N(x_A\ln x_A + x_B\ln x_B)$ for large N. The first term gives the ideal expression, the second term gives no contribution. The third term is negligible for large N.

Exercise 16.9.2

From the treatment of thermal vacancies we obtained $\Delta S_m/R =$ $-\ln(1 - y_v) - [y_v/(1 - y_v)] \cdot \ln y_v$. Compare with the expression obtained for a binary alloy A–B if we identify B with vacancies.

Hint

Examine for what size of a system each expression is defined.

Solution

$\Delta S_m/R = -x_A \ln x_A - x_B \ln x_B = -(1 - x_B)\ln(1 - x_B) - x_B \ln x_B$ holds for one mole of an alloy. The quantity y_v is the site fraction but x_B is also a site fraction because there is only one kind of site and in the A–B alloy they are all filled by atoms. We could thus get complete agreement between the two cases by expressing ΔS_m for the case with vacancies per mole of sites, i.e., by multiplying with $1 - y_v$, the number of atoms per site.

Exercise 16.9.3

An A–B–C alloy is prepared by mixing x_C moles of pure C with $1 - x_C$ moles of an A–B alloy. Calculate (a) the entropy of mixing in $1 - x_C$ moles of the initial A–B alloy, (b) the entropy due to the mixing of C and the A–B alloy and (c) the sum of the two contributions.

Hint

The A content of the initial alloy can be expressed as $x_A/(x_A + x_B)$ if x_A and x_B refer to the final, ternary alloy. The effect of mixing C with a mixture of A + B is the same as the effect of mixing C with a single element which could very well consist of two isotopes. Remember that $x_A + x_B = 1 - x_C$.

Solution

$\Delta S(a) = (1 - x_C) \cdot \{x_A/(x_A + x_B) \cdot \ln[x_A/(x_A + x_B)] + x_B/(x_A + x_B) \cdot \ln[x_B/(x_A + x_B)]\} = x_A \ln x_A + x_B \ln x_B - (x_A + x_B)\ln(x_A + x_B);$
$\Delta S(b) = x_C \ln x_C + (1 - x_C)\ln(1 - x_C) = x_C \ln x_C + (x_A + x_B)\ln(x_A + x_B);$
$\Delta S(c) = x_A \ln x_A + x_B \ln x_B + x_C \ln x_C.$

16.10   *Restricted random mixture*

Many crystalline phases have more than one family of sites. It is convenient to describe such phases with the use of sublattices and the state may be defined by giving the site fractions (see Section 3.11).

$$y^t_j = N^t_j/N^t$$

If the atoms in one sublattice are mixed with each other at random, they give the following contribution to the entropy

$$\Delta S^t = - kN^t \sum_i \frac{N^t_i}{N^t} \ln \frac{N^t_i}{N^t}$$

Since entropy is an extensive property which obeys the law of additivity, one could add the contributions from the individual sublattices,

$$\Delta S_m = \Sigma \Delta S^t = - R \sum_s a^s \sum_i y^s_i \ln y^s_i$$

This expression holds for one mole of atoms if there are no vacant sites. a^s is then defined as N^s/N, the fraction of all sites belonging to the s sublattice, and Σa^s is unity. One may instead like to consider one mole of formula units where the formula is written with integers for all a^s. Then Σa^s is the number of atoms per formula unit.

This may be regarded as an ideal solution model for the particular type of crystalline structure. If all the elements can go into all the sublattices with the same probability, this model reduces to the previous ideal solution model because y^s_i is then identical to x_i.

Exercise 16.10.1

Consider a so-called Laves phase with two components each on two sublattices, $(A,B)_1(C,D)_2$. Calculate the entropy of mixing assuming random mixing within each sublattice for equal amounts of A and B and also for C and D. Compare with the ideal entropy of mixing when all four components are mixed randomly with each other.

Hint

Consider 3 moles of atoms, 0.5 of A, 0.5 of B, 1 or C and 1 of D.

Solution

$\Delta S/R = - 1 \cdot (0.5 \ln 0.5 + 0.5 \ln 0.5) - 2 \cdot (0.5 \ln 0.5 + 0.5 \ln 0.5) = + 3 \ln 2 = 2.08$. For ideal solution we get

$$\Delta S/R = 3[(1/6)\ln(1/6) + (1/6)\ln(1/6) + (2/6)\ln(2/6) + (2/6)\ln(2/6)] = 3.99.$$

16.11   # *Crystals with stoichiometric vacancies*

A binary crystal can vary in composition even if each component is restricted to its own sublattice. An example is wüstite which has separate sublattices for Fe and O. The Fe sublattice can have vacant sites and we should thus write the formula as $(Fe,Va)_1O_1$. We shall call such vacancies **stoichiometric vacancies** because in the simplest case, their number is fixed by the stoichiometric imbalance between the amounts of the elements. The entropy of disorder in a crystal with any kind of vacancies is given by the expression already discussed, if the vacancies are distributed at random on a sublattice. The vacancies are then treated as the atoms of any element and one must define their site fraction, y_{Va}^s. On the other hand, the vacancies are not included in the mole fractions x_i which give the composition of the crystal. Consequently, when evaluating x from y, one should not include the vacancies in the summation. We can give the equation as

$$x_j = \sum_s a^s y_j^s \bigg/ \left(\sum_s a^s - \sum_s a^s y_{Va}^s \right)$$

On the other hand, it is not always possible to calculate the y values from the overall composition given by the set of x values. If each element is dissolved in one sublattice only, then we can still evaluate the site fractions from the composition of the crystal provided that three is at least one sublattice without vacancies. First we can identify that sublattice as the one having the largest value of $\Sigma x_i^s/a^s$. Let us denote that sublattice by r. The site fractions in any sublattice t are then given by

$$y_j^t = x_j a^r / a^t \sum_i x_i^r$$

If Σy_j^t in any sublattice is less than unity then the difference from unity gives the site fraction of vacancies

$$y_{Va}^t = 1 - \sum_i y_i^t$$

This relation illustrates why they are called stoichiometric vacancies. The amount of stoichiometric vacancies does not change directly with the temperature but there may be an indirect effect if the phase is in equilibrium with another phase. The composition may then vary by an exchange of atoms between the phases. The amount of vacancies can also vary if the composition of the other phase varies by an action from the outside. A typical example is an oxide in equilibrium with an atmosphere of variable P_{O_2}.

Exercise 16.11.1

At 1300 K the molar content of Fe in wüstite can vary between 0.467 and 0.488 and the O content from 0.533 to 0.512. What is the range of variation of the vacancy content?

Hint

Wüstite has the NaCl structure with $a^t = a^r$. Evidently the O sublattice is filled. We have Fe vacancies. It is convenient to use y fractions.

Solution

(a) $y^t_{Fe} = 0.467 \cdot 1/1 \cdot 0.533 = 0.876$; $y^t_{Va} = 0.124$.

(b) $y^t_{Fe} = 0.488 \cdot 1/1 \cdot 0.512 = 0.953$; $y^t_{Va} = 0.047$.

Exercise 16.11.2

We know the composition of an oxide by chemical analysis as $x_U = 0.252$, $x_{Pu} = 0.094$ and $x_O = 0.654$. Calculate the site fractions under the assumption that the oxide is a so-called stoichiometric phase with only a small deviation from stoichiometry.

Hint

Suppose that U and Pu occupy one sublattice and O another. The number of sites can then be estimated from the number of atoms supposing there are no vacancies. Vacancies can then be assumed on one sublattice by stoichiometric reasons.

Solution

$x_U + x_{Pu} = 0.346$; $x_O/(x_U + x_{Pu}) = 0.654/0.346 = 1.89$. This is close to 2. We may thus assume that the formula is $(U,Pu)_1(O,Va)_2$ and we find

$y_O = 0.654 \cdot 1/2 \cdot 0.346 = 0.945$; $y_{Va} = 0.055$.

16.12   *Interstitial solutions*

Phases with two or more sublattices are often called compounds or intermediary phases, or intermetallic phases when appropriate. Such compounds may be strictly stoichiometric or may show a deviation from stoichiometry, caused by defects. One such defect is the vacancy. A related case is the interstitial solution where some solute atoms dissolve in a crystalline solvent by going into interstitial sites that are initially empty. These sites form a sublattice. An example is the solution of carbon in the bcc modification of iron, so-called ferrite or α-Fe. The formula can be written as $Fe_1(Va,C)_3$. The entropy of interstitial solutions can be treated with the method discussed in the preceding sections. A more complicated case occurs when an element goes mainly into ordinary lattice sites but some of its atoms go into interstitial sites.

In Section 3.12 we found that it was possible to derive an expression for the chemical potential of a compound in a stoichiometric phase but not for the chemical potential of an element. The situation is different for a phase with vacancies. For the interstitial solution of C in α-Fe we can by standard methods derive expressions for μ_{FeC_3} and μ_{FeVa_3}. We can thus evaluate μ_C from

$$\frac{1}{3}(\mu_{FeC_3} - \mu_{FeVa_3}) = \frac{1}{3}\mu_{Fe} + \mu_C - \frac{1}{3}\mu_{Fe} - \mu_{Va} = \mu_C - \mu_{Va} = \mu_C$$

because we can usually assume that the chemical potential of vacancies is zero (see Exercise 16.2.2). Assuming random mixing of carbon atoms and vacancies we obtain the following contributions to the entropy of mixing, the Gibbs energy of mixing and the chemical potential of carbon

$$\Delta S_m = -3R(y_{Va}\ln y_{Va} + y_C\ln y_C)$$
$$\Delta G_m = 3RT(y_{Va}\ln y_{Va} + y_C\ln y_C)$$

$$\Delta\mu_C = \Delta\mu_C - \Delta\mu_{Va} = \frac{1}{3}\left(\Delta G_m + \frac{\partial\Delta G_m}{\partial y_C} - \Sigma y_i\frac{\partial\Delta G_m}{\partial y_i} - \Delta G_m - \frac{\partial\Delta G_m}{\partial y_{Va}}\right.$$

$$\left. + \Sigma y_i\frac{\partial\Delta G_m}{\partial y_i}\right) = \frac{1}{3}\left(\frac{\partial\Delta G_m}{\partial y_C} - \frac{\partial\Delta G_m}{\partial y_{Va}}\right) = \frac{1}{3}\cdot3RT\ln\frac{y_C}{y_{Va}} = RT\ln\frac{y_C}{1 - y_C}$$

Interstitial solutions will be further discussed in Section 18.2.

It is important to notice that the deviation from stoichiometry of a compound may also be caused by some atoms going into sites of the 'wrong' sublattice (in that connection called anti-sites) or into interstitial sites or by stoichiometric vacancies.

Exercise 16.12.1

The interstitial solution of carbon in fcc- and bcc-iron can be represented with the formula $(Fe)_1(Va,C)_1$ and $(Fe)_1(Va,C)_3$ respectively. The martensitic transformation from fcc to bcc is very rapid and carbon is not able to take advantage of the additional interstitial sites. Estimate how much larger the driving force for the martensitic transformation would have been if carbon could be distributed among all the sites available in bcc-Fe. Make a numerical calculation for an Fe–C alloy with a molar content of 0.02 C.

Hint

For simplicity, suppose that the redistribution of carbon atoms has an effect on the Gibbs energy through the ideal entropy term only. Consider a system with one mole of Fe atoms and thus $z_C = 0.02/0.98 = 0.0204$ moles of C atoms.

Solution

$\Delta G = 3RT[(z_C/3)\ln(z_C/3) + (1 - z_C/3)\ln(1 - z_C/3)] - RT[z_C\ln z_C + (1 - z_C)\ln(1 - z_C)] = RT(- 0.1219 + 0.0996) = - 0.223RT$. The driving force increases with $0.223RT$.

Exercise 16.12.2

Consider the interstitial solution of C and N in bcc-Fe. Assume random mixing of them and derive an expression for the contribution to the chemical potential of C from the mixing.

Hint

Derive μ_C from ΔG_m, using $\mu_C = \mu_C - \mu_{Va}$. Use $y_N = 1 - y_{Va} - y_C$; $\partial y_N/\partial y_{Va} = \partial y_N/\partial y_C = - 1$ if y_N is chosen as a dependent variable.

Solution

$\Delta G_m = 3RT(y_{Va}\ln y_{Va} + y_C\ln y_C + y_N\ln y_N)$ per mole of Fe.
$\Delta\mu_C = (1/3)(\partial\Delta G_m/\partial y_C - \partial G_m/\partial y_{Va}) = RT(y_C/y_C + \ln y_C - y_{Va}/y_{Va} - \ln y_{Va}) = RT\ln(y_C/y_{Va}) = RT\ln[y_C/(1 - y_C - y_N)]$

Exercise 16.12.3

Express the contribution to the chemical potential of carbon from random mixing in the interstitial sublattice in bcc-Fe in terms of the ordinary molar content, x_C.

Hint

The problem is to express y_C in terms of x_C. Start with the definition of y_C.

Solution

$y_C = x_C/3x_{Fe} = x_C/3(1 - x_C)$; $1 - y_C = (3 - 4x_C)/3(1 - x_C)$; $\Delta\mu_C = RT\ln[y_C/(1 - y_C)] = RT\ln[x_C/(3 - 4x_C)]$.

Mathematical modelling of solution phases

Ideal solution

The thermodynamic properties of some solutions were illustrated graphically in Section 6.1 and some mathematical expressions were also given. We shall now give a more thorough discussion. In Section 15.7 we mentioned the possibility of modelling a special physical effect, p, in a substance and defining the remaining part of the Gibbs energy as the property of a hypothetical state, h, which does not have that physical effect,

$$G_m = G_m^h + G_m^p$$

This approach can also be applied to solution phases. The most important application is the treatment of the thermodynamic effects of mixing the atoms in a solution. The hypothetical state would then be a so-called mechanical mixture of the pure components at the same temperature and pressure,

$$G_m^h = \Sigma x_i \, {}^\circ G_i$$

The physical effect, G_m^p would here be the contribution due to the intimate mixing of the atoms in a solution. It is usually denoted by $^M G_m$, which is thus defined by

$$G_m = \Sigma x_i \, {}^\circ G_i + {}^M G_m$$

The mechanical mixture may be regarded as a reference for the properties of a solution and $^M G_m$ gives the solution behaviour. It is called Gibbs energy of mixing but a better name would have been 'Gibbs energy of solution' because it represents the effect of forming a solution from a mechanical mixture. For a binary system the reference is a straight line in the molar Gibbs energy diagram, in a ternary system it is a plane, etc. Earlier, in Fig. 6.1, it was demonstrated that one can use this straight line in a binary system as the line of reference in a molar Gibbs energy diagram and the concept of Gibbs energy of mixing was introduced.

The simplest model for the intimate mixing of atoms in a solution is based on the assumption of random mixing and no particular interactions between atoms of different kinds. For that case we have

$$G_m = \Sigma x_i \, {}^\circ G_i - T \cdot {}^M S_m^{ideal}$$

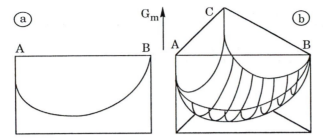

Figure 17.1 Characteristic shapes of the molar Gibbs energy curves in (a) binary and (b) ternary systems, caused by the entropy of mixing.

and expressions for the ideal entropy of mixing in phases with different structures were derived in Sections 16.9–16.12. A solution obeying such an equation may be called an ideal solution. For an ideal substitutional solution we have

$$^M S_m^{ideal} = - R\Sigma x_i \ln x_i$$
$$G_m = \Sigma x_i {}^\circ G_i + RT\Sigma x_i \ln x_i$$

and we may also define a partial ideal entropy of mixing

$$^M S_i^{ideal} = - R\ln x_i$$

Sometimes one explicitly requires that the molar volume of an ideal solution should be equal to a weighted average of the values for the pure components. It is easy to see that the last equation for G_m satisfies this requirement. The derivative with respect to P yields

$$V_m = \partial G_m/\partial P = \Sigma x_i \partial {}^\circ G_m/\partial P = \Sigma x_i {}^\circ V_i$$

The term $- T \cdot {}^M S_m^{ideal}$ is important even in real, non-ideal solutions and it gives the Gibbs energy its characteristic shape of a hanging rope for a binary system and a canopy for a ternary. This has already been illustrated in a number of diagrams in Sections 6.1 and 6.10 and is again shown in Fig. 17.1.

Exercise 17.1.1

Suppose one knows that a binary solution is not ideal because the molar volume obeys the relation: $V_m = x_A {}^\circ V_A + x_B {}^\circ V_B + x_A x_B \cdot |{}^\circ V_A - {}^\circ V_B|$. Examine the effect on G_m if ${}^\circ V_A - {}^\circ V_B = 2 \cdot 10^{-6}\,m^3/mol$.

Hint

Represent G_m with $x_A {}^\circ G_A + x_B {}^\circ G_B - T \cdot {}^M S_m^{ideal} + x_A x_B L$ and evaluate the effect on L.

Solution

Compare with $V_m = \partial G_m/\partial P = x_A{}^\circ V_A + x_B{}^\circ V_B + x_A x_B \partial L/\partial P$. Thus, $\partial L/\partial P = ({}^\circ V_A - {}^\circ V_B)$ and, neglecting the pressure dependence of ${}^\circ V_A - {}^\circ V_B$, we find $L = L_o + P \cdot ({}^\circ V_A - {}^\circ V_B)$. At 1 bar ($10^5$ Pa) this effect would increase L by $10^5 \cdot 2 \cdot 10^{-6} = 0.2$ J/mol. This is negligible.

17.2 *Mixing quantities*

The value of any molar quantity in a solution can be compared with the weighted average of the values for the pure components. We may thus generalize the equation defining the Gibbs energy of mixing

$$A_m = \Sigma x_i{}^\circ A_i + {}^M A_m$$

where ${}^M A_m$ is any **molar quantity of mixing**. The additional requirement for an ideal solution, mentioned in the preceding section, may thus be formulated by stating that the volume of mixing, ${}^M V_m$, must be zero.

Partial quantities of mixing can be defined relative to the value for the pure component,

$${}^M A_i = A_i - {}^\circ A_i$$

It is evident that all relations derived for A_m and A_i in Chapter 3 also apply to ${}^M A_m$ and ${}^M A_i$ because the references ${}^\circ A_i$ will drop out from all such relations. We get, for instance,

$${}^M A_m = A_m - \Sigma x_i{}^\circ A_i = \Sigma x_i A_i - \Sigma x_i{}^\circ A_i = \Sigma x_i (A_i - {}^\circ A_i) = \Sigma x_i{}^M A_i$$

We find, as for A_m in Section 3.5,

$${}^M A_j = {}^M A_m + \frac{\partial {}^M A_m}{\partial x_j} - \Sigma x_i \frac{\partial {}^M A_m}{\partial x_i}$$

Exercise 17.2.1

Show that the partial enthalpy of mixing can be calculated from ${}^M H_j = \partial({}^M G_j/T)/\partial(1/T)$.

Hint

We know $H_j = \partial(G_j/T)/\partial(1/T)$. Use ${}^M H_j = H_j - {}^\circ H_j$ and ${}^M G_j = G_j - {}^\circ G_j$.

Solution

$$\partial(^M G_j/T)/\partial(1/T) = \partial(G_j/T)/\partial(1/T) - \partial(^\circ G_j/T)/\partial(1/T) = H_j - {}^\circ H_j = {}^M H_j$$

17.3 *Excess quantities*

The various mixing quantities were defined relative to the mechanical mixture of the components. In the same way one may define **excess quantities** relative to an ideal solution, e.g.

$$G_m = \Sigma x_i {}^\circ G_i - T \cdot {}^M S_m^{ideal} + {}^E G_m$$

The excess quantities represent the deviation from ideal behaviour and are thus subject to direct study and modelling. In this chapter we shall examine mathematical models of the excess Gibbs energy and in the next chapter we shall discuss some very simple physical models.

From the excess Gibbs energy one may define partial excess Gibbs energies and obtain, by standard procedures,

$$
\begin{aligned}
{}^E G_j &= {}^E G_m + \partial^E G_m/\partial x_j - \Sigma x_i \partial^E G_m/\partial x_i \\
{}^E G_i &= G_i - {}^\circ G_i + T \cdot {}^M S_i^{ideal} \\
{}^E G_m &= \Sigma x_i {}^E G_i \\
{}^E S_i &= S_i - {}^\circ S_i - {}^M S_i^{ideal} \\
{}^E H_i &= H_i - {}^\circ H_i = {}^M H_i
\end{aligned}
$$

We note that it is not necessary to introduce the concept of excess enthalpy because all enthalpy of mixing is in excess of the ideal solution behaviour. For a substitutional solution we get

$$
\begin{aligned}
{}^E G_i &= G_i - {}^\circ G_i - RT \ln x_i \\
{}^E S_i &= S_i - {}^\circ S_i + R \ln x_i
\end{aligned}
$$

Exercise 17.3.1

Show that the excess partial enthalpy of mixing can be calculated from ${}^E H_j = \partial(^E G_j/T)/\partial(1/T)$.

Hint

We know $H_j = \partial(G_j/T)/\partial(1/T)$. Use ${}^E H_j = H_j - {}^\circ H_j$ and for a substitutional solution ${}^E G_j = G_j - {}^\circ G_j - RT \ln x_j$.

Solution

$\partial(^E G_j/T)/\partial(1/T) = \partial(G_j/T)/\partial(1/T) - \partial(^\circ G_j/T)/\partial(1/T) - \partial(R\ln x_j)/\partial(1/T) = H_j - {}^\circ H_j = {}^E H_j$ because $\partial(R\ln x_j)/\partial(1/T) = 0$. The relation is thus correct at least for this particular choice of $^M S_i^{ideal}$. In fact, it is correct also for any other expression of the ideal entropy.

Exercise 17.3.2

Using a Maxwell relation at constant T and P we get $\partial G_i/\partial N_j = \partial^2 G/\partial N_i \partial N_j = \partial^2 G/\partial N_j \partial N_i = \partial G_j/\partial N_i$. Using this relation, show that $\partial^E G_i/\partial N_j = \partial^E G_j/\partial N_i$.

Hint

$G_i = {}^\circ G_i + RT\ln x_i + {}^E G_i; \quad \partial x_i/\partial N_j = -x_i/N; \quad \partial x_j/\partial N_i = -x_j/N.$

Solution

$\partial G_i/\partial N_j = RT(1/x_i)\cdot(-x_i/N) + \partial^E G_i/\partial N_j = -RT/N + \partial^E G_i/\partial N_j; \quad \partial G_j/\partial N_i = -RT/N + \partial^E G_j/\partial N_i$. If the two left-hand sides are equal, we obtain $\partial^E G_i/\partial N_j = \partial^E G_j/\partial N_i$.

17.4   *Empirical approach to substitutional solutions*

We shall now discuss an empirical approach to the modelling of the excess Gibbs energy of solution phases. We have already seen that the difference between the properties of a real solution and an ideal solution may be represented by a quantity called the excess Gibbs energy, $^E G_m$. For a substitutional solution we can immediately see that it must go to zero for the pure components and a convenient way of representing its composition dependence in a binary system is to introduce the factor $x_A x_B$ which goes to zero for pure A as well as pure B. We shall thus express $^E G_m$ as $x_A x_B I$. We shall allow I to be a function of composition. It may for instance be represented by a power series, often called the Redlich–Kister (1948) polynomial. Figure 17.2 illustrates the properties of the various terms in $^E G_m$.

$$I = {}^0 L + {}^1 L(x_A - x_B) + {}^2 L(x_A - x_B)^2 + \ldots$$

$$^E G_m = x_A x_B \sum_{k=0}^{n} {}^k L(x_A - x_B)^k$$

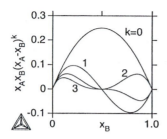

Figure 17.2 Properties of Redlich–Kister terms.

The parameter I may be regarded as a representation of the interaction between the two components. It is convenient to give the two components as an index to I and write it as I_{AB}. When I is a constant, independent of composition as well as temperature, one talks about a **regular solution** and ^0L may thus be called the regular solution parameter. When it appears alone, it is usually denoted by L. ^1L may be called the subregular solution parameter and ^2L the subsubregular solution parameter. However, it should be mentioned that a regular solution is sometimes defined as a solution where ^EG_m is independent of temperature but may have any composition dependence. On the other hand, in recent years there is a tendency to call a solution regular as soon as I is independent of composition, whether or not I is independent of temperature.

Positive values of the regular solution parameter result in a tendency of demixing. If not interrupted by other reactions a miscibility gap will form when the temperature is lowered. It is easy to calculate its spinodal from the condition for the stability limit. For a constant I we get

$$g_B = dG_m/dx_B = (x_A - x_B)^0L + RT(x_B/x_B + \ln x_B - x_A/x_A - \ln x_A)$$
$$g_{BB} = d^2G_m/dx_B^2 = (-1-1)^0L + RT(1/x_B + 1/x_A) = 0$$
$$T_{sp} = x_A x_B \cdot 2^0L/R$$
$$x_A x_B = RT_{sp}/2^0L$$

The spinodal will thus be a parabola in this simple case and its maximum will fall at $x_A = x_B = 0.5$ and be a consolute point.

$$T_{cons} = {}^0L/2R \text{ and } x_A x_B = T_{sp}/4T_{cons}$$

If I is not a constant, one would also have to calculate the consolute composition. A second equation is required and it is obtained through the condition of a critical point.

$$d^3G_m/dx_B^3 = 0$$

For further reference it is convenient here to derive expressions for the partial Gibbs energies from the Redlich–Kister polynomial. Using the standard method presented in Section 3.5, we can evaluate the last two terms in the following general expression

$$G_l = {}^0G_l - T \cdot {}^MS_l^{ideal} + {}^EG_l$$

With the power series representation of $^{E}G_{m}$ we first obtain a general expression for the effect of an interaction between two components, l and i,

$$^{E}G_{l} = {}^{0}L_{li}x_{i}(1 - x_{l}) + \sum_{k=1}^{n} {}^{k}L_{li}x_{i}(x_{l} - x_{i})^{k-1}[(k + 1)(1 - x_{l})(x_{l} - x_{i}) + kx_{i}]$$

For a binary A–B system we can replace $1 - x_{A}$ by x_{B} and $1 - x_{B}$ by x_{A} and obtain

$$^{E}G_{A} = x_{B}^{2}\{{}^{0}L_{AB} + \sum_{k=1}^{n} {}^{k}L_{AB}(x_{A} - x_{B})^{k-1}[(2k + 1)x_{A} - x_{B}]\}$$

$$^{E}G_{B} = x_{A}^{2}\{{}^{0}L_{AB} + \sum_{k=1}^{n} {}^{k}L_{AB}(x_{A} - x_{B})^{k-1}[x_{A} - (2k + 1)x_{B}]\}$$

It should be noticed that in our notation the sign of $^{k}L_{AB}$ changes for odd k if the order between A and B is reversed. It is also interesting to note that the expression for the partial excess Gibbs energy of one component contains the square of the molar content of the other. The partial excess Gibbs energy of a component will thus go to zero asymptotically as the pure component is approached. In this respect the very dilute solutions are ideal, a result which is usually formulated in Raoult's law. It will be discussed in the next section.

For a solution with more than two components we should consider interactions within each combination of two components and possibly also interactions between more than two components, so-called ternary, quaternary, etc., interactions. By limiting the present discussion to binary interactions we obtain

$$G_{m} = \Sigma x_{i}{}^{o}G_{i} + RT\Sigma x_{i}\ln x_{i} + \sum_{i}\sum_{j>i} x_{i}x_{j}I_{ij}$$

and $^{E}G_{l}$ will be the sum of contributions from all I_{li} but also from I_{ij}. We obtain

$$^{E}G_{l} = \sum_{i \neq l}\left\{{}^{0}L_{li}x_{i}(1 - x_{l}) + \sum_{k=1}^{n} {}^{k}L_{li}x_{i}(x_{l} - x_{i})^{k-1}[(k + 1)(1 - x_{l})(x_{l} - x_{i}) + kx_{i}]\right\}$$

$$- \sum_{i \neq l}\sum_{j \neq l,>i} x_{i}x_{j}\left[{}^{0}L_{ij} + \sum_{k=1}^{n} {}^{k}L_{ij}(x_{i} - x_{j})^{k}(k + 1)\right]$$

For a ternary system with constant interaction energies (i.e. for the regular solution model) we can write the result as follows by omitting the superscript 0 in ^{0}L,

$$G_{A} = {}^{o}G_{A} + RT\ln x_{A} + x_{B}(x_{B} + x_{C})L_{AB} - x_{B}x_{C}L_{BC} + x_{C}(x_{C} + x_{B})L_{CA}$$
$$G_{B} = {}^{o}G_{B} + RT\ln x_{B} + x_{A}(x_{A} + x_{C})L_{AB} + x_{C}(x_{C} + x_{A})L_{BC} - x_{C}x_{A}L_{CA}$$
$$G_{C} = {}^{o}G_{C} + RT\ln x_{C} - x_{A}x_{B}L_{AB} + x_{B}(x_{B} + x_{A})L_{BC} + x_{A}(x_{A} + x_{B})L_{CA}$$

Exercise 17.4.1

It has been suggested that the temperature dependence of the regular solution parameter L should be described with $A + B/T$ instead of $a + bT$ because the latter expression may make L change sign at some high temperature which is claimed to be unrealistic. Of course, the same objection can be raised against the new suggestion because it may change sign at some low temperature. Anyway, accept the new suggestion and evaluate the enthalpy and entropy parts of L.

Hint

Write $L = L_H - L_S T$.

Solution

By definition $L_S = -\partial L/\partial T = B/T^2$ and $L_H = \partial(L/T)/\partial(1/T) = A + 2B/T$.

Exercise 17.4.2

Show that the expressions given for $^E G_A$ and $^E G_B$ in a binary system obey the equation $G_m = \Sigma x_i G_i$.

Hint

We know that for an ideal solution $G_m = \Sigma x_i G_i$. Now it is thus sufficient to show that $^E G_m = \Sigma x_i {}^E G_i$.

Solution

$\Sigma x_i {}^E G_i = x_A x_B^2 \{^0 L_{AB} + \Sigma^k L_{AB}(x_A - x_A)^{k-1}[(2k+1)x_A - x_B]\}$
$+ x_B x_A^2 \{^0 L_{AB} + \Sigma^k L_{AB}(x_A - x_B)^{k-1}[x_A - (2k+1)x_B]\} = x_A x_B \{^0 L_{AB}(x_A + x_B)$
$+ \Sigma^k L_{AB}(x_A - x_B)^{k-1}[x_B(2k+1)x_A - x_B^2 + x_A^2 - x_A(2k+1)x_B]\} = x_A x_B[^0 L_{AB} +$
$\Sigma^k L_{AB}(x_A - x_B)^k] = {}^E G_m$

Exercise 17.4.3

Mo has the same structure as α-Fe (bcc) and it is thus possible to combine them into the same model, covering the whole range of composition in the Fe–Mo phase diagram. Determine the G_m expression for this model from the solubilities at 1300 °C which are 0.16 Mo in Fe and 0.075 Fe in Mo.

Hint

Since two pieces of information are given, we can only determine two parameters. The excess term will thus be written as $x_{Fe}x_{Mo}[^0L + {}^1L(x_{Fe} - x_{Mo})]$ where ^0L and ^1L will be considered as independent of T. At equilibrium $G_{Fe} - {}^0G_{Fe}$, i.e. $RT\ln x_{Fe} + {}^EG_{Fe}$ would have the same value in both phases and so would $RT\ln x_{Mo} + {}^EG_{Mo}$. For a binary solution we find ${}^EG_A = x_B^2[^0L + {}^1L(3x_A - x_B)]$ and ${}^EG_B = x_A^2[^0L + {}^1L(x_A - 3x_B)]$.

Solution

$RT\ln x_{Fe} + {}^EG_{Fe} = RT\ln 0.84 + (0.16)^2[^0L + {}^1L(3\cdot0.84 - 0.16)] = RT\ln 0.075 + (0.925)^2[^0L + {}^1L(3\cdot0.075 - 0.925)]$;

$RT\ln x_{Mo} + {}^EG_{Mo} = RT\ln 0.16 + (0.84)^2[^0L + {}^1L(0.84 - 3\cdot0.16)] = RT\ln 0.925 + (0.075)^2[^0L + {}^1L(0.075 - 3\cdot0.925)]$. The numerical result is $^0L = 34\,500$ and $^1L = -4500$ J/mol.

Exercise 17.4.4

The Fe–Cr phase diagram has a two-phase region $\gamma(fcc) + \alpha(bcc)$ which cannot be described by applying the thermodynamics for dilute solutions because it has a minimum (at about 1123 K). Try to explain this behaviour with the regular solution model. At the same time evaluate the 'lattice stability' of fcc-Cr as a function linear in T, i.e. $^0G_{Cr}^\gamma - {}^0G_{Cr}^\alpha = a + bT$. The quantity $^0G_{Fe}^\gamma - {}^0G_{Fe}^\alpha$ is known and is tabulated here together with information on the $\gamma + \alpha$ region. The tabulated compositions are taken in the middle of $\gamma + \alpha$ where one may assume that γ and α have the same Gibbs energy.

T(K)	x_{CR}°	$^0G_{Fe}^\gamma - {}^0G_{Fe}^\alpha$ (J/mol)
1170	0.012	10.54
1170	0.121	10.54
1600	0.057	− 30.12

Hint

From the table we get three conditions through $G_m^\alpha = G_m^\gamma$. Two parameters are already defined, a and b. Let the third one be $L^\gamma - L^\alpha$.

Solution

For each phase $G_m = x_{Fe}^\circ G_{Fe} + x_{Cr}^\circ G_{Cr} - TS_m^{ideal} + x_{Fe}x_{Cr}L$. At $x_{Cr}^\alpha = x_{Cr}^\gamma = x_{Cr}^\circ$ where $G_m^\alpha = G_m^\gamma$: $0 = x_{Fe}^\circ(^\circ G_{Fe}^\gamma - ^\circ G_{Fe}^\alpha) + x_{Cr}^\circ(^\circ G_{Cr}^\gamma - ^\circ G_{Cr}^\alpha) + x_{Fe}^\circ x_{Cr}^\circ(L^\gamma - L^\alpha)$. The data in the table will give:

$$0.988 \cdot 10.54 + 0.012(a + 1170b) + 0.988 \cdot 0.012(L^\gamma - L^\alpha) = 0;$$
$$0.879 \cdot 10.54 + 0.121(a + 1170b) + 0.879 \cdot 0.121(L^\gamma - L^\alpha) = 0;$$
$$0.943 \cdot (-30.12) + 0.057(a + 1600b) + 0.943 \cdot 0.057(L^\gamma - L^\alpha) = 0.$$

We find $^\circ G_{Cr}^\gamma - ^\circ G_{Cr}^\alpha = a + bT = 3476 + 2.417T$ J/mol; $L^\gamma - L^\alpha = -7259$ J/mol.

Exercise 17.4.5

For a binary solution with a constant positive L in $G_m = x_A{}^\circ G_A + x_B{}^\circ G_B + RT(x_A \ln x_A + x_B \ln x_B) + Lx_Ax_B$ there will be a symmetric miscibility gap. Show that close to the critical point the width of the gap is $\sqrt{3}$ times the width of the spinodal curve.

Hint

We have seen that the spinodal curve for a symmetric system is described by $x_Ax_B = T/4T_{cons} = RT/2L$. Let Δx be the width of the spinodal. By symmetry reasons $x_B' = (1 - \Delta x)/2$; $x_B'' = (1 + \Delta x)/2$, and Δx can be expressed in terms of $(T_{cons} - T)/T_{cons}$. Derive a similar equation for the binodal curve using the condition that dG_m/dx_B should have the same value for the phases in equilibrium, i.e. for $x_B' = (1 - \Delta x)/2$ and for $x_B'' = (1 + \Delta x)/2$. Then introduce the approximation $\ln(1 + \Delta x) = \Delta x - \Delta x^2/2 + \Delta x^3/3$.

Solution

Spinodal: $T/T_{cons} = 4(1 + \Delta x)(1 - \Delta x)/4 = 1 - \Delta x^2$; $\Delta x^2 = (T_{cons} - T)/T_{cons}$.
Binodal: $dG_m/dx_B = -^\circ G_A + ^\circ G_B + RT\ln(x_B/x_A) + L(x_A - x_B)$ should have the same value on both sides: $RT\ln[(1 - \Delta x)/(1 + \Delta x)] + L(1 + \Delta x - 1 + \Delta x)/2 = RT\ln[(1 + \Delta x)/(1 - \Delta x)] + L(1 - \Delta x - 1 - \Delta x)/2$. This gives $RT(-\Delta x +$

$\Delta x^2/2 - \Delta x^3/3 - \Delta x - \Delta x^2/2 - \Delta x^3/3) + L\Delta x \cong RT(\Delta x - \Delta x^2/2 + \Delta x^3/3 + \Delta x + \Delta x^2/2 + \Delta x^3/3) - L\Delta x$ and $1 + \Delta x^2/3 = L/2RT = T_{\text{cons}}/T; \ \Delta x^2 = 3[(T_{\text{cons}}/T) - 1] = 3(T_{\text{cons}} - T)/T \cong 3(T_{\text{cons}} - T)/T_{\text{cons}}.$

17.5 *Real solutions*

As discussed already in Section 6.1, one often represents the properties of real solutions with the activity a_i and activity coefficient f_i, defined through

$$\mu_i = \mu_i^{\text{REF}} + RT\ln a_i = \mu_i^{\text{REF}} + RT\ln x_i + RT\ln f_i$$

It is common to choose as reference the state of pure i at the same temperature and pressure as the state under consideration.

The activity coefficient is often intended for use in dilute solutions only. It is interesting to examine its variation with composition by applying the regular solution model for a binary substitutional solution. Figure 17.3 shows three curves for the activity obtained with $^0L/RT = 2, 0$ and -2. For high B contents x_A is small and the regular solution model yields

$$a_B = x_B \exp(^0L x_A^2/RT) \cong x_B$$
$$f_B \cong 1$$

As a consequence, all the curves approach asymptotically the diagonal in the diagram. This is **Raoult's law** and its validity is very general. We have already seen that $^E G_B$ has the factor x_A^2 independent of what power series has been chosen. This factor originates from the factor $x_A x_B$ introduced in order to make $^E G_m$ go to zero for pure B and for pure A.

For low B contents where x_A may be approximated by 1 we obtain

$$a_B = x_B \exp(^0L x_A^2/RT) \cong x_B \exp(^0L/RT)$$
$$f_B \cong \exp(^0L/RT)$$

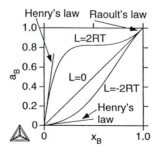

Figure 17.3 The properties of a binary solution according to the regular solution model with three different values of the regular solution parameter.

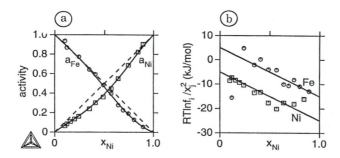

Figure 17.4 The properties of liquid Fe–Ni solutions at 1852 K according to direct measurements and a subregular solution model with $^0L_{Fe,Ni} = -10$ and $^1L_{Fe,Ni} = 5\,kJ/mol$.

This gives the slope of the tangent to the curve at the origin. It is self-evident that at low B contents the alloys fall close to this tangent. This is often formulated as **Henry's law**. It must be emphasized that this law says nothing about the slope of the tangent, contrary to Raoult's law. It is interesting to mention that Raoult's law for component A can be derived by applying Henry's law to component B.

A convenient method of studying the properties of a binary solution is based upon the presence of the factor x_A^2 in the expression for the partial excess Gibbs energy of B in the empirical model. We obtain

$$^0L_{AB} + \sum_{k=1}^{n} {}^kL_{AB}(x_A - x_B)^{k-1}[x_A - (2k+1)x_B] = {}^EG_B/x_A^2 = RT\ln(a_B/x_B)/x_A^2$$
$$= RT\ln f_B/x_A^2$$

If a plot of experimental values of $RT\ln f_B/x_A^2$ versus x_B can be represented by a straight line then the properties can be represented by two parameters, $^0L_{AB}$ and $^1L_{AB}$. If experimental data are available for both components, then the two-parameter representation requires that one finds two straight lines, one for each component and such that they yield the same set of ^0L and ^1L values. An example for liquid Fe–Ni at 1852 K is given in Fig. 17.4. In Fig. 17.4(a) the two lines have the same slope, representing $^1L_{Fe,Ni}$, and the same value at $x_{Ni} = 0.25$ for the Ni line and $x_{Fe} = 0.25$ for the Fe line, representing $^0L_{Fe,Ni}$. The experimental scatter is considerable in Fig. 17.4(b), especially at low Ni contents for Fe and at low Fe contents for Ni because of the very small values then taken by the factors x_{Ni}^2 and x_{Fe}^2 in the denominators.

It should be emphasized that there are many binary systems where the power series representation is not very convenient. Figure 17.5 shows an example from liquid Bi–Mg at 973 K where a very large number of terms would be needed for a satisfactory representation of the data. It is evident that some particular physical effect occurs at the centre of the system and it would be difficult to represent such data mathematically without identifying that effect and representing it with an adequate model. Such models will be described later on.

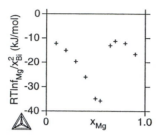

Figure 17.5 Experimental data from liquid Bi–Mg alloys at 973 K. This behaviour cannot be well represented with a power series.

Exercise 17.5.1

Consider the gaseous mixture (solution) of H_2 and O_2 at 1 bar and a temperature high enough for the reaction $2H_2 + O_2 \rightarrow 2H_2O$ to go to equilibrium but still low enough to make the formation of H_2O practically complete. Examine how the activity of O_2 would vary across the binary H_2–O_2 system as a function of x_{O_2}.

Hint

The binary system has two components and it is evident that they are defined as H_2 and O_2 and that x_{O_2} is thus defined with no regard for the formation of H_2O. Suppose that the gas is actually an ideal gas mixture of H_2, O_2 and H_2O. At constant T and P we then have $RT\ln a_{O_2} = \mu_{O_2} - {}^\circ\mu_{O_2} = RT\ln y_{O_2}$ where the definition of y_{O_2} takes into account the presence of H_2O.

Solution

Between $x_{O_2} = 0$ and $\frac{1}{3}$ there are practically no molecules of O_2. All oxygen goes into H_2O. Between $x_{O_2} = \frac{1}{3}$ and 1 there are, by the same reason, practically no H_2 molecules and $y_{H_2O} + y_{O_2} = 1$ and $x_{O_2} = (y_{O_2} + 0.5y_{H_2O})/(y_{O_2} + 1.5y_{H_2O})$ and thus $a_{O_2} = y_{O_2} = (3x_{O_2} - 1)/(x_{O_2} + 1)$. It is evident that it is impossible to describe the deviation of this solution from the ordinary ideal solution with a power series. Any reasonable model must recognize that the interaction between the two components is here so strong that large quantities of a new kind of molecule form.

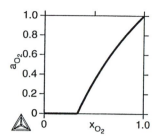

Solution 7.5.1

Exercise 17.5.2

It is common to define the activity of a component i in a solution with reference to pure i at the same temperature. The chemical potential of i in the solution is then written as $\mu_i = {}^\circ G_i(T) + RT\ln a_i$. Suppose that one would instead like to define an activity by referring to pure i at absolute zero as reference. Derive a relation between the two kinds of activity.

Hint

The chemical potential of i in the solution, μ_i, is an absolute quantity just like volume, in spite of the fact that it has no natural zero point. μ_i is thus independent of the choice of reference.

Solution

μ_i is the same in the two equations:
$\mu_i = {}^\circ G_i(T) + RT\ln a_i(T)$ and $\mu_i = {}^\circ G_i(0) + RT\ln a_i(0)$. Let us take the difference:
$0 = {}^\circ G_i(T) - {}^\circ G_i(0) + RT\ln[a_i(T)/a_i(0)]; \quad a_i(0) = a_i(T)\exp[({}^\circ G_i(T) - {}^\circ G_i(0))/RT]$.
Thus $a_i(0) < a_i(T)$ since $G_i(T) < {}^\circ G_i(0)$.

Exercise 17.5.3

Show that for a multicomponent solution $\partial\ln f_j/\partial x_i - \Sigma x_k \partial\ln f_j/\partial x_k = \partial\ln f_i/\partial x_j - \Sigma x_k \partial\ln f_i/\partial x_k$.

Hint

From Exercise 17.3.2 we know that $\partial^E G_j/\partial N_i = \partial^E G_i/\partial N_j$ and thus $\partial\ln f_j/\partial N_i = \partial\ln f_i/\partial N_j$. Now consider $\ln f_i$ and $\ln f_j$ as functions of x_i, etc., and remember that $\partial x_i/\partial N_i = (1 - x_i)/N$ and $\partial x_i/\partial N_j = - x_i/N$.

Solution

$\partial \ln f_j / \partial N_i = (\partial \ln f_j / \partial x_i) \cdot (1 - x_i / N) + \sum_{k \neq i} (\partial \ln f_j / \partial x_k) \cdot (- x_k / N) = [\partial \ln f_j / \partial x_i -$

$\Sigma x_k \partial \ln f_j / \partial x_k] / N$; $\partial \ln f_i / \partial N_j = [\partial \ln f_i / \partial x_j - \Sigma x_k \partial \ln f_i / \partial x_k] / N$. The two right-hand sides must be equal because the two left-hand sides are.

17.6   *Applications of the Gibbs–Duhem relation*

At constant T and P the Gibbs–Duhem relation reduces to

$$\Sigma x_i dG_i = 0 \text{ or } \Sigma x_i d\ln a_i = 0$$

We have here chosen the notation G_i instead of μ_i since we shall consider a single solution phase. By introducing partial excess Gibbs energies or activity coefficients through

$$G_i = {}^{\circ}G_i + RT\ln x_i + {}^{E}G_i = {}^{\circ}G_i + RT\ln x_i + RT\ln f_i$$

we get

$$\Sigma x_i d^{E}G_i = 0 \text{ or } \Sigma x_i d\ln f_i = 0$$

because

$$\Sigma x_i d\ln x_i = \Sigma x_i dx_i / x_i = \Sigma dx_i = 0$$

In a binary system it is thus possible to evaluate f_1 from measurements of f_2 by integration from pure 1.

$$\ln f_1 = - \int_0^{x_2} \frac{x_2}{x_1} \frac{d\ln f_2}{dx_2} dx_2$$

For graphical or numerical integration Wagner (1952) suggested that this equation should first be transformed by integration by parts, yielding

$$\ln f_1 = \int_0^{x_2} \frac{\ln f_2}{x_1^2} dx_2 - \frac{x_2 \ln f_2}{x_1}$$

One can also evaluate ${}^{E}G_m$ from the information on f_2. The last equation gives

$$x_1 \ln f_1 + x_2 \ln f_2 = x_1 \int_0^{x_2} \frac{\ln f_2}{x_1^2} dx_2$$

and this is identical to ${}^{E}G_m / RT$.

As an introduction to a discussion of ternary systems, it may be useful to repeat the last derivation by starting from the well-known expression for a binary system in Section 6.1, when x_1 is regarded as a dependent variable.

$$^EG_2 = {}^EG_m + (1 - x_2)\mathrm{d}^EG_m/\mathrm{d}x_2$$

This can be rearranged into

$$^EG_2 = (1 - x_2)^2 \cdot \mathrm{d}[^EG_m/(1 - x_2)]/\mathrm{d}x_2$$

and integration from $x_2 = 0$ where $^EG_m = 0$, yields

$$^EG_m = (1 - x_2)\int_0^{x_2} [^EG_2/(1 - x_2)^2]\mathrm{d}x_2$$

which is identical to the previous result.

For a ternary system one may derive the same equations if the ratio x_1/x_3 is regarded as the second independent variable. However, when integrating from $x_2 = 0$ one now starts from a binary 1–3 alloy and its ^EG_m is not zero. We should thus write the result as

$$^EG_m = \left\{ ^EG_m(x_2 = 0) + (1 - x_2)\int_0^{x_2} [^EG_2/(1 - x_2)^2]\mathrm{d}x_2 \right\}_{x_1/x_3}$$

This equation can be useful if one has measured $^EG_2$ (i.e. $RT\ln f_2$) in sections of constant x_1/x_3.

Exercise 17.6.1

The diagram shows experimental values of a_{Al} in Al–Ag alloys, represented through the quantity $\ln f_{Al}/x_{Ag}^2$. The solid line represents a reasonable curve across the whole

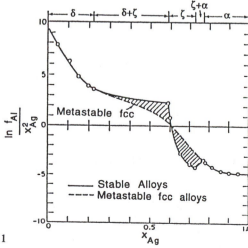

Exercise 17.6.1

system. Both Al and Ag are fcc and one can imagine a gradual change of this phase from pure Al to pure Ag provided that the ζ phase can be prevented. A corresponding curve (dashed line) has been drawn tentatively but obeying the rule that the two shaded areas must be equal. Prove that rule.

Hint

Imagine that ^EG_m for an alloy with a low value of x_{Ag} is calculated from $\ln f_{Al}$ by integration from pure Ag, following two different routes in the middle of the system.

Solution

$^EG_m/RT = x_{Ag}\ln f_{Ag} + x_{Al}\ln f_{Al} = x_{Ag}\int(\ln f_{Al}/x_{Ag}^2)dx_{Al}$. The integral represents the area under the curve one follows on integration. The difference between the two shaded areas represents the difference in integrated area if we follow the two different paths. In order to yield the same result after crossing the two-phase field, the two shaded areas must be equal.

Exercise 17.6.2

Derive Raoult's law from Henry's law for a binary solution. Then examine if one can go the other way.

Hint

We assume that f_2 is constant and thus $d\ln f_2/dx_2 = 0$ from $x_2 = 0$ up to some small value.

Solution

When calculating $\ln f_1$ by integration we find that the integrand is zero. Thus $\ln f_1 = 0$ and $f_1 = 1$ in agreement with Raoult's law. On the other hand, if we start with Raoult's law, $f_1 = 1$, then $d\ln f_1/dx_1 = 0$ but when calculating $\ln f_2$ by integration we now find that the integrand is indeterminate, $(d\ln f_1/dx_1)/x_2 = 0/0$. It is not possible to predict the slope of the line representing Henry's law.

Exercise 17.6.3

Show that $G_2 = G_m + (1 - x_2) \cdot (\partial G_m/\partial x_2)_{x_1/x_3}$ in a ternary system by starting from $G = NG_m(x_2, x_1/x_3)$.

Hint

Since $x_1/x_3 = N_1/N_3$ we find $(\partial[x_1/x_3]/\partial N_2)_{N_1, N_3} = 0$.

Solution

$x_2 = N_2/(N_1 + N_2 + N_3)$; $(\partial x_2/\partial N_2)_{N_1, N_3} = (N - N_2)/N^2 = (1 - x_2)/N$; $G_2 \equiv (\partial G/\partial N_2)_{N_1, N_3} = G_m + N \cdot (\partial G_m/\partial x_2)_{x_1/x_3} \cdot (\partial x_2/\partial N_2)_{N_1, N_3} + N \cdot (\partial G_m/\partial[x_1/x_3])_{x_2} \cdot (\partial[x_1/x_3]/\partial N_2)_{N_1/N_3} = G_m + N \cdot (\partial G_m/\partial x_2)_{x_1/x_3} \cdot (1 - x_2)/N + N \cdot (\partial G_m/\partial[x_1/x_3])_{x_2} \cdot 0 = G_m + (1 - x_2) \cdot (\partial G_m/\partial x_2)_{x_1/x_3}$

Exercise 17.6.4

Show that for a binary solution phase

$$2\frac{\mathrm{d}\ln f_B}{\mathrm{d}(x_A^2)} = 2\frac{\mathrm{d}\ln f_A}{\mathrm{d}(x_B^2)} = \frac{\mathrm{d}\ln(f_B/f_A)}{\mathrm{d}x_A} = \frac{-1}{RT}\cdot\frac{\mathrm{d}^2 {}^E G_m}{\mathrm{d}x_A^2}$$

Hint

Take the derivative of ${}^E G_m = x_A {}^E G_A + x_B {}^E G_B$ remembering $x_A + x_B = 1$.

Solution

$\mathrm{d}^F G_m/\mathrm{d}x_A = {}^E G_A - {}^E G_B = RT\ln(f_A/f_B)$ since $\Sigma x_i \mathrm{d}^E G_i = 0$. Thus we get, by again using the Gibbs–Duhem relation, $(-1/RT) \cdot \mathrm{d}^2 {}^E G_m/\mathrm{d}x_A^2 = -\mathrm{d}\ln(f_A/f_B)/\mathrm{d}x_A = -(\mathrm{d}\ln f_A - \mathrm{d}\ln f_B)/\mathrm{d}x_A = -[\mathrm{d}\ln f_A + (x_A/x_B)\mathrm{d}\ln f_A]/\mathrm{d}x_A = -(1/x_B) \cdot \mathrm{d}\ln f_A/\mathrm{d}x_A = 2\mathrm{d}\ln f_A/\mathrm{d}(x_B^2)$.

Exercise 17.6.5

By studying the ratio of vapour pressures over A–B alloys one can measure how the ratio of activities a_A/a_B varies with composition. Show how one can evaluate a_A and a_B from such information.

Hint

Use the Gibbs–Duhem relation in the form $x_A \mathrm{d}\ln a_A + x_B \mathrm{d}\ln a_B = 0$. Replace $\ln a_B$ by $\ln(a_B/a_A) + \ln a_A$.

Solution

$0 = x_A \mathrm{d}\ln a_A + x_B \mathrm{d}\ln(a_B/a_A) + x_B \mathrm{d}\ln a_A = (x_A + x_B)\mathrm{d}\ln a_A + x_B \mathrm{d}\ln(a_B/a_A) = \mathrm{d}\ln a_A + x_B \mathrm{d}\ln(a_B/a_A)$; $\ln a_A = - \int x_B \mathrm{d}\ln(a_B/a_A)$

17.7   *Dilute solution approximations*

When discussing dilute solutions Wagner (1952) suggested that one should consider the concentration dependence of the activity coefficient. For a dilute solution of B in A he wrote

$$(G_B - {^\circ}G_B)/RT = \ln a_B = \ln f_B + \ln x_B = \ln{^\circ}f_B + \ln x_B + \varepsilon_B x_B$$

This may be compared with the expression we obtained for a regular solution model with a constant parameter

$$G_B = {^\circ}G_B + RT \ln x_B + (1 - 2x_B + x_B^2)L_{AB}$$

The two formalisms are identical if the x_B^2 term can be neglected, i.e. for dilute solutions. We identify parameters as follows if the same reference state is used in both cases.

$$\ln{^\circ}f_B = L_{AB}/RT$$
$$\varepsilon_B = - 2L_{AB}/RT$$

However, a different reference state is often used in the ε formalism. As an example, the infinite-dilution reference state is defined in such a way that it makes $\ln{^\circ}f_B = 0$ and gives the relation

$${^\circ}G_B^{\mathrm{inf.dil.}} = {^\circ}G_B + L_{AB}$$
$$G_B = {^\circ}G_B^{\mathrm{inf.dil.}} + RT \ln x_B + (- 2x_B + x_B^2)L_{AB}$$

For a ternary system, where small amounts of B and C are dissolved in A, Wagner gave the expressions

$$(G_B - {}^\circ G_B)/RT = \ln a_B = \ln f_B + \ln x_B = \ln {}^\circ f_B + \ln x_B + \varepsilon_B^B x_B + \varepsilon_B^C x_C$$
$$(G_C - {}^\circ G_C)/RT = \ln a_C = \ln f_C + \ln x_C = \ln {}^\circ f_C + \ln x_C + \varepsilon_C^B x_B + \varepsilon_C^C x_C$$

This may be compared with the expressions obtained for the regular solution model with constant parameters,

$$G_B = {}^\circ G_B + RT\ln x_B + (1 - 2x_B + x_B^2 - x_C - x_B x_C)L_{AC} +$$
$$(- x_C + x_B x_C + x_C^2)L_{AB} + (x_C - x_B x_C)L_{BC}$$
$$G_C = {}^\circ G_C + RT\ln x_C + (- x_B + x_B^2 + x_B x_C)L_{AB} +$$
$$(1 - x_B - x_B x_C - 2x_C + x_C^2)L_{AC} + (x_B - x_B x_C)L_{BC}$$

The two formalisms are still identical if the second-order terms are neglected and if the ε parameters are identified as

$$\varepsilon_B^B = - 2L_{AB}/RT$$
$$\varepsilon_B^C = (L_{BC} - L_{AB} - L_{AC})/RT = \varepsilon_C^B$$
$$\varepsilon_C^C = - 2L_{AC}/RT$$

We note that ε_B^C and ε_C^B are equal.

The regular solution model demonstrates that one must go to the second-order power in order to see a deviation from Raoult's law for the solvent, A.

$$G_A = {}^\circ G_A + RT\ln x_A + (x_B^2 + x_B x_C)L_{AB} + (x_B x_C + x_C^2)L_{AC} - x_B x_C L_{BC}$$

With the ε formalism with its first-power terms one would not see any deviation from Raoult's law. One would simply get

$$(G_A - {}^\circ G_A)/RT = \ln x_A$$

and $\ln {}^\circ f_A = 0$, ${}^\circ f_A = 1$.

One may object to the ε formalism because the three expressions for G_A, G_B and G_C do not satisfy the Gibbs–Duhem relation. However, this can be accomplished by adding some particular terms, here represented by K,

$$(G_A - {}^\circ G_A)/RT = \ln a_A = \ln f_A + \ln x_A = \ln x_A + K$$
$$(G_B - {}^\circ G_B)/RT = \ln a_B = \ln f_B + \ln x_B = \ln {}^\circ f_B + \ln x_B + \varepsilon_B^B x_B + \varepsilon_B^C x_C + K$$
$$(G_C - {}^\circ G_C)/RT = \ln a_C = \ln f_C + \ln x_C = \ln {}^\circ f_C + \ln x_C + \varepsilon_C^B x_B + \varepsilon_C^C x_C + K$$

One can show that K must have the following quadratic form

$$- K = \tfrac{1}{2}\varepsilon_B^B x_B^2 + \varepsilon_B^C x_B x_C + \tfrac{1}{2}\varepsilon_C^C x_C^2$$

The quadratic formalism originates from Darken (1967) but has recently been discussed in more detail (Pelton and Bale, 1986). It should be emphasized that it cannot apply over the whole compositional range. It may be described as a method to make the solution look like a regular solution by introducing new reference states for the solutes in a dilute solution, ${}^\circ G_i + M_i$ instead of ${}^\circ G_i$. The molar Gibbs energy for a multicomponent solution in a solvent 1 would thus be

$$G_m = x_1{}^\circ G_1 + \sum_{i>1} x_i{}^\circ G_i + RT\Sigma x_i \ln x_i + \sum_i \sum_{j>i} x_i x_j I_{ij}$$

The partial Gibbs energies are obtained from an expression derived in Section 17.4. For the solvent we obtain, by using only the first coefficients ^0L and denoting them by L for simplicity,

$$G_1 = {}^\circ G_1 + RT\ln x_1 + \sum_{i \neq 1} L_{1i}x_i(1 - x_1) - \sum_{i \neq 1}\sum_{j>i} x_i x_j L_{ij}$$

For the solutes we obtain

$$G_l = {}^\circ G_l + M_l + RT\ln x_l + \sum_{i \neq 1} L_{li}x_i(1 - x_l) - \sum_{i \neq l}\sum_{j>i, \neq l} x_i x_j L_{ij}$$

For dilute solutions in component 1 it is advantageous to insert $1 - \Sigma x_i$ instead of x_1. After some manipulations we obtain

$$G_1 = {}^\circ G_1 + RT\ln x_1 + \tfrac{1}{2}\sum_{i>1}\sum_{j>1} x_i x_j(L_{1i} + L_{1j} - L_{ij})$$

$$G_l = {}^\circ G_l + M_l + L_{1l} + RT\ln x_l + \sum_{i>1}(L_{il} - L_{1i} - L_{1l})x_i$$

$$+ \tfrac{1}{2}\sum_{i>1}\sum_{j>1} x_i x_j(L_{1i} + L_{1j} - L_{ij})$$

where all $L_{ii} = 0$ and $L_{ij} = L_{ji}$. This is identical to the quadratic modification of the ε formalism if

$$\varepsilon_j^i = \varepsilon_i^j = (L_{ij} - L_{1i} - L_{1j})/RT$$
$$\varepsilon_j^j = -2L_{1j}/RT$$
$$\ln^\circ f_j = (M_j + L_{1j})/RT = M_j/RT - 0.5\varepsilon_j^j$$

or, turned the other way,

$$L_{1j} = -0.5RT\varepsilon_j^j$$
$$L_{ij} = L_{ji} = 0.5RT(2\varepsilon_j^i - \varepsilon_i^i - \varepsilon_j^j)$$
$$M_j = RT(\ln^\circ f_j + 0.5\varepsilon_j^j)$$

Exercise 17.7.1

Derive the above expression for K in the quadratic formalism.

Hint

Apply the Gibbs–Duhem relation in the x_B direction and in the x_C direction, respectively, in a ternary system, after first expressing x_A in terms of x_B and x_C.

Solution

$(1 - x_B - x_C)\mathrm{d}\ln a_A + x_B\mathrm{d}\ln a_B + x_C\mathrm{d}\ln a_C = 0$; $\partial/\partial x_B$ of this gives $(1 - x_B - x_C)[-1/(1 - x_B - x_C) + \partial K/\partial x_B] + x_B[1/x_B + \varepsilon_B^B + \partial K/\partial x_B] + x_C[\varepsilon_C^B + \partial K/\partial x_B] = 0$; $\partial K/\partial x_B = -x_B\varepsilon_B^B - x_C\varepsilon_B^B - x_C\varepsilon_C^B$. In the same way we find $\partial K/\partial x_C = -x_C\varepsilon_C^C - x_B\varepsilon_B^C$. These conditions are satisfied with $K = -(\varepsilon_B^B x_B^2 + 2\varepsilon_B^C x_B x_C + \varepsilon_C^C x_C^2)/2$.

Exercise 17.7.2

In the ε formalism we may define ε_B^C as $\partial^E G_B/\partial x_C$ and ε_C^B as $\partial^E G_C/\partial x_B$. Examine a ternary system. Under what conditions would we have $\varepsilon_B^C = \varepsilon_C^B$.

Hint

First apply the Gibbs–Duhem relation for excess quantities to variations in x_B and take the derivative with respect to x_C. Then do it the other way and compare the results. Notice that x_A must be replaced by $1 - x_B - x_C$.

Solution

Gibbs–Duhem in this form $(1 - x_B - x_C)\partial^E G_A/\partial x_B + x_B\partial^E G_B/\partial x_B + x_C\partial^E G_C/\partial x_B = 0$ would give us $(1 - x_B - x_C)\partial^{2E} G_A/\partial x_B\partial x_C - \partial^E G_A/\partial x_B + x_B\partial^{2E} G_B/\partial x_B\partial x_C + \partial^E G_C/\partial x_B + x_C\partial^{2E} G_C/\partial x_B\partial x_C = 0$.
Gibbs–Duhem in the form $(1 - x_B - x_C)\partial^E G_A/\partial x_C + x_B\partial^E G_B/\partial x_C + x_C\partial^E G_C/\partial x_C = 0$ would give us $(1 - x_B - x_C)\partial^{2E} G_A/\partial x_C\partial x_B - \partial^E G_A/\partial x_C + x_B\partial^{2E} G_B/\partial x_C\partial x_B + \partial^E G_B/\partial x_C + x_C\partial^{2E} G_C/\partial x_C\partial x_B = 0$.
From the difference we obtain $\partial(^E G_C - {}^E G_A)/\partial x_B = \partial(^E G_B - {}^E G_A)/\partial x_C$. It is thus necessary to have $\partial^E G_A/\partial x_B = \partial^E G_A/\partial x_C$. This is fulfilled in the ε formalism because these derivatives are both zero. It is not fulfilled in the quadratic formalism. We may conclude that $\varepsilon_B^C = \varepsilon_C^B$ holds for dilute solutions only.

Exercise 17.7.3

Show for a ternary system under what conditions $RT\varepsilon_B^C = g_{BC}$, which is a notation for $\partial^2 G_m/\partial x_B\partial x_C$.

Hint

From the definition of ε_B^C we have $RT\varepsilon_B^C = \partial G_B/\partial x_C$ which can be calculated from $G_B = G_m + (1 - x_B)\partial G_m/\partial x_B - x_C\partial G_m/\partial x_C$ where G_m is regarded as a function of x_B and x_C.

Solution

$RT\varepsilon_B^C = \partial G_B/\partial x_C = \partial G_m/\partial x_C + (1 - x_B)\partial^2 G_m/\partial x_B\partial x_C - \partial G_m/\partial x_C - x_C\partial^2 G_m/\partial x_C^2 = \partial^2 G_m/\partial x_B\partial x_C - x_B\partial^2 G_m/\partial x_B\partial x_C - x_C\partial^2 G_m/\partial x_C^2$. The second term may be neglected at small x_B but for convenience we shall keep the part which comes from the ideal entropy of mixing, RTx_B/x_A. The third term may be approximated by the part coming from the ideal entropy of mixing $RTx_C(1/x_A + 1/x_C)$. We thus obtain $RT\varepsilon_B^C = \partial^2 G_m/\partial x_B\partial x_C - RT/x_A$. The last term is approximately $-RT$ and it can be neglected only when $|\varepsilon_B^C| \gg 1$.

17.8 *Predictions for solutions in higher-order systems*

Taking all binary interactions into account we obtain the following expression for the excess Gibbs energy in a multicomponent system.

$$^E G_m = \sum_i \sum_{j>1} x_i x_j I_{ij}$$

If this model applies, all the I_{ij} coefficients can be determined experimentally on the respective binary systems and the properties of the higher-order system can be predicted by combination. On the other hand, if a composition-dependent I is required in order to represent the experimental information on a binary system, then there is no simple physical model predicting the properties of the higher-order system. It may be stated that a composition-dependent I implies that the interaction energy is not determined completely by the pair-wise interaction of atoms. If I in a binary system is equal to $^0L + (x_A - x_B)^1 L$, then the excess Gibbs energy is described by

$$^E G_m = x_A x_B{}^0L + x_A x_A x_B{}^1L - x_A x_B x_B{}^1L$$

where the last two terms seem to originate from interactions within groups of three atoms. If that is the case, one should expect similar effects in the higher-order system, for instance an interaction between an A, a B and a C atom given by $x_A x_B x_C I_{ABC}$. Of course, there is no way by which this interaction can be predicted from the binary systems where that group of atoms does not occur.

When the binary interaction energies depend upon the composition, it should be advisable to introduce I_{ABC} in the description of the ternary system. However, it must be realized that there is no unique way of defining such a parameter because there is no unique way of predicting how the properties of a binary system contribute to the properties of the higher-order system unless I_{AB} is a constant. For variable I_{AB} several expressions are used

$$I_{AB} = {}^0Lx_A^2 + {}^1Lx_Ax_B + {}^2Lx_B^2$$
$$I_{AB} = {}^0L + {}^1Lx_B + {}^2Lx_B^2$$
$$I_{AB} = {}^0L + {}^1L(2x_B - 1) + {}^2L(6x_B^2 - 6x_B + 1)$$
$$I_{AB} = {}^0L + {}^1L(x_A - x_B) + {}^2L(x_A^2 - 4x_Ax_B + x_B^2)$$
$$I_{AB} = {}^0L + {}^1L(x_A - x_B) + {}^2L(x_A - x_B)^2$$

These expressions are quite equivalent when applied to a binary system because they can be transformed into each other by the use of $x_A + x_B = 1$. When applied to a higher-order system the expressions give different results because $x_A + x_B$ is no longer unity. For practical reasons it may be important to select a particular expression and at present there is a strong preference for the last one. It was first suggested by Redlich and Kister (1948) that this particular form for the binary systems should be used for representing the binary contributions in a multicomponent system.

The same kind of problem appears when one wants to predict the properties of a quaternary solution from the four ternaries. A general method may be based upon the observation that all the expressions, listed for the binary interaction energy I_{AB}, become identical if x_A is replaced by $x_A + (1 - x_A - x_B)/2$ and x_B by $x_B + (1 - x_A - x_B)/2$. In fact, all the expressions then become identical to the Redlich–Kister polynomial. Generalizing this method we find that x_A should be replaced by $x_A + (1 - x_A - x_B - x_C)/3$, x_B by $x_B + (1 - x_A - x_B - x_C)/3$ and x_C by $x_C + (1 - x_A - x_B - x_C)/3$ in the ternary interaction energy I_{ABC}. This method can easily be extended to higher-order terms (Hillert, 1980).

Exercise 17.8.1

Show that $I_{AB} = {}^0L + {}^1Lx_B + {}^2Lx_B^2$ becomes identical to $I_{AB} = {}^0L + {}^1L(x_A - x_B) + {}^2L(x_A - x_B)^2$ if x_B is replaced by $x_B + (1 - x_A - x_B)/2$. Show how the second set of L parameters can be evaluated from the first one.

Hint

The expression by which we shall replace x_B is equal to $1/2 - (x_A - x_B)/2$.

Solution

$$I_{AB} = {}^0L + {}^1L/2 - {}^1L/2 \cdot (x_A - x_B) + {}^2L/4 + {}^2L/4 \cdot (x_A - x_B)^2 - 2^2L/4 \cdot (x_A - x_B).$$

Comparison shows that: New $^0L = {}^0L + {}^1L/2 + {}^2L/4$; New $^1L = -({}^1L + {}^2L)/2$; New $^2L = {}^2L/4$.

Exercise 17.8.2

Suppose the properties of a certain phase are known in the binary systems A–B, B–C and C–A. When the ternary parameter I_{ABC} was evaluated from some ternary information one found that it was practically zero. In that assessment one described the binary A–B properties with an expression $x_A x_B({}^0L_B + {}^1L_B x_B)$. Suppose that the assessment is repeated using an expression $x_A x_B({}^0L_A + {}^1L_A x_A)$. What value would one then obtain for the ternary parameter?

Hint

We must suppose that the two binary expressions are identical in the binary case, $^0L_B + {}^1L_B(1 - x_A) = {}^0L_A + {}^1L_A x_A$. Thus, $^0L_A = {}^0L_B + {}^1L_B$ and $^1L_A = - {}^1L_B$.

Solution

Omit all terms from B–C and C–A because they are the same in both cases. The result of the second assessment is written in the form: $x_A x_B({}^0L_A + {}^1L_A x_A) + x_A x_B x_C I_{ABC}$. By the use of $x_B = 1 - x_A - x_C$, transform the result of the first assessment to this form: $x_A x_B({}^0L_B + {}^1L_B x_B) = x_A x_B({}^0L_B + {}^1L_B - {}^1L_B x_A - {}^1L_B x_C) = x_A x_B({}^0L_B + {}^1L_B - {}^1L_B x_A) + x_A x_B x_C(- {}^1L_B)$. Comparison gives $^0L_A = {}^0L_B + {}^1L_B$ and $^1L_A = - {}^1L_B$ which we knew before but we also find $I_{ABC} = - {}^1L_B$.

17.9 *Numerical methods of predictions for higher-order systems*

As an alternative to the analytical methods described in the preceding section, several numerical methods have been suggested. They allow the properties of a ternary alloy to be estimated from binary alloys without first assessing all the information available in the binary systems. However, it should be emphasized that they are nevertheless based upon some assumptions regarding the properties. The expression for the Gibbs energy of a ternary alloy of the composition x_A, x_B, x_C must contain the terms $x_A {}^\circ G_A + x_B {}^\circ G_B + x_C {}^\circ G_C$ and what remains is the Gibbs energy of mixing, $^M G_m$. The first aim is to be able to predict $^M G_m$ for the ternary alloy as some average of $^M G_m$ for a few

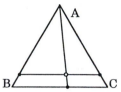

Figure 17.6 Selection of binary alloys (filled circles) for the estimation of G_m of a ternary alloy (open circle).

binary alloys which have been studied experimentally. Bonnier and Caboz (1960) suggested that three binary alloys should be chosen as demonstrated in Fig. 17.6. Two of the binary alloys have the same A content as the ternary alloy and their compositions can thus be given as $x_A, 1 - x_A$. Bonnier further suggested that their ^MG_m values should be weighted with factors representing the relative distances from the ternary point, i.e. $x_B/(x_B + x_C)$ and $x_C/(x_B + x_C)$. The third alloy has the same ratio B to C as the ternary alloy. Its composition is thus $x_B/(x_B + x_C), x_C/(x_B + x_C)$. Bonnier suggested that it should be given a weighting factor of $1 - x_A$. This method of prediction can thus be represented by the equation

$$^MG_m^{ABC}(x_A, x_B, x_C) = \frac{x_B}{x_B + x_C} \, ^MG_m^{AB}(x_A, 1 - x_A)$$

$$+ \frac{x_C}{x_B + x_C} \, ^MG_m^{AC}(x_A, 1 - x_A) + (1 - x_A)^MG_m^{BC}\left(\frac{x_B}{x_B + x_C}, \frac{x_C}{x_B + x_C}\right)$$

For a substitutional solution one should certainly like the ideal entropy expression to come out right and that is indeed the case,

$$RT\left\{\frac{x_B}{x_B + x_C} \cdot [x_A \ln x_A + (1 - x_A)\ln(1 - x_A)]\right.$$

$$+ \frac{x_C}{x_B + x_C} \cdot [x_A \ln x_A + (1 - x_A)\ln(1 - x_A)]$$

$$\left. + (1 - x_A) \cdot \left[\frac{x_B}{x_B - x_C} \ln \frac{x_B}{x_B - x_C} + \frac{x_C}{x_B + x_C} \ln \frac{x_C}{x_B + x_C}\right]\right\}$$

$$= RT(x_A \ln x_A + x_B \ln x_B + x_C \ln x_C)$$

In addition, one may hope that the same method will give a reasonable result for systems deviating from ideal behaviour. However, for the leading terms in the excess Gibbs energy, $\Sigma\Sigma x_i x_j \,^0L_{ij}$, the method would give

$$\frac{x_B}{x_B + x_C} x_A(1 - x_A)^0L_{AB} + \frac{x_C}{x_B + x_C} x_A(1 - x_A)^0L_{AC} + (1 - x_A)\frac{x_B}{x_B + x_C} \cdot \frac{x_C}{x_B + x_C} \,^0L_{BC}$$

$$= x_A x_B \,^0L_{AB} + x_A x_C \,^0L_{AC} + \frac{x_B x_C}{x_B + x_C} \,^0L_{BC}$$

The last term does not come out right. As a consequence, Toop (1965) suggested that the weighting factor for the third binary alloy should be taken as $(1 - x_A)^2$ which is equal to $(x_B + x_C)^2$. However, it is then impossible to reproduce the ideal entropy expression. Toop's method should thus be applied only to the excess Gibbs energy, and it must be assumed that the ideal entropy expression can be applied directly.

Toop's method is asymmetric because one component is treated differently from the other two. Three symmetric methods by Kohler (1960), Colinet (1967) and Muggianu, Gambino and Bros (1975), which also apply to the excess Gibbs energy, deserve special mention.

Kohler
$$^EG_m = \Sigma(x_i + x_j)^2 \, ^EG_m^{ij}(x_i/[x_i + x_j]; x_j/[x_i + x_j])$$

Colinet
$$^EG_m = \Sigma \left\{ \frac{x_j/2}{1 - x_i} \, ^EG_m^{ij}(x_i; [1 - x_i]) + \frac{x_i/2}{1 - x_j} \, ^EG_m^{ij}([1 - x_j]; x_j) \right\}$$

Muggianu
$$^EG_m = \Sigma \frac{4x_i x_j}{(1 + x_i - x_j)(1 + x_j - x_i)} \, ^EG_m^{ij}([1 + x_i - x_j]/2; [1 + x_j - x_i]/2)$$

The three methods are compared in Fig. 17.7.

All the weighting factors have been selected in such a way that the methods reproduce the terms $x_i x_j \, ^0L_{ij}$. Colinet and Muggianu *et al.* also reproduce the terms $x_i x_j (x_i - x_j)^1 L_{ij}$. In addition, Muggianu *et al.* reproduce all the higher-power terms if they are written in the form $x_i x_j (x_i - x_j)^{k \cdot k} L_{ij}$. As a consequence, the numerical method by Muggianu *et al.* and the analytical method based upon the Redlich–Kister polynomial give the same result and may be recommended for general use.

The numerical methods are often described as geometrical methods because the compositions of the selected binary alloys are found by some simple geometrical construction. It should be emphasized that they only apply to integral quantities. On the other hand, the mathematical expressions defining these methods have given rise to analytical methods by the introduction of analytical expressions for the binary systems.

Exercise 17.9.1

Refer to Fig. 17.7 and show that the binary A–B alloy used in Muggianu's method has an A content of $(1 + x_A - x_B)/2$.

Hint

Draw lines through the ternary alloy, parallel to sides A–C and B–C. The intercepts on the A–B side have lengths equal to x_B, x_C and x_A.

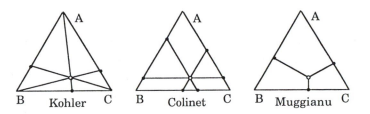

Figure 17.7 Selection of binary alloys (filled circles) according to three symmetric methods of predicting ternary properties.

Solution

The distance of the binary alloy from the B corner is $x_C/2 + x_A$. Using the relation $x_C = 1 - x_A - x_B$ one can transform this expression into $1/2 + x_A/2 - x_B/2$.

Exercise 17.9.2

Prove that Muggianu's method for a ternary system correctly reproduces a term written in the form $x_i x_j (x_i - x_j)^{k.k} L_{ij}$ in the binary ij system.

Hint

Start by evaluating the value of $^E G_m^{ij}$ for the binary alloy used in the method. Then multiply by the weight recommended by Muggianu.

Solution

$x_i^{bin} = (1 + x_i - x_j)/2; \quad x_j^{bin} = (1 + x_j - x_i)/2;$

$^E G_m^{ij} = (1 + x_i - x_j)/2 \cdot (1 + x_j - x_i)/2 \cdot [(1 + x_i - x_j - 1 - x_j + x_i)/2]^{k.k} L_{ij} = (1/4) \cdot$
$(1 + x_i - x_j)(1 + x_j - x_i) \cdot (x_i - x_j)^{k.k} L_{ij};$

$^E G_m = \Sigma \dfrac{4 x_i x_j}{(1 + x_i - x_j)(1 + x_j - x_i)} \cdot (1/4) \cdot (1 + x_i - x_j)(1 + x_j - x_i)(x_i - x_j)^{k.k} L_{ij}$

$= \Sigma x_i x_j (x_i - x_j)^{k.k} L_{ij}.$

Exercise 17.9.3

The excess Gibbs energy for a binary system is represented with the following expression according to the subregular solution model, $^E G_m =$

$x_A x_B [^0 L_{AB} + {}^1 L_{AB}(x_A - x_B)]$. Suppose this expression is included in the Gibbs energy for a ternary A–B–C system. Derive expressions for the corresponding contributions to the partial Gibbs energies for A and C, respectively.

Hint

Use $^E G_j = {}^E G_m + \partial G_m / \partial x_j - \Sigma x_i \partial G_m / \partial x_i$. Note that $x_B = 1 - x_A$ cannot be used in the ternary system.

Solution

$$^E G_A = {}^0 L_{AB}(x_A x_B + x_B - x_A x_B - x_B x_A) + {}^1 L_{AB}(x_A^2 x_B - x_A x_B^2 + 2x_A x_B - x_B^2 - x_A \cdot 2 x_A x_B + x_A x_B^2 - x_B x_A^2 + x_B \cdot 2 x_A x_B) = {}^0 L_{AB} x_B (1 - x_A) + {}^1 L_{AB}[x_B^2 (2x_A - 1) + x_B (1 - x_A) \cdot 2 x_A]; {}^E G_C = {}^0 L_{AB}(x_A x_B - x_A x_B - x_B x_A) + {}^1 L_{AB}(x_A^2 x_B - x_A x_B^2 - 2 x_A^2 x_B + x_A x_B^2 - x_A^2 x_B + 2 x_A x_B^2) = {}^0 L_{AB}(- x_A x_B) + {}^1 L_{AB}[- 2 x_A x_B (x_A - x_B)] = - x_A x_B [^0 L_{AB} + 2 {}^1 L_{AB}(x_A - x_B)]$$

Solution phases with sublattices

Sublattice solution phases

In the substitutional solutions discussed in Section 17.9 all lattice sites were equivalent and a solution was formed from a pure substance by substituting new kinds of atoms for the initial one. However, relatively few crystalline phases belong to this class. The great majority has different kinds of lattice sites and they can be described by using two or more sublattices. Examples of such phases will be discussed in this chapter. It will be demonstrated that a great variety of such phases can be modelled in a very direct way using an approach often called the compound energy model. It is a crude model in the sense that it assumes random mixing within each sublattice. Even so, the set of equations from which one can calculate the state of equilibrium can easily get very complicated. However, it should be realized that actual calculations of equilibria, and even whole phase diagrams, can now be carried out with sophisticated computer programs which only require that the expression for the molar Gibbs energy of each phase is defined.

In Chapter 17 an expression was derived for the entropy assuming random mixing of all the components present in each sublattice. The result was expressed in terms of the site fraction variable, y. We shall first consider the rather simple case where there is only one component, M, in one sublattice, and a number of components, $i, j, \ldots$, in another sublattice. It is then convenient to consider a formula unit with one mole of atoms in the second sublattice $(M)_b(i, j, \ldots)_1$. For 1 mole of such formula units we get

$$S_m^{ideal} = - R\Sigma y_i \ln y_i$$

The deviation from ideal solution behaviour may be represented by the interactions between the components in the second sublattice. Using the Redlich–Kister type of power series we have, for the interaction between components i and j, when the first sublattice is filled with M,

$$^E G_m = y_i y_j \sum_{k=0}^{n} {}^k L_{ij}^M (y_i - y_j)^k$$

For $y_i = 1$ the phase is identical to a compound $M_b i$ and the whole expression for the molar Gibbs energy will thus be

$$G_m = \Sigma y_i {}^{\circ}G_{Mbi} + RT\Sigma y_i \ln y_i + \sum_i \sum_{j>i} y_i y_j \sum_{k=0}^{n} {}^k L_{ij}^M (y_i - y_j)^k$$

It should be noticed that y_j is at the same time the site fraction of j in the second sublattice and the molar content of the compound $M_b j$ among all $M_b i$ compounds. From G_m we may derive expressions for the partial Gibbs energies of the compounds, using an equation derived in Section 3.5.

$$G_{Mbl} = {}^{\circ}G_{Mbl} + RT\ln y_l + \sum_{i \neq l} \left\{ {}^0 L_{li}^M y_i (1 - y_l) + \sum_{k=1}^{n} {}^k L_{li}^M (y_l - y_i)^{k-1} \right.$$

$$\left. \cdot [(k+1)(1 - y_l)(y_l - y_i) + k y_i] \right\} - \sum_{i \neq l} \sum_{j \neq l, > i} y_i y_j \left\{ {}^0 L_{ij}^M + \sum_{k=1}^{n} {}^k L_{ij}^M (y_i - y_j)^k (k+1) \right\}$$

The expressions are thus analogous to those for a substitutional solution in Section 17.4, except that y is substituted for x. A ternary phase of this type would behave as a binary, substitutional phase. It is sometimes called a **quasi-binary** or **pseudo-binary** phase. By the same reason one may call a quaternary phase of this type a quasi-ternary phase.

Exercise 18.1.1

High-temperature measurements have shown complete miscibility in the solid phase of Al_2O_3–Cr_2O_3. Information from lower temperatures is less certain but there is some report of a miscibility gap with a maximum at about $T_{crit} = 2000$ K. Model this solution phase.

Hint

Introduce a constant regular solution parameter. Express the ideal entropy contribution with regard to the size of the formula unit chosen.

Solution

Define the unit as $M_1O_{1.5}$. Then we get $G_m = y_{Al} {}^{\circ}G_{AlO_{1.5}} + y_{Cr} {}^{\circ}G_{CrO_{1.5}} + RT(y_{Al}\ln y_{Al} + y_{Cr}\ln y_{Cr}) + y_{Al}y_{Cr}L$. Because we have chosen a symmetric description, the miscibility gap will be symmetric and its maximum will be found at $y_{Al} = y_{Cr} = 0.5$ and $0 = d^2G_m/dy_{Cr}^2 = RT/y_{Al}y_{Cr} - 2L$; $L = 2RT_{crit} = 2R \cdot 2000 = 33\,000$ J/mol.

18.2   ***Interstitial solutions***

In Chapter 17 we discussed the ideal entropy of mixing in an interstitial solution. In fact, it may be regarded as the special case of a solution with two sublattices obtained by allowing vacant sites on one of the sublattices. For 1 mole of sites in that sublattice we can write the formula as $(M)_b(Va,i,j,\ldots)_1$. All the equations derived for the phase with two sublattices in the preceding section can be applied if the vacancies are included as a component. For a binary M–C system we write the interstitial solution as $M_b(Va,C)_1$ and obtain for 1 mole of formula units,

$$G_m = y_C{}^\circ G_{M_bC} + y_{Va}{}^\circ G_{M_bVa} + RT(y_C\ln y_C + y_{Va}\ln y_{Va}) + y_Cy_{Va}\sum_{k=0}^{n} {}^kL_{CVa}(y_C - y_{Va})^k$$

$$G_{M_b} = {}^\circ G_{M_bC} + RT\ln y_C + y_{Va}^2\left\{{}^0L_{CVa} + \sum_{k=1}^{n} {}^kL_{CVa}(y_C - y_{Va})^{k-1}[(2k+1)y_C - y_{Va}]\right\}$$

$$G_{M_bVa} = {}^\circ G_{M_bVa} + RT\ln y_{Va} + y_C^2\left\{{}^0L_{CVa} + \sum_{k=1}^{n} {}^kL_{CVa}(y_C - y_{Va})^{k-1}[y_C - (2k+1)y_{Va}]\right\}$$

Since all the sites in the second sublattice are vacant in the compound M_bVa, it is identical to b moles of pure M. G_M is thus obtained by dividing the last equation by b. G_C can be obtained by taking the difference between the two equations because

$$G_{M_bC} - G_{M_bVa} = bG_M + G_C - bG_M - G_{Va} = G_C - G_{Va}$$

and it may be assumed that the chemical potential of vacancies is zero at equilibrium. We thus obtain

$$G_C = G_{M_bC} - G_{M_bVa} = {}^\circ G_{M_bC} - b{}^\circ G_M + RT\ln(y_C/y_{Va}) + {}^0L_{CVa}(y_{Va} - y_C)$$

$$+ \sum_{k=1}^{n} {}^kL_{CVa}(y_C - y_{Va})^{k-1}[2ky_Cy_{Va} - (y_C - y_{Va})^2]$$

Actually, we can obtain $G_{M_bC} - G_{M_bVa}$ and thus G_C directly from G_m by

$$G_{M_bC} - G_{M_bVa} = \frac{\partial G_m}{\partial y_C} - \frac{\partial G_m}{\partial y_{Va}}$$

in view of the rule of calculating partial quantities given in Section 3.5.

It is important to notice that pure C cannot exist in the interstitial sublattice without M in the other sublattice. As a consequence, the reference state for C must be chosen using another phase. If pure C in a form α is chosen and y_{Va} is replaced by $1 - y_C$, then we obtain

$$RT\ln a_C = G_C - {}^\circ G_C^\alpha = {}^\circ G_{M_bC} - b{}^\circ G_M - {}^\circ G_C^\alpha + RT\ln[y_C/(1 - y_C)]$$
$$- {}^0L_{CVa}\cdot(1 - 2y_C) + {}^1L_{CVa}\cdot(-1 + 6y_C - 6y_C^2) - {}^2L_{CVa}\cdot(1 - 10y_C + 24y_C^2 - 16y_C^3) + \ldots$$

The activity coefficient for very dilute solutions would be given by

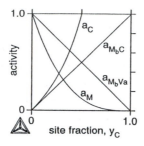

Figure 18.1 The variation of different activities in an ideal interstitial solution of C in M with the formula $M_{1/3}(Va,C)_1$.

$$^\circ f_C = \exp(^\circ G_{M_bC} - b^\circ G_M - {^\circ}G_C^\alpha + {^0}L_{CVa} - {^1}L_{CVa} + {^2}L_{CVa} + \ldots)/RT$$

The deviation from Henry's law for less dilute solutions depends strongly upon the choice of composition variable. In order to get a constant activity coefficient for C over the whole system, when all the L parameters are zero, one must use $y_C/(1 - y_C)$ as the composition variable. If y_C is used one will find that a_C goes to infinity at $y_C = 1$. This is demonstrated in Fig. 18.1. On the other hand, in this type of diagram the activities of M_bVa and M_bC show an ideal behaviour when all the parameters are zero. The diagrams were drawn for $b = 1/3$ and all L parameters equal to zero.

Exercise 18.2.1

In the plot of a_M versus y_C for an interstitial solution of C in M it may seem surprising in view of Raoult's law that the curve for a_M does not approach the diagonal close to $y_C = 0$. Find the reason and calculate the initial slope according to Raoult's law.

Hint

Raoult's law says that a_M should approach x_M, i.e. $1 - x_C$, not $1 - y_C$. Evaluate x_M from y_C using the formula $M_b(Va,C)_1$ with $b = 1/3$ (the value for bcc).

Solution

For a binary interstitial solution we have
$x_M = b/(b + y_C) = 1/(1 + y_C/b) \cong 1 - y_C/b = 1 - 3y_C$. The initial slope should be -3 and not -1.

18.3   *Reciprocal solution phases*

With a stoichiometric phase one usually means a phase with a constant composition. This may, for instance, be caused by a crystalline structure which is composed of different sublattices, one for each component. An example is cementite, Fe_3C. Such a phase is also described as a compound. When a further component is added, it may go into one of the existing sublattices, an example being manganese-alloyed cementite, $(Fe,Mn)_3C$. The composition of such a phase may thus vary along a line in the ternary phase diagram and it is sometimes described as a line compound or a quasi-binary phase. Such phases were described in Section 18.1. If still another component is added, two alternatives result, examples being $(Fe,Mn,Cr)_3C$ and $(Fe,Mn)_3(C,N)_1$. The latter type of phase is sometimes called a reciprocal solution phase because the central alloy can be regarded as a solution between either Fe_3C and Mn_3N or Fe_3N and Mn_3C. Both kinds of phases have the restriction to the variation in composition which we have called stoichiometric constraint. The composition of a reciprocal solution phase is represented by two sets of site fractions, one set for each sublattice, $y_A + y_B = 1$ and $y_C + y_D = 1$.

Accepting the stoichiometric constraint it is logical to consider the binary compounds as the components of the system. They were introduced in Section 3.12 and are called component compounds. Let us discuss the reciprocal case represented by $(A,B)_b(C,D)_c$. This is a quaternary phase but its composition can only be varied with two degrees of freedom instead of three due to the stoichiometric constraint. All possible compositions can thus be represented on a plane just like a ternary system. As shown in Fig. 12.11, a square diagram is now the natural shape and each corner is an end-member and represents a component compound. Perpendicular to that plane we may plot the Gibbs energy (see Fig. 18.2).

The diagram demonstrates that it is, in general, impossible to place a plane through the four points representing the Gibbs energy values of the four component compounds. The question is then what surface of reference one should use when giving the Gibbs energy value for a composition inside the system. The most natural choice seems to be the curved surface shown in the diagram and for a simple case that choice may be supported by the random mixing version of the nearest-neighbour bond energy

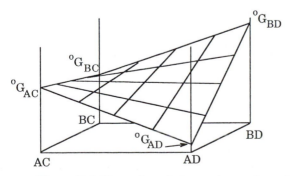

Figure 18.2 The surface of reference for a reciprocal system.

model to be described in the next chapter. This surface of reference is accepted in the **compound energy model** (Andersson *et al.* 1986) and it yields the following expression for the Gibbs energy.

$$G_{\mathrm{m}} = y_A y_C {}^{\circ}G_{A_bC_c} + y_A y_D {}^{\circ}G_{A_bD_c} + y_B y_C {}^{\circ}G_{B_bC_c} + y_B y_D {}^{\circ}G_{B_bD_c}$$
$$+ RT[by_A \ln y_A + by_B \ln y_B + cy_C \ln y_C + cy_D \ln y_D] + {}^{E}G_{\mathrm{m}}$$

This expression may be regarded merely as a definition of the excess term ${}^{E}G_{\mathrm{m}}$ but it becomes very important if one, as a first approximation, neglects the excess term. G_{m} refers to 1 mole of formula units M_bN_c. The simplest type of power series representation of the excess term makes use of the remaining two second-power terms, $y_A y_B$ and $y_C y_D$. However, in order to allow different behaviour on two opposite sides of the composition square we must go to third-power terms,

$$^{E}G_{\mathrm{m}} = y_A y_B y_C I_{\mathrm{AB:C}} + y_A y_B y_D I_{\mathrm{AB:D}} + y_C y_D y_A I_{\mathrm{A:CD}} + y_C y_D y_B I_{\mathrm{B:CD}}$$

The colon in the subscript is used to separate sublattices. For any side of the system this expression reduces to the expression discussed in the preceding section and all these interaction energies can thus be evaluated from the properties of the side systems. As an example, for the A_bC_c–B_bC_c side of the system we have $y_D = 0$, $y_C = 1$ and the excess Gibbs energy expression reduces to $y_A y_B I_{\mathrm{AB:C}}$.

For more complicated cases one may express each one of the parameters $I_{\mathrm{AB:C}}$, etc., with a power series of the Redlich–Kister type using site fractions. We have already done this for the simple case of a phase with sublattices considered previously. In this way we can introduce a large number of higher-power terms but it should be noticed that for each such term all the y_i except for one come from one sublattice and all the coefficients can be evaluated from the side systems. In order to adjust a description to information from inside the composition square we need a term like $y_A y_B y_C y_D I_{\mathrm{AB:CD}}$.

In our discussion on constituents in Section 3.11 we saw that the chemical potentials of the compounds can be evaluated from the following equation

$$G_{A_bC_c} = bG_A + cG_C = G_{\mathrm{m}} + \partial G_{\mathrm{m}}/\partial y_A + \partial G_{\mathrm{m}}/\partial y_C - \Sigma y_i \partial G_{\mathrm{m}}/\partial y_i$$

We thus obtain, for instance,

$$G_{A_bC_c} = {}^{\circ}G_{A_bC_c} + y_B y_D \cdot \Delta^{\circ}G_{\mathrm{AD+BC}} + bRT\ln y_A + cRT\ln y_C + {}^{E}G_{A_bC_c}$$
$$G_{B_bC_c} = {}^{\circ}G_{B_bC_c} - y_A y_D \cdot \Delta^{\circ}G_{\mathrm{AD+BC}} + bRT\ln y_B + cRT\ln y_C + {}^{E}G_{B_bC_c}$$

where the quantity

$$\Delta^{\circ}G_{\mathrm{AD+BC}} = {}^{\circ}G_{A_bD_c} + {}^{\circ}G_{B_bC_c} - {}^{\circ}G_{A_bC_c} - {}^{\circ}G_{B_bD_c}$$

is the Gibbs energy for the reciprocal reaction, $A_bC_c + B_bD_c \rightarrow A_bD_c + B_bC_c$. For constant interaction energies, to be denoted by L, we find

$$^{E}G_{A_bC_c} = y_B(y_D y_A + y_B y_C)L_{\mathrm{AB:C}} + y_D(y_D y_A + y_B y_C)L_{\mathrm{A:CD}}$$
$$+ y_B y_D(y_D - y_C)L_{\mathrm{B:CD}} + y_B y_D(y_B - y_A)L_{\mathrm{AB:D}}$$
$$^{E}G_{B_bC_c} = y_A(y_D y_B + y_A y_C)L_{\mathrm{AB:C}} + y_D(y_D y_B + y_A y_C)L_{\mathrm{B:CD}}$$

$$+ y_A y_D (y_D - y_C) L_{A:CD} + y_A y_D (y_A - y_B) L_{AB:D}$$

It should be noticed that the quantity $\Delta^\circ G_{AD+BC}$ is evaluated from information on the four pure component compounds and does not even concern the quasi-binary sides. It often has a dominating influence on the properties of alloys inside the quaternary system. One may regard $\Delta^\circ G_{AD+BC}$ as a representation of the difference in interaction between nearest neighbours, i.e. usually between atoms in different sublattices. The L values, on the other hand, which enter in the excess Gibbs energy and control the behaviour of the quasi-binary sides, represent the interactions between atoms in the same sublattice, i.e. next-nearest neighbours in most cases, and may thus be of secondary importance.

It is worth noting that the partial Gibbs energies of the four component compounds are not independent of each other. From Sections 3.10 and 3.12 it is evident that

$$G_{A_bD_c} + G_{B_bC_c} - G_{A_bC_c} - G_{B_bD_c} = bG_A + cG_D + bG_B + cG_C - bG_A - bG_C - bG_B - cG_D = 0$$

If one, by some reason, wants to consider the partial Gibbs energies of the elements, then it can be done relative to the value of one of them, e.g. G_A,

$$cG_C = G_{A_bC_c} - bG_A$$
$$cG_D = G_{A_bD_c} - bG_A$$
$$bG_B = G_{B_bC_c} - G_{A_bC_c} + bG_A = G_{B_bD_c} - G_{A_bD_c} + bG_A$$

However, G_A is indeterminate unless a second phase is present. The same phenomenon is illustrated for binary and ternary systems in Figs. 6.5 and 6.6 but it cannot be easily illustrated for a phase with four components.

One can introduce Redlich–Kister polynomials to describe the composition dependence of the interaction energies. When calculating $^E G_{M_bi_c}$ we must then evaluate several kinds of contributions if there are many components. We find contributions of the following forms from interactions on the second sublattice

$$\Delta_1^E G_{M_bi_c} = {}^0 L_{M:ij} y_j (y_i + y_M - 2y_i y_M)$$

$$+ \sum_{k=1}^{n} {}^k L_{M:ij} (y_i - y_j)^{k-1} y_j \{ (y_i - y_j)[y_M(1 + k)(1 - y_i) + y_i - y_M y_i] + k y_M y_j \}$$

$$\Delta_2^E G_{M_bi_c} = {}^0 L_{N:ij} y_N y_j (1 - 2y_i)$$

$$+ \sum_{k=1}^{n} {}^k L_{N:ij} (y_i - y_j)^{k-1} y_N y_j \{ (y_i - y_j)[(1 + k)(1 - y_i) - y_i] + k y_j \}$$

$$\Delta_3^E G_{M_bi_c} = {}^0 L_{M:lj} y_l y_j (1 - 2y_M) + \sum_{k=1}^{n} {}^k L_{M:lj} (y_l - y_j)^k y_l y_j (1 - 2y_M - k y_M)$$

$$\Delta_4^E G_{M_bi_c} = {}^0 L_{N:lj} y_N y_l y_j (-2) + \sum_{k=1}^{n} {}^k L_{N:lj} (y_l - y_j)^k y_N y_l y_j (-2 - k)$$

Equivalent terms would come from the interactions on the first sublattice, $L_{MN:i}$, $L_{MN:j}$, $L_{KN:i}$ and $L_{KN:j}$.

The model for reciprocal phases, which has been discussed here, can be generalized to several sublattices and several components on each one. For three sublattices we find

$$G_m = \Sigma\Sigma\Sigma y_i^1 y_j^2 y_k^{3\,\circ}G_{i_a j_b k_c} + RT[a\Sigma y_i \ln y_i + b\Sigma y_j \ln y_j + c\Sigma y_k \ln y_k] + {}^E G_m$$

It should be repeated that one can make computer calculations by simply defining the G_m expression. The complicated expressions for partial excess quantities given here are seldom needed.

Exercise 18.3.1

Examine how $G_{B_b D_c}$ varies along the $A_b C_c$–$B_b D_c$ diagonal if all the L parameters are zero. Compare with a binary substitutional solution A–B.

Hint

On the diagonal $y_B = y_D$ and $y_A = y_C$.

Solution

The model gives $G_{B_b D_c} = {}^\circ G_{B_b D_c} + y_A y_C \Delta^\circ G + bRT \ln y_B + cRT \ln y_D = {}^\circ G_{B_b D_c} + y_A^2 \Delta^\circ G + (b + c)RT \ln y_B$. The expression holds for one mole of atoms if $b + c = 1$ and it then resembles the regular solution model if ${}^\circ L = \Delta^\circ G = {}^\circ G_{A_b D_c} + {}^\circ G_{B_b C_c} - {}^\circ G_{A_b C_c} - {}^\circ G_{B_b D_c}$. This result may be guessed directly by inspecting Fig. 18.2.

Exercise 18.3.2

Suppose ${}^E G_m$ for a reciprocal solution $(A,B)_r(C,D)_s$ contains a term ${}^1 L_{AB}^C \cdot y_A y_B y_C \cdot (y_A - y_B)$. What is the corresponding term in ${}^E G_{A,C_s}$?

Hint

Use either the expression derived for $\Delta_1{}^E G_{Mb_ic}$ where M must be identified with C and i with A or start from the basic equation for ${}^E G_m$.

Solution

$^EG_{A,C_s} = \Delta_1{}^EG_{Mbic} = {}^1L_{AB:C}\cdot y_B\{(y_A - y_B)[y_C\cdot2\cdot(1 - y_A) + y_A - y_Cy_A] + y_Cy_B\} = y_B(2y_Ay_C + y_A^2 - 3y_A^2y_C - y_By_C - y_Ay_B + 3y_Ay_By_C).$
The basic equation gives $^EG_{A,C_s} = {}^1L_{AB:C}\cdot(y_A^2y_By_C - y_Ay_B^2y_C + 2y_Ay_By_C - y_B^2y_C + y_A^2y_B - y_Ay_B^2 - 2y_A^2y_By_C - y_A^2y_By_C - y_A^2y_By_C + y_Ay_B^2y_C + 2y_Ay_B^2y_C + y_Ay_B^2y_C).$ This can be simplified to the same expression.

18.4

Combination of interstitial and substitutional solution

The compound energy model, used to describe a reciprocal phase, can also be used for the case where there are two substitutionally mixed elements and one interstitial element. If we use C to represent the interstitial element then D represents the vacant interstitial sites. A and B represent the two substitutional elements. The compound A_bD_c will thus simply represent b atoms of A and B_bD_c represents b atoms of B. The difference between A_bC_c and A_bD_c or the difference between B_bC_c and B_bD_c can be used to represent c atoms of pure C. By the methods used in our preceding discussion on interstitial solutions we now obtain

$$G_A = {}^\circ G_A - y_By_C\Delta^\circ G_{AD+BC}/b + RT\ln y_A + RT\frac{c}{b}\ln(1 - y_C) + {}^EG_A$$

$$G_B = {}^\circ G_B + y_Ay_C\Delta^\circ G_{AD+BC}/b + RT\ln y_B + RT\frac{c}{b}\ln(1 - y_C) + {}^EG_B$$

$$G_C = y_A({}^\circ G_{A_bC_c} - {}^\circ G_{A_bD_c})/c + y_B({}^\circ G_{B_bC_c} - {}^\circ G_{B_bD_c})/c$$
$$+ RT\ln[y_C/(1 - y_C)] + {}^EG_C$$

where

$$b^EG_A = y_By_C(L_{AB:D} - L_{AB:C} - L_{B:CD} + L_{A:CD}) + y_B^2L_{AB:D} + y_C^2L_{A:CD}$$
$$+ 2y_B^2y_C(L_{AB:C} - L_{AB:D}) + 2y_By_C^2(L_{B:CD} - L_{A:CD})$$
$$b^EG_B = y_Ay_C(L_{AB:D} - L_{AB:C} + L_{B:CD} - L_{A:CD}) + y_A^2L_{AB:D} + y_C^2L_{B:CD}$$
$$+ 2y_A^2y_C(L_{AB:C} - L_{AB:D}) + 2y_Ay_C^2(L_{A:CD} - L_{B:CD})$$
$$c^EG_C = y_Ay_B(L_{AB:C} - L_{AB:D}) + (1 - 2y_C)(y_AL_{A:CD} + y_BL_{B:CD})$$

The expression for G_C has been made symmetric for A and B. Alternatively, we can modify the expression by considering A as the base metal.

Exercise 18.4.1

By considering the Fe–Mn–S melt as a reciprocal solution $(Fe,Mn)_1(Va,S)_1$, estimate the ε_S^{Mn} parameter at 1900 K from the following binary information, $^\circ G_{MnS} - {}^\circ G_{Mn} - {}^\circ G_S = -139.4$ and $^\circ G_{FeS} - {}^\circ G_{Fe} - {}^\circ G_S = -62.4$ kJ/mol.

Hint

The model is actually the combination of a substitutional and interstitial model. Neglecting binary interaction energies we obtain for S, which in the model plays the role of an interstitial element, $G_S = {}^\circ G_{FeS} - {}^\circ G_{Fe} + RT\ln[(y_S/(1 - y_S)] + y_{Mn}\Delta^\circ G$. The ε_S^{Mn} parameter is defined from $(G_S - {}^\circ G_S)/RT = \ln{}^\circ f_S + \ln x_S + \varepsilon_S^{Mn} x_{Mn}$ but for low S and Mn contents we may approximate x by y.

Solution

By comparing the two expressions we find $RT\varepsilon_S^{Mn} \cong \Delta^\circ G = {}^\circ G_{MnS} - {}^\circ G_{Mn} - {}^\circ G_{FeS} + {}^\circ G_{Fe} = -139.4 + 62.4 = -77.0$ kJ/mol; $\varepsilon_S^{Mn} = -77\,000/1900R = -4.9$.

18.5   *Phases with variable order*

So far we have discussed reciprocal phases where each element can go into one sublattice, only. However, there are many cases where an element can go into two or more sublattices although it energetically prefers a particular one. The distribution on various sublattices will then vary with temperature, with the highest degree of order found at the lowest temperature at which the atoms are still able to move. To illustrate the case we shall consider a phase with the formula $(A,B)_b(B,A)_c$ where A prefers the first sublattice and B the second one. For simplicity we shall choose $b + c = 1$. The four component compounds will be A_bB_c, A_bA_c, B_bB_c and B_bA_c. We can apply all the equations already derived for a reciprocal phase but now the site fractions are not fixed by the composition. However, there is a relation between composition and site fractions and we can write it in two different ways,

$$x_A = by'_A + cy''_A = y'_A y''_A + by'_A y''_B + cy'_B y''_A$$

There is still a degree of freedom among the site fractions and we may define an internal variable. In general it represents the progress of the ordering reaction. Its equilibrium value is determined by minimizing the Gibbs energy at any given T,P,x_A.

In order to eliminate $y'_A y''_A$ and $y'_B y''_B$, let us first transform the expression for G_m derived for a reciprocal phase, using the above relation and a corresponding one for x_B,

$$x_B = by'_B + cy''_B = y'_B y''_B + by'_B y''_A + cy'_A y''_B$$

By neglecting the excess Gibbs energy we obtain

$$
\begin{aligned}
G_m &= y'_A y''_A {}^\circ G_{A_bA_c} + y'_A y''_B {}^\circ G_{A_bB_c} + y'_B y''_A {}^\circ G_{B_bA_c} + y'_B y''_B {}^\circ G_{B_bB_c} \\
&\quad + RT(by'_A \ln y'_A + by'_B \ln y'_B + cy''_A \ln y''_A + cy''_B \ln y''_B) \\
&= x_A {}^\circ G_{A_bA_c} + x_B {}^\circ G_{B_bB_c} + [y'_A y''_B({}^\circ G_{A_bB_c} - b \cdot {}^\circ G_{A_bA_c} - c \cdot {}^\circ G_{B_bB_c}) \\
&\quad + y'_B y''_A({}^\circ G_{B_bA_c} - b \cdot {}^\circ G_{B_bB_c} - c \cdot {}^\circ G_{A_bA_c})] + RT(by'_A \ln y'_A + by'_B \ln y'_B \\
&\quad + cy''_A \ln y''_A + cy''_B \ln y''_B)
\end{aligned}
$$

${}^\circ G_{A_bA_c}$ is the Gibbs energy of 1 mole of A in this structure and it may be denoted by ${}^\circ G_A$. By the same reasoning ${}^\circ G_{B_bB_c}$ will be denoted by ${}^\circ G_B$. We can introduce the following two quantities, which represent the Gibbs energy of formation of the compounds A_bB_c and B_bA_c.

$$
\begin{aligned}
\Delta^\circ G_{A_bB_c} &= {}^\circ G_{A_bB_c} - b \cdot {}^\circ G_{A_bA_c} - c \cdot {}^\circ G_{B_bB_c} = {}^\circ G_{A_bB_c} - b \cdot {}^\circ G_A - c \cdot {}^\circ G_B \\
\Delta^\circ G_{B_bA_c} &= {}^\circ G_{B_bA_c} - b \cdot {}^\circ G_{B_bB_c} - c \cdot {}^\circ G_{A_bA_c} = {}^\circ G_{B_bA_c} - b \cdot {}^\circ G_B - c \cdot {}^\circ G_A
\end{aligned}
$$

This is why the model is sometimes called the Compound Energy Model. The sum of the two quantities is identical to the Gibbs energy of the reciprocal reaction introduced in Section 18.3,

$$
\Delta^\circ G_{A_bB_c} + \Delta^\circ G_{B_bA_c} = {}^\circ G_{A_bB_c} + {}^\circ G_{B_bA_c} - {}^\circ G_{A_bA_c} - {}^\circ G_{B_bB_c} = \Delta^\circ G_{AB+BA}
$$

We now obtain

$$
\begin{aligned}
G_m &= x_A {}^\circ G_A + x_B {}^\circ G_B + y'_A y''_B \Delta^\circ G_{A_bB_c} + y'_B y''_A \Delta^\circ G_{B_bA_c} \\
&\quad + RT(by'_A \ln y'_A + by'_B \ln y'_B + cy''_A \ln y''_A + cy''_B \ln y''_B)
\end{aligned}
$$

By choosing y'_A as the independent internal variable we get $dy'_B = -dy'_A$ and $dy''_B = -dy''_A = (b/c)dy'_A$ under constant x_A, and the equilibrium value of y'_A under constant composition and any T can thus be calculated from

$$
\begin{aligned}
c \cdot \partial G_m / \partial y'_A &= (cy''_B + by'_A)\Delta^\circ G_{A_bB_c} - (cy''_A + by'_B)\Delta^\circ G_{B_bA_c} \\
&\quad + bc \cdot RT(1 + \ln y'_A - 1 - \ln y'_B - 1 - \ln y''_A + 1 + \ln y''_B) = 0
\end{aligned}
$$

However, it is easier to solve for T as a function of y'_A,

$$
bcRT = \frac{(cy''_A + by'_B)\Delta^\circ G_{B_bA_c} - (cy''_B + by'_A)\Delta^\circ G_{A_bB_c}}{\ln(y'_A y''_B / y'_B y''_A)}
$$

When the two sublattices are equivalent, $b = c = 0.5$ and $\Delta^\circ G_{A_bB_c} = \Delta^\circ G_{B_bA_c} = 0.5\Delta^\circ G_{AB+BA}$, and the equilibrium condition is

$$
\frac{RT}{-\Delta^\circ G_{AB+BA}} = \frac{y'_A - y'_B - y''_A + y''_B}{\ln(y'_A y''_B / y'_B y''_A)}
$$

This may be called the symmetric case. The disordered state, $y'_A = y''_A = x_A$ is a solution to the equation at all temperatures and compositions but in order to examine where it is a stable or unstable equilibrium, we must study the second-order derivative,

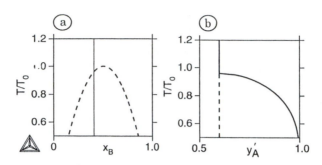

Figure 18.3 Second-order transition in a binary system. (a) Phase diagram. (b) Variation of degree of order with temperature for $x_A = 0.6$.

$$c \cdot \partial^2 G_m / \partial y_A'^2 = (b + b + b + b) \cdot 0.5 \Delta°G_{AB+BA} + bRT(c/y_A' + c/y_B' + b/y_A'' + b/y_B'') = 2b\Delta°G_{AB+BA} + bRT/x_A x_B$$

For positive values of $\Delta°G_{AB+BA}$ this is always positive and the disordered state is always the stable state. There is no ordering tendency. For negative values of $\Delta°G_{AB+BA}$ it is positive at higher temperatures and negative at lower and there is a transition temperature where it is zero, corresponding to the criterion $g_{\xi\xi} = 0$ in Section 14.2,

$$\frac{RT_{tr}}{-\Delta°G_{AB+BA}} = 2x_A x_B$$

The disordered state is thus a state of stable equilibrium above this critical temperature but an unstable equilibrium below. Two new solutions to the equation appear there, representing stable ordered states. This is demonstrated for a composition of $x_A = 0.6$ in Fig. 18.3(b) and the dashed line represents the unstable, disordered state below the transition point. The phase diagram is shown in Fig. 18.3(a). The whole curve is there drawn with a dashed line because it is not a phase boundary but a transition line, the transition being of second order. The site fraction y_A' approaches the x_A value without any jump, as demonstrated in Fig. 18.3(b). We have thus managed again to model the second-order type of ordering transition but this time we have used a model containing parameters of some physical significance. In Chapter 14 we followed Landau's approach and simply worked with coefficients in a power series expansion.

When $\Delta°G_{A_bB_c}$ and $\Delta°G_{B_bA_c}$ are not equal, the result will be quite different. This is demonstrated in Fig. 18.4(a) for $b = c = 0.5$, $x_A = 0.5$, and

$$\Delta°G_{A_bB_c} - \Delta°G_{B_bA_c} = -0.1(\Delta°G_{A_bB_c} + \Delta°G_{B_bA_c}) = -0.1\Delta°G_{AB+BA}$$

This may be called the asymmetric case. As soon as the two compound energies differ, the completely disordered state will never be stable and the ordered region in the phase diagram does not show an order–disorder transition.

The variation with composition is shown at three temperatures in Fig. 18.4(b). The results for $x_A = 0.5$ can be read on the diagonal between the upper left and lower right corners.

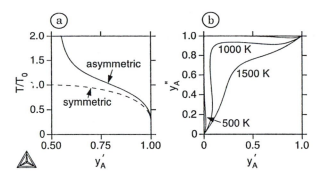

Figure 18.4 Model calculation of an ordered phase without a transition to a disordered state at any temperature. (a) Result for $x_A = 0.5$ compared with the result for a symmetric case showing a transition (dashed line). (b) Variation with composition.

Exercise 18.5.1

Where in Fig. 18.4(a) would a line for $\Delta^\circ G_{A_bB_c} - \Delta^\circ G_{B_bA_c} = +0.1\Delta^\circ G_{AB + BA}$ fall?

Hint

Do not try to solve this problem by looking at the equations. The answer should be based upon a more basic consideration.

Solution

A change of sign of $\Delta^\circ G_{AB + BA}$ means that the other sublattice will be preferred by the A atoms but otherwise the effects will be the same as before. The effect of $\Delta^\circ G_{AB + BA}$ is to accentuate the order, independent of what sublattice the A atoms prefer. We will thus get the same shape of curve as in Fig. 18.4(a) but starting from $y_A' = 0$ at $T = 0$ and approaching $y_A' = 0.5$ from the left.

18.6   ## *Ionic solid solutions*

In this chapter we have modelled various types of phases with sublattices without actually discussing the nature of the atoms (or 'species' to be more general). The compound energy model also applies to ionic substances but there are some complications due to the requirement of electroneutrality which should now be discussed.

Let us first consider solid solutions between NaCl, KCl, NaBr and KBr which all have the same crystalline structure. All the elements are ionized and we could give the formula as $(Na^{+1},K^{+1})_1(Cl^{-1},Br^{-1})_1$. One can vary the composition freely within the composition square because all the ions are univalent and the condition of electroneutrality is automatically fulfilled over the whole square by the fact that the two sublattices have the same number of sites. The compound energy model can be applied with no additional complications in this case.

Let us next consider the solution of $CaCl_2$ in $NaCl$. A complication is then caused by Ca being divalent and, in order to compensate for this, some of the cation sites will be vacant. The formula would thus be $(Na^{+1},Ca^{+2},Va^0)_1(Cl^{-1})_1$. This seems to resemble a ternary system with Na_1Cl_1, Ca_1Cl_1 and Va_1Cl_1 as the components. However, electroneutrality requires that $y_{Ca} = y_{Va}$ and one can only vary the composition along a straight line in the composition triangle. In order to model the properties of solutions on this line we shall apply the ordinary expression for the compound energy model but with the additional condition of electroneutrality. Since $y_{Cl} = 1$ we get, by neglecting the excess Gibbs energy,

$$G_m = y_{Na}{}^°G_{NaCl} + y_{Ca}{}^°G_{CaCl} + y_{Va}{}^°G_{VaCl} + RT\Sigma y_i \ln y_i$$

Here we have two quantities which cannot be studied experimentally, $^°G_{CaCl}$ and $^°G_{VaCl}$, because they are defined for charged compounds. However, due to the auxiliary condition $y_{Ca} = y_{Va}$, they always appear in the neutral combination, $(^°G_{CaCl} + ^°G_{VaCl})$. The properties of this combination can be studied by studying the neutral solutions. Instead of introducing this combination in the equation, it may be recommended to keep the original form and apply the condition of electroneutrality in the final expression one wants to use. When listing the parameter values for an ionic system one could select one of the charged compounds and give all the other charged compounds relative to that one. As an example, one may like to express the chemical potential of $CaCl_2$ in the solution. It is obtained as

$$\mu_{CaCl_2} \equiv G_{CaCl} + G_{VaCl} = ^°G_{CaCl} + RT\ln y_{Ca} + ^°G_{VaCl} + RT\ln y_{Va}$$
$$= (^°G_{CaCl} + ^°G_{VaCl}) + RT\ln(y_{Ca}^2)$$

since $y_{Ca} = y_{Va}$. The neutral combination $(^°G_{CaCl} + ^°G_{VaCl})$ can be given a numerical value.

Let us now consider the opposite case, the solution of $NaCl$ in $CaCl_2$. Electroneutrality may there be satisfied by the formation of vacant sites on the anion sublattice, $(Ca^{+2},Na^{+1})_1(Cl^{-1},Va^0)_2$, and the condition of electroneutrality is now $1 \cdot y_{Na} = 2 \cdot y_{Va}$. This is a reciprocal system and the compound energy model yields

$$G_m = y_{Ca}y_{Cl}{}^°G_{CaCl_2} + y_{Ca}y_{Va}{}^°G_{CaVa_2} + y_{Na}y_{Cl}{}^°G_{NaCl_2}$$
$$+ y_{Na}y_{Va}{}^°G_{NaVa_2} + RT\Sigma y_i \ln y_i + {}^EG_m$$

The chemical potential of $NaCl$ is obtained as

$$2\mu_{NaCl} = G_{NaCl_2} + G_{NaVa_2}$$
$$= {}^°G_{NaCl_2} - y_{Ca}y_{Va}\Delta^°G + RT\ln y_{Na} + 2RT\ln y_{Cl} + {}^EG_{NaCl_2}$$
$$+ {}^°G_{NaVa_2} + y_{Ca}y_{Cl}\Delta^°G + RT\ln y_{Na} + 2RT\ln y_{Va} + {}^EG_{NaVa_2}$$

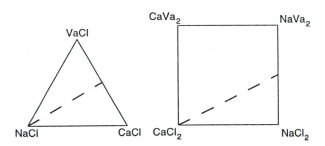

Figure 18.5 Neutral line in two cases of ionic solutions.

$$= {}^\circ G_{\text{NaCl}_2} + {}^\circ G_{\text{NaVa}_2} + (1 - y_{\text{Na}})^2 \Delta^\circ G + 2RT\ln[0.5 y_{\text{Na}}^2(1 - 0.5 y_{\text{Na}})]$$
$$+ {}^E G_{\text{NaCl}_2} + {}^E G_{\text{NaVa}_2}$$

${}^\circ G_{\text{NaCl}_2} + {}^\circ G_{\text{NaVa}_2}$ and $\Delta^\circ G$ (which is equal to ${}^\circ G_{\text{CaCl}_2} + {}^\circ G_{\text{NaVa}_2} - {}^\circ G_{\text{CaVa}_2} - {}^\circ G_{\text{NaCl}_2}$ and may be written as $\Delta^\circ G_{\text{CaNa}+\text{ClVa}}$) represent neutral combinations and can be given numerical values. Figure 18.5 illustrates the neutral line in the two cases. How the model may describe the properties outside these lines is of no practical consequence.

Many ionic compounds are non-stoichiometric, an example being CeO_2. It may be modelled by assuming that some of the Ce ions are only trivalent, yielding the formula $(Ce^{+4}, Ce^{+3})_1(O^{-2}, Va^0)_2$.

$$G_m = y_{\text{Ce}^{+4}} y_O {}^\circ G_{\text{Ce}^{+4}O_2} + y_{\text{Ce}^{+4}} y_{\text{Va}} {}^\circ G_{\text{Ce}^{+4}Va_2} + y_{\text{Ce}^{+3}} y_O {}^\circ G_{\text{Ce}^{+3}O_2}$$
$$+ y_{\text{Ce}^{+3}} y_{\text{Va}} {}^\circ G_{\text{Ce}^{+3}Va_2} + RT\Sigma a y_i \ln y_i + {}^E G_m$$

The deviation from the stoichiometric composition depends upon the oxygen potential in the surroundings. In order to derive an expression for the oxygen potential one must find a reaction formula for the formation of oxygen which balances atoms as well as charges.

$$4Ce^{+4} + 2O^{-2} \rightarrow 4Ce^{+3} + O_2$$

Then the reaction formula must be expressed in terms of the component compounds in the model. Instead of $2O^{-2}$ we thus write $Ce^{+3}O_2\text{–}Ce^{+4}Va_2$. With all charges shown we write the reaction as

$$5(Ce^{+4})_1(O^{-2})_2 \rightarrow (Ce^{+4})_1(Va^0)_2 + 4(Ce^{+3})_1(O^{-2})_2 + O_2$$

We thus obtain

$$2\mu_O = \mu_{O_2} = 5G_{\text{Ce}^{+4}O_2} - G_{\text{Ce}^{+4}Va_2} - 4G_{\text{Ce}^{+3}O_2}$$
$$= 5{}^\circ G_{\text{Ce}^{+4}O_2} + 5y_{\text{Ce}^{+3}} y_{\text{Va}} \Delta^\circ G_{\text{Ce}^{+4}Va+\text{Ce}^{+3}O} + 5RT\ln y_{\text{Ce}^{+4}} + 10RT\ln y_O$$
$$+ 5{}^E G_{\text{Ce}^{+4}O_2} - {}^\circ G_{\text{Ce}^{+4}Va_2} + y_{\text{Ce}^{+3}} y_O \Delta^\circ G_{\text{Ce}^{+4}Va+\text{Ce}^{+3}O} - RT\ln y_{\text{Ce}^{+4}}$$
$$- 2RT\ln y_{\text{Va}} - {}^E G_{\text{Ce}^{+4}Va_2} - 4{}^\circ G_{\text{Ce}^{+3}O_2} + 4y_{\text{Ce}^{+4}} y_{\text{Va}} \Delta^\circ G_{\text{Ce}^{+4}Va+\text{Ce}^{+3}O}$$
$$- 4RT\ln y_{\text{Ce}^{+3}} - 8RT\ln y_O - 4{}^E G_{\text{Ce}^{+3}O_2}$$
$$= 5{}^\circ G_{\text{Ce}^{+4}O_2} - {}^\circ G_{\text{Ce}^{+4}Va_2} - 4{}^\circ G_{\text{Ce}^{+3}O_2} + RT\ln[y_{\text{Ce}^{+4}}^4 y_O^2 / y_{\text{Ce}^{+4}}^4 y_{\text{Va}}^2]$$

$$+ (y_{Ce^{+3}} + 4y_{Va})\Delta^\circ G_{Ce^{+4}Va+Ce^{+3}O} + 5^E G_{Ce^{+4}O_2} - 4^E G_{Ce^{+3}O_2} - {}^E G_{Ce^{+4}Va_2}$$

where the factor $y_{Ce^{+3}} + 4y_{Va}$ can be written as $8y_{Va}$ because the condition of electroneutrality for $(Ce^{+4}, Ce^{+3})_1(O^{-2}, Va^0)_2$ is

$$1 \cdot [4(1 - y_{Ce^{+3}}) + 3y_{Ce^{+3}}] = 2 \cdot 2(1 - y_{Va}) \text{ or } y_{Ce^{+3}} = 4y_{Va}$$

For low deviations from stoichiometry, $y_{Ce^{+4}} \cong 1$ and $y_O \cong 1$ and we may neglect all excess terms, so that

$$RT\ln P_{O_2} = \mu_{O_2} - {}^\circ\mu_{O_2}$$
$$= 5^\circ G_{Ce^{+4}O_2} - 4^\circ G_{Ce^{+3}O_2} - {}^\circ G_{Ce^{+4}Va_2} - {}^\circ\mu_{O_2} - RT\ln(256y_{Va}^6)$$

This model will thus predict that the vacancy content is proportional to $P_{O_2}{}^{-1/6}$. Experimental data indicates that the true value may rather be $-1/5$ at low values of y_{Va}. The model may thus require some modification. One possibility is to introduce a strong association between a vacancy and a cation with abnormal valency, in this case Ce^{+3}.

It may again be emphasized that a computer calculation only requires that the G_m expression is defined. On the other hand, the complicated expressions derived in this section are needed for the kind of analytical examination of the model presented here.

Exercise 18.6.1

It is possible to dissolve Al and O in Si_3N_4. Show in a composition square Si_3N_4–SiO_2–Al_2O_3–AlN where you would expect to find such a solid solution phase.

Hint

Since these materials are strongly covalent, the vacancy concentration is likely to be low. As a first approximation we may thus neglect vacancies. On the other hand, the ordinary condition of electroneutrality may be applied.

Solution

The general formula would be $(Si^{+4}, Al^{+3})_3(N^{-3}, O^{-2})_4$. The requirement of electroneutrality gives $3[4(1 - y_{Al}) + 3y_{Al}] = 4[3(1 - y_O) + 2y_O]$ and thus $3y_{Al} = 4y_O$. For the highest Al content, $y_{Al} = 1$, we find $y_O = 0.75$. The solution phase thus falls on the join Si_3N_4–$Al_3N_1O_3$. How close to $Al_3N_1O_3$ one can actually get experimentally depends on the competition with other phases.

Exercise 18.6.2

In order for the solution phase $(Si^{+4},Al^{+3})_3(N^{-3},O^{-2})_4$ to extend to the AlN and Al_2O_3 corners of the Si_3N_4–SiO_2–Al_2O_3–AlN diagram (which does not really happen) it would be necessary to introduce vacancies. Even though such vacancies are not very likely to form, show how many vacancies would be needed in the two cases.

Hint

Introduce vacancies into the formula to an amount required by electroneutrality if there are no O^{-2} in the first case and no N^{-3} in the second.

Solution

$(Al^{+3})_3(N^{-3}{}_{1-x},Va_x)_4$ gives $3 \cdot 3 = 4 \cdot 3(1-x);\ \ x = 1/4.$
$(Al^{+3}{}_{1-x},Va_x)_3(O^{-2})_4$ gives $3 \cdot 3(1-x) = 4 \cdot 2;\ \ x = 1/9.$

19

Physical solution models

19.1 *Concept of nearest-neighbour bond energies*

The modelling of solution phases described in Chapters 17 and 18 was based on the proper expression for the ideal entropy of mixing assuming random mixing within each sublattice. The rest of the modelling was purely mathematical and was not related directly to any physical effects. The present chapter is devoted to models based on more physical considerations. In particular, the interaction energies between atoms will be considered. A very simple and useful way of modelling the thermodynamic properties of a binary solution is based upon the assumption that the energy of the whole system is the sum of the bond energies between neighbouring atoms. In the simplest case one only considers the energies of pairs of nearest neighbours. The formation of the solution from the pure components can then be regarded as a chemical reaction between different kinds of bonds, similar to a reaction between molecules

$$A-A + B-B \Leftrightarrow A-B + B-A$$

We have here chosen to distinguish between an A–B bond and a B–A bond although they are of course quite equivalent if all the lattice sites are equivalent.

The reaction gives a change of energy which we may denote by $2v$ and regard as an **exchange energy**. Since our aim is to construct an expression for the Gibbs energy of the solution we shall consider the Gibbs energies of the bonds rather than the internal energies. This actually means that we take into account not only the bond energies but also their temperature dependence.

$$2l = g_{AB} + g_{BA} - g_{AA} - g_{BB}$$

In some cases g_{AB} and g_{BA} may be different and it may thus be useful to define two different quantities from the beginning,

$$v_{AB} = g_{AB} - g_{AA}/2 - g_{BB}/2$$
$$v_{BA} = g_{BA} - g_{AA}/2 - g_{BB}/2$$

where $v_{AB} + v_{BA} = 2v$. All bonds of each kind are assumed to have the same energy independent of the local composition. The Gibbs energy contribution from the bond energies can thus be evaluated by counting the number of bonds, N_{AA}, N_{BB}, N_{AB} and N_{BA}.

$$\Delta G = \sum_i \sum_j N_{ij} g_{ij}$$

This is the mathematical definition of the nearest-neighbour bond energy model. In the next section we shall evaluate N_{ij} and add the contribution due to the entropy of configurational disorder.

The counting of the number of bonds of each kind can be done with different degrees of ambition. In the simplest treatment, which is called the Bragg–Williams model, one assumes that the atoms are placed at random on the sites in the crystal and it leads to an expression which is identical to the so-called regular solution model. It may thus be used to justify the regular solution model. In more ambitious treatments one tries to calculate how the v value influences the number of bonds. A positive v value indicates that the A and B atoms do not like to mix with each other and, if they have been mixed with each other in a solution, they should at least try to arrange themselves in such a way that there are less A–B bonds than in a random arrangement. A negative v value, on the other hand, would favour arrangements where the A atoms are surrounded by more B atoms than in a random arrangement. Such effects will be considered later in this chapter using an approximation called the quasi-chemical approach. It is primarily based on a random mixture of the nearest neighbour bonds. In Kikuchi's cluster variation method one considers the random mixture of larger clusters. In principle, one should get an exact description of the configurational entropy by going to clusters of infinite size but that is not practically possible, nor is it necessary. A sufficiently good result is probably obtained by including just a few clusters. It is interesting to note that in the cluster variation method one estimates the energy of a cluster as a sum of its bond energies (also called pair energies), assuming that each kind of pair energy is a constant, independent of the local and global composition.

The concept of nearest neighbour bond energies is closely related to the concept of molecules with a Gibbs energy of formation for each kind of molecule but it is much more difficult to justify. In a substance with molecules the atoms are actually present as groups of atoms bound together tightly and it is often a good approximation to neglect interactions between atoms in different molecules. However, the splitting up of the total energy of a crystal into a large number of bond energies is quite arbitrary and one may, for instance, choose to consider or neglect next-nearest neighbour bonds and to consider bond energies related to pairs of atoms or to larger groups of atoms, sometimes called clusters. Even if one decides to consider only pairs of atoms or larger groups of atoms, the energy of the different kinds of bonds is rather arbitrary unless one has information relating to different types of ordering. This being so, it is doubtful whether a rather random distribution of atoms can be described with cluster energies evaluated from ordered arrangements. A very crude but useful way of improving the pair energy model would be to assume that only part of the excess Gibbs energy in a disordered state is of such short-range character that it can affect short- and long-range order. That approach would give an additional adjustable parameter to be used in the description of thermodynamic and configurational information.

The justification of the nearest-neighbour bond energy model has to come from its success in explaining experimental facts. It has been found very useful in giving qualitative explanations of many phenomena in alloy systems but less successful in accounting for experimental data in detail. There are many modifications of the basic treatment but we shall first consider the simplest possible approach, the random mixing model of Bragg and Williams.

Exercise 19.1.1

Suppose one has found experimentally that v is a constant across a binary system. This result may be interpreted by assuming that all the bond energies are independent of composition. However, suppose one has some theoretical reason to expect that g_{AA} and g_{BB} should vary linearly across the system. Would it then be possible to explain the experimental result?

Hint

Let $g_{AA} = g^\circ_{AA} + ax_B$ and $g_{BB} = g^\circ_{BB} + bx_A$ in the second interpretation. Suppose in the first interpretation $g^\circ_{AB} = g^\circ_{BA}$ and in the second one $g_{AB} = g_{BA}$.

Solution

We can achieve $g^\circ_{AB} + g^\circ_{BA} - g^\circ_{AA} - g^\circ_{BB} = 2v = g_{AB} + g_{BA} - g_{AA} - g_{BB} = 2g_{AB} - g^\circ_{AA} - ax_B - g^\circ_{BB} - bx_A$ if $2g_{AB} = g^\circ_{AA} + ax_B + g^\circ_{BB} + bx_A + 2g^\circ_{AB} - g^\circ_{AA} - g^\circ_{BB} = (2g^\circ_{AB} + b) + (a - b)x_B$. Yes, it is possible.

19.2   *Random mixing model for a substitutional solution*

A solid phase where atoms of different components can substitute for each other, i.e. occupy the same kind of lattice sites, is called a substitutional solution. The composition of such a solution is conveniently described with the molar contents of the atoms, x_A and x_B in a binary solution. In order to describe the arrangement of the atoms it may be convenient to introduce the fractions of bonds, p_{AA}, p_{AB}, p_{BA} and p_{BB}. The notation p is chosen because the fraction of a certain kind of bond is equal to the probability of finding that kind of bond. Of course, $\Sigma p_{ij} = 1$. The fractions are also related through the composition and for the simple case where all the atoms have the same number of bonds this condition can be written in two ways

$$x_A = p_{AA} + (p_{AB} + p_{BA})/2$$
$$x_B = p_{BB} + (p_{AB} + p_{BA})/2$$

In the case of random mixing we get

$$p_{AA} = x_A^2$$
$$p_{AB} = x_A x_B = p_{BA}$$
$$p_{BB} = x_B^2$$

Let us assume that all atoms have the same number of nearest neighbours. We can thus introduce a single z value as the coordination number. The total number of bonds in a system containing one mole of atoms is thus $zN^A/2$ since each bond is shared between two atoms. N^A is Avogadro's number. One will thus obtain, for instance, $N_{AA} = p_{AA} \cdot zN^A/2$ and for one mole of pure A we obtain

$$^\circ G_A = g_{AA} \cdot 1 \cdot zN^A/2$$

The Gibbs energy contribution from all the bond energies is

$$\Delta G_m = \sum_i \sum_j N_{ij} g_{ij} = (p_{AA} g_{AA} + p_{AB} g_{AB} + p_{BA} g_{BA} + p_{BB} g_{BB}) \cdot zN^A/2$$

By inserting v_{AB} and v_{BA} we obtain

$$\Delta G_m = [p_{AB} v_{AB} + p_{BA} v_{BA} + g_{AA}(p_{AA} + p_{AB}/2 + p_{BA}/2) + g_{BB}(p_{BB} + p_{AB}/2$$
$$+ p_{BA}/2)] \cdot zN^A/2 = x_A {}^\circ G_A + x_B {}^\circ G_B + [p_{AB} v_{AB} + p_{BA} v_{BA}] \cdot zN^A/2$$

When A and B atoms are mixed at random in the solution, there will be a configurational contribution to the entropy and we get, by substituting $x_A x_B$ for p_{AB} and p_{BA} and inserting $2v = v_{AB} + v_{BA}$,

$$G_m = x_A {}^\circ G_A + x_B {}^\circ G_B + vzN^A x_A x_B + RT(x_A \ln x_A + x_B \ln x_B)$$

It is evident that a positive value of v gives a positive deviation from the ideal solution behaviour and thus a tendency for demixing. The consequences of this tendency were discussed in Section 17.4. Negative values give a tendency of ordering of the atoms in such a way that unlike atoms prefer to be nearest neighbours. The result may be short-range order and even long-range order. These phenomena will be discussed in the following sections.

Exercise 19.2.1

Compare the final G_m expression with the corresponding expression according to the regular solution model.

Hint

According to Section 17.4, the regular solution model gives $^{E}G_{m} = x_{A}x_{B}{}^{0}L$.

Solution

The random bond energy model yields $^{E}G_{m} = vzN^{A}x_{A}x_{B}$. The two models can be exactly translated into each other by the relation $^{0}L = vzN^{A}$.

19.3   *Deviation from random distribution*

In order to describe non-random solutions we shall make extensive use of the bond probabilities, p_{ij}. In order to describe both long- and short-range order we need two independent internal variables and they can be defined in many ways. A convenient choice is the following,

$$K = (p_{AB} + p_{BA})/2$$
$$L = (p_{AB} - p_{BA})/2$$

and it yields

$$p_{AB} = K + L; \quad dp_{AB} = dK + dL$$
$$p_{BA} = K - L; \quad dp_{BA} = dK - dL$$
$$p_{AA} = x_{A} - (p_{AB} + p_{BA})/2 = x_{A} - K; \quad dp_{AA} = dx_{A} - dK$$
$$p_{BB} = x_{B} - (p_{AB} + p_{BA})/2 = x_{B} - K; \quad dp_{BB} = - dx_{A} - dK$$

Long-range order can only be described with the use of two or more sublattices. The situation in an ordered alloy with two sublattices can be described with the site fractions which, in turn, can be expressed through the bond probabilities. Assuming that all the bonds go between atoms in two different sublattices we find

$$y'_{A} = p_{AA} + p_{AB} = x_{A} + L; \quad dy'_{A} = dx_{A} + dL$$
$$y'_{B} = p_{BB} + p_{BA} = x_{B} - L; \quad dy'_{B} = - dx_{A} - dL$$
$$y''_{A} = p_{AA} + p_{BA} = x_{A} - L; \quad dy''_{A} = dx_{A} - dL$$
$$y''_{B} = p_{BB} + p_{AB} = x_{B} + L; \quad dy''_{B} = - dx_{A} + dL$$

Long-range order means that element A prefers one sublattice and B the other. It is conveniently defined as

$$\text{l.r.o.} = y'_{A} - y''_{A} = p_{AA} + p_{AB} - p_{AA} - p_{BA} = 2L$$

Short-range order means that the atoms with given site fractions do not arrange themselves at random within each sublattice. Random distribution would yield the following probabilities of finding various bonds between the two sublattices and we shall still assume that there are no bonds within a sublattice.

$$p_{AA} = y'_A y''_A$$
$$p_{AB} = y'_A y''_B$$
$$p_{BA} = y'_B y''_A$$
$$p_{BB} = y'_B y''_B$$

Short-range order may be defined as the deviation from this arrangement

$$\text{s.r.o.} = [p_{AB} - y'_A y''_B + p_{BA} - y'_B y''_A]/2$$
$$= [2K - (x_A + L)(x_B + L) - (x_B - L)(x_A - L)]/2 = K - L^2 - x_A x_B$$

We know from Section 16.10 that a random distribution within each sublattice would yield the following configurational entropy for 1 mole of atoms,

$$-S_m/R = [y'_A \ln y'_A + y'_B \ln y'_B + y''_A \ln y''_A + y''_B \ln y''_B]/2$$

This does not account for short-range order. In an attempt to treat that case, one could start by considering a random distribution of the bonds,

$$-S_m/R = (z/2)[p_{AA}\ln p_{AA} + p_{AB}\ln p_{AB} + p_{BA}\ln p_{BA} + p_{BB}\ln p_{BB}]$$

However, this does not reduce to the previous expression when the random values for the four p_{ij} are inserted. This condition can be satisfied if we divide each p_{ij} under an ln sign by its random value, which will make the whole expression go to zero for a random case, and then add the previous expression, which should be the correct one for the random case.

$$-S_m/R = (z/2)[p_{AA}\ln(p_{AA}/y'_A y''_A) + p_{AB}\ln(p_{AB}/y'_A y''_B) + p_{BA}\ln(p_{BA}/y'_B y''_A)$$
$$+ p_{BB}\ln(p_{BB}/y'_B y''_B)] + [y'_A \ln y'_A + y'_B \ln y'_B + y''_A \ln y''_A + y''_B \ln y''_B]/2$$

The contribution from the bond energies was derived in the preceding section and is

$$\Delta G_m = x_A{}^\circ G_A + x_B{}^\circ G_B + [p_{AB}v_{AB} + p_{BA}v_{BA}]\cdot zN^A/2$$

The complete expression will thus be

$$G_m = x_A{}^\circ G_A + x_B{}^\circ G_B + [p_{AB}v_{AB} + p_{BA}v_{BA}]\cdot zN^A/2$$
$$+ RT(z/2)[p_{AA}\ln(p_{AA}/y'_A y''_A) + p_{AB}\ln(p_{AB}/y'_A y''_B) + p_{BA}\ln(p_{BA}/y'_B y''_A)$$
$$+ p_{BB}\ln(p_{BB}/y'_B y''_B)] + RT[y'_A \ln y'_A + y'_B \ln y'_B + y''_A \ln y''_A + y''_B \ln y''_B]/2$$

We shall now apply this general expression to several special cases. However, it should be remembered that the expression is valid only under the assumption that there are two sublattices, that they contain the same number of bonds and that all bonds go between atoms in different sublattices. The simple bcc structure (A2) can order in this way, yielding B2.

Exercise 19.3.1

Demonstrate that the general equation for ordering reduces to the model for random mixing in a substitutional solution without short-range order.

Hint

In a substitutional solution there is no long-range order and thus $y'_A = y''_A = x_A$ and $y'_B = y''_B = x_B$ and without short-range order $p_{ij} = y'_i y''_j = x_i x_j$. Furthermore, v_{AB} and v_{BA} must be equal in order to prevent long-range order.

Solution

The first part of the entropy contribution reduces to zero and the second one to $RT[x_A \ln x_A + x_B \ln x_B]$. In the energy part $p_{AB} v_{AB} + p_{BA} v_{BA}$ reduces to $2 x_A x_B v$. Thus, $\Delta G_m = x_A{}^\circ G_A + x_B{}^\circ G_B + z N^A x_A x_B + RT[x_A \ln x_A + x_B \ln x_B]$.

19.4   *Short-range order*

At a high enough temperature long-range order will disappear if $v_{AB} = v_{BA} = v$ and only short-range order remains. For this case, $y'_A = y''_A = x_A$ and $y'_B = y''_B = x_B$ yield $L = 0$ and $p_{AB} = p_{BA}$. We obtain

$$G_m = x_A{}^\circ G_A + x_B{}^\circ G_B + vzN^A p_{AB} + RT(z/2)[p_{AA} \ln(p_{AA}/x_A^2)$$
$$+ 2p_{AB} \ln(p_{AB}/x_A x_B) + p_{BB} \ln(p_{BB}/x_B^2)] + RT[x_A \ln x_A + x_B \ln x_B]$$

We have only one internal variable $K = (p_{AB} + p_{BA})/2 = p_{AB}$. Its equilibrium value under constant composition is obtained from

$$(\partial G_m/\partial K)_{L, x_A} = vzN^A + RT(z/2)[-\ln(p_{AA}/x_A^2) - p_{AA}/p_{AA} + 2\ln(p_{AB}/x_A x_B)$$
$$+ 2p_{AB}/p_{AB} - \ln(p_{BB}/x_B^2) - p_{BB}/p_{BB}]$$
$$= vzN^A + RT(z/2)\ln[(p_{AA})^2/p_{AA}p_{BB}] = 0$$

$$\frac{(p_{AB})^2}{p_{AA}p_{BB}} = \exp(-2v/kT)$$

This resembles the law of mass action for a chemical reaction between molecules and the method of correcting the entropy expression is thus called the quasi-chemical method. It should be emphasized that p_{AB} is here defined as half the number of AB bonds because the other half is counted as p_{BA}.

It is worth noting that the quasi-chemical method of correcting the entropy expression is valid only for small deviations from randomness, i.e. for low values of v/kT. It is immediately evident that very large values of v/kT will produce unreasonable results. An infinite value of v/kT will make $p_{AB} = p_{BA} = 0$ and thus $p_{AA} = x_A$ and $p_{BB} = x_B$. This implies a separation of the system into two parts, one containing all the A atoms and the other all the B atoms. The configurational entropy should thus be zero but our entropy expression will yield

$$S_m = - R(z/2)[x_A\ln(1/x_A) + 0 + x_B\ln(1/x_B)] - R[x_A\ln x_A + x_B\ln x_B]$$
$$= (z/2 - 1)R[x_A\ln x_A + x_B\ln x_B]$$

This yields the correct value (zero) for $z = 2$, which applies when all the atoms are arranged in a string. For all realistic z values (e.g. $z = 8$ for bcc and $z = 12$ for fcc) the entropy expression yields large negative values. This result emphasizes that the quasi-chemical method should be used only for low values of v/kT, i.e. for low deviations from the ideal entropy. There, one can simplify the expression by series expansions,

$$\exp\varepsilon = 1 + \varepsilon + \varepsilon/2$$
$$\sqrt{1 + \varepsilon} = 1 + \varepsilon/2 - \varepsilon^2/8 + \varepsilon^3/16$$

yielding

$$p_{AB} = p_{BA} = x_A x_B[1 - 2x_A x_B v/kT - 2x_A x_B(x_A - x_B)^2(v/kT)^2]$$
$$G_m = x_A{}^\circ G_A + x_B{}^\circ G_B + RT(x_A\ln x_A + x_B\ln x_B)$$
$$+ vzN^A x_A x_B[1 - x_A x_B v/kT - (2/3)x_A x_B(x_A - x_B)^2(v/kT)^2]$$

It is self-evident that the Gibbs energy will decrease due to short-range order, otherwise it would not form. The presence of short-range order will thus stabilize the disordered state and depress the temperature for the transition to a state with long-range order. This conclusion holds independent of the sign of v. Positive v will result in a miscibility gap at low temperatures if that reaction is not prevented by other reactions. Already above the miscibility gap a positive v will favour A–A bonds and B–B bonds and result in clusters. This tendency grows very strong as the consolute point is approached on cooling of a system with the right composition. The tendency to actually separate into two phases will thus decrease and the consolute point of the miscibility gap will be depressed to lower temperatures. Also, the consolute point will be flatter than calculated from the regular solution model because the effect will be strongest close to the consolute point. This effect is quite noticeable in the liquid state where the presence of clusters may give rise to opalescence close to the consolute point. In the solid case the effect is counteracted by coherency stresses due to the difference in atomic sizes.

In order to treat this effect properly it is not enough to consider nearest neighbours. It is not even enough to extend the consideration to larger clusters. It has to be treated with a mathematical technique called the renormalization group approach. The resulting shape of the miscibility gap is non-analytical, especially close to the maximum.

Exercise 19.4.1

In the next section we shall find that long-range order is predicted to occur below $T_{\text{tr}} = 2x_A x_B(-vz/k)$ if the effect of short-range order is neglected. Evaluate the short-range order in a 50/50 bcc alloy at this temperature in order to test if the neglect of short-range order is serious.

Solution

$p_{AA}p_{BB} = (p_{AB}{}^2\exp(2v/kT_{\text{tr}})) = (p_{AB})^2\exp[2v/2x_A x_B(-vz)] = (p_{AB})^2\exp(-4/z)$; $(0.5 - p_{AB})^2 = 0.606(p_{AB})^2$; $p_{AB} = 0.2811$; s.r.o. $= K - L^2 - x_A x_B = 0.2811 - 0 - 0.25 = 0.0311$. This is not negligible. The value of T_{tr} given above may be regarded as a rough estimate but short-range order makes the disordered state more stable and it should depress T_{tr}.

19.5 *Long-range order*

Let us again consider negative exchange energies. There is always a tendency for short-range order but as an introduction to the more general case it may be illustrative first to consider long-range order but neglect short-range order which is done by taking all $p_{ij} = y_i' y_j''$. The general equation then only contains y_i variables. The first part of the entropy contribution reduces to zero and with $v_{AB} + v_{BA} = 2v$ we find

$$G_{\text{m}} = x_A{}^\circ G_A + x_B{}^\circ G_B + (y_A' y_B'' + y_B' y_A'')\cdot vzN^A/2$$
$$+ RT[y_A'\ln y_A' + y_B'\ln y_B' + y_A''\ln y_A'' + y_B''\ln y_B'']/2$$

With a fixed composition there is only one independent internal variable and we may choose any one of the four y_i or the long-range order parameter L. Using the relations between L and the y variables in Section 19.3 we thus find the equilibrium from

$$(\partial G_{\text{m}}/\partial L)_{x_A} = (y_B'' + y_A' - y_A'' - y_B')\cdot vzN^A/2 + RT[y_A'/y_A' + \ln y_A'$$
$$- y_B'/y_B' - \ln y_B' - y_A''/y_A'' - \ln y_A'' + y_B''/y_B'' + \ln y_B'']/2 = 0$$

$$\frac{kT}{-vz} = \frac{y_A' + y_B'' - y_A'' - y_B'}{\ln(y_A' y_B''/y_A'' y_B')}$$

using $R = kN^A$. It is evident that $y_A' = y_A'' = x_A$ (and thus $y_B' = y_B'' = x_B$) is a solution for all T values and it represents a completely random distribution. However, below a particular T value another kind of solution appears and it represents long-range order. That temperature has to be calculated as the limiting value of T where y_A' and y_A'' of the new kind of solution approach x_A since the numerator and denominator are both zero. It is obtained by taking the ratio of their derivatives,

$$\frac{kT_{tr}}{-vz} = \frac{1+1+1+1}{1/y'_A + 1/y''_B + 1/y''_A + 1/y'_B} = \frac{4}{2/x_A + 2/x_B} = 2x_Ax_B$$

This relation defines the transition line, i.e. the boundary of the ordering region in the T,x_B phase diagram, see Fig. 18.3(a). It has a parabolic shape just as the spinodal curve for positive v values. Its maximum is found at $x_A = x_B = 0.5$ and $T_{tr}^{max} = -vz/2k$. The present result is identical to the result of the compound energy model in Section 18.5 and the two models are related by $\Delta°G_{AB+BA} = vzN^A$. The Bragg–Williams model thus provides a physical interpretation of the model parameter in this simple application of the compound energy model.

Above the ordering region the present model predicts complete disorder and the expression for G_m degenerates to

$$G_m = x_A°G_A + x_B°G_B + vzN^Ax_Ax_B + RT(x_A\ln x_A + x_B\ln x_B)$$

because $y'_A = y''_A = x_A$ and $y'_B = y''_B = x_A$. This expression was derived in Section 19.2 and it was then emphasized that vzN^A is identical to the regular solution parameter $°L$. Within the limitations of the nearest-neighbour bond energy model it would thus be possible to predict the ordering behaviour of a binary solution at low temperatures from the value of the regular solution parameter at high temperatures. However, as mentioned in Section 19.1, it should be realized that part of the energy may be of a more long-range character then nearest-neighbour interactions. It is thus possible that the ordering tendency should be represented by a v value which is not quite equal to $°L/zN^A$.

Exercise 19.5.1

We have derived an expression for the transition line for ordering in a binary system, taking into account a gradual increase of long-range order with decreasing temperature. Now, formulate a more primitive theory by considering only two possible states, complete order and complete disorder. Calculate the transition temperature and compare with the result just obtained for the critical temperature. Limit the calculation to a 50/50 alloy.

Hint

One can directly calculate the Gibbs energy contributions from configurational entropy and bond energies for the two states. Do not introduce the exchange energy v until the two states are compared.

Solution

In the disordered state the configurational entropy for 1 mole of atoms is $R\ln2$

and the energy due to bonds is $(x_A^2 g_{AA} + 2x_A x_B g_{AB} + x_B^2 g_{BB})zN^A/2 = (g_{AA} + 2g_{AB} + g_{BB}) \cdot zN^A/8$ for $x_A = x_B = 0.5$. In the ordered state there is no configurational entropy and the bonds give an energy of $g_{AB}zN^A/2$. At equilibrium: $-RT\ln2 + (g_{AA} + 2g_{AB} + g_{BB})zN^A/8 = g_{AB}zN^A/2$; $RT\ln2 = g_{AA} - 2g_{AB} + g_{BB})zN^A/8 = -vzN^A/4$; $T = -vz/4k\ln2$. This should be compared with $T_{tr} = -vz/2k$. The ratio is $T/T_{tr} = 1/2\ln2 = 0.72$. It is interesting to note that the same result was obtained in a similar treatment of a magnetic transition in Section 16.7.

Exercise 19.5.2

Derive an expression for the curved line in Fig. 18.3(b), i.e. for the equilibrium distribution in an ordered 50/50 alloy.

Hint

Express all y_i in terms of y'_A for $x_A = 0.5$. Then insert them in $kT/(-vz) = (y'_A + y''_B - y''_A - y'_B)\ln(y'_A y''_B/y''_A y'_B)$.

Solution

$y'_A + y''_A = 2x_A = 1$; $y''_A = 1 - y'_A$; $y'_A + y'_B = 1$; $y'_B = 1 - y'_A$; $y''_B = 1 - y''_A = y'_A$;
$kT/(-vz) = (4y'_A - 2)/\ln[(y'_A)^2/(1 - y'_A)^2] = (2y'_A - 1)/\ln[y'_A/(1 - y'_A)]$.
Using $kT_{tr}/(-vz) = 2x_A x_B$ we get $T/T_{tr} = (2y'_A - 1)/\ln[y'_A/(1 - y'_A)]/2x_A x_B = 2(2y'_A - 1)/\ln[y'_A/(1 - y'_A)]$

Exercise 19.5.3

Apply the condition of the stability limit and verify that the transition temperature falls at the limit of stability as it should for a second-order transition.

Hint

The condition can here be written as $(\partial^2 G_m/\partial L^2)_{x_A} = 0$. We have already an expression for $(\partial G_m/\partial L)_{x_A}$.

Solution

$(\partial^2 G_m/\partial L^2)_{x_A} = (1 + 1 + 1 + 1)vzN^A/2 + RT[1/y'_A + 1/y'_B + 1/y''_A + 1/y''_B]/2$. For the disordered state $y'_A = y''_A = x_A$ we get $(\partial^2 G_m/\partial L^2)_{x_A} = 2vzN^A + RT(1/x_A + 1/x_B) = 2vzN^A + RT/x_Ax_B = 0$ yielding $T_{limit} = -2vzx_Ax_B/k$ in agreement with the expression for T_{tr}.

19.6   *Long- and short-range order*

In the general case there will be both long- and short-range order and the situation will be described with two independent internal variables. We shall use K and L and with the relations of the various y and p quantities to L and K, derived in Section 19.3, we obtain

$$(\partial G_m/\partial K)_L = (v_{AB} + v_{BA})\cdot zN^A/2 + RT(z/2)[-\ln(p_{AA}/y'_Ay''_A) - 1$$
$$+ \ln(p_{AB}/y'_Ay''_B) + 1 + \ln(p_{BA}/y'_By''_A) + 1 - \ln(p_{BB}/y'_By''_B) - 1]$$
$$(\partial G_m/\partial L)_K = (v_{AB} - v_{BA})\cdot zN^A/2 + RT(z/2)[-p_{AA}/y'_A + p_{AA}/y''_A + \ln(p_{AB}/y'_Ay''_B)$$
$$+ 1 - p_{AB}/y'_A - p_{AB}/y''_B - \ln(p_{BA}/y'_By''_A) - 1 + p_{BA}/y'_B + p_{BA}/y''_A$$
$$+ p_{BB}/y'_B - p_{BB}/y''_B] + RT[\ln y'_A + 1 - \ln y'_B - 1 - \ln y''_A - 1$$
$$+ \ln y''_B + 1]/2$$
$$= (v_{AB} - v_{BA})\cdot zN^A/2 + RT(z/2)[-(y'_A - p_{AB})/y'_A + (y''_A$$
$$- p_{BA})/y''_A - p_{AB}/y'_A$$
$$- p_{AB}/y'_B + p_{BA}/y'_B + p_{BA}/y''_A + (y'_A - p_{BA})/y'_B - (y''_B - p_{AB})/y''_B$$
$$+ \ln(p_{AB}y'_By''_A/p_{BA}y'_Ay''_B)] + RT[\ln(y'_Ay''_B/y'_By''_A)]/2$$
$$= (v_{AB} - v_{BA})\cdot zN^A/2 + RT(z/2)[\ln(p_{AB}y'_By''_A/p_{BA}y'_Ay''_B)]$$
$$+ RT[\ln(y'_Ay''_B/y'_By''_A)]/2$$
$$= (v_{AB} - v_{BA})\cdot zN^A/2 + (RTz/2)\ln(p_{AB}/p_{BA})$$
$$- (z - 1)(RT/2)[\ln(y'_Ay''_B/y'_By''_A)] = 0$$

The two conditions can be written as

$$\frac{p_{AB}p_{BA}}{p_{AA}p_{BB}} = \exp[-(v_{AB} + v_{BA})\cdot z/kT]$$

$$z\ln(p_{AB}/p_{BA}) = (z - 1)\ln(y'_Ay''_B/y'_By''_A) - (v_{AB} - v_{BA})/kT$$

It is easy to make numerical calculations from these two equations.

In this chapter we have so far discussed ordering of different kinds of atoms. This phenomenon is often called **chemical ordering** or **configurational ordering**. A completely different kind of ordering occurs when a liquid solidifies and the atoms arrange themselves in a regular pattern, the crystalline structure. In this connection the liquid is said to be topologically disordered. The solid → liquid reaction can thus be regarded as an order–disorder transition and, evidently, it is a first-order transition. In the first-order transition, illustrated in Fig. 14.4, the superheated, ordered state reaches a stability limit at some high temperature and the undercooled, disordered state reaches a stability limit at some low temperature. The question whether such limits of stability exist for the solid–liquid transition has attracted some attention. The very simple and crude model described in Section 16.3 may be used to model a continuous transition from solid to liquid and it would predict limits of stability. On the other hand, cooling experiments with liquid alloys have revealed that the high entropy of the liquid state, as compared to the solid, disappears on cooling, and extrapolation of the high-temperature data seems to indicate that it approaches the entropy of the solid at some low temperature. In the same range of temperature the viscosity increases drastically and the liquid transforms into a viscous, amorphous state, which has an entropy similar to the solid. It seems that the amorphous state has only a low topological disorder although it is enough to prevent the topological long-range order found in crystalline solids. It seems that the topological disorder in the amorphous-liquid phase increases gradually with temperature and never shows a transition point. This is somewhat similar to the behaviour of the asymmetric case of chemical ordering illustrated in Fig. 18.4. From the mechanical point of view, one defines the drastic increase of the viscosity as a glass transition. Thermodynamically, one could define a related point where extrapolated data predict that the entropy of liquid and solid should be equal. However, one should not really expect that the entropy of the liquid reaches that of the crystal and then suddenly starts to follow the value of the crystal. Most probably, there is no thermodynamic transition point between the liquid and the amorphous states.

Due to the topological disorder in a liquid, it would not be possible to observe chemical long-range order in liquid alloys. Nevertheless, there are many cases with very strong chemical ordering in liquids, the most typical case is a molten salt where the electric charges make the cations tend to surround themselves with anions and vice versa. A very realistic model of a molten salt is thus based on the assumption of two sublattices, one for cations and one for anions. That would be to assume complete long-range order and the effect of short-range order could then be added to the model. A problem with such a model is that the coefficients in the chemical formula (the stoichiometric coefficients) will vary with composition if there are cations of different valencies or anions of different valencies.

Another difficulty appears when one wants to model the change in chemical order in a liquid from a high value at a low temperature and to a low value at a high temperature. As demonstrated in Section 19.4 the quasi-chemical approach to short-range order becomes unrealistic at large degrees of short-range order. It has been proposed that this difficulty can be overcome by using $z = 2$, for which the quasi-chemical approach does not break down at high degrees of short-range order. However, that would make the model less physical. Another possibility would be to use a two-sublattice model of the asymmetric type illustrated in Fig. 18.4(a). It can predict a gradual change from very low to very high degrees of order and without a transition point. Another possibility is to mimic the chemical ordering in a liquid by the formation of molecular-like clusters of atoms, so-called associates. Finally, in an attempt to develop the two-sublattice model to a model applicable to many different types of systems and to intermediate cases, the two-sublattice model with complete long-range order has been manipulated in such a way that it can describe high as well as low degrees of ordering.

Exercise 19.6.1

Calculate the degree of short-range order when the disordered state is just becoming unstable in a second-order transition.

Hint

There are two internal degrees of freedom, K and L. The limit of stability is thus obtained as $g_{KK} \cdot g_{LL} - g_{KL} \cdot g_{LK} = 0$ according to Section 5.8. However, L is still zero at that point, $y'_A = y''_A = x_A$ and $p_{AB} = p_{BA}$. Then it is easily shown that $g_{KL} = g_{LK} = 0$ and $g_{KK} > 0$. The condition is thus $g_{LL} = 0$.

Solution

From $\partial G_m/\partial L$ we get, using the relation between dL and dp_{ij} and dy_i^s in Section 19.3: $g_{LL} = (\partial^2 G_m/\partial L^2)_K = 0 + (zRT/2)(1/p_{AB} + 1/p_{BA}) - (z - 1)(RT/2)(1/y'_A + 1/y''_B + 1/y''_A + 1/y'_B) = 0$; $p_{AB} = x_A x_B z/(z - 1) = p_{BA} = K$; s.r.o. $= K - L^2 - x_A x_B = x_A x_B[z/(z - 1) - 1] = x_A x_B/(z - 1)$.

19.7   The compound energy model with short-range order

In Section 19.2 it was assumed that all atoms have the same z value. In Section 19.3 it was further assumed that all bonds go between atoms in two different sublattices. These assumptions were carried over to the treatment of order in Sections 19.3–19.6. However, the treatment can be generalized to cases with different coordination numbers, z' and z'', if it is still assumed that all bonds go between atoms in two different sublattices but that requires that the number of sites, b and c, are related by $bz' = cz''$. The total number of bonds in a system containing one mole of atoms is $bz'N^A = cz''N^A$ if the sum of b and c is made equal to unity. Thus $N_{AA} = p_{AA} \cdot bz'N^A$ and $°G_A = g_{AA} \cdot bz'N^A$ and

$$x_A = p_{AA} + bp_{AB} + cp_{BA}; \quad p_{AA} = x_A - bp_{AB} - cp_{BA}$$
$$x_B = p_{BB} + bp_{BA} + cp_{AB}; \quad p_{BB} = x_B - bp_{BA} - cp_{AB}$$

Using these expressions for p_{AA} and p_{BB} we obtain, for the Gibbs energy contribution from all the bond energies,

$$\Delta G_m = \sum_i \sum_j N_{ij} g_{ij} = [p_{AA} g_{AA} + p_{AB} g_{AB} + p_{BA} g_{BA} + p_{BB} g_{BB}] \cdot bz' N^A$$
$$= [x_A g_{AA} + x_B g_{BB} + p_{AB}(g_{AB} - bg_{AA} - cg_{BB})$$
$$+ p_{BA}(g_{BA} - bg_{BB} - cg_{AA})] \cdot z' bN^A/2$$

We find, by generalizing the definitions of the v quantities,

$$v_{AB} = g_{AB} - bg_{AA} - cg_{BB}$$
$$v_{BA} = g_{BA} - bg_{BB} - cg_{AA}$$
$$\Delta G_m = x_A {}^\circ G_A + x_B {}^\circ G_B + bz' N^A \cdot [p_{AB} v_{AB} + p_{BA} v_{BA}]$$

The expression for the configurational entropy in Section 19.3 will also be modified by replacing $z/2$ by bz' or cz'' and $1/2$ by b or c, yielding, for the total Gibbs energy,

$$G_m = x_A {}^\circ G_A + x_B {}^\circ G_B + bz' N^A \cdot [p_{AB} v_{AB} + p_{BA} v_{BA}]$$
$$+ RTbz'[p_{AA} \ln(p_{AA}/y'_A y''_A) + p_{AB} \ln(p_{AB}/y'_A y''_B) + p_{BA} \ln(p_{BA}/y'_B y''_A)$$
$$+ p_{BB} \ln(p_{BB}/y'_B y''_B)] + RT[by'_A \ln y'_A + by'_B \ln y'_B + cy''_A \ln y''_A + cy''_B \ln y''_B]$$

This may be regarded as a generalization of the final equation in Section 19.3. However, the definitions of v_{AB} and v_{BA} cannot be reconciled with the simple concept of bond energies presented in Section 19.1. The new definition is closely related to the definition of $\Delta^\circ G_{A_b B_c}$ and $\Delta^\circ G_{B_b A_c}$ in the compound energy model in Section 18.5, the only difference being the numerical factor $bz' N^A$. However, the compound energy model goes one step further. It postulates that phases, with some bonds between atoms in the same sublattice, can be treated with the same formalism. That may not be correct but may be remedied by the use of the I parameters in the excess terms in Section 18.3.

Using the G_m expression just derived one can formally account for the effect of short-range order with the compound energy model. The G_m equation from Section 18.5 would be modified to

$$G_m = x_A {}^\circ G_A + x_B {}^\circ G_B + p_{AB} \Delta^\circ G_{A_b B_c} + p_{BA} \Delta^\circ G_{B_b A_c}$$
$$+ RTbz'[p_{AA} \ln (p_{AA}/y'_A y''_A) + p_{AB} \ln(p_{AB}/y'_A y''_B) + p_{BA} \ln(p_{BA}/y'_B y''_A)$$
$$+ p_{BB} \ln(p_{BB}/y'_B y''_B)] + RT[by'_A \ln y'_A + by'_B \ln y'_B + cy''_A \ln y''_A + cy''_B \ln y''_B]$$

For a reciprocal solution, i.e., a quaternary phase where each sublattice dissolves only two of the four elements, we would obtain

$$\Delta G_m = x_A {}^\circ G_A + x_B {}^\circ G_B + x_C {}^\circ G_C + x_D {}^\circ G_D$$
$$+ p_{AC} \Delta^\circ G_{A_b C_c} + p_{AD} \Delta^\circ G_{A_b D_c} + p_{BC} \Delta^\circ G_{B_b C_c} + p_{BD} \Delta^\circ G_{B_b D_c}$$
$$+ RTbz'[p_{AC} \ln(p_{AC}/y'_A y''_C) + p_{AD} \ln(p_{AD}/y'_A y''_D) + p_{BC} \ln(p_{BC}/y'_B y''_C)$$
$$+ p_{BD} \ln(p_{BD}/y'_B y''_D)] + RT[by'_A \ln y'_A + by'_B \ln y'_B + cy''_C \ln y''_C + cy''_D \ln y''_D]$$

This would be a case of short-range order at complete long-range order. Again we have four p_{ij} related by $\Sigma p_{ij} = 1$ but their relation to the composition now yields two independent equations instead of one at the beginning of Section 19.2. In this case it is most convenient to use the y_i variables which are here fixed by the composition,

$$p_{AC} + p_{AD} = y'_A$$
$$p_{AC} + p_{BC} = y''_C$$

Consequently, now there is just one independent internal variable, one of the four p_{ij}, instead of two, and it represents short-range order because the long-range order is here perfect. It may be convenient to define the short-range order in a way similar to that in Section 19.3. However, since each kind of bond can now go in one direction, only, we write

$$\text{s.r.o.} = p_{AC} - y'_A y''_C$$

We shall denote this variable by s and by combination with the relations of p_{ij} to the composition we find

$$p_{AC} = y'_A y''_C + s; \quad dp_{AC} = ds$$
$$p_{AD} = y'_A y''_D - s; \quad dp_{AD} = -ds$$
$$p_{BC} = y'_B y''_C - s; \quad dp_{BC} = -ds$$
$$p_{BD} = y'_B y''_D + s; \quad dp_{BD} = ds$$

The equilibrium value of s is obtained from

$$(\partial G_m/\partial s)_{y'_A y''_C} = \Delta^\circ G_{A_b C_c} - \Delta^\circ G_{A_b D_c} - \Delta^\circ G_{B_b C_c} + \Delta^\circ G_{B_b D_c}$$
$$+ RTbz'[1 + \ln(p_{AC}/y'_A y''_C) - 1 - \ln(p_{AD}/y'_A y''_D) - 1 - \ln(p_{BC}/y'_B y''_C)$$
$$+ 1 + \ln(p_{BD}/y'_B y''_D)] = \Delta^\circ G_{AC+BD} + RTbz'\ln(p_{AC}p_{BD}/p_{AD}p_{BC}) = 0$$

$$\frac{p_{AC}p_{BD}}{p_{AD}p_{BC}} = \exp(-\Delta^\circ G_{AC+BD}/RTbz')$$

This is a result characteristic of the quasi-chemical approach. The notation $\Delta^\circ G_{AC+BD}$ was introduced in Section 18.3 and used in Section 18.5.

Exercise 19.7.1

Consider a solution phase defined by the formula $(A,B)_b(C,D)_c$ with $N_A = N_B$ and $N_C = N_D$. Apply the compound energy model and estimate how much the short-range order decreases the Gibbs energy if changes in the configurational entropy are neglected. Suppose $z' = 8$, $b = c = 0.5$ and $\Delta^\circ G_{AC+BD} = 8RT$.

Hint

According to the instruction, we should study:
$p_{AC}\Delta^\circ G_{A_b C_c} + p_{AD}\Delta^\circ G_{A_b D_c} + p_{BC}\Delta^\circ G_{B_b C_c} + p_{BD}\Delta^\circ G_{B_b D_c}$ which can be changed to
$y'_A y''_C \Delta^\circ G_{A_b C_c} + y'_A y''_D \Delta^\circ G_{A_b D_c} + y'_B y''_C \Delta^\circ G_{B_b C_c} + y'_B y''_D \Delta^\circ G_{B_b D_c} + s(\Delta^\circ G_{A_b C_c} - \Delta^\circ G_{A_b D_c} + \Delta^\circ G_{B_b C_c} - \Delta^\circ G_{B_b D_c})$. All site fractions are fixed ($= 0.5$). The change is thus given by the last term and it can be written as $s \cdot \Delta^\circ G_{AC+BD}$.

Solution

$\Delta^\circ G_{AC+BD}/RTbz' = 2$. The quasi-chemical equation yields $(\frac{1}{4} + s)^2/(\frac{1}{4} - s)^2 = \exp(-2)$; $s = -0.1155$; $s \cdot \Delta^\circ G_{AC+BD} = -0.1155 \cdot 8RT = -0.92RT$. This is an appreciable part of $\Delta^\circ G_{AC+BD}$.

19.8   *Interstitial ordering*

Interstitial solutions can also show ordering. Two effects may be recognized. The first effect is a tendency for the interstitial atoms to avoid occupying sites which are nearest neighbour to each other. In the extreme case, all the sites which are nearest neighbours to an interstitial atom will be excluded from occupancy. 'Excluded sites' models have been developed for this case. As an example, we may consider fcc-Fe where the interstitial sites form their own fcc sublattice. Each interstitial site is surrounded by 12 nearest-neighbour interstitial sites. According to the model, each interstitial atom in a dilute solution excludes its own site and 12 neighbouring interstitial sites from being occupied. The lattice would thus behave as if it will be completely filled at a value of $y_C = 1/13$ where C represents the interstitial component. The partial entropy of C would thus be

$$S_C = - R\ln\frac{y_C}{1 - 13y_C} \cong - R(\ln y_C + 13y_C)$$

In the ideal case, i.e. neglecting all other interactions, this model will predict a very strong positive deviation from Henry's law. Experimental information on the carbon activity in fcc-Fe–C does show a strong positive deviation but not enough to satisfy the excluded-sites model. This is demonstrated by Fig. 19.1 where $\ln(a_C/x_C)$ has been plotted versus x_C. The slope is about 8 but should have been 14 according to the excluded-sites model because $y_C/(1 - 13y_C)$ is equal to $x_C/(1 - 14x_C)$ when $a = 1$ for both sublattices.

In order to obtain better agreement for carbon in fcc-Fe one has to relax the condition of absolute exclusion to one in which a neighbouring site can be occupied at the expense of a certain energy v. Such models are sometimes called statistical models. A treatment can also be based upon the quasi-chemical approach which will formally apply to interactions between atoms in the interstitial sublattice just as well as to the ordinary lattice in a substitutional solution. By substituting y for x we obtain

$$G_m = y_C{}^\circ G_{MbC} + y_{Va}{}^\circ G_{MbVa} + RT(y_C\ln y_C + y_{Va}\ln y_{Va})$$
$$+ vzN^A y_C y_{Va}[1 - y_C y_{Va}v/kT - (2/3)y_C y_{Va}(y_C - y_{Va})^2(v/kT)^2]$$

With the rule of calculation given for G_C we obtain, for concentration-independent v,

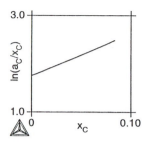

Figure 19.1 The activity coefficient of C in fcc-Fe at 1400 K as function of the carbon content.

$$RT\ln a_C = G_C - {}^\circ G_C^\alpha = {}^\circ G_{M_bC} - a^\circ G_M - {}^\circ G_C^\alpha + RT\ln[y_C/(1 - y_C)]$$
$$+ vzN^A(y_{Va} - y_C)[1 - 2y_Cy_{Va}v/kT + 8y_Cy_{Va}(y_Cy_{Va} - 1/6)(v/kT)^2]$$

For small y_C the last term can be approximated by $vzN^A\{1 - 2y_C[1 + v/kT + (2/3)(v/kT)^2]\}$. By comparing with the power series treatment of a random interstitial solution in Section 18.2 we see that vzN^A corresponds to ${}^0L_{CVa}$ if v is independent of composition. For $v/kT = 0$ we thus have complete agreement with that expression. For small v/kT there will be a difference of $-2y_CvzN^A\cdot[v/kT + (2/3)(v/kT)^2]$. This difference is initially negative for positive as well as negative v which is natural because the random mixing model always results in higher Gibbs energies than a model which allows the arrangement of the atoms to be adjusted to minimize the Gibbs energy. For a given value of v, or ${}^0L$, the quasi-chemical model will thus predict a slightly weaker deviation from Henry's law.

At high enough values of the interstitial content and of $-v/kT$ there will be long-range order, an example being the γ' phase in the Fe–N system. It has a nitrogen content corresponding to Fe_4N and the Fe atoms have the same arrangement as in fcc-Fe. It may be regarded as an ordered interstitial solution of N in fcc-Fe. It is interesting to note that no N atoms occupy nearest neighbour sites. This phase is thus a perfect example of the excluded-sites principle. The high N content, compared to the limiting value of $y_C = 1/13$ given previously, is due to the fact that each excluded site is now nearest neighbour to not only one interstitial atom but four which does not happen in a dilute solution.

Exercise 9.8.1

What value of ${}^0L_{CVa}$ in the random mixing model or v/kT in the quasi-chemical model is required in order to explain the slope of 8 in the plot of $\ln(a_C/x_C)$ versus x_C for carbon in fcc-Fe?

Hint

The expression for G_C in Section 18.2 gives
$$\ln[y_C/(1 - y_C)] = \ln[x_C/(1 - 2x_C)] \cong \ln x_C + 2x_C. \text{ Remember } z = 12.$$

Solution

The random mixing model for an interstitial solution gives $RT\ln(a_C/x_C) \cong$ const $+ 2RTx_C - {}^0L \cdot 2y_C \cong$ const $+ (2RT - 2{}^0L)x_C$, yielding a slope of $2RT - 2{}^0L = 8RT$; ${}^0L/RT = -3$. The quasi-chemical model gives, for small y_C, $RT\ln(a_C/x_C) \cong$ const $+ 2RTx_C - zvN^A(-2y_C)[1 + (v/kT) + (2/3)(v/kT)^2 + ...] \cong$ const $+ \{2RT - 2zvN^A[1 + (v/kT) + (2/3)(v/kT)^2]\} \cdot x_C$, yielding a slope of $2RT - 2zvN^A[1 + (v/kT) + (2/3)(v/kT)^2] = 8RT$; $-24(v/kT)[1 + (v/kT) + (2/3)(v/kT)^2] = 8 - 2$; $v/kT = -0.339$, corresponding to ${}^0L/RT = zv/kT = 12 \cdot (-0.342) = -4.07$.

19.9   *Composition dependence of physical effects*

In Section 15.7 it was mentioned that one should sometimes model the effect of a special physical phenomenon separately. A typical example would be the effect of magnetic ordering in a ferromagnetic material. That effect will now be used to demonstrate how the composition dependence of such an effect can be taken into account.

Figure 19.2 shows (a) the magnetic contribution to the heat capacity and (b) the decrease of the magnetic entropy from a disordered state of bcc-Fe at very high temperature. The peak temperature for the heat capacity and the inflexion point for the magnetic entropy represent the Curie temperature, T_C, i.e. the transition point between the high-temperature paramagnetic state which has only magnetic short-range order and the low-temperature state with long-range magnetic order.

A very simple model of magnetic ordering was presented in Section 16.7. It takes into account only long-range order. It would be possible also to include short-range order with a quasi-chemical type of approach. However, it would still be a very primitive model. For reasonably accurate modelling it seems necessary to rely on experimental measurements and some simple assumption of the dependence on composition. There have been several suggestions of mathematical expressions to be used. The first was made by Inden (1977) and it is still the most widely used model. He described the heat capacity with a logarithmic expression but in order to derive an analytical expression for the contribution to the Gibbs energy it was later found necessary to use a series expansion. With truncation after the third term it yields the following expressions above and below the Curie temperature.

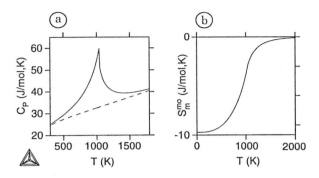

Figure 19.2 The effect of magnetic ordering on the heat capacity and entropy of bcc-Fe.

$$G_m^{mo} = -RT\ln(\beta + 1)\left(\frac{\tau^{-5}}{10} + \frac{\tau^{-15}}{315} + \frac{\tau^{-25}}{1500}\right)\bigg/\left[\frac{518}{1125} + \frac{11\,692}{15\,975}\left(\frac{1}{p} - 1\right)\right]$$

for $\tau > 1$

$$G_m^{mo} = RT\ln(\beta + 1)\left\{1 - \left[\frac{79\tau^{-1}}{140p} + \frac{474}{497}\left(\frac{1}{p} - 1\right)\left(\frac{\tau^3}{6} + \frac{\tau^9}{135} + \frac{\tau^{15}}{600}\right)\right]\right\}$$

$$\bigg/\left[\frac{518}{1125} + \frac{11\,692}{15\,975}\left(\frac{1}{p} - 1\right)\right] \text{ for } \tau < 1$$

p is a constant defined as the fraction of the total disordering enthalpy which is absorbed above the critical temperature. It was given as 0.28 for fcc metals and 0.40 for bcc metals. τ is a normalized temperature defined as T/T_C. The superscript 'mo' refers to magnetic order. The equations contain two material constants, the Curie temperature T_C and the Bohr magneton number β. The equations could hopefully be applied to solutions by inserting experimental or estimated expressions for their composition dependence.

The two diagrams in Fig. 19.2 were calculated from these equations after an assessment of the magnetic properties of bcc-Fe. However, experimental measurements indicate that C_P goes to infinity at T_C from both sides (see Fig. 16.4) and that result was originally modelled by Inden. It was abandoned for practical reasons when a Gibbs energy expression was derived by integration. This approximation may seem as a bad one from a theoretical point of view but for the calculation of phase equilibria it seems to have a comparatively small effect. An exception may be the position of the tri-critical point in a binary diagram like Fig. 14.14. According to the crude treatment of the effect of C_P^p, presented in Section 14.5, the tri-critical point should approach the side of the binary system if the magnetic C_P goes to infinity at T_C.

In order to extend the magnetic model to alloys it is necessary to make the model parameters composition-dependent. So far, investigators have studied the effect of composition-dependent β and T_C but kept p constant.

Exercise 19.9.1

Use the C_P curve in Fig. 19.2(a) for an estimate of p, the fraction of the total disordering enthalpy which has not yet been absorbed on heating to T_C because of the remaining short-range order.

Hint

Divide the area between the solid and the dashed curve into two, one above and one below T_C and make a crude graphical integration of these two areas.

Solution

The two areas are proportional to the two parts of ΔH. Thus
$p = A_1/(A_1 + A_2) = 1/(1 + A_2/A_1) \cong 1/(1 + 1.5) = 0.4$.

Exercise 19.9.2

Show schematically how the solubility of phosphorus (P) in bcc-Fe can be expected to vary with temperature around the intersection with the magnetic transition line, the Curie line. Consider three different conditions: (a) No magnetic effect. (b) A magnetic effect but no separation into regions of different composition. (c) A magnetic effect and separation into regions of different composition, resulting in a miscibility gap with a tri-critical point.

Hint

Consult Section 14.5 and, in particular, Fig. 14.15 and Exercise 14.5.1.

Solution

(a) Without a magnetic effect we expect that $\ln(x_P)$ should vary linearly with $1/T$ (see Exercise 10.4.1). (b) The slope should be smaller where C_P is larger due to the magnetic effect. (c) See diagram.

Solution 19.9.2

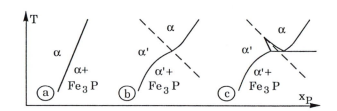

References

von Alkemade, A. C. van Ryn (1893) *Z. Phys. Chem.*, Vol. 11, p. 289

Andersson, J.-O., Fernández Guillermet, A., Hillert, M., Jansson, B. and Sundman, B. (1986) *Acta Metall.*, Vol. 34, p. 437

Bonnier, E. and Caboz, R. (1960) *CR. Hebd. Séances Acad. Sci.*, Vol. 250, p. 527

Cahn, J. W. (1961) *Acta Metall.*, Vol. 9, p. 795

Cahn, J. W., Pan, J. D. and Balluffi, R. W. (1979) *Scripta Metall.*, Vol. 13, p. 503

Callen, H. B. (1988) *Thermodynamics and an Introduction to Thermostatistics*, John Wiley and Sons, New York

Carnot, S. (1960) *Reflections on the Motive Power of Fire*, Dover, New York

Colinet, C. (1967) D.E.s., Fac.des Sci., Univ. Grenoble

Darken, L. S. (1948) *J. Amer. Chem. Soc.*, Vol. 70, p. 2046

Darken, L. S. and Gurry, R. W. (1953) *Physical Chemistry of Metals*, McGraw-Hill, London

Darken, L. S. (1967) *Trans. AIME*, Vol. 239, p. 90

De Donder, Th. (1922) *Bull. Acad. Roy. Belg.*, (Cl. Sc.), Vol. 7(5), p. 197,205

Ehrenfest, P. (1911) *Z. Phys. Chem.*, Vol. 77, p. 227

Gibbs, J. W. (1876) *Trans. Conn. Acad.*, Vol. 3, p. 108

Gibbs, J. W. (1948) *The Collected Works*, Vol. I, Yale Univ. Press, New Haven, Conn.

Goodman, D. and Cahn, J. W. (1981) *Bull. Alloy Phase Diagr.*, Vol. 2, p. 29

Guggenheim, A. E. (1952) *Mixtures*, Clarendon Press, Oxford

Gupta, H., Morral, J. E. and Nowotny, H. (1986) *Scripta Metall.*, Vol. 20, p. 889

Guthrie, F. (1884) *Phil. Mag.*, Vol. 17, p. 462

Hillert, M. (1961) *Acta Metall.*, Vol. 9, p. 525

Hillert, M. (1980) *CALPHAD*, Vol. 4, p. 1

Hillert, M. (1985) *Intern. Metals Rev.*, Vol. 30, p. 45

Hultgren, A. (1947) *Trans. ASM*, Vol. 39, p. 915

Inden, G. (1976) *Pro. CALPHAD V*, Max-Planck Inst. Eisenforschung, Düsseldorf, pp. 1–13

Kauzmann, W. (1948) *Chem. Rev.*, Vol. 43, p. 219

Kohler, F. (1960) *Monatsh. Chemie*, Vol. 91, p. 738

Konovalov, D. (1881) *Ann. Phys.*, Vol. 14, p. 34

Landau, L. D. and Lifshitz, E. M. (1958) *Statistical Physics*, Chapter XIV, Addison-Wesley, Reading, Mass.

Le Chatelier, H. (1984) *Comptes Rendus*, Vol. 99, p. 786

Masing, G. (1949) *Ternäre Systeme*, Leipzig, Akad. Verlag

Morey, G. W. (1936) in *Commentary on the Scientific Writings of J. Willard Gibbs*, (ed. F. G. Donnan and A. Haas) pp. 233–93, Yale Univ. Press, New Haven, Conn.

Morey, G. W. and Williamson, E. D. (1918) *J. Amer. Chem. Soc.*, Vol. 40, p. 59

Muggianu, Y.-M., Gambino, M. and Bros, J.-P. (1975) *J. Chimie Physique*, Vol. 72, p. 83

Münster, A. (1970) *Classical Thermodynamics*, Wiley-Interscience, London

Murnaghan, F. D. (1944) *Proc. Nat. Acad. Sci.* (USA), Vol. 30, p. 244

Palatnik, L. S. and Landau, A. I. (1964) *Phase Equilibria in Multicomponent Systems*, New York, Holt, Rinehart and Winston

Pelton, A. D. and Bale, C. W. (1986) *Metall. Trans.*, Vol. 17A, p. 1211

Pelton, A. D. (1992) *J. Chim. Phys.*, Vol. 89, p. 1931

Prigogine, I. and Defay, R. (1958) *Chemical Thermodynamics*, London, Longmans

Redlich, O. and Kister, A. T. (1948) *Ind. Eng. Chem.*, Vol. 40, p. 345

Scheil, E. (1935/36) *Arch. Eisenh. W.* Vol. 9, p. 571

Schreinemakers, F. A. H. (1911) *Die Ternäre Gleichgewichte*, Braunschweig Vol. III, Part I, p. 72

Schreinemakers, F. A. H. (1912) *Die Ternäre Gleichgewichte*, Braunschweig Vol. III, Part II

Schreinemakers, F. A. H. (1915) *Proc. K. Akad. Wetensch, Amsterdam*, (*Section on Sciences*), Vol. 18, p. 116

Toop, G. W. (1965) *Trans. AIME*, Vol. 233, p. 850

Wagner, C. (1952) *Thermodynamics of Alloys*, Addison–Wesley, Cambridge

Weiss, R. J. (1963) *Solid State Physics for Metallurgists*, Pergamon, Oxford

Zener, C. M. (1948) *Elasticity and Anelasticity*, Univ. Chicago Press, Chicago

Index